Indian and World Geography

Indian and World Geography

Physical, Social and Economic

H.M. Saxena

RAWAT PUBLICATIONS
Jaipur • New Delhi • Bangalore • Guwahati • Kolkata

ISBN 978-81-316-0743-5 (HB)
ISBN 978-81-316-0744-2 (PB)

Published by
Prem Rawat for **Rawat Publications**
Satyam Apts, Sector 3, Jawahar Nagar, Jaipur 302 004 (India)
Phone: 0141 265 1748 / 7006 Fax: 0141 265 1748
E-mail: info@rawatbooks.com
Website: www.rawatbooks.com

New Delhi Office
4858/24, Ansari Road, Daryaganj, New Delhi 110 002
Phone: 011 2326 3290

Also at *Bangalore, Guwahati* and *Kolkata*

Typeset by Rawat Computers, Jaipur
Printed at Nice Printing Press, New Delhi

Contents

Section II
WORLD: Physical, Social & Economic Geography

Preface

Physical, Social and Economic Geography of India and the World is one of the most important component of Paper I – General Studies in UPSC Civil Services Preliminary test. The present book is an attempt to assist and facilitate all those students who prepare themselves for Preliminary Examination of Civil Services as well as for other competitive examinations.

The book is organized into two sections spread over 30 chapters. Section-I deals with Indian Geography, while Section-II is related to the Geography of the World. In Section-I there are 15 chapters, in which brief analysis of the various facets of physical, social and economic geography of India have been discussed. It includes geological structure, physiography, drainage, climate, natural vegetation, soils, social and demographic structure and economic aspects of India. The remaining 15 chapters of the book focus on Geography of the World covering all three aspects – physical, social and economic.

The topics of the book have been designed keeping in mind the changing nature of examination and to make this book as the standard material which students can turn to for all their needs related to Indian and the World Geography. It is written in clear and simple style and supported with maps and diagrams. More emphasis has been given on factual informations, which is a basic requirement in competitive examinations. At the end of each section multiple choice questions related to each topic have been given which will be very useful to the students in their preparation for the examination. The book will not only be useful for the students appearing for competitive examinations but also to those interested in gernal geography of the world and India.

Many scholars and friends have had a role, either direct or indirect, in making of this book. The author expresses his sincere thanks to all authors whose works have been consulted. I am also thankful to the entire team of Rawat Publications for this nice publication.

My wife Nirmal has, as ever, been my greatest source of strength; without her active cooperation and moral support of other members of my family, this book would have been much harder to write. I am thankful to all.

H.M. Saxena

Section I

INDIA

Physical, Social & Economic Geography

1

India: An Introduction

India, also called *'Bharat'* is a unique country where geographical factors not only direct its growth and development but also determine the whole socio-cultural pattern. The word 'India' has its origin in the Greek meaning, the land of 'Indoi'; the people living near the *Indos* (latin Indus). The country is the best example of 'Unity in Diversity', which is evident in its physical, economic, social, cultural as well as in political set-up. The territorial limits of Indian mainland extend between 8°4´ and 37°6´ north latitudes and 68°7´ and 97°25´ east longitudes. The Tropic of Cancer passes through almost the centre of the country and crosses 8 states of India, viz. Gujarat, Rajasthan, Madhya Pradesh, Chhattisgarh, Jharkhand, West Bengal, Tripura and Mizoram. Thus, India lies entirely in the North-Eastern hemisphere.

The total area of India is 32,87,263 sq km and in terms of area, it ranks the seventh largest country in the world after Russia, Canada, USA, China, Australia and Brazil. The land area of the country is about 2.4 per cent of the world land area. But in terms of population, it is second to China having about one-sixth (17.4%) of the world population. The northernmost point is the 'Indra Col' which lies in the state of Jammu & Kashmir. The southernmost point of India is Indira Point (formerly known as Pygmalion Point) located in the Nicobar Islands. The maximum length of the country from north to south is 3,214 km and from east to west is 2,933 km. The land boundary of India measures 15,200 km and the total length of the coastline of mainland is 6,100 km but, when length of island coastlines is included, it becomes around 7,516 km.

Space Relationship of India with Neighbouring Countries

India shares 15,200 km land boundary with seven countries. The countries having common land boundary with India are Pakistan, China, Nepal, Bhutan, Bangladesh, Myanmar and Afghanistan. The longest border of India is with Bangladesh with a length of 4,096 km; while its shortest border is 80 km with Afghanistan. Table 1.1 indicates the length of India's border's with the neighbouring countries.

The *Indo-Pakistan Boundary* is the result of the partition of the country in 1947. The 3,310 km long boundary is also known as *Redcliffe Line* after the name of Sir Cyril Radclifffe who was the chairman of the Boundary Commission. The boundary starts from the Rann of Kachchh in Gujarat, and traverses through the states of Rajasthan, Punjab and Jammu & Kashmir.

Table 1.1 *Length of India's Border with the Neighbouring Countries*

Country	Length of Border (km)	Percentage of Total Length of Border
1. Bangladesh	4,096	26.95
2. China	3,917	25.77
3. Pakistan	3,310	21.78
4. Nepal	1,752	11.53
5. Myanmar	1,458	9.89
6. Bhutan	587	3.86
7. Afghanistan	80	0.57

The *Indo-China Border* is 3,917 km long, which is prominently separated by the Himalayas. Five Indian states, viz. Jammu & Kashmir, Himachal Pradesh, Uttarakhand, Sikkim and Arunachal Pradesh touch the Indian boundary with China. The Indo-China border line is divided into three sectors:

1. *The Western Sector* is 2,152 km long and separates Jammu & Kashmir state of India from Sinkiang province of China.
2. *The Central Sector* is about 625 km long and runs along the watershed from Ladakh to Nepal. The two Indian States – Himachal Pradesh and Uttarakhand touch this border.
3. *The Eastern Sector* of Indo-China border is 1,140 km long and also known as *McMahon Line* after Henry McMahon, a British representative who signed the 1913-14 Shimla Convention. This line normally runs along the crest of the Himalayas between Bhutan and Myanmar. Although China has not yet accepted this boundary fully and laid claims to some areas of India, this has become a cause of boundary dispute between India and China.

The *India-Bangladesh Border* is the longest with a length of 4,096 km. The Indian states adjoining this boundary is West Bengal, Assam, Meghalaya, Tripura and Mizoram. The Indo-Bangladesh boundary line criss-crosses the Ganga-Brahmaputra delta and runs through an entirely flat country.

The *India-Nepal Border* stretches up to 1,752 km. Five Indian States, viz. Uttarakhand, Uttar Pradesh, Bihar, West Bengal and Sikkim touch the Nepalese border with India.

The *India-Bhutan Border* is 587 km long. The four Indian States, viz. Sikkim, West Bengal, Assam and Arunachal Pradesh touch the Indo-Bhutan border.

The *India-Myanmar Border* is 1,458 km long and runs from India-China-Myanmar tri-junction in the north to the southern tip of Mizoram.

The *India-Afghanistan Border* is the shortest one between the two countries and spreads only in 80 km. The Indian state that touches this border is Jammu & Kashmir.

Coastline of India

India is having a long coastline of about 6,100 km. The mainland of India is washed on three sides of the country by the Indian Ocean and its two seas, viz. The Arabian Sea and Bay of Bengal. Apart from this, if coastlines of Andaman and Nicobar Islands and Lakshadweep are included, the total length of Indian's coastline becomes 7,517 km. This long coastline helps India it its oceanic trade and commerce as well as in exploiting the marine resources. The coastal states of India are: Gujarat, Maharashtra, Goa, Karnataka, Kerala, Tamil Nadu, Andhra Pradesh, Odisha and West Bengal.

The Indian Neighbour – Sri Lanka

Apart from the neighbours of India located in the west, north-west, north and north-east, the other important neighbouring country is Sri Lanka, which is separated from India by a narrow and shallow Sea called Palk Strait. The two points Dhanushkodi on Tamil Nadu coast in India and Talaimanar in Jaffna peninsula in Sri Lanka are only 32 km away from each other. In between these two, Adam's Bridge *(Ram Setu)* is located. The maritime boundary between India and Sri Lanka passes through Palk Strait as Palk Bay touching Dhanushkodi.

India and the Indian Ocean

'India is the keystone of the arch of the Indian Ocean' (O.H.K. Spate, *India and Pakistan,* 1967: 4). In fact, India's location in Indian Ocean is so important that its name is given after India. The Indian Ocean in bounded by South Asia in the north, Indonesian islands and Australia in the east and Africa in the west. Of the great oceans of the world, the Indian Ocean alone, is an embayed ocean. The vital feature which differentiates the Indian Ocean from Pacific and the Atlantic is not the two sides but the sub-continent of India which juts out far into the sea for a thousand kilometres to its tapering end at Cape Comorin. It is the geographical position of India that changes the character of the Indian Ocean.

The Indian Ocean has great strategic and economic importance for India. The 'landlocked' nature of the Indian Ocean has given India, a commanding position. The several islands such as Mauritius, Diego Garcia, Seychelles, Socotra, Lakshadweep, Chagos, Andaman and Nicobar, Maldives, Cocos, Chrismas, etc. are having great strategic importance in world geopolitics. Indian Ocean provides a link not only between Europe and Asia but also between the countries of Africa, South-East Asia, Australia and New Zealand. The Indian Ocean is endowed with a rich variety of natural resources. Right from the ancient times, Indian Ocean always plays a vital role in Indian economy and politics.

Unity in Diversity

India is an example of 'Unity in Diversity'. Because of the vastness of the country, there are several elements of diversity which can be categorized as:

1. Physical diversities
2. Economic diversities
3. Socio-cultural and demographic diversities

1. *Physical Diversities*
   - (i) Structural
   - (ii) Topographical
   - (iii) Drainage pattern
   - (iv) Climatic
   - (v) Soil
   - (vi) Vegetation
   - (vii) Water resources
2. *Economic Diversities*
   - (i) Agricultural
   - (ii) Means of irrigation
   - (iii) Power resources
   - (iv) Minerals
   - (v) Industrial

(vi) Transportation
(vii) Communication

3. *Socio-cultural and Demographic Diversities*
   (i) Racial
   (ii) Language
   (iii) Religion
   (iv) Tribes
   (v) Food habits, etc.
   (vi) Human Resources

Unifying Elements

In spite of the above mentioned diversities, India is united because of geo-economic and socio-cultural elements also.

The factors of 'Unity in Diversity' are:

1. Physiographical integration.
2. Monsoon climate acts as the most powerful unity factor.
3. Cultural values based on mutual understanding.
4. The basic philosophy of all the religions is unity based on co-operation.
5. The philosophical cultural base is *'Vasudhave Kutumbkam'* (वसुधैव कुटुम्बकम).
6. Agricultural base of economy.
7. Spatial linkages through transportation and communication.
8. Apart from all, the Constitution of India is the most unifying element of the country.

Administrative Set-up

At present, there are 29 states and 7 union territories in India. Table 1.2 indicates the states, union territories, the areas and capital cities.

Table 1.2 *States and Union Territories of India*

	State	*Area (sq km)*	*Capital*
1.	Andhra Pradesh	1,60,200	Hyderabad
2.	Arunachal Pradesh	83,743	Itanagar
3.	Assam	78,438	Dispur
4.	Bihar	94,163	Patna
5.	Chhattisgarh	1,36,034	Raipur
6.	Goa	3,702	Panaji
7.	Gujarat	1,96,024	Gandhi Nagar
8.	Haryana	44,212	Chandigarh
9.	Himachal Pradesh	55,673	Shimla
10.	Jammu & Kashmir	2,22,236	Srinagar
11.	Jharkhand	79,714	Ranchi
12.	Karnataka	1,91,791	Bangalore (Bengaluru)

Cont'd...

...Cont'd

13.	Kerala	38,863	Thriuvananthapuram
14.	Madhya Pradesh	3,08,000	Bhopal
15.	Maharashtra	3,07,713	Mumbai
16.	Manipur	22,727	Imphal
17.	Meghalaya	22,429	Shillong
18.	Mizoram	21,081	Aizawl
19.	Nagaland	16,579	Kohima
20.	Odisha	1,55,707	Bhubaneshwar
21.	Punjab	50,362	Chandigarh
22.	Rajasthan	3,42,239	Jaipur
23.	Sikkim	7,096	Gangtok
24.	Tamil Nadu	1,30,058	Chennai
25.	Tripura	10,492	Agartala
26.	Uttar Pradesh	2,38,566	Lucknow
27.	Uttarakhand	53,484	Dehradun
28.	West Bengal	88,752	Kolkata
29.	Telangana	1,14,800	Hyderabad
	Union Territories		
1.	Andaman and Nicobar Islands	8,249	Port Blair
2.	Chandigarh	114	Chandigarh
3.	Dadra and Nagar Haveli	491	Silvassa
4.	Daman and Diu	112	Daman
5.	Delhi	1,483	Delhi
6.	Lakshadweep	32	Kavaratti
7.	Pondicherry	492	Pondicherry
	INDIA	32,87,263	New Delhi

It becomes clear from Table 1.2 that Rajasthan with an area of 3,42,239 sq km is the largest state of India followed by Madhya Pradesh and Maharashtra having an area of 3,08,000 and 3,07,713 sq km respectively. In fact, these three states each are having an area more than three lakh sq km. On the other end, Goa is the smallest state of India with an area of 3,702 sq km only. The five smallest states of India in ascending order are Goa, Sikkim, Tripura, Nagaland and Mizoram. Lakshadweep is the smallest union territory having an area of only 32 sq km; while Andaman and Nicobar Islands with an area of 8,249 sq km is the largest union territory.

Administratively, India has 29 States, 7 Union Territories, 640 Districts, 5,924 Sub-divisions (Tehsils), 7,936 Towns and 6.41 lakh Villages.

Figure 1.1 *Political Map of India*

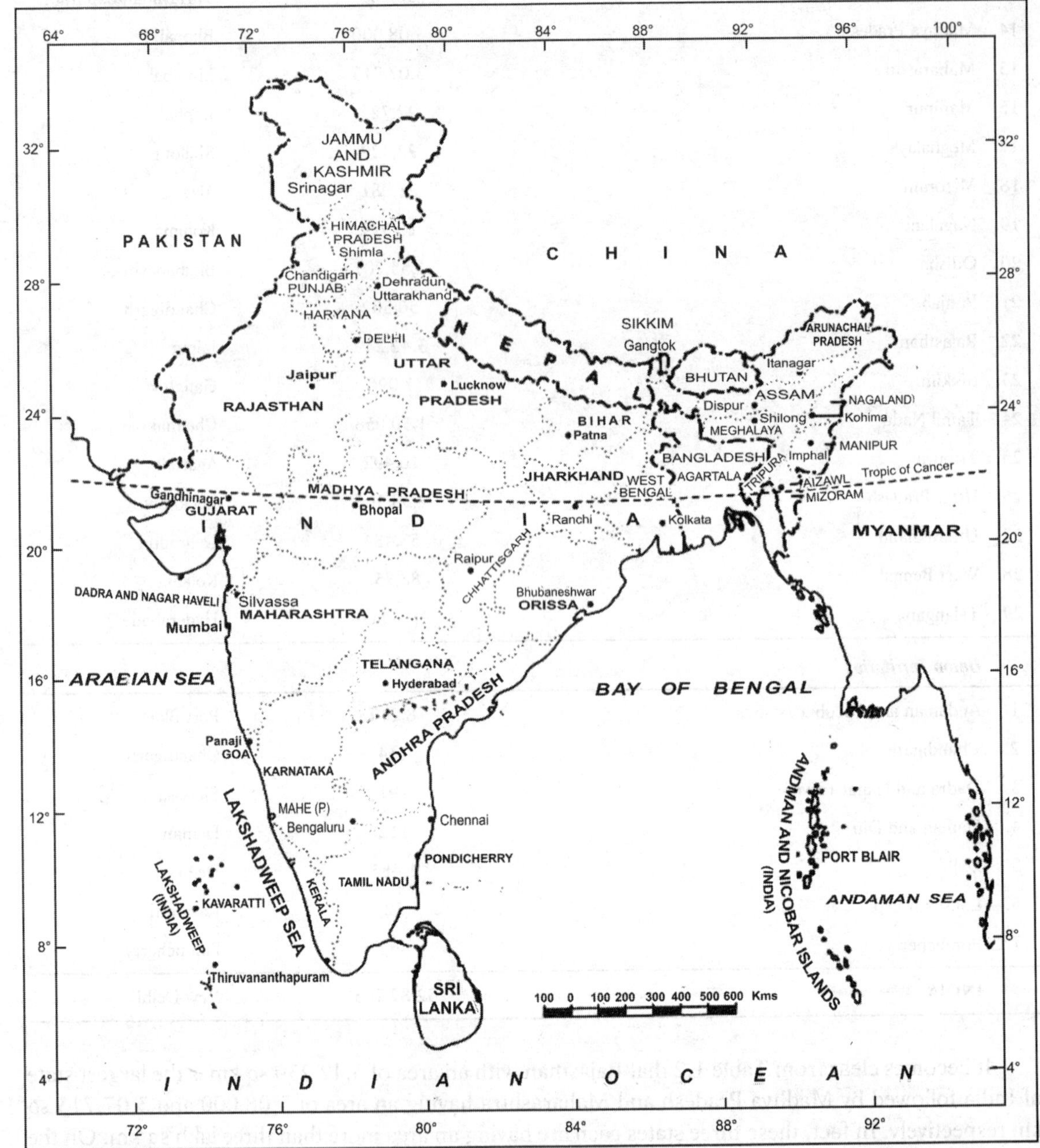

2

Geological Structure

The geological structure of India is very complex and unique. The rock formation in different parts of the country varies from most ancient to recent. Indian Peninsula is the oldest landmass and was a part of the Gondwana land, which included India, Australia, South Africa and South America. The split and deft of Indo-Australian Plate from Gondwana land resulted in the existence and location of India. The northward movement of the Indian Plate and its collision with Eurasian Plate was responsible for large scale folding of the sedimentary rocks of Tethys and has given rise to the Himalayan mountain system in Tertiary period which is geologically of a recent origin. In between the Peninsular India and the Himalayas, aggradational formation of the Indo-Gangetic Plain has occurred during the Pleistocene period. Geologically, India is divided into three regions, viz. (i) the Peninsular region, (ii) the extra Peninsular region (Himalaya), and (iii) The Indo-Gangetic Plain.

Geological History/Rock Formations

The history of rock formation of India ranges from Pre-Cambrian (600 million years ago) to the recent times. Sir T. Holland of the Geological Survey of India has traced the geological history of India and classified the four major divisions of the rock formation. These are:

1. The Archaean rock system
2. The Purana rock system
3. The Dravidian rock system
4. The Aryan rock system

The Archaean Rock System

The Archaean rock system is the oldest in the world. The word 'Archaean', first used by J.D. Dana, for rock structure belongs prior to Cambrian system. It includes two groups of rocks, i.e. gneisses and schists and Dharwar system.

1. *The Archaean system – Gneisses and Schists:* These oldest rocks were formed in Pre-Cambrian period. In India, about two-third of the peninsular rock is the gneiss in which mineral composition varies from granite to garbo. The schists are mostly crystalline rocks, and include mica, talc, hornblende, chlorite, etc. These rocks are found in north-east peninsula, large areas of Odisha, Madhya Pradesh, Chhattisgarh, Jharkhand (Chotanagpur Plateau) and Meghalaya.

2. *The Dharwar System:* The Dharwar system according to Wadia is the most ancient metamorphosed sedimentary rock-system of India. The rocks belonging to this system spread in Dharwar-Bellary-Mysore belt of Karnataka, Ranchi, Hazaribagh (Jharkhand), Balaghat and Rawa (Madhya Pradesh), Bastan, Dantewara and Kankar (Chhattisgarh), Sundergarh and Keonjhar (Orrisa) and in the Aravallis between Jaipur and Palampur. Meghalaya plateau also have such rocks. In rocks of Dharwar system, several useful minerals such as iron-ore, manganese, copper, lead, gold, quartzite, slates, mica, etc. are found.

The Purana Rock System

With reference to Proterozoic, the word 'Purana' has been used for the period between 1,400–600 million years ago. The two sub-divisions of this system are Cuddaph and Vindhyan system.

The Cuddaph system is made of shales, slates, limestone and quartzite and named after the district of Cuddapah in Andhra Pradesh. This occurs in the (i) Cuddapah and Kurnool districts of Andhra Pradesh, (ii) Chhattisgarh, (iii) Rajasthan-Delhi to the south of Alwar, and (iv) The Lesser Himalayas.

These rocks are having sandstone, shales, limestones, quartzites, slates and inferior quality of iron-ore, manganese, asbestos, copper, nickel, cobalt, marble, jasper, etc.

The Vindhyan System

The Vindhyan system derives its name from the Vindhyan mountains located between Ganga plain and the Deccan plateau. Vindhyan system stretches from Sasaram and Rohtas in Bihar to Chittorgarh in Rajasthan in an extensive area of more than one lakh sq km. These rocks are also found in Chhattisgarh and Andhra Pradesh. These rocks are well known for sandstone, and other building materials, conglomerates, diamondiferous and raw materials for cement, lime, glass and chemical industries. The three important series of Vindhyan system are Bhander series, Bijwar series and Kaimur series.

The Dravidian Rock System (Palaeozoic-Cambrian to Carboniferous Period)

The rocks of this system came into being about 600–300 million years ago. The Palaeozoic era includes the Ordovician, Silurian, Devonian, Carboniferous and the Permian periods of the geological time scale. Most rocks of this system are found in the extra peninsular region. These formations exist in the Salt Range, Pir-Punjal, Handwara, Liddes Valley, Anantnag (Jammu & Kashmir), Spiti, Kangra and Shimla region of Himachal Pradesh and Garhwal and Kumaon of Uttaranchal. The Cambrian rocks include shales, sandstone, clays, quartzites, shells, salts and marble. Coal formation started in the carboniferous period.

The Aryan Rock System

This system comprises rock formations from the upper Carboniferous to recent periods. The Mesozoic era includes three periods. Triassic, Jurassic and Cretaceous extends from the upper Carboniferous up to the beginning of the Cenozoic era.

The Gondwana system is considered to be the most important because most of the coal formation has occurred in this phase. The main areas of Gondwana rocks in the peninsula are along the Damodar valley in Jharkhand, along the Mahanadi river valley in Chhattisgarh and Orissa, in the southern part of Madhya Pradesh and a series of troughs along the Godawari from Nagpur to the delta. The other regions of these rocks are found in Kashmir, Darjeeling and Sikkim. India's 98 per cent coal deposits are found in these rocks.

The Triassic and Jurassic systems have limited areas in India, but Cretaceous system is widely distributed.

The Cretaceous System (Deccan Trap)

The Cretaceous system extends from about 146 million to 65 million years ago and represented by a wide variety of rocks, deposited in the land, sea, estuaries and lakes. No other system is so widely distributed in India as this system is, both in peninsular and the extra-peninsular parts.

From the end of the Cretaceous till the beginning of Eocene, stupendous volcanic outburst overwhelmed a vast area of peninsular India. These volcanic deposits are termed as 'Deccan Trap' or 'Deccan Lava'. The Deccan Trap has been divided into Upper, Middle and Lower Trap. The Upper Trap is mostly found in Maharashtra having layers of volcanic lava of about 450 m of thickness. While the Middle Trap is about 1,200 m thick and spreads over central India and Malwa. On the other hand, the Lower Trap is only 150 m thick and found in central India and Tamil Nadu. The weathering of these rocks for a long time has given birth to black cotton soil known as *'regur'*. Basalt is the main rock of Deccan Trap but dolarite, rthyolite, gabro and many other rocks are also found. The other minerals found in these areas are bauxite, magnetite, agate, magnesium, carbonate, potash, phosphates and semi-precious stones.

The Tertiary System

The Tertiary is the most significant period in India's geological history because in this period Himalayas were born. The rocks of Tertiary system were formed from Eocene to Pliocene, about 60 to 70 million years ago. The Tertiary period, is also known as the 'Age of Mammals' because of the abundance of fossil remains of these animals in the deposits of this period.

The two most significant events of the Tertiary Period are (i) the final breaking up of old Gondwana continent, and (ii) the rise of Himalayas from the Tethys geosyncline. The three phases of the upliftment of Himalayas are:

I Phase during Eocene (about 65 million years ago) up to Oligocene, resulted in upheaval of greater Himalayas.

II Phase during mid-Miocene (about 45 million years ago), which resulted in formation of Lesser Himalayas.

III Phase during post-Pliocene period (about 1.4 million years ago), resulted in formation of Shiwaliks or the outer Himalayas.

According to Geologists, the process of rising of Himalayas is still going on. Apart from Himalayan region, Tertiary System also occurred on coast of Kachchh, Konkan, Malabar, Nilgiris and the Eastern Ghats.

The Pleistocene and Recent (Quaternary) Period

This period is of nearly one million years and said to have just begun. The northern plains of India camc into existence during this period. Pleistocene period is marked by glaciations, while withdrawal of glaciations has started in recent period. During Quaternary period, the ice-sheets descended to as low as about 1,500 metres in altitude and resulted in large scale deposits of the moraine, which is responsible for the formation of Indo-Gangetic and Brahmaputra plain. Glaciation also lead to the formation of a number of glacial lakes in Himalayan region such as Kailash Kund, Gulmarg, Sheshnag, Sanasar, Kaunsarmag, etc. There is no evidence of glaciations in peninsular part of India.

3

Physiography

Physiographically, India is a land of great diversity. This diversity is evident in its topography or relief features. India's present day relief features have been superimposed on three basic structural units: the Himalayas in the north, the Deccan Plateau in the south and the Indo-Gangetic Plain between the two. These three divisions form the basis of the physiographic divisions, but some scholars treat coastal plains and islands as separate entity. Thus, India can be divided into following five physiographic regions:

1. The Himalayan Mountains
2. The Great Plains of India
3. The Peninsular Plateau
4. The Coastal Plains
5. The Islands

The Himalayan Mountains

The mountain ranges of the Great Himalayas are stretching over the northern borders of India. The Himalayas are also known as the *Himadri, Himavan* or *Himachal*. These are the youngest and the loftiest mountain systems of the world. The main Himalayas alone stretch for a distance of over 2,400 km and its width varies from 500 km in Kashmir to 200 km in Arunachal Pradesh. Pamir is the central part of Himalayas, also known as 'roof of the world'. From the Pamir, Himalayan ranges spread eastward as well as in west and south-westward directions. The Himalayan mountains are also known for the several lofty peaks. There are 14 mountain peaks over 8,000 metres high and another 20 peaks are above 7,500 metres high. The number of peaks over 7,000 metres is more than hundred. In fact, Himalayan ranges are the most extensive and consist of three parallel ranges.

Regarding the origin of Himalayas, several scholars have expressed their views. The views of O.H.K. Spate, D.N. Wadia, M.S. Krishnan, S. Burrard, E.H. Pasco, G.E. Pilgrim, De Terra, T.T. Paterson, T. Hagen, Anden, A. Heim, A. Gansser, etc. are significant to some extent. There is no need to go into the details of the origin of Himalayas, because the main concern of this chapter is the relief features, which have had direct impact on the geo-economy of India.

All the scholars have a general agreement that Himalayas have come out of the geosyncline known as *Tethys Sea*, located between Gondwana land in the south and Angaraland in the north. Deposition of sediments in this sea during several geological periods and powerful compression either southward movement of the Angaraland or due to northward movement of Gondwana land has resulted into folding of rocks, which is responsible for the origin of Himalayas.

Another acceptable theory is the *Plate Tectonic* put forward by W.J. Morgon. This theory put forward in 1967 is based on the concept of 'Sea-Floor Spreading' given by H.H. Hess. Plate tectonics is a theory of global dynamics in which lithosphere is believed to be broken into series of plates that move in response to convection in the upper mantle. In case of Himalayas, about 65–30 million years ago, Indian plate came very close to the Asian plate and started subducting under the Asian plate. This caused lateral compression due to which the Tethys sediments were squeezed and folded into three parallel ranges of the Himalayas. Since northward movement of the Indian plate is still continuing, the height of the Himalayan peaks is increasing.

Figure 3.1 *India: Physiographic Divisions*

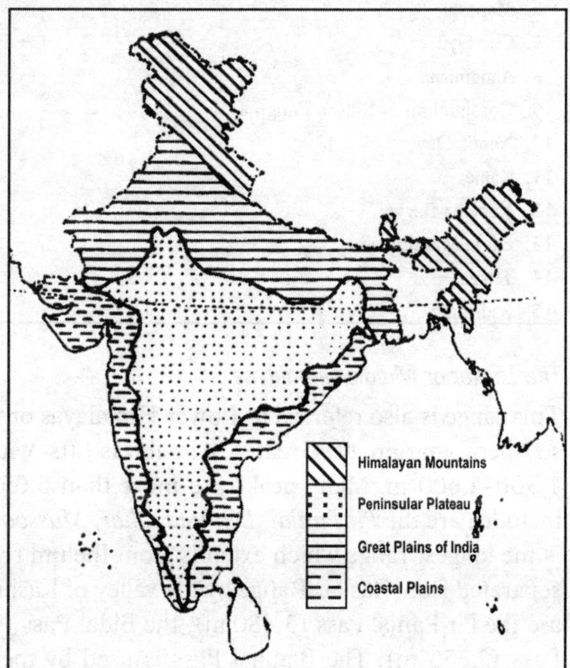

Physiographic Divisions of the Himalayas

Physiographically, Himalayas can be divided into the following divisions:

1. The Great Himalayas or Inner Himalayas
2. The Lesser or Middle Himalayas
3. The Shiwaliks or Outer Himalayas
4. The Trans Himalayas

The Great Himalayas or Inner Himalayas

The Great Himalayas are about 25 km wide with an average height above 6,100 metres and these mountains rise abruptly like a wall north of the lesser Himalayas. Most of the lofty peaks of Himalayas *Mt. Everest, Kanchanjunga, Nangaparbat, Gasherbrun, Manaslu, Dhaulagiri, Annapurna, Gosainthan, Cho-Oyu, Nanda Devi, Kamet, Badrinath,* etc. lie in this division. The highest peak not only of Himalayas but of the world is Mount Everest, which is 8,850 m above mean sea level (earlier the height of Mt. Everest was considered 8,848 m. But, according to the measurements made by the National Geographic Society using GPS satellite equipment on May 2000, the height of the peak is 8,850 metres). Its Nepalese name, is '*Sagarmatha*' means '*The Goddess of the Sky*', while Tibetans call it '*Chomlungma*'.

Some important peaks of the Great Himalayas are as follows:

Name of the Peak	*Height above Sea Level (metres)*
1. Mount Everest	8,850
2. Kanchenjunga	8,598
3. Lhotse	8,501
4. Makalu	8,481
5. Dhaulagiri	8,172

Cont'd...

...Cont'd

6. Manaslu	8,156
7. Cho Oyu	8,153
8. Annapurna	8,078
9. Gosainathan or Shisha Pangma	8,013
10. Nanda Devi	7,817
11. Kamet	7,756
12. Namcha Barwa	7,756
13. Gurla Mandhata	7,728
14. Trisul	7,140
15. Badrinath	7,138

The Lesser or Middle Himalayas

This range is also referred as Lower Himalayas or the Himachal Range. This range extends along the southern margin of Greater Himalayas. Its width is about 80 km with an average height of 1,300–4,600 m. Many peaks are more than 5,050 m above mean sea level. The important ranges included are the *Pir Panjal, Dhalola Dhar, Mussoorie and Nag Tiba*. The Pir Panjal range in Kashmir is the longest range which extends from Jhelum river to the upper Beas river for 300–400 km and is separated from Zaskar Range by the valley of Kashmir. The well-known passes of the Pir Panjal range are the Pir Panjal Pass (3,480 m), the Bidal Pass (4,270 m), Golabghar Pass (3,812 m), and Banihal Pass (2,835 m). The Banihal Pass is used by the Jammu–Srinagar highway. The famous valley of Kashmir is located between Pir Panjal and Zaskar Range over a distance of 135 km. Other famous valleys are Kangra and Kulu Valley. Many Himalayan hill resorts like Shimla, Mussoorie, Nainital, Ranikhet, Almora, Lansdowne, Darjeeling, etc. are located here.

The Shiwaliks or Outer Himalayas

The Shiwaliks are the outermost range of Himalayas, which run parallel to the Lesser Himalayas for a distance of about 2,400 km from Potwar Plateau to the Brahmaputra valley. The width of these ranges varies from 50 km in Himachal Pradesh to less than 15 km in Arunachal Pradesh with an average height of about 900 to 1,500 metres above mean sea level. There are several longitudinal valleys located between Shiwaliks and the Lesser Himalayas known as *'Doons'* or *'Duns'*. Important Duns are Dehra Dun, Potli, Kothri, Kathmandu, Chumbi and Kyarda.

The Trans Himalayas

The Trans Himalayas are the northern most ranges located north of the Greater Himalayas. This part of Himalayan ranges is also known as the 'Tibetan Himalaya' because most of it lies in Tibet. The *Zaskar, Ladakh, Kailash* and *Karakoram* are the main ranges of Trans Himalayas. Its length is about 1,000 km in east-west direction and average elevation is 3,000 m above mean sea level.

Regional/Longitudinal Divisions of Himalayas

Sir S. Burrard divided the Himalayan ranges into four regional divisions, namely (i) The Punjab Himalaya (ii) The Kumaon Himalaya (iii) The Nepal Himalaya (iv) The Assam Himalaya

Another division has been suggested by S.P. Chatterjee (1973). He divided these ranges into five transverse divisions from west to east. These are: (i) The Kashmir Himalayas, (ii) The Himachal Himalayas, (iii) The Kumaon Himalayas, (iv) The Central Himalayas, and (v) The Eastern

Himalayas. Both the above divisions are based on their regional location from west to east and accordingly their names have been suggested.

Passes of Himalayas

A characteristic feature of the Himalayan topography is a number of passes located in these ranges. The main passes are as follows:

Name	*Location/Elevation*
Aghil Pass	Karakoram-Ladakh in J&K, 5,000 m.
Banihal Pass	J&K, Jawahar Tunnel, Joins Jammu with Srinagar, 2,835 m
Bara Lancha	J&K, Manali-Leh National Highway pass through it, 4,843 m
Bomdi La	Arunachal Pradesh, 4,331 m.
Burzail Pass	J&K, 5,000 m.
Chang – La Pass	J&K, 5,270 m.
Dibsa Pass	Himachal Pradesh, 5,270 m
Dihang Pass	Arunachal Pradesh, 4,000 m.
Diphu Pass	Arunachal Pradesh, Shortest access to Myanmar.
Imis La	J&K, 4,500 m
Khardung La	J&K, joins Leh with Siachin glacier, above 6,000 m.
Jelepa La	Sikkim, connects Sikkim with Lhasa, 4,538 m.
Likhapani	Arunachal Pradesh, above 4,000 m.
Lipu Lekh	Uttarakhand, connects Uttarakhand with Tibet. The pilgrims for Mansarover Lake travel through this pass.
Nathu La	Sikkim, located on Indo-China border, 4,310 m.
Pir-Panjal Pass	J&K, connects Jammu to Kashmir valley with shortest route.
Rohtang Pass	Himachal Pradesh, connects Kullu with Lahul and Spiti valley, 3,979 m.
Shipki-La	Himachal Pradesh, 4,300 m.
Thang-La	Ladakh (J&K), 5,359 m.
Zoji La	J&K, joins Srinagar with Kargil and Leh, 3,850 m.

Apart from the above mentioned passes, there are several other passes located in the Himalayan ranges.

Significance of the Himalayas

Himalayas are not only a distinct feature of the geography of India but always remain an integral part of the socio-cultural history of India. Himalayas are considered as abode of God and the body and soul of India. Himalayas have direct impact on physical, social, economic and political structure of India. The significance of Himalayas can be summarized in the following points:

1. Controller of Indian monsoon as well as climate of India
2. Strategic importance with reference to China and Nepal
3. Source of several rivers
4. Potentialities of hydro power
5. Rich forest wealth
6. Rich in biodiversity
7. Number of tourist resorts are located. Tourist and adventure sports can be developed.

8. Several pilgrimage centers such as Kailash, Amarnath, Badrinath, Vaishno Devi, Jwalaji, Uttarkashi, Gangotri, Yamnotri, etc. are located in Himalayan region.
9. Possibilities of some minerals such as coal, copper, lead, zinc, nickel, cobalt, antimony, tungsten, gold, gypsum are also predicted.
10. Source of fertile soil.

In fact, 'There are no mountain ranges anywhere in the world which have so much shape the life of a country as the Himalayas have in respect of India' (K.M. Panikkar, 1959). Panikkar further states that 'in a very special measure the Himalayas is India's national mountain – the very description of India which has come down to us by tradition is '*Himavat-Setu Paryantam*' from the 'Himalayas to Rameswaram'.

The Great Plains of India

The Great Plains of India is located in a transitional zone between the Shiwalik hills of Himalayas and the Peninsular India. This is an aggradational plain formed by the alluvial deposits of Indus, Ganga, Brahmaputra rivers and their tributaries. The plain stretches for about 2,400 km in Indian sector and its width varies from 150 to 300 km. Regarding the origin of this plain, divergent views have been expressed but it is universally accepted that this plain has been formed as a result of filling of deep depression between Peninsular and Himalayan region by the depositional work of the rivers coming from these to land masses.

Physiographic Divisions of the Great Plain

Although, the Great Plain of India is monotonous and has more or less a homogenous surface and featureless, but due to variations in depositional features, this region can be divided into following sub-regions:

The Bhabar Plain is a narrow belt about 8–16 km wide running in east-west direction along the foothills of Shiwaliks from Indus to the Tista. This area consists of gravel and unassorted sediments deposited by the rivers descending from the Himalayas and the Shiwalik. The porosity is so high that most of the small streams disappear in this fact. In this region, only big trees with long roots can survive and the region is not suitable for agriculture.

The Tarai: This is a 15–30 km wide belt located south of Bhabar tract. Due to re-emergence of streams, the area becomes an ill drained marshy land. The Tarai lands of Punjab, Uttar Pradesh and Uttaranchal have been reclaimed and turned as agricultural land.

The Bhangar or Bangar: These are older alluvial plains of the Middle Pleistocene age. The Bhangar land lies above the flood limits of the rivers. This region is dominated by nodules of impure calcium carbonate or *'kankar'* and also having small patches of saline and alkaline soil known as *'Reh'*, *'Kallar'* or *'Thur'*.

The Khadar Plains are the newer alluvium and form the flood plains along the river banks. The Khadar tracts are rich in alluvium soil deposited every year. Khadar lands are now used for the cultivation of various crops.

Delta plains: These plains are the extension of Khadar plains and cover an area of about 1.9 lakh sq km in lower reaches of Ganga River. These are marshy lands. The uplands of the delta region are called '*Chars*' and the marshy lands are known as '*Bils*'.

Apart from the above mentioned lands of the northern plains, in Upper Ganga-Yamuna Doab, *'Bhur'* denotes an elevated piece of land situated along the banks of river.

Regional Divisions: The Great Plain of India is divided into the following regional divisions:

1. The Rajasthan Plain
2. The Punjab-Haryana Plain
3. The Ganga Plain
4. The Brahmaputra Plain

These regions are different geo-climatically as well as in topography.

The Peninsular Plateau

The Peninsular India is a plateau which spreads southwards of the North Indian Plain in the form of a triangle with its broad base to the north and the narrow apex to the south. This region is surrounded on three sides by water and thus is a peninsular plateau. The Bay of Bengal, the Arabian Sea and the Indian Ocean flank it on its three sides. The entire plateau covers about 16 lakh sq km, and thus forms the largest physiographic division of India. It is bounded by the Aravalli range in north-west, the Vindhya, Satpura hills in the north and Hazaribagh and Rajmahal hills in the northeast while its western and eastern boundaries are formed by the Western Ghats and the Eastern Ghats. The average height of the plateau is 600–900m above mean sea level although many parts are above 1,000 m. The Peninsular plateau is an ancient tableland composed mostly by Archaean gneisses and schists. The physiography of this region is varied and has several highlands and plateaus. A brief description of these major topographical features is as follows:

The Plateaus and Uplands of the Peninsular India:

1. *The Marwar upland:* The upland of eastern Rajasthan lies in east of Aravalli range having an average height of 250–500 above mean sea level.
2. *The Central Highland* also known as *'Madhya Bharat Pathar'* is located east of Marwar uplands and drained by Chambal river.
3. *The Bundelkhand Upland:* These are located between Madhya Bharat Pathar and the Vindhyan Scarp lands. The average height of this region is 300–600 m above sea level.
4. *The Malwa Plateau:* It is bounded by Aravallis in the north, the Vindhyan Range in the south and the Bundelkhand in the east. It has two drainage systems-Narmada and Mahi towards the Arabian Sea and Chambal, Sind, Betwa and Ken towards the Bay of Bengal.
5. *The Baghelkand or Vindhyanchal Plateau:* This region is located east of the Maikal Range and includes plateau of Satna, Rewa (MP) and Mirzapur (UP). The region has a general elevation of 150 m to 1,200 m and uneven relief. Narmada-Son trough (rift valley) and scarps of Vindhyan are the characteristic features.
6. *The Chotanagpur Plateau:* It spreads over parts of West Bengal, Jharkhand, Chhattisgarh, Odisha and north eastern part of Andhra Pradesh. The Hazaribagh Plateau, Ranchi Plateau and Rajmahal hills are the part of this plateau. The plateau is drained by numerous rivers, among them Damodar, Subarnrekaha, North Koel and South Koel and Barkar are important.
7. *The Meghalaya Plateau:* This part of plateau has been detached from Indian Peninsula by Malda Gap and includes Garo, Khasi, Jantia hills and outlying Mikir and Rengma hills.
8. *The Deccan Plateau:* Spread over an arca of about five lakh sq km, this triangular plateau is bounded by the Satpura and the Vindhya in the north-west, Mahadev and Maikal in the north, Western Ghats in the west and Eastern Ghats in the east. The average height of the region is 600 m and its general slope is from west to east. Mahanadi, Godawari, Krishna, Cauvery are the main rivers flowing in this region. Further subdivisions of this region are *Maharashtra Plateau, Karnataka Plateau, Telangana Plateau, Mahanadi Basin, Chhattisgarh Plain, Dandakaranya, Tamil Nadu Upland,* etc.

Hills of the Peninsular India

Several hills are also located in this region. Among them Aravalli, Vindhyan, Satpura ranges and Western and Eastern Ghats are prominent.

The Aravalli Range is also included in hills of Peninsular India. Aravalli runs from north-east to south-west for about 800 km between Delhi to Palampur in Gujarat. These are the oldest mountains of the world. *Guru Shikar* is the highest peak of 1,722 metres. Other peaks of Aravalli range are *Ser* (1,597 m), *Delwara* (1,442 m), *Jarga* (1,431 m), *Achalgarh* (1,380 m), *Mt. Abu* (1,295 m), *Kumbalgarh* (1,224 m), etc.

The Vindhyan Range: These ranges are more or less parallel to Narmada Valley in east-west direction from Jobat in Gujarat to Sasaram in Bihar at a distance of over 1,200 km. Bharner and Kaimur hills are their eastward extensions. The average height of these ranges is 300 to 600 m.

The Satpura Range: The Satpura range is a series of seven mountains. It runs in an east-west direction south of the Vindhyas and in between Narmada and Tapi rivers. *Rajpipla hills, Mahadev hills* and *Maikala range* are parts of these hills. *Dhupgarh* (1,350 m) is the highest peak located near Pachmarhi on Mahadev hills. *Astamba Dongarr* (1,325 m) and *Amarkantak* (1,127 m) are other important peaks.

The Western Ghats: These are rocky hills extending in a narrow belt over about 1,600 km in north-south direction from the mouth of Tapi river (Maharashtra) to Kanya Kumari (Tamil Nadu). These hills are also called '*Sahyadris*'. The average width of these Ghats varies from 50 km in north to 300 km in south. The slope of the Ghats is steeper towards west, while the eastern slope is gentle. Important rivers of Peninsular India like Godawari, Krishna and Kaveri rise from the Western Ghats. *The Gersoppa (Jog Falls)* on Sharvati river is the highest waterfall in India. These Ghats form a physical barrier to movement from coastal areas to the interior. The *Palghat, Borghat* and *Thalghat* are important passes. The important peaks of the Western Ghats are *Kundermuch (1,892 m), Pushpagiri* (1,714 m), *Kalsubai* (1,646 m), *Salher* (1,567 m) *Mahabaleshwar* (1,438 m) and *Harishchandra* (1,424 m). The Nilgiri hills are the southernmost part of Western Ghats, which joins them with Eastern Ghats. At the meeting points of these two is located the *Anaimudi* (2,695 m) peak, which is the highest in Peninsular India. The other important peaks are *Doda Betta* (2,637 m) and *Makurti* (2,554 m). From the nodal point Anaimudi, three ranges radiate, these are – *Anaimalai* (1,800–2,000 m) to North, *Palni* (900–1,200 m) to the north-east and the *Cardamom Hills* (Ealaimalai) to the South.

Figure 3.2 *Relief of Peninsular India*

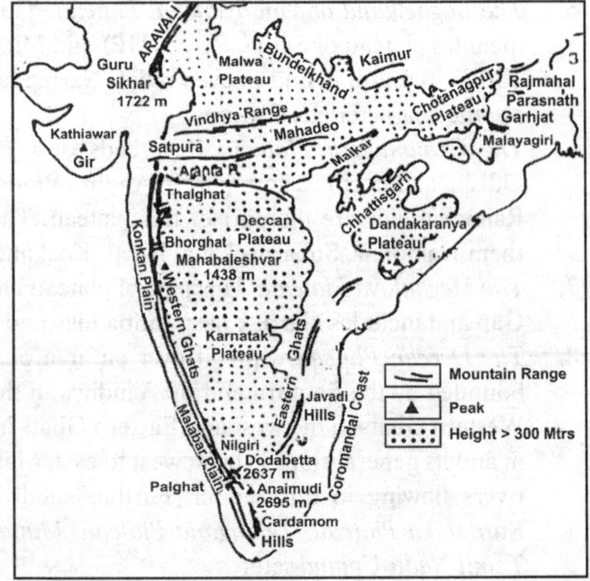

The Eastern Ghats: The Eastern Ghats form the eastern boundary of Deccan plateau and run almost parallel to the east coast of India. In contrast to the Western Ghats, the Eastern Ghats are not a continuous range but it is a chain of broken and detached hills. The average height of the Eastern Ghats is about 600 m. These Ghats extend along the eastern coast of India over a length of 800 km from

Mahanadi in the north to the Nilgiri hills in the south. The Mahanadi, Godawari, Krishna and Kaveri rivers flow eastward through the gaps in the Ghats. *Aroya-Konda* (1,680 m) is the highest peak of the Eastern Ghats located in Andhra Pradesh. Other important peaks are *Dewodi-Munda* (1,598 m), *Singa-Raju* (1,516 m), *Nimalgiri* (1,515 m) and *Mahendragiri* (1,501).

The Coastal Plains

The coastal plains of India stretch along the coasts of Bay of Bengal and the Arabian sea. The two plains, western and eastern differ from each other in respect of their geomorphologic characteristics and width. Although, both have formed depositional actions of rivers as well as erosional and depositional actions of the sea waves.

The West Coastal Plain stretches from Rann of Kachchh in the north to Kanyakumari in the south. The length of the plain from Gujarat coast in the north to Kerala coast in the south is about 1,500 km and average width is about 65 km. This plain is known with different names in various parts. The northernmost part of this plain is called the *Gujarat Coastal Plain,* while the southernmost is known as the *Kerala Coastal Plain.* In between these two lies *Konkan Coastal Plain* in the north and *Malabar Coastal Plain* in the south.

The Kachchh Peninsula is an extensive tract of tidal mudflats and is known as 'Rann of Kachchh'. The salt in the soil makes this low-lying marshy area almost barren and unproductive. The entire coastal area is characterized by sandy beaches, coastal sand-dunes, mud-flats, lagoons, alluvial tracts along rivers, laturite-platforms and residual hills. There are several shallow lagoons and backwaters, known as *Kayals* and *Teris*.

The Eastern Coastal Plain: This coastal plain lies between the Eastern Ghats and the Bay of Bengal and stretches along the coasts of West Bengal, Odisha, Andhra Pradesh and Tamil Nadu. The major part of the plain is formed by the alluvial deposits by the rivers Mahanadi, Godawari, Krishna and Cauvery. The deltas of these rivers are also located in this plain. The Eastern Plain is divided into three regions: *Utkal Plain, Andhra Plain* and *Tamil Nadu Plain*. The southern part is also known as *'Coromandal Coast'*.

The Utkal Plain is the coastal area of Odisha. It includes the Mahanadi delta. *Chilka Lake* is located in this area, which is the biggest lake in the country with an area of 780 sq km in winter and 1,144 sq km in the monsoon months.

The Andhra Plain is located south of the Utkal plain and extends upto Pulicat Lake, some 40 km north of Chennai. This lake has been barred by a long sand spit known as *Sriharikota Island,* on which is located the satellite launching station of the Indian Space Research Organization. The deltas of Godawari and Krishna are located in this area.

The Tamil Nadu Plain stretches from Pulicat lake to Kanyakumari along the coast of Tamil Nadu in about 675 km. Cauvery delta is the characteristic feature of this plain.

The main difference between the eastern and western coasts is that eastern coastline is primarily a coastline of emergence, while the western coastline is the coastline of submergence. Another major difference is that in eastern plains, several deltas have been formed, while in western plains, there is an absence of delta.

The Islands

The Islands located in Indian territories are distinct physiographic units. The Indian islands can be grouped as (i) Islands of Bay of Bengal, and (ii) Islands of the Arabian sea.

Islands of Bay of Bengal

The two large groups of islands located in the Bay of Bengal are – Andaman and Nicobar. These two groups form the Union Territory of India. The Andaman group has 325 islands which cover an area of 6,710 sq km; while the Nicobar group has only 24 islands with an area of 1,765 sq km and the two is separated by the *Ten Degree Channel* from the Nicobar group in the south. The Andaman is a closely knit group of 325 islands. It is 260 km long and 30 km wide with a total area of 6,596 sq km. This group of islands is divided into three major groups, viz. North Andaman, Middle Andaman and South Andamans. Little Andaman is separated from the Great Andaman by a 50 km wide Duncan passage. Most of the islands in Andaman group are formed by sandstone, limestone and shale rocks. But two of them – Narcondam island and Barren island are of volcanic origin. Saddle Peak in North Andaman is the highest peak with a height of 737 m.

The Nicobar group includes 24 islands, among them, the Great Nicobar Island is the largest. The Car Nicobar is the northernmost island of this group. The southernmost point of India, the *Indra Point* is located on the southern tip of the Great Nicobar Island. Some of the islands of Nicobar group are of coral origin.

Islands of Arabian Sea

Lakshadweep and *Minicoy* are the two groups of Arabian Sea islands. The Lakshdweep Islands is a group of 25 small islands. They are widely scattered over an area of 108.78 sq km. The northern group of islands in the Lakshadweep is called the *Amindivi Island* and the southernmost part, the Minicoy island. The *Nine Degree Channel* lies between Minicoy Islands. These two islands are of coral origin. These islands are also a Union Territory of India.

Offshore Islands

A number of islands are located in the delta region of Ganga, the Gulf of Mannar and on western coast. These are *New Moore, Ganga Sagar* and *Sagar* (Ganga Delta), *Short, Wheeler* (Mahanadi-Brahmani mouth), *Sri Harikota* (mouth of Pulicat Lake), *Vypin* near Kochi, *Pamban, Crocodile, Adunda* (Gulf of Mannar), *Bhatkal, Pegioncook, St. Marry* (Mangalore Coast), *Butches, Elephanta, Karanja, Cross* (Near Mumbai), *Khadiabet, Aliabet* (Narmada-Tapi mouths), *Karunbhar, Diu, Vaida, Nora, Pirotan* (Kachchh coast), *Piram, Bhaisala* (Kathiawar) etc.

4

Drainage (Rivers), Lakes and Waterfalls

Drainage (Rivers)

The drainage pattern or river system of India is determined by its geological structure and topography. The two broad drainage systems of India are: (i) Himalayan river system including Indus, Ganga and Brahmaputra rivers and their tributaries, and (ii) Rivers of Peninsular India and their tributaries; the main rivers are Mahanadi, Godavari, Krishna, Kaveri, Narmada and Tapi (Tapti).

The area drained by main river and all its tributaries is known as its 'river basin'. India has one hundred and thirteen river basins, of which 14 are large (catchment area of more than 20,000 sq km), 44 medium (catchment area between 2,000 to 20,000 sq km) and 55 minor (catchment area less than 2,000 sq km) river basins. The major river basins and their basin area are given in Table 4.1

Table 4.1 *Major River Basins of India*

River Basins	*Basin Area (sq km*)*	*Percentage Area*
Ganga	8,61,404	26.2
Indus	3,21,284	9.8
Godavari	3,12,812	9.5
Krishna	2,58,948	7.9
Brahmaputra	2,58,008	7.8
Mahanadi	1,41,589	4.3
Narmada	98,795	3.0
Kaveri	87,900	2.7
Tapi	65,150	2.0
Pennar	55,213	1.7
Brahamani	39,033	1.2
Mahi	34,481	1.0
Subarnrekha	19,296	0.6
Sabarmati	21,895	0.7
Medium and Minor Rivers	7,11,833	23.6
Total	32,87,697	100.00

Source: S.P. Das Gupta, 1989 * Basin area in India.

A brief description of the rivers of India is as follows:

The Rivers of North India

The north Indian river system is also known as 'Himalayan Drainage System', because most of the rivers of this system have their origin from the glaciers of the Himalayas. The Himalayan rivers are although perennial but have seasonal fluctuation in the form of floods during rainy seasons. These rivers are not only eroding agents but also act as depositing agents in the plains and deltas; thus, creating fertile plains of north India.

The Indus River System

The three river systems of the north India are the Indus, the Ganga and the Brahmaputra.

River Indus also known as 'Sindhu' along with its tributaries is one of the biggest river systems of the world. The length of the river is 2,880 km, of which 709 km lies in India. The total catchment area of Indus is about 11,65,000 sq km out of which about 3,21,284 sq km is in India. River Indus rises from the Bokhar Chu in northern slopes of Kailash Range in the Tibetan Plateau. A number of tributaries such as *Gilgit, Gortang, Dras, Shiger, Hunza* joined it as Himalayan tributaries. But its most important tributaries are *Sutlej, Chenab, Ravi, Vyas* and *Jhelum.* After flowing through the territories of India and Pakistan, it falls into Arabian Sea. The following table shows the details of the Indus drainage system.

Table 4.2 *The Indus Drainage System*

Name of the River	*Source*	*Length (km)*
Indus	Near Mansarovar Lake at 5,182 m elevation	2,880 (709 in India)
Jhelum	Verinag at 4,900 m altitude	724
Chenab	Bara Lancha Pass	1,180
Ravi	Near Rohtang Pass	725
Beas	Near Rohtang pass at 4,062 m height	460
Sutlej	Mansarovar-Raks Lakes at 4,570 m height	1,450 (1,050 in India)

River Jhelum (Vitasa): It rises from a spring at Verinag at an altitude of 4,900 m in the south-eastern part of Kashmir valley. It flows north-westward from its source to Wular Lake for about 110 km and further down south-westward. Its tributaries Lidar, Sind and Pohru, rise in Kashmir and join the main river. It joins Chenab river at Trimmu.

River Chenab (Asikni): It originates from near the Bara Lancha pass in the LahulSpiti part of the Zaskar Range. Two streams Chandra and Bhaga join it. This united river is called Chandrabhaga in Himachal Pradesh. After flowing parallel to the Pir Panjal range, it enters Jammu & Kashmir as Chenab river. Salal, Baghliar and Dulshi are the important hydel projects located on river Chenab.

River Ravi (Parushni or Iravati): The Ravi originates from Kullu and from the triangle formed by the Pir Panjal and Dhauladhar ranges, flows west, entering the Punjab plains near Madhopur. Three hundred and seventy six kilometers from its origin, 26 km from Amritsar, it enters Pakistan near Lahore.

River Beas (Vipas or Argikiya): The river Beas originates near the north Rohtang Pass in Kullu (4,062 metres altitude). Flowing down the steep slopes, it forms a meandering course and crossing the Shiwalik ranges, reaches Hoshiarpur, flowing in the western direction. Forming a deep valley, it reaches Talwara from Larji and enters the Punjab plains. After flowing for 40 km, it flows into the Sutlej near Horike.

River Sutlej (Satadru or Satudri): The origin of Sutlej is at an altitude of 4,630 metres near the Darma pass in the Zaskar range of Tibet, flowing in a north-west direction; it flows over the northern slopes of the Kailash mountains. Then it turns south and reaches the Punjab plains near Ropar. Then flowing west upto Ferozepur, it forms the border between India and Pakistan and enters Pakistan near the Suliman mountains. After flowing for a stretch of 1,440 km, it meets the Chenab in Pakistan.

Ghaggar: The source of river Ghaggar is in Dagshai, at an altitude of 1,927 metres near Shimla in Himachal Pradesh. Throughout its length, the river flows in a south-west direction. Ten kilometers from its origin, near Kalka, the river enters Haryana. It winds in way along the borders of Punjab and Haryana. It receives the waters of the Tangari and Markanda rivers through the river Saraswati, which flows in from its left, near the Shatrana village. Further on the stream is called Hakra which get lost near Hanumangarh. In fact, river Ghaggar is an ephemeral stream coming into flow during rainy season. This river is also known as *'Dead River'*.

The Ganga River System

River Ganga and its large number of tributaries form the Ganga river system of North India. Table 4.3 indicates the drainage system of river Ganga.

Table 4.3 *The Drainage System of River Ganga*

Name of the River	*Source*	*Length (km)*
Ganga	Gangotri Glacier at 7,010 m	2,525
Yamuna	Yamnotri at 6,330 m	1,376
Chambal	Near Mhow (MP)	1,050
Ramganga	Garhwal district at 3,110 m	596
Ghaghara	Near Gurla Mandhota peak, south of Manasarovar	1,080
Gandak	Tibet-Nepal border at 7,620 m	425 in India
Kosi	Sikkim-Nepal-Tibet Himalaya	730 in India

River Ganga: The Ganga is the most important river of India and considered as a holy river of the country. It is an international river as it passes through Bangladesh before merging into the Bay of Bengal. The origin of this stream is at Gomukhi glacier, situated at an altitude of 7,016 km. After flowing turbulently for 25 km through the mountain caves/valleys, the Ganga enters the plain near Rishikesh. Flowing further downstream for 30 km, the Ganga reaches Haridwar, which is a holy centre of pilgrimage.

From its origin till Devprayag, the Gomukhi Ganga is known by the names of *Mandakini*, *Jahnavi* and *Bhagirathi*. It is on this river that the Tehri dam is being constructed. At Rudra Prayag, another mountain stream, *Alaknanda* meets the comparatively muddy and warm stream of the *Mandakini* and this combined stream is known as Bhagirathi. Beyond Haridwar, the Ganga assimilates the Ramganga before Allahabad. At Allahabad, its principal tributary, the Yamuna flows into it from the right, which is called the Triveni confluence, because here, the confluence of three rivers, Ganga, Yamuna and Saraswati occurs. The popular belief is that the third river Saraswati has disappeared.

After Allahabad, flowing east for 250 km, the Ganga reaches that great holy pilgrimage centre of Varanasi. After Varanasi, the river *Gomati* flows into the Ganga. Crossing Varanasi, the middle stretch of the Ganga's course begins, when forming the borders of Uttar Pradesh and Bihar for 100 km, it enters Bihar from the north-west. Here many important rivers flow into the Ganga from the north. Among these, the *Ghaghra* near Chhapra, *Gandak* near Patna, *Burhi Gandak* near the town of Munger and the *Kosi* near Kursda, are among the major rivers that flow into the Ganga. From the south, the rivers flowing into the Ganga are the *Karamnasa* near Chausa, *Sone* near *Koilwar* and *Punpun* which flows into the Ganga 25 km east of Patna. The Ganga, flowing east, touches the Rajmahal hills and ahead of Manihari Ghat enters the borders of West Bengal. It is said that the delta of the Ganga begins from Farakka itself, 40 km past Farakka, the stream of the Ganga gets bifurcated. The first branch, known by the name of *Padma*, enters Bangladesh flowing eastwards. The second branch flows in West Bengal, known by the name of *Bhagirathi*. In the north, the river *Mahanadi* flowing in from Darjeeling gets divided into two streams. The first branch meets the Ganga beyond Farakka and the second reaches Bangladesh to vanish into the stream of the Padma. Flowing south-east in Bangladesh, the Padma first meets the Brahmaputra and then combining with Meghna, flowing south, it gets divided into many branches and sub-branches ending up in the Bay of Bengal.

From its origin to its estuary, the total length of the river Ganga is 2,525 km which includes the length of the *Bhagirathi* and the Hooghly. Of this length, 1,450 km is in Uttar Pradesh, 110 km forms the borders of Uttar Pradesh and Bihar, 445 km is in Bihar and 550 km lies in West Bengal.

The Ganga has many tributaries, several of whom are quite famous These are the *Ramganga, Yamuna, Tons, Gomati, Ghaghra, Sone, Gandak, Punpun, Kiul, Burhi Gandak, Kuri* and the *Mahendra*. Along the course of the *Bhagirathi* and the *Hooghly*, the main tributaries of the Ganga are the *Dwarika*, *Ajoy*, *Damodar*, *Dwarkeshwari*, *Rupnarayan* and *Haldi/Ildi*. Brief details of important tributaries are as follows:

River Yamuna: The Yamuna is the most important and biggest tributary of the Ganga which flows into it from its right. Its origin is at an altitude of 6,330 metres in the Yamnotri glacier in the Tehri-Garhwal Himalaya in Uttar Pradesh. This river forms a border between Uttar Pradesh and Haryana for 320 km, then coursing through Delhi enters Uttar Pradesh covering a distance of 970 km till the confluence. The biggest tributary of the river Yamuna in the Himalayan region, the Tons, flows into it near Kalsi.

Another Himalayan river, *Gauri* originates near Shimla and flows into the Yamuna near Paonta Sahib. The Yamuna descends from the hills near Tajewala and enters the plains reaching Delhi via Muzaffarnagar. Here, the waters of the *Sahibi river* flows into the Yamuna through the Sahibi drain. After Delhi, the Yamuna flows for 203 km in a southward direction till Mathura. The *Hindon* flows into it near Dhankaur from its left. After Mathura, the river Yamuna flows in a south-east direction and flowing through the Agra, Mainpuri, Etah, Kanpur and Allahabad districts meets the Ganga at the Triveni confluence near Prayag. Meanwhile, several small rivers flow into the Yamuna from its left of which the *Karen*, *Sanger* and *Rind* are the major ones. The rivers flowing in from the right are more important of which the *Chambal*, *Sindh*, *Betwa*, *Dhasan*, and *Caine* are the major ones. The origin of all the rivers flowing in from the south is in the Vindhyan ranges. From its origin till its confluence with the Ganga, the total length of the Yamuna is 1,376 km.

Chambal: The Chambal is the main tributary of the Yamuna. Its length is 960 km. It originates from the Vindhyachal Hills near Mhow in Indore district of Madhya Pradesh. The Chambal flows for about 320 km in a northern direction till the borders of Madhya Pradesh and Rajasthan. From here, it enters Rajasthan and after flowing for 30 km, turns right and then flows in a north-east direction. After flowing for 520 km in a north-east direction, it assimilates the waters of its tributaries, the *Kali-Sindh* and *Kural* and forms the borders of Rajasthan and Madhya Pradesh. Near Rameshwar Village here, one of its principal tributaries, the *Banas* flows into it from its left. From here the Chambal river forming the borders of Madhya Pradesh and Uttar Pradesh for 117 km, flows in a north-east direction till village Pinhat. Then it turns right and flowing in a south-east direction enters Uttar Pradesh. After flowing for 46 km in Uttar Pradesh, the river Chambal meets the Yamuna near Sahan in the Etawah district.

Ramganga: Flowing in from the left, the Ramganga is the Ganga's first major tributary. From its origin till its confluence with the Ganga, its length is 596 km. It originates from an altitude of 3,110 metres near Lohaba Ganga in the Garhwal region of Uttar Pradesh. This river flows into the Ganga near Kannauj (UP).

Gomati: The origin of the Gomati is at an altitude of 200 meters, 3 km east of the Pilibhit town in the Pilibhit district of Uttar Pradesh. In the beginning, it waters the region between the Ramganga and Sharada and further downstream, flows through the Ganga and Ghaghra. The total length of this river in UP is 940 km.

River Ghaghra: Famous by the names of *Manchu* and *Karnali*, the Ghaghra originates at an altitude of 4,800 metres near the Mansarovar Lake. In the beginning, it flows for about 72 km in the south-east direction, and then, the river enters Nepal. It has a meandering course till *Chandi Ghat*. Its important tributaries are the *Sarda*, the *Sarju* on the bank of which Ayodhya is located and the *Rapti*. Ghaghara joins the Ganga a few kilometres downstream Chopra in Bihar after a 1,080 km long journey from its source.

River Son originates from the Amarkkantak Plateau of the Maikal Range at an elevation of about 600 m. After flowing for some distance to the north, turns its course to the north-east. It joins the Ganga about 16 km upstream of Danapur in Patana district of Bihar after flowing for a distance of 784 km from its source.

River Gandak originates near the Tibet-Nepal border at a height of 7,620 and receives a number of tributaries in Nepal Himalaya, the main are Kali Gandak, Mayangadi, Barj and Trishuli. It debouches into the plains at Tribeni and pours into the Ganga at Hajipur in Bihar after flowing for a distance of 425 km in India.

River Kosi: Kosi River consists of seven streams, viz. *Sut Kosi*, *Tamba Kosi*, *Talkha Doodh Kosi*, *Botia Kosi*, *Arun* and *Tamber*. Arun is the main stream which originates from the northern slopes of Mt. Everest in Tibet (China). Coming out of the Saptakosi region, the streams flow in south-west direction and the seven streams mingle with each other to form three streams named Tumar, Arun and Sun Kosi. They unite at Triveni north of Mahabharat Range to Kosi. The Kosi flows for a distance of 730 km in India and joins the Ganga near Kursela. The Kosi River is called as the '*Sorrow of Bihar*', because the river gets flooded every year and also changes its course as a result of which, people have to face many difficulties.

River Damodar: The Damodar river rises in the hills of Chotanagpur Plateau and flows through a rift valley. Every year through its floods, it used to destroy thousands of hectares of cultivable land and homes in the Burdwan and Bankura districts of West Bengal, thus this is known as '*Sorrow of West Bengal*'. The main tributary of Damodar is Barakar. After flowing through Bankura and Burdwan districts, it joins Hugli and then flows into the Bay of Bengal.

The other tributaries of River Ganga are *Tons*, *Punpun*, *Kiul*, *Burhi Gandak*, *Bagmati*, *Kamla*, *Mahananda*, *Rupnarayan*, *Ken*, *Karnali*, *Kali*, etc.

The Brahmaputra River System

The Brahmaputra river system is the drainage system controlled by river Brahmaputra and its tributaries. It covers North-Eastern States of Assam, Meghalaya, Manipur, Tripura, Nagaland, Arunachal Pradesh, Mizoram and the northern part of West Bengal.

Brahmaputra: The origin of Brahmaputra meaning the son of God Brahma, is near Mansarovar in Tibet from the Kong Shu Lake, which is situated at a height of 5,150 metres. Bearing its Tibetan name *Tsangpo*, this river flows 1,770 km to the east, parallel to the Himalayan ranges. Flowing across the Himalayas in the north, this river collects the water of its tributaries like the *Rega*, *Tsangpo Naghcu*, *Kyuchu* and *Gramda Chu*. Turning north from a place called PE (3,000 metres), the Brahmaputra encircles the mountains of Gyala Paro and Namcha Barua (7,756 metres) and then turning towards south and south-west, enters the Indian territory through the foothills of Arunachal Pradesh. In this region, it is initially called *Siang* and then *Dihang*. Flowing south-west from here, it enters Assam from Sadia Nagar. Here, it meets its two main tributaries, *Dibang* and *Lohit*. Here onwards, it assumes its popular name Brahmaputra. This huge river creates the famous Assam valley flowing west from east. While flowing along its course, tributaries like the *Jiyadhal*, *Kameng*, *Dhanshree*, *Puthimari*, *Pagladia*, *Manas*, *Champamati*, *Sarbhanga* and *Sankokh* flow into it. Among its tributaries which flow in from the south, the *Nav Dihing Old* (*Burhi*) *Dihing*, *Disang*, *Dirbat*, *Jhanji*, *Dhanshri* (South) and *Kopili* deserve mention. From here, turning from the Garo Hills, the Brahmaputra enters Bangladesh near Golpara. Here, many of its tributaries flow in from North Bengal among which the *Teesta*, *Jal Dhaka* (*Dharia*) *Torsa*, *Kalyanjki* and *Raidak* are the major ones. All these river streams meet the Brahmaputra inside the territory of Bangladesh. Flowing south from here, the Brahmaputra meets the Ganga at Golando 250 km downstream. The combined streams of Ganga and Brahmaputra are known as *Padma*. After flowing for 105 km from here, one of its major tributaries called the *Meghna* flows into it from the left. Meghna has its origin in the Assam Himalayas. Beyond this confluence, the Meghna Brahmaputra gets divided into many streams and forming a huge estuary, flows into the Bay of Bengal. The major tributaries of Brahmaputra are:

Subanshri: Its origin is in the plateau of Tibet. In the early stages, this river is called *Sukanchi*. Its name changes after the confluence with *Sarichu*, and this combined stream is known as *Subanshri*. The total length of the Subanshri river is 442 km of which 192 km is in Arunachal Pradesh and 100 km in Assam.

Kameng: The origin of Kameng is across the Ulla and EK hills of the Himalayan ranges in Tibet. It flows for over 190 km through the hilly tract falling in Kameng district of Arunachal Pradesh. It meets Brahmaputra 11 km before Tejpur. The total length of the river is 265 km.

Manas: This river has been formed by the confluence of the Tongsa and Kar streams. And it originates in Bhutan near the Sukali village. The Manas river after covering 8 km in the north and in the south reaches the Barpeta district of Assam. After flowing for about 30 km towards the south-west, it reaches the Golpara district. From the Angroka village here, forming a meandering course, the Manas flows into Brahmaputra near Jogigopa. The total length of Manas river is 376 km of which 104 km is in India.

Burhi Dihing: This river has been formed by the confluence of the Namphuk and Namchuk rivers, which originate in the Patkai mountain ranges. It flows into a stream of Brahmaputra 32 km from Dibrugarh. The total length of Burhi Dihing along with Namchuk is 362 km.

Dhansiri: The origin of Dhanshri is on the south-west corner of Nagaland near the aisang peak. In the beginning, it flows towards the north-west for a distance of 37 km. Then turning north-east, it flows ahead for 76 km and reaches Dimapur. Beyond Dimapur, the flow of Dhanshri is generally northwards. Reaching Golaghat, it suddenly changes it course towards the west and meets the Brahmaputra at a place called Dhanshrimukh. The total length of this river is 353 km.

Kopili: The origin of Kopili is the Malangha village in the mountain ranges of Assam between Mikir and Cachar. It flows for a distance of 60 km through the Mikir and North Chachar districts adjoining the Khasi and Jaintiya hills. It meets the Brahmaputra near Raja Mayang. The total length of the Kopili river is 256 km of which 78 km forms the boundary of Nagaland and Assam and the rest of the 178 km is in the state of Assam.

Teesta: The largest river of Sikkim and North Bengal is the Teesta. It originates in the Himalayan ranges of North Sikkim. Flowing south through deep valleys for a distance of 138 km, the Teesta enters the plains of the Jalpaiguri district near Sewak. From here, it reaches the Jalpaiguri town flowing south-east, beyond which it deeps on changing its course. The main tributaries of the Teesta river are *Ramgiri, Rangit, Rali, Lish, Gish* and *Chel*. After flowing for about 309 km, the Teesta meets the Brahmaputra near Rangpur in Bangladesh.

Jaldhaka: The origin of Jaldhaka too is in Sikkim. Its length is 186 km. It reaches Bangladesh flowing through Bhutan, Darjeeling, Jalpaiguri and Cooch Behar and flows into the Brahmaputra at a place called Alipur in the Rangpur district.

Torsa: The river flows from the Chumbi valley in Tibet, where it is called *Maochu*. After flowing for 113 km through Tibet, it enters Bhutan and is referred to by the name of *Amachu*. After flowing for 145 km, it enters India at Tetapara (District Jalpaiguri) in West Bengal. Before meeting the Brahmaputra, the river Torsa flows for about 160 km in the Cooch Behar region of North Bengal. Among its tributaries are the *Hulang*, *Gargaria*, *Kaijani*, *Raidhak* and *Sankosh*.

Barak: Originating from the Manipur mountains, south of Mhow near the Kohima-Imphal road, the Barak river forms a deep valley toward the south-west and reaches Hitmairnukha. The length of the river from the point of its origin till Lakhimpur is 408 km and along with Kushiara till the boundaries of Assam, it is 564 km near Bhairav Bazaar in Bangladesh. After covering a distance of 902 km, it flows into the Meghna. Its main tributaries are the *Giri*, *Chin*, *Harong*, *Katkhal* and *Longat*.

Gornti: Its origin is in the mountains of Tripura. Combining with the streams of Surma and Rajma, it takes the form of a river. It flows towards the west throughout its course.

River Imphal: It flows creating a separate basin from the Brahmaputra. Descending from the hills of north Manipur, it flows through the capital Imphal. Its tributary Iril flows in from the south and the Thombal flows in from the east while the Khuga flows in from south-east. The waters of the Loktak lake, situated in the midst of Manipur, flows into Imphal through the Kordak river.

Rivers of Penisular India

The important rivers of this region are the Narmada and the Tapti which flow towards the west and the Suvarna Rekha, Vaitarni, Brahmani, Mahanadi, Godavari, Krishna, Penang and Cauvery which flow eastward. This region incorporates all the states of south like Andhra Pradesh, Telangana, Karnataka, Tamil Nadu and Kerala as also portions of Odisha, Maharashtra, Gujarat and Madhya Pradesh. Table 4.4 indicates the main features of the rivers of Penisular India.

River Mahanadi: The Mahanadi flows from a reservoir near village Faresha in the Raipur District of Chhattisgarh and for 200 km, flows in a western and north-western direction. In the beginning, many streams flow into it. But it is near Bilaspur, 13 km before Shivrinarayan, one of its main tributaries, the Shivnath flows into it. Flowing for 200 km towards the east, the Mahanadi enters Odisha and is met by the Ewe river. Other rivers which meet in Mahanadi are *Aung* and *Tel*. The river enters the region of Eastern Ghats and flows through a deep valley called Satkosia.

Table 4.4 *Important Rivers of the Peninsular India*

	Name of the River	Source	Length (km)	Main Tributaries
1.	Mahanadi	Dandakarnya near Sihawa in Raipur District of Chhattisgarh	857	Ib, Mand, Hasdo, Seonath, Ong, Jonk and Tel
2.	Godawari	Trimbak Plateau near Nasik	1,465	Penganga, Wardha, Wainganga, Indrawati, Sabari, Pranhita
3.	Krishna	Near Mahabaleshver	1,400	Koyna, Ghataprabha, Bhima, Malprabha, Tungbhadra, Musi, Muneru
4.	Kaveri (Cauvery)	Tall Kaveri (Western Ghats)	800	Herangi, Hemavati, Lokpavani, Shimsha, Arkavati, Lakshmanatirath, Kabani, Suvanavati, Amaravati.
5.	Narmada	Amarkantak Plateau	1,310	Hiran, Orsang, Barna, Kalar, Burhner, Banjar, Shar, Tawa, Kundi
6.	Tapi (Tapti)	Multain in Betul District	730	Purna, Betul, Patkai, Ganjal, Dhatranj, Bokad, and Amravati

Coming out of the valley, the Mahanadi enters its delta region near Neraj. Past Neraj, its stream gets divided into the branches, *Katjuri* and *Birupa*. Further downstream, the Mahanadi gets divided into many small streams and finally it flows into the Bay of Bengal near the False Point. The total length of this river is 851 km of which 357 km is in Madhya Pradesh and the remaining 494 km is in Odisha.

River Godavari: The Godavari flowing across the southern region from the Western Ghats to the Eastern Ghats, covers a distance of 1,465 km. The origin of Godavari in at an altitude of 1,067 m in the Nasik district of Maharashtra of which about 693 km is in Maharashtra and for the rest of the 772 km, it flows in Andhra Pradesh in a south-east direction. This river then flows into the Bay of Bengal. The main tributaries of the Godavari in Maharashtra are the *Darha*, *Kadwa*, *Parwara*, *Onula*, *Purna* and in Andhra Pradesh, the river *Manira* flows in from the south.

After Rajahmundhry, the Godavari gets divided into many streams. The eastern branch is called the *Vashistha Godavari*. They create a big delta region. Of all the rivers in the peninsula, the Godavari is the biggest and the most important. Only the Ganga and the Indus surpass in terms of piousness, utility and grandeur. Therefore, it is also referred to as the *Old Ganga* or the *Ganga of the South*.

River Krishna: The river Krishna too is a big river of South India and like the Godavari and the Cauvery, flows from east to west across the peninsula. Its origin is near Mahabaleshwar, at an altitude of 1,337 m in the Western Ghats. Its main tributaries are *Bhima* and *Tungabhadra* and others are *Meria*, *Verma*, *Panchang*, *Doodhganga*, *Musi*, *Puleru*, *Menerruth*, *Ghatprabha* and *Malprabha*. The total length of the river is 1,400 km of which 369 km is in Maharashtra, 484 km in Karnataka and 611 km in Andhra Pradesh.

River Cauvery: The origin of Cauvery is at an altitude of 1,341 m at Taal Cauvery in the Brahmagiri hills in the Coorg district of Western Ghats. In the beginning, the river has a meandering course in the Coorg district and while it cruises down the slopes, it is met by the Harangi, its principal tributary. The other tributaries of the river are *Himawati*, *Lakshman Tieth*, *Suvarnavati*, *Bhawani Noel*, *Amaravati* and *Vinnar*. Finally, the Cauvery as a small and sluggish stream flows into the Bay of Bengal.

River Pennar: Among the peninsular rivers flowing east, the Pennar is only next in importance to the Cauvery. Its origin is in the Chenna Keshav hills in the Nadidurg ranges in the state of Karnataka. Having flown through the Kolar and Tumkur districts of Karnataka in a north-west direction for 490 km, it enters Andhra Pradesh. Its main tributaries are *Kumudavalli*, *Jaimangli*, *Kunderru Papagni*, *Sangileru*, *Vyonpalli*, *Ghiyaru*. The total length of the river is 597 km of which 61 km is in Karnataka and 536 km in Telangana and Andhra Pradesh.

River Narmada: The Narmada flows from a place called Amarkantak in the Maikal ranges in the Shahdol district of Madhya Pradesh. It is the biggest river to flow westward in the peninsula. There are many water falls in this river of which Jabalpur at 404 km, Gandhar at 806 km, Thari at 853 km and Maheshwar at 966 km are located, where the waterfalls are respectively at a height of 15 mtr, 12 mtr and 6 mtr respectively. Near Jabalpur, the Narmada, at Bhedaghat, reaches the higher plains through the Dhualadhar falls. The main tributaries fo Narmada are *Bamer, Banjer, Kiran, Sher, Shakkar, Tava, Chhoti Tava, Barna, Uri, Curzan* and *Orsansh*.

Tapi or Tapti: Thee second biggest river to flow west in the peninsula is the Tapti. Its origin is in the Betui district of Madhya Pradesh near Multai at an altitude of 752 m. For 241 km, this river flows through open and fertile plains. From its origin, to the sea, the total length of the river is 724 km of which 228 km is in Madhya Pradesh, 228 km in Maharashtra, 214 in Gujarat and the remaining 54 km is in the form of Maharasthra-Madhya Pradesh border. Among the main tributaries of the Tapti are the *Purna, Vadhur, Gima, Bori, Panjchrr, Bure, Amer, Arunavati* and the *Gornai*.

Other Rivers of Central and South India

The other important rivers are *Suvarna Rekha, Vaitarni, Brahmani*. The rivers flowing towards west are – *Luni, Machuchu, Rupen, Saraswati, Banas, Bhader, Shatranji, Sabarmati, Mahi, Ambika, Vaitarna, Ulhas, Savitri, Mandvi, Kalinadi, Gangawali, Sharavathy, Netravathi, Cheliar, Bharatpuzha, Periyar* and *Panib*. While the rivers flowing towards east are *Burha Balang, Risikulva, Bahunda, Vanshdhara, Nagvalli, Sharda, Varaha, Tanrwa, Vogarivupu, Gundlakamma, Mousi, Pulleru, Maneru, Kunlru, Swarnamnkhi, Arnaiver, Kotalvar, Kooam, Advar, Palar, Ginjel, Ponnaiyyar, Velar, Tiruchi-ellar, Varshali, Valgai, Gunder, Vaipar* and *Tamraparani*.

Lakes

India has large number of lakes which differ from each other in their origin, size and other characteristics. The lakes are classified variously such as permanent, seasonal, freshwater, salt lakes, but their basic classification is on the basis of origin. The following are the types of lakes in India.

1. *Tectonic Lakes:* Those lakes which are formed due to the fractures and faults in the earth's crust. Tso Moriri and Pangong Tso of Ladakh and Nainital and Bhimtal of Uttarkhand are its examples.
2. *Crater Lakes* are of volcanic origin and are formed when craters are filled with water, such as Lonar Lake of Buldhana in Maharashtra.
3. *Glacial Lakes* are formed by the erosional action of glaciers. Glacial lakes in India are found in Jammu & Kashmir, Himachal Pradesh and Uttarakhand. Gangobal Lake of Kashmir is an example.
4. *Fluvial Lakes* are formed by the action of water both erosional and depositional. Such lakes are common in Upper, Middle and Lower course of River Ganga and river Brahmaputra.
5. *Aeolian Lakes* are formed by wind action. Such lakes have been formed in western Rajasthan.
6. *Dissolution Lakes* are caused by underground water action on limestone rocks. Such lakes are found in Meghalaya.
7. *Lagoons:* These lakes are formed along sea coast by the deposition of sand bars. Chilka lake of Odisha, Pulicat of Andhra Pradesh, Vembanad, Asthamndi and Kayals of Kerala are the lagoon lakes.
8. *Landslide Lakes* are formed when an obstruction in the course of stream is caused due to landslide. The Gohna lake of Garhwal is an example.

Table 4.5 indicates the main lakes of India and their locations.

Table 4.5 *Important Lakes of India*

	Lake	*State*
1.	Asthamudi	Kerala
2.	Bhimtal	Uttarakhand
3.	Boj Wetland	Madhya Pradesh (Near Bhopal)
4.	Chandra Tal	Himachal Pradesh
5.	Chembarambakkam Lake	Tamil Nadu
6.	Chilka Lake	Odisha
7.	Dal Lake	Jammu & Kashmir
8.	Dhebar Lake	(Jaisamand) Rajasthan
9.	Himayat Sagar	Near Hyderabad
10.	Hussain Sagar	Hyderabad, Andhra Pradesh/Telangana
11.	Kaliveli Lake	Tamil Nadu
12.	Khajjiar Lake	Himachal Pradesh
13.	Khecheopalri Lake	Sikkim
14.	Kolleru Lake	Andhra Pradesh
15.	Loktak Lake	Manipur
16.	Nako Lake	Himachal Pradesh
17.	Osmal Sagar	Hyderabad, Andhra Pradesh/Telangana
18.	Pongonng TSO	Ladakh (J&K)
19.	Pulicat Lake	Andhra Pradesh
20.	Pushkar Lake	Rajasthan
21.	Renuka Lake	Himachal Pradesh
22.	Roop Kund	Uttarakhand
23.	Sambhar Lake	Rajasthan
24.	Sasthamkotta Lake	Kerala
25.	Satta or Sat Tal	Uttarakhand
26.	Pichola	Rajasthan
27.	Tawa Reservoir	Madhya Pradesh
28.	Tsongmo Lake	Sikkim
29.	Veranmam Lake	Tamil Nadu
30.	Vemnanad Lake	Kerala
31.	Veeran puzha Lake	Kerala
32.	Vembanattu	Kerala
33.	Wular Lake	Jammu & Kashmir

Apart from above mentioned there are several other small lakes located in various parts of India. *Chilka Lake* is the largest lake of India and a saline lake; while *Wular Lake* in Jammu & Kashmir is the largest freshwater lake in India. Jaisamand (Dhaber lake) near Udaipur in Rajasthan is the largest artificial lake of India.

Water Falls

Waterfall is a vertical or near-vertical fall of water or series of step-like falls, developed where a river course is interrupted by marked break of gradient. There are many waterfalls in India. The main waterfalls of India are as follows:

Table 4.6 *Main Waterfalls of India*

Name of the Waterfall	*Location*
Agaya Gangai	Kolli Hills, Eastern Ghats, Tamil Nadu
Ayyanar	Virudhungar District, Tamil Nadu
Bareshipani	Budhabalanga River, Mayurbanj, Odisha
Barkana	Sita River, Shimoga, Karnataka
Dudhsagar	Mandovi River, Goa
Duduma	Machhkund River, Odisha
Bundia	Bundia Stream, Himachal Pradesh
Chulia	Chambal River, Rajasthan
Chechai	Bihad River, Madhya Pradesh
Catherine	Tamil Nadu
Birthi	Pancholi Peak, Uttarakhand
Dhuandhar	Narmada River, Madhya Pradesh
Dasam	Subernarekha River, Jharkhand
Gokak	Ghataprabha River, Karnataka
Jog (Gersoppa)	Sharavathi River, Karnataka
Khandahar	Korapari nala, Sundararh, Odisha
Kalhatta Giri	Karnataka
Kune	Maharashtra
Kynrem	Meghalaya
Kiliyur	Servary Hill, Tamil Nadu
Kurtalam	Tirunelveli District, Tamil Nadu
Kapildhara	Narmada River, Madhya Pradesh
Langshian	Meghalaya
Hundru	Subernarekha River, Jharkhand
Lodh	Budh River, Jharkhand
Elephant	Meghalaya
Magod	Bedti River, Karnataka
Shivasamndram (Cauvery Fall)	Caveri River, Karnataka
Siruvani	Siruvani River, Western Ghats, Tamil Nadu
Talakonce	Tirupati, Andhra Pradesh
Tirathgarh	Mugabahar River, Chhattishgarh
Paiaruvi	Kerala
Rajat Pratap	Pachmarhi, Madhya Pradesh
Nokhalikai	Meghalaya
Thalaiyer (Rat Tail Fall)	Near Kodaikanal, Tamil Nadu
Vasudhara	Near Badrinath, Uttarakhand
Vattaparia	Pazhayar River, Tamil Nadu
Vazhachai	Thrissur District, Kerala

There are several other waterfalls spread throughout the country and are of local and regional importance.

5

Climate

The climate of India is complex and varies from region to region as Blandford has rightly observed that 'We may speak of the climates but not the climate of India, for the world itself affords no greater contrast than is to be met with at one and the same time within its limits'. The climatic conditions of India ranges from tropical in the south to temperate in the north as well as arid in Thar desert to alpine in the Himalayas. But in spite of these diversities, India is basically a tropical country although its northern part is situated in the temperate belt. In fact, India is per excellence, a tropical monsoon country.

Factors Influencing the Climate of India

The Indian climate is directed and determined by the following factors:

1. *Location:* India is located between 8°4´ and 37°6´ north latitudes and the Tropic of Cancer passes through the middle of the country. Thus, areas south of the Tropic of Cancer are having higher temperature, while areas north to it have warm temperate location.
   Another locational factor is the location in terms of ocean and sea. The location of India with reference to Indian ocean and Bay of Bengal and Arabian Sea has a direct control over the climate. The whole mechanism of monsoon is controlled by the Indian Ocean. Similarly, the coastal areas of India have maritime climate, while interior areas of the country experience continental climate.
2. *Physiography:* The country's physiographical features such as mountains, plateau, plains, desert, coasts, etc. have a direct impact on temperature, pressure, winds and rainfall system. The Himalayan mountain ranges act as an effective barrier for rain bearing south-west monsoon and also protect India from cold and dry winds during winters. In fact, Himalayas act as a climatic divide of the Indian sub-continent and Central Asia. The Western Ghats of Peninsular India control the rainfall distribution in South India and are the cause of lesser rainfall in states located on its eastern side. The other smaller mountain ranges of central and north-east India and Eastern Ghats also have their impact on climate regionally.
   Apart from mountains, the Thar desert influences the climatic conditions of the western Rajasthan and its adjoining areas. The Ganga and Brahmaputra Valley provides a passage to the monsoon winds originated from the Bay of Bengal. The western and eastern coasts of India are having moderate climate due to the maritime impact.

3. *Wind System:* The Indian climate is the result of 'monsoon winds' in which there is a complete reversal of winds during summer and winter season which is the main cause of the seasonal change in climatic conditions.
4. *Tropical Cyclones and Western Disturbances:* The tropical cyclones generated in the Bay of Bengal and the Arabian Sea influence the large part of Peninsular India and north-east India. Similarly, western disturbances originated over Mediterranean Sea travel eastward, influence the climate of western and northern plains of India and often a cause of winter rainfall in these areas.
5. *Upper Air circulation:* The upper air circulation in the form of Jet Streams over Indian landmass influences the climate of India. Westerly Jet Streams blown at a very high speed during winter exercises a significant influence on winter conditions of India while Easterly Jet streams during summer is considered as a main cause of the onset of the south-west monsoon in India.
6. *El-Nino and La-Nina Effect:* El-Nino is a narrow warm current which occasionally appears of the coast of Peru and is a cause of widespread climatic change in tropical regions of the world. When this creates warming of tropical pacific waters, it affects the global paths of pressure and wind system including the Indian monsoon system.

La-Nina appears only after the El-Nino when weather conditions return to normal. Sometimes when trade winds become so strong, they cause abnormal accumulation of cold water in central and eastern pacific region. This event is called La-Nina and is the cause of heavy rainfall in India.

Indian Monsoon and Its Origin

The word 'monsoon' is derived from the Arabic word *'mausam'*, meaning season; the term originally referred to the winds of Arabian sea, which blow for about six months from the north-east and for another six months from the south-west. In fact, monsoon is the climate of contrasting season and reversal of wind system or rhythm is the keynote of monsoon climate, which is the characteristic of the climate of India. Some scholars tend to treat the monsoon winds as land and sea breezes on a large scale. The whole climatic pattern of India is controlled by monsoon, therefore, it is necessary to understand the mechanism of its origin.

Several scholars have given their views regarding the origin of monsoon, but still this mystery has not been fully resolved. Although much of its mechanism now has been explained by scientists. The concepts regarding origin of monsoon can be divided into two categories, viz. traditional or classical and modern concepts. The brief review of main concepts is as follows:

The Thermal Concept of Halley: In 1686, Sir Edmund Halley has given the classical thermal concept regarding origin of the monsoon. He stated that Indian monsoon is the result of thermal contrasts between continents and oceans due to their differential heating. During summer season, when sun's rays are vertical over the Tropic of Cancer, the huge landmass of Asia heats quickly resulting in high temperature and low pressure in Central Asia, while the pressure is high over Indian ocean. Thus, it results in air flow from seas to land and brings heavy rainfall in India. While in winter season, the conditions become reverse due to the vertical sunshine over Tropic of Capricorn. The Central Asian region and the north-western part of India grow colder and develop a high pressure area and in the Arabian Sea and the Bay of Bengal, a low pressure area has developed resulting in the flow of monsoon winds from land to sea.

The Halley's thermal concept was criticized on the ground that only differential heating is not a cause of the origin of monsoon but the shape of the continents, orography and conditions of upper air circulation, etc. also have significant role.

The Dynamic Concept by Flohn: In 1951, Flohn has put forward the dynamic concept of the origin of monsoon and suggested that the tropical monsoon of Asia is simply a modification of the planetary winds. The Inter-Tropical Convergence Zone (ITCZ) is formed due to the convergence of north-east and south-east trade winds near equator. The northern and southern boundaries of the ITCZ are called NITC (Northern Inter-Tropical Convergence) and SITC (Southern Inter-Tropical Convergence). The seasonal shift of the ITCZ is the main cause of monsoon winds. The equatorial westerlies become south-west monsoon winds. Due to the earth's rotation, the trade winds of Southern Hemisphere after crossing the equator turn towards their right. Thus ITCZ becomes a zone of clouds and heavy rainfall.

Similarly, the north-east or winter monsoon originates due to low pressure in the Southern Hemisphere. Due to the southward movement of the Sun at the time of winter solstice, the north-east trade winds become winter monsoon.

In Flohn's concept, emphasis has ben given on shifting of Inter-Tropical Convergence Zone but he has not taken into consideration the upper air atmospheric circulation.

Recent Concepts

Several scientists have expressed their views regarding the origin of monsoon. Among them, M.T. Yin, Hwang, Takahasi, E. Palmen, P. Koteswaram, Krishnamurti, Krishanan, Raman, Ramanathan, Rama Ratta, Ramaswami, Anant Krishnan are notable. In recent concepts, emphasis has been given on upper air circulation (Jet stream), role of Tibetan Plateau and El-Nino. A brief explanation of these concepts is as follows:

Jet Stream and Indian Monsoon

Jet stream is a high-altitude wind blowing at approximately 9–15 km above the earth's surface. It has a high velocity with normal maximum speed of 180–270 km/hr, but may occupationally reach over 500 km/hr. The Jet streams are high altitude westerly winds blowing between middle latitudes (Summer 35°N – 45°N; Winter 20°N–35°N) in Northern Hemisphere and have direct impact on surface weather conditions, specially on Indian monsoon.

A westerly subtropical jet stream blows over India during winter causing a high pressure over northern parts of the country. It causes winds to blow from north to south in the form of north-east monsoon while during summer with northward shift of the vertical rays of the sun, the jet stream also shifts northwards and an easterly jet is established over central India. The easterly tropical jet over central India leads to the creation of a low pressure over northern plains. This helps in creation of the southwest monsoon over India during summer. In this process the role of high temperature over the Tibetan plateau in summer is also an important factor leading to the formation of tropical easterly jet over India in summer. A timely northward shift of the sub-tropical westerly jet from over India in the beginning of the summer season is critical to the onset of southwest monsoon. If this shift is delayed, the delay in the onset of southwest monsoon will occur. Similarly, an early withdrawal of this jet stream heralds an early monsoon.

Tibetan Plateau and Indian Monsoon

P. Koteswaram, the Director General of the Indian Meteorological Observatories has expressed his views on 'The Monsoons of the world' in an International Symposium in 1958, that the summer time heating of the Tibetan Plateau is the most important cause of monsoonal circulation.

This view was confirmed by the Monsoon Expedition (Monex) in 1973. The Monex was a joint venture of the Soviet Union and India, under which four Russian and two Indian ships equipped with modern scientific instruments have been used to investigate the mechanism of Indian monsoon. The scientists of both the countries reached to the conclusion that the Tibet Plateau plays a crucial role in initiating the Indian monsoon.

The Tibet Plateau is an ellipsoidal plateau at an attitude of about 4,000 m above sea level with an area of about 4.5 million sq km. This plateau is surrounded by mountain ranges which rise 6,000 to 8,000 m above sea level. It gets heated in summer and is 2°C to 3°C warmer than the air over the adjoining areas. Because Tibet Plateau is a source of heat for atmosphere, it generates an area of rising air. During its ascent the air spreads outwards and gradually sinks over equatorial part of the Indian Ocean. The ascending air is deflected in an anti-clockwise direction leading to anticyclone conditions in upper troposphere over Tibet. It finally approaches the west coast of India as a return current from a south-westerly direction and picks up moisture from Indian Ocean and causes rainfall in India.

El-Nino Effect on Monsoon

Recent studies in the field of monsoon revealed that there is a link between Indian monsoon and a long distance event known as El-Nino and the Southern Oscillation. El-Nino is a warm ocean current which appears along the coast of Peru in December. El-Nino is a Spanish word which means '*Child Christ*' because it appears around Christmas. Under normal conditions, the Peru is a cold water current. The appearance of El-Nino reverses the conditions and develops warm conditions over the eastern Pacific and cold conditions in western Pacific. Under these conditions, coastal areas of South America receive more rainfall while the eastern coasts of Australia and Indonesia record drought conditions. The occurrence of El-Nino results into weak monsoon. This warming of tropical Pacific waters affects the global pattern of pressure and wind systems including the Indian monsoon.

On the contrary, when weather conditions return to normal after El-Nino, sometimes, trade winds become so strong that they cause abnormal accumulation of cold water in the central and eastern Pacific region. This event is known as La-Nina, which in effect is the complete opposition of El-Nino. In India, the presence of La-Nina has a positive impact in the form of heavy monsoon showers.

Southern Oscillation (SOI) is another event which affects the Indian monsoon. Southern Oscillation is a climatic change often observed between the Pacific and the Indian Ocean. Whenever the surface level pressure is high over the Indian Ocean, there is low pressure over the Pacific Ocean and vice-versa. This interrelation of high and low pressure over the Pacific and the Indian Ocean is called the Southern Oscillation. When the winter pressure is high over the Pacific Ocean and low over the Indian Ocean, the south-west monsoon in India tend to be stronger. In the reverse, the monsoon is most likely to be weaker.

The El-Nino and Southern Oscillation have close relationship and the two are jointly referred as an ENSO event. The Southern Oscillation also gives rise to vertical circulation along the equator with its rising limb over low pressure area and descending limb over high pressure area. This is known as *Walker Circulation.*

Progress of the Indian Monsoon

The Gradual northward movement of monsoon or its progress is a special feature because it determines the distribution of rainfall in India. In India, normal dates of the arrival of monsoon varies in different parts of the country. The normal date of the arrival of monsoon is 20 May in Andaman and Nicobar Islands located in the Bay of Bengal. Thereafter the southwest monsoon reaches the Malabar Coast of Kerala by 1 June. This is actually the beginning of monsoon in the mainland of India. The start of the monsoon is sudden and accompanied by thunder, lightning and heavy downpour. This sudden onset of rain is termed as 'burst of monsoon'. The subsequent progress of the monsoon in the country is divided into two branches, viz. the Arabian Sea branch and the Bay of Bengal branch. The Arabian Sea branch of monsoon gradually advances northward reaching Mumbai by 10 June and spread over Maharashtra, parts of Gujarat and Central India by 15 June. Its further advances in Rajasthan, Haryana, and Punjab, Himachal Pradesh by the 1st week of July. Meanwhile, the Bay of

Bengal branch moves northward into the central Bay of Bengal by the third week of May and it spreads rapidly over Meghalaya, Assam and other states of north-east India. The normal date of monsoon arrival at Kolkata is 7 June. On reaching foothills of the Himalayas, the Bay of Bengal branch deflects westward and advances in the northern plain of India through the Ganga Valley covering Bihar, Jharkhand and Uttar Pradesh states. Meanwhile the two branches, i.e. the Arabian Sea, and the Bay of Bengal merge with each other and gradually extend to west Uttar Pradesh, Haryana, Punjab, Rajasthan, Himachal Pradesh and Jammu & Kashmir. By the end of June, the whole country is under the grip of monsoon winds. Figure 5.1 shows the normal dates of onset of south-west monsoon in India.

Figure 5.1 *Progress of Monsoon in India*

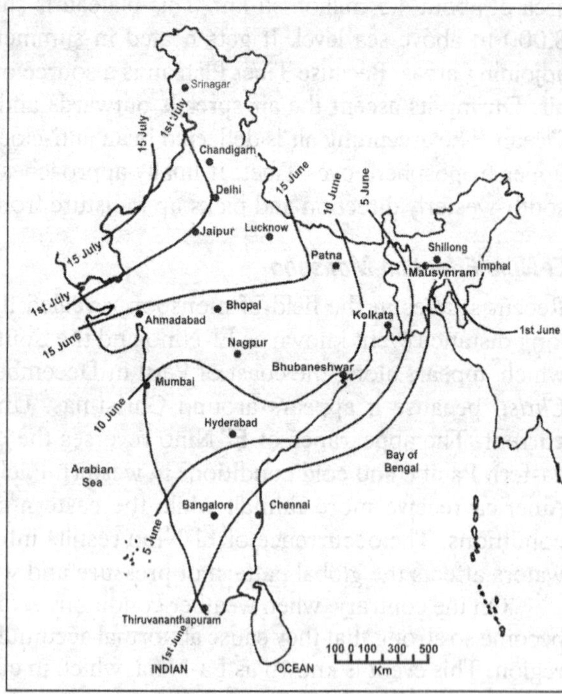

Retreat of Monsoon

The South-west monsoon begins to retreat in the third week of September from northern India. This retreat is steady and gradual. By the end of September, it retreats from Himachal Pradesh, Punjab, Haryana, Western Uttar Pradesh and Rajasthan. By mid-October, monsoon withdraws from three-fourth of India. Only the southernmost part of the country remains, that too has a minor effect. Figure 5.2 depicts the general pattern of the withdrawal of south-west monsoon.

Figure 5.2 *Withdrawal of South-West Monsoon*

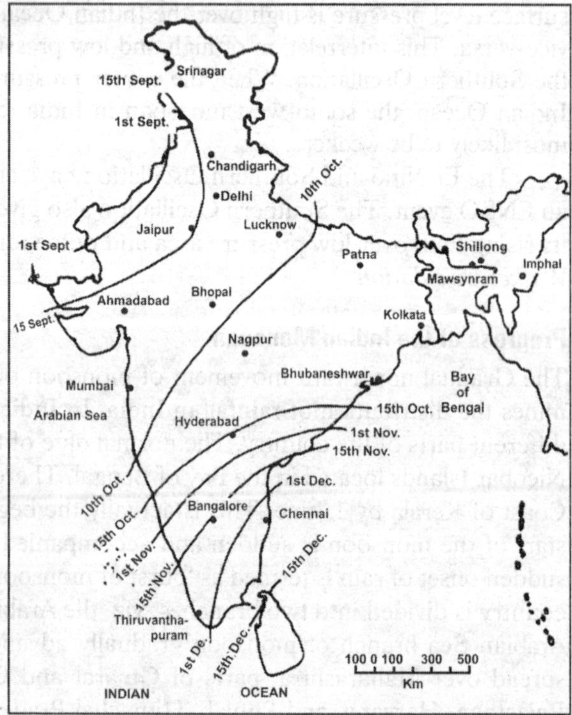

Distribution of Rainfall in India

The temporal and spatial distribution of annual rainfall in India is very uneven. The greater parts of India receive over 85 per cent of their total rainfall during rainy season, i.e. from June to September through south-west monsoon. The Arabian Sea branch of monsoon causes rainfall all along the Malabar-Coast, Konkan Coast, Western Ghats, Maharashtra, Gujarat and parts of Madhya Pradesh.

Mumbai records about 190 cm of rainfall while leewardside of Western Ghats receive lesser rainfall. The sub-branch of Arabian Sea

Seasons in India

The characteristic features of the Indian climate is the seasons. The Meterological Department of India divides the seasons of India into following four seasons:

1. Seasons of north-east Monsoon
   (i) Winter Season (mid December to mid March)
   (ii) Hot Weather Season (mid March to May)
2. Seasons of south-west Monsoon
   (iii) Rainy Season (June to September)
   (iv) Season of Retreating Monsoon (October to mid December)

current which moves northward through Kachchh, Gujarat and Rajasthan gives little rainfall to these regions due to parallel alignment of the Aravalli. The Bay of Bengal branch of south-west monsoon clashes towards North-east India and also westward towards the Ganga Valley. All the states of northern and central India used to receive rainfall from this branch.

Winter rainfall occurs in Tamil Nadu, Southern Andhra Pradesh, south-east Karnataka due to north-east monsoon. The highest season's rainfall of about 75 cm between October and December, occurs along the south-eastern coast of Tamil Nadu and its adjoining areas. During winter western disturbances also cause some rains in Rajasthan, Punjab and Haryana as well as in Gangetic Plains.

The spatial distribution of rainfall in India is highly uneven. On one side, large parts of Meghalaya receive about 1,000 cm annual rainfall, while on the other side, the average annual rainfall in Rajasthan is less than 25 cm. The variations in rainfall distribution are common throughout the India. The following regions have been identified in India according to the annual average rainfall:

1. *Regions of very High Rainfall:* Areas receiving an annual rainfall of 200 cm and above are included in this category. Such areas are Assam, Meghalaya, Nagaland, Mizoram, Arunachal Pradesh, Sikkim, and parts of Manipur, north-western west Bengal. Meghalaya is the wettest part of the country with Mawsynram and Cherrapunji are getting 1,221 and 1,102 cm annual rainfall respectively.
2. *Areas of High Rainfall:* These areas receive 100 to 200 cm rainfall. Major parts of the Northern plain, Odisha, Madhya Pradesh, Chhattisgarh, Jharkhand, Andhra Pradesh, Tamil Nadu and eastern slopes of western ghats are included in this category.
3. *Areas of Low Rainfall:* Those areas which receive 50 to 100 cm average annual rainfall are large parts of Gujarat, Maharashtra, Western Madhya Pradesh, Andhra Pradesh, Karnataka, eastern Rajasthan, Punjab, Haryana, and parts of Uttar Pradesh.

Figure 5.3 *Annual Average Rainfall Distribution in India*

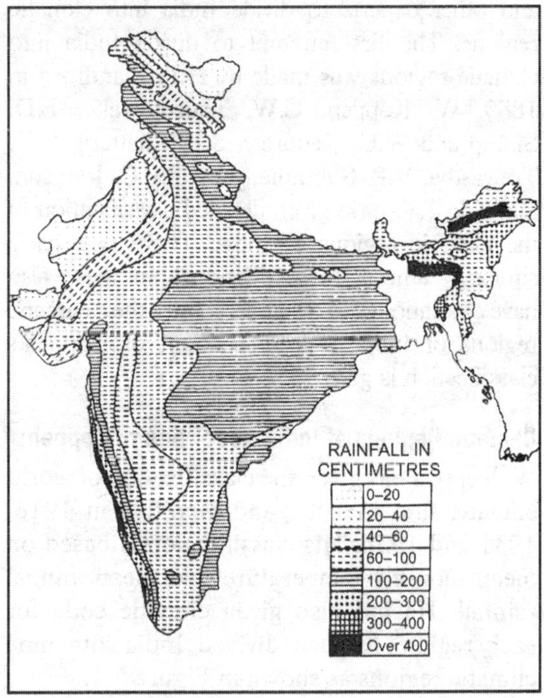

Regional Terms Related with Climate of India

Kalbaishakhi (Norwester): These are tornado-like dust storms. The direction of squalls is mainly from the north-west, therefore, they are called 'Norwester'. These are prevelant in West Bengal and the adjoining areas of Jharkhand, Odisha and Assam. In West Bengal, such storms are locally known as *'Kalbaishakhi'* because the period of their occurrence is mostly the month of *'Vaisakh'*.

Mangoshowers: At the end of summer season the coastal areas of Kerala and Karnataka receive rains from thunderstorms, which helps in ripening of mangoes, thus, known as 'Mangoshowers'. Such rains have also occurred in Tamil Nadu and Karnataka.

Cherry Blossoms: During summers, the cyclonic rains that occur in Karnataka have salutary effect on the coffee plantations known as 'Cherry Blossoms'.

Loo: The hot, dusty winds blown during summer in Rajasthan, Haryana, Punjab and their adjoining areas is called *'Loo'*.

Mavath: In north-western region of the sub-continent, winter precipitation is caused by the depressors that are associated with the westerly disturbances. Such winter rainfall is known as *'mavath'* in north-west India. This rainfall is highly beneficial to 'Rabi' crops.

4. *Areas of very Low Rainfall:* Where annual average rainfall is less than 50 cm; the western Rajasthan, Kachchh and Ladakh region are included in this category.

The annual average rainfall distribution is shown in Figure 5.3.

Climatic Regions of India

India is a vast country having too much regional variations in climatic conditions. Several attempts have been made by climatologists, geographers and other experts to divide India into climatic regions. The first attempt to divide India into climatic regions was made by H.E. Blandford in 1889. W. Koppen, C.W. Thornthwaite, L.D. Stamp and W.G. Kendrew, S.P. Chatterji, G.T. Trewastha, V.P. Subramanyam, B.L.C. Johnson, and K.L. Rao have given detailed classification of the climatic regions of India. There is a basic similarity among these divisions but these also have variations. To understand the climatic regions of India, a brief account of Koppen's classification is given here.

Climatic Regions of India According to Koppen:

W. Koppen has given the classification of world climate, first in 1901 and modified in 1918, 1931 and 1936. His classification is based on mean monthly temperature and mean annual rainfall. He has also given climatic code for each region. Koppen divided India into nine climatic regions as shown in Figure 5.4.

Figure 5.4 *Koppen's Climatic Regions of India*

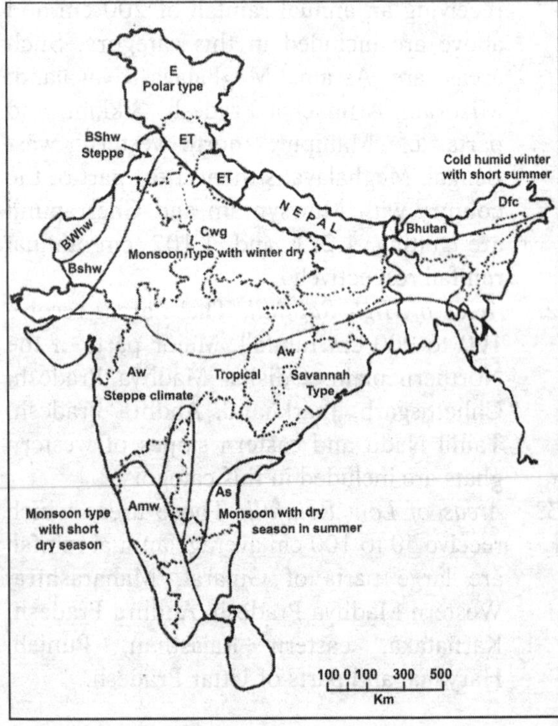

1. *Amw (Monsoon type with short dry winter season):* This area receives over 250 cm annual rainfall from south-west monsoon. This climate is found in parts of Konkan, Malabar coasts and adjoining areas of Western Ghats.
2. *As (Tropical Moist Climate):* The mean monthly temperature is more than 18°C and annual rainfall varies between 75 to 100 cm. It is found in coastal Tamil Nadu and adjoining areas of Andhra Pradesh.
3. *Aw (Tropical Savannah type):* Temperature of the hottest month May mean maximum temperature is 40°C and mean minimum is 20°C, while the coldest month temperature is around 18°C. Average annual rainfall is about 75 cm. Such climate is found in most of the areas of Jharkhand, Chhattisgarh, Odisha, Andhra Pradesh, Telangana, Maharashtra and southern parts of West Bengal.
4. *BShw (Semi-Arid Steppe Climate):* This is a semi-arid type of climate in which rainfall varies from 12 to 25 cm and winters are dry. Large parts of Rajasthan, rainshadow areas of Western Ghats, Haryana and Gujarat are having this type of climate.
5. *BWhw (Hot Desert type):* It covers the Thar desert of India, specially the western Rajasthan. The summer temperature often crosses 45°C and in winter may fall below 5°C and the rainfall is less than 15 cm.
6. *Cwg (Monsoon type with dry winters):* This type of climate is found in most parts of the Gangetic plain, eastern Rajasthan, Assam and Madhya Pradesh. Summer temperature rises up to 40°C and falls to 27°C in winters. Winters are dry and rainfall occurs in summer.
7. *Dfc (Cold Humid Winter type):* It is found in Sikkim, Arunachal Pradesh and parts of Assam. Winters are cold with mean temperature of about 10°C and summers are short but humid.
8. *ET (Tundra Type):* It occupies the higher altitudes of Ladakh, Kashmir, Himachal Pradesh and Uttarakhand. The average temperature varies between 0°C and 10°C.
9. *E (Polar type):* These areas are caused with snow for most part of the year. The higher areas of Jammu & Kashmir and Himachal Pradesh experience polar climate.

6

Natural Vegetation and Biodiversity in India

The naturally growing plant cover without human aid is called natural vegetation. The natural vegetation in India primarily comprises of forests. Several major factors such as soil, topography, temperature and rainfall have influenced the natural vegetation of India to a large extent.

As of 2010, the Food and Agriculture Organization of United Nations estimates India's forest cover to be about 68 million hectares. The 2013 Forest Survey of India states its forest cover increased to 69.8 million hectares by 2012. According to Forest Survey of India Report 2013, the various types of forest covers in India are as given in Table 6.1.

Table 6.1 *Forest Cover in India – 2013*

Class	Area (sq km)	Percentage Out of Total Geographical Area
Forest Cover		
(a) Very Dense Forest	83,503	2.54
(b) Moderately Dense Forest	3,18,745	9.7
(c) Open Forest	2,95,651	8.99
Total Forest Cover (includes 4,629 sq km under mangroves)	6,97,898	21.23
Scrub	41,383	1.26
Non Forest	25,47,982	77.51
Total Geographical Area	32,87,263	100

The 2007 Forest Census Data published by Government of India shows that the five states with largest area under forest cover are Madhya Pradesh, Arunachal Pradesh, Chhattisgarh, Odisha and Maharashtra. While according to India's 2013 Forest Survey Report the forest cover in top five states has increased with exception to Arunachal Pradesh. This fact is clear from Table 6.2.

Table 6.2 *Forest Area in Top Five States of India*

	States	Forest Area in Million Hectares	
		2007 Forest Survey	2013 Forest Survey
1.	Madhya Pradesh	7.64	7.75
2.	Arunachal Pradesh	6.8	6.73
3.	Chhattisgarh	5.6	5.6
4.	Odisha	4.83	5.03
5.	Maharashtra	4.68	5.06

In terms of percentage of forest cover with reference to total geographical area of the state, Mizoram with 90.38 per cent has the highest forest cover followed by Lakshadweep (84.56%), Andaman & Nicobar Islands (81.36%), Arunachal Pradesh (80.39%), Nagaland (78.68%), Meghalaya (77.08%), Manipur (76.10%) and Tripura (75.01%). The states having less than 5 per cent forest cover area are Punjab 3.52 per cent, and Haryana 3.59 per cent. At present 15 states and UTs have above 33 per cent of geographical area under forest cover. Out of these states and UTs, eight states and UTs, namely, Mizoram, Lakshadweep, A&N Islands, Arunachal Pradesh, Nagaland, Meghalaya, Manipur and Tripura have more than 75 per cent forest cover.

Classification Types of Natural Vegetation

India has great variety of natural vegetation due to climatic and physiographical regional variations. The first classification of Indian vegetation was given by H.G. Champion in 1936, afterwards G.S. Puri (1960), Legris (1963), Chamion and S.K. Seth (1968) and S.S. Negi (1990) have also classified vegetation of India.

The four broad categories of Indian natural vegetation are:

1. Tropical Forests
2. Sub-Tropical Forests
3. Temperate Forests
4. Alpine Forests

Tropical Forests

There are two main sub-types of tropical forests: (i) Moist Forest and (ii) Dry Forest.

Moist Forests

Includes Tropical Wet Evergreen, Tropical Semi-evergreen, Tropical Moist Deciduous, Littoral and Swamps.

Tropical Wet Evergreen Forests

These are found in areas having more than 250 cm of annual rainfall The trees found in these areas are Cedar, Mesua, Jamun, Nopea, etc. In India such forests are found in north-east India, Andaman & Nicobar Islands, western side of the Western Ghats, etc.

Tropical Semi-evergreen Forests

The forests are found in areas having 200 to 250 cm annual rainfall. Regions having such forests are western coast, lower slopes of Eastern Himalayas, Odisha and Andamans. Semul, Kadam, Rosewood, Aini, Kusum, etc. tree species are common in these forests.

Tropical Moist Deciduous Forests

In areas having 100 to 200 cm rainfall these forests are found. The species of trees in these forests include Teak, Sal, Laurel, White Chuglam, Mahua, etc. In India these forests are mostly found in Madhya Pradesh, Chhattisgarh, parts of Odisha and West Bengal and along Shiwalik ranges.

Littoral and Swamp Forests

These forests are found in area of deltas, estuaries and creeks. Trees in these areas can survive both in fresh and brackish water. Common trees in such forests are Sundari, Agar, Bhandi, Aemur, etc. Ganga, Brahmaputra delta are famous for such forests.

Dry Forests

Three types of dry forests are found in India These are Tropical Dry Evergreen Forests, Tropical Dry, Deciduous Forests and Tropical Thorn Forests.

Tropical Dry Evergreen Forests are found in coasts of Tamil Nadu. The tree species found are Khirni, Jamun, Tamarind, Neem etc.

Tropical Dry, Deciduous Forests are similar to evergreen forests but their trees shed their leaves in dry season. Teak, Aylewood, Tendu Palas, Bel, etc. are common trees of these forests.

Tropical Thorn Forests are mostly found in western Rajasthan, western Haryana, south-western Punjab and Kachchh area of Gujarat. The common tree species of these forests are Khair, Neem, Babul, Cacti, Palas, etc.

Figure 6.1 *Distribution of Natural Vegetation in India*

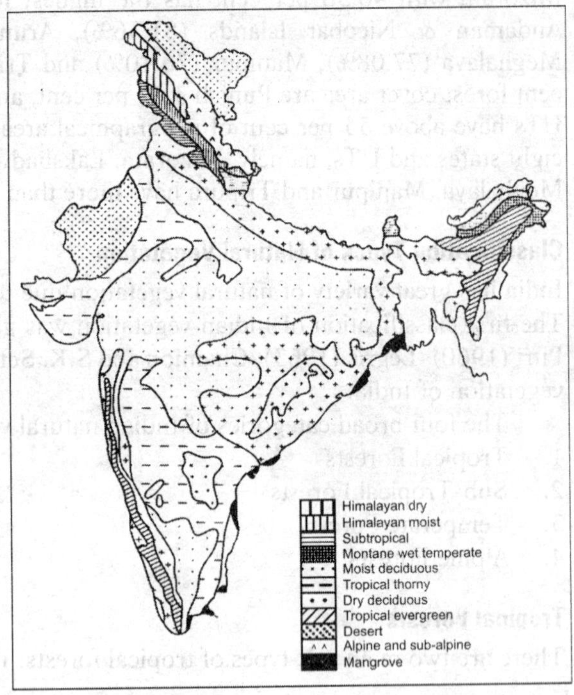

Sub-Tropical Forest

The three types of sub-tropical forests are Sub-Tropical Broad leaved, Hill Forest, Sub-Tropical Moist Pine Forests and Sub-Tropical Dry Evergreen Forests.

Sub-Tropical Broad-leaved Forests are found in Eastern Himalayas as well as in the Nilgiri Hills between 1,070 to 1,525 m altitude. The common trees of these forests are Oak, Sal, Pine, Chestnuts, etc.

Sub-Tropical Moist Pine Forests are found in western Himalayas between 1,000 to 2,000 m altitudes. The most common tree is Chir.

Sub-Tropical Dry Evergreen Forests are found in Shiwalik hills, Western Himalayas and in Bhabar area. The dominant trees are Olive, Acacia, Modesta and Pistacia.

Temperate Forests

In India temperate forests are mostly found in Himalayan regions, but their nature varies according to height. Between 1,800 to 3,000 m altitude Montane wet temperate forests are found. Such forests are also found in higher hills of Tamil Nadu and Kerala. The common trees of these forests are Deodar, Chilaum, Chestnut, Birchi, Blue-pine, etc. whereas between 1,500 to 3,300 m altitude Himalayan moist temperate forests are found with trees of Pine, Cedar, Silvas, Fir, Spruiucem, etc.

Alpine Forests

The Alpine forests are found all along the Himalayas at altitudes ranging between 2,500 to 3,500 m. The trees found in this zone are Kail, Spruce, Yew, Firs, Birch, Honeyshckle, Artemesia, Potentilla and small scrubs.

Mangroves

Mangroves are special types of forests found along sheltered sea-coasts in tropical and sub-tropical regions. These forests are included in tidal forest category. In India these forests are found along the coastal areas of Bay of Bengal and also in some areas of west coast. The main areas of mangroves are *Sundarbans* (West Bengal), *Mahanadi and Bhitakanika* (Odisha), *Godawari and Krishna delta (Andhra Pradesh), Pichavarma Muthupur, Point Calimere (Tamil Nadu), Gulf of Kachchh and Gulf of Khambat* (Gujarat), *Goa, Kundapur* (Karnataka), *Achra Ratnagi* (Maharashtra), *Vembrud* (Kerala) and *Andaman and Nicobar Islands.* The delta of Sundarbans is covered by Sundari *(Heritiera minor)* tree. The trees found in mangroves areas are used for fuelwood and hard durable timber for construction and boat making.

Administrative Classification of Forests

In India, forests have been classified as Reserved, Protected and Unclassified Forests.

Reserved Forests: are under the direct supervision of government and entry in these forests is not allowed for collection of timber or grazing of cattle. About 53 per cent of the total forest areas of India is in this category.

Protected Forests: are those which are looked after by the government, but local people are allowed to collect fuel wood and graze their cattle without causing any damage to the forest. About 25 per cent forests of the country are included in protected forests.

Unclassified Forests: About 18 per cent of the country's forest area is under unclassified category. In these forests there is no restriction on cutting of trees, etc.

New Forms of Forestry

Social Forestry

Social forestry refers to the forests planted by the people of the society. This is a forestry of the people, by the people and for the people. Its objective is to manage and protect forests as well as afforestation of barren lands, which helps in reducing pressure on traditional forest area by developing plantations of fuelwood, fodder and grasses.

Agro Forestry

It involves raising of trees and agricultural crops either on the same land or in close association in such a way that all land and waste patches are put for tree plantation. This enables the former to get food, fodder, fuel, fruits and other products.

Community Forestry

It is a part of social forestry. It involves raising of trees on community lands for the benefit of the community as a whole. Generally the plants and seedlings are provided by forest department and their protection is done by the community. In states like Gujarat, Haryana, Himachal Pradesh, Karnataka, Kerala, Odisha, Punjab, Rajasthan, Tamil Nadu, Uttarakhand and Uttar Pradesh, community forestry is a big success.

Commercial Farm Forestry

It involves growing of trees on private land for the direct commercial purpose. This will reduce pressure of natural forests and is also a source of employment and earning.

Biodiversity In India

India is one of the world's 12 megadiversity countries. India contains 8.1 per cent of the world's biodiversity on 2.4 per cent of the earth's surface. It is estimated that 47,000 wild species of plants and over 89,450 wild species of animals are found in our country. India has a tremendous range of ecosystem, species and genetic diversity. The location of the Indian subcontinent at the confluence of three biogeographic realms has led to the presence of elements of African, European and Chinese and Indo-Malayan characteristics in the flora and fauna. The diversity is also due to the fact that India has almost every major type of habitat and climatic condition – from alpine heights to coasts and plains from areas of the heaviest rainfall to dry deserts.

India is a bio-rich country having great variety of plants and animals many of which are not found elsewhere. (Table 6.3)

Table 6.3 *Biodiversity of India and Its World Ranking*

	India's World Ranking	*Number of Species in India*
Mammals	8th	350
Birds	8th	1,200
Reptiles	5th	453
Amphibians	15th	182
Angiosperms	15th–20th	14,500

India has 350 different mammals (rated the eighth highest in the world), 1,200 species of birds (eighth in the world), 453 species of reptiles (fifth in the world) and 45,000 plant species, of which most are angiosperms, (fifteenth in the world). These include especially high species diversity of ferns (1,022 species) and orchids (1,082 species). India has 50,000 known species of insects, including 13,000 butterflies and moths. It is estimated that the number of unknown species could be several times higher.

It is estimated that 18 per cent of Indian plants are endemic to the country and found nowhere else in the world. Among the plant species, the flowering plants have a much higher degree of endemism, a third of these are not found elsewhere in the world. Among amphibians found in India, 62 per cent are unique to this country. Among lizards, of the 153 species recorded, 50 per cent are endemic. High endemism has also been recorded for various groups of insects, marine worms, centipedes, mayflies and freshwater sponges.

India can be conveniently divided into ten major regions, based on the geography, climate and pattern of vegetation seen and the communities of mammals, birds, reptiles, amphibians, insects and other invertebrates that live in them. Each of these regions contains a variety of ecosystems such as forests, grasslands, lakes, rivers, wetlands, mountains and hills, which have specific plant and animal species.

India's Biogeographic Zones are as follows:

1. The cold mountainous snow covered Trans-Himalayan region of Ladakh
2. The Himalayan ranges and valleys of Kashmir, Himachal Pradesh, Uttarakhand, Assam and other North-eastern states
3. The Terai, the lowland where the Himalayan rivers flow into the plains
4. The Gangetic and Brahmaputra plains
5. The Thar desert of Rajasthan
6. The semi-arid grassland region of the Deccan plateau, Gujarat, Maharashtra, Andhra Pradesh, Karnataka and Tamil Nadu
7. The North-eastern states of India

8. The Western Ghats in Maharashtra, Karnataka and Kerala
9. The Andaman and Nicobar Islands
10. The long western and eastern coastal belt along sandy beaches, forests and mangroves

Loss of Biodiversity

The extinction of several species or loss of biodiversity is a major problem of the international world and India is no exception to it. India's rich biological diversity is rapidly receding. At least 10 per cent of its recorded flora, and possibly a large fraction of its wild fauna, is threatened. Many species have been lost and many others are on the verge of extinction. In the last few decades, India has lost at least 15 per cent of its forests. More than 150 of the known species of medicinal plants in India have already become extinct due to unsustainable methods of harvesting. Of the species of flora and fauna that remains, over 10 per cent of flowering plants, 21 per cent of mammals and 5 per cent of birds are believed to be under threat. India's domesticated biodiversity is also under threat. Hundreds of crop varieties have disappeared. The main causes of the loss of biodiversity are:

1. Habitat destruction
2. Introduction of invasive species or exotic species
3. Overexploitation
4. Pollution
5. Climatic change
6. Hybridization
7. Loss of traditional knowledge
8. Changing agricultural and forestry practices
9. Ineffective legal system, etc.

Conservation of Biodiversity

The two categories of the methods of the conservation of biodiversity are: (i) *In-situ* conservation and (ii) *Ex-situ* conservation.

In-situ Conservation is done within the ecosystem and natural habitats, while in Ex-situ Conservation a controlled situation is created such as botanical garden for plants or zoological parks for animals, etc.

India has taken several steps towards in-situ conservation of biodiversity.

India's major attempt to conserve in-situ wild biodiversity has been through a network of protected areas throughout the country. Two kinds of protected areas are recognized by the legal system in India: National Parks and Wildlife Sanctuaries. National parks are highly protected by law. No human habitation, private landholding or traditional human activity such as firewood collection or grazing is allowed within the park. Sanctuaries are also protected, but certain types of activities such as collection of firewood and grazing of cattle are permitted within these areas.

The first such initiative was the establishment of the Corbett National Park in 1936. India has 99 national parks and 513 wildlife sanctuaries occupying 1,56,000 sq km, or over 5 per cent of the country's area. These protected areas have helped in conserving habitats and their biodiversity Several special projects have also been launched to save certain animal species which have been identified as needing concerted protection effort. These projects are designed to protect the species in-situ, by protecting and conserving their natural habitat. Project Tiger and Project Elephant are two such major initiatives. Other species conservation projects include Project Crocodile Breeding and Management, which was started in 1976 and now operates in 16 sanctuaries; and the Gir Lion Sanctuary Project in Gujarat, which aims at saving the Asiatic lion which once roamed over a wide stretch of the northern and central parts of the Indian subcontinent, but is today one of the most threatened species of animals in India. Project Vulture, Project Snow Leopard, Project Rhino, Project Hangul, Project Red Panda, Project Maipur, Thamin and Project Olive Ridley Turtles are also working.

To conserve medicinal plants, the forest departments of Andhra Pradesh, Karnataka, Kerala, Maharashtra and Tamil Nadu, with the help of the Foundation for the Revitalization of Local Health Traditions (FRLHTs), have set aside 54 forest patches as Medicinal Plants Conservation Areas (MPCAs) measuring 200 to 500 ha each. These MPCAs represent all the forest types and climatic zones of the region and harbour many species of medicinal plants threatened with extinction.

Wildlife Sanctuaries and National Parks of India

There are 589 Protected Areas in India, of which 99 are national parks and 513 are wildlife sanctuaries. They include a variety of ecosystems and habitats. Some have been created in order to protect highly endangered species of wild plants and animals found nowhere else in the world.

The Great Himalayan National Park is the largest sanctuary in this ecosystem and is one of the last homes of the beautiful snow leopard. The Dachigam Sanctuary is the only place where the rare Hangul or Kashmir Stag is found. There are several sanctuaries in the terai region; the Kaziranga National Park is the most famous which has elephant, gaur, wild boar, swamp deer, and hog deer, in large numbers, as well as tigers and leopards. Its bird life is extremely rich and includes ducks, geese, pelicans and storks. The Manas Sanctuary, in addition to the above terai species, also includes the rare golden langur and the very rare pygmy hog, the smallest wild boar in the world. The florican is found only in a few undisturbed grasslands in the terai sanctuaries.

In the sal forests of Madhya Pradesh, there are several PAs (Protected Areas). *Kanha* offers a wonderful opportunity to observe wild tigers from elephant back. It is the only PA in which a sub-species of the barasingha is found.

Bharatpur is one of the most famous water-bird sanctuaries in the world. Thousands of ducks, geese, herons, and other wading birds can be seen here. This is the only home of the very rare Siberian crane, which migrates to India every winter. During the last 20 years, the 30 or 40 Siberian cranes have dwindled to only 2 or 3. During 2002–03, no cranes were seen and it is possible that this beautiful bird will never again come to India.

In the Thar Desert, the wildlife is protected in the Desert National Park. Here, large numbers of blackbuck, nilgai and chinkara can be seen. The Great Indian Bustard lives in these arid lands.

Ranthambore was the most well-known sanctuary for observing tigers in the wild.

The Great and the Little Rann of Kachchh have been made into sanctuaries to protect the very rare wild ass, the flamingo, the star tortoise and the desert fox.

In Gujarat, the Gir Sanctuary protects the last population of the majestic Asiatic lion. This thorn and deciduous forest is also the home of large herds of chital, sambhar, and nilgai.

The sanctuaries of the Western Ghats and associated hill ranges protect some of the most diverse forest types in the country. The few examples of highly-threatened species include the Malabar giant squirrel, the flying squirrel and a variety of hill birds, several species of amphibians, reptiles and insects. These regions are also rich in highly endemic plant life. Sanctuaries such as, Bhimashankar, Koyana, Chandoli and Radhanagari preserve this rich flora in Maharashtra, Bandipur, Bhadra, Dandeti, Nagarhole, etc. in Karnataka; and Eravikulam, Parambikutam, Periyar, and the Silent Valley in Kerala.

In the Nilgiri Hills the rich forest sanctuaries protect some of the last pockets of the Indian elephant in South India. Examples include *Bandipur, Mudumalai, Wynad* and *Bhadra*. During the last 10 years, a large number of the great tusker elephants of this region have been ruthlessly killed for their ivory. Now, very few of these magnificent animals are left in these jungles.

Two important sanctuaries meant for preservation of coastal ecosystems are the *Chilka Lake* in Odisha and Point Calimere in Tamil Nadu. The *Sunderbans* protect the largest mangrove delta in India. The Marine National Park in Gujarat protects shallow areas in the sea, islands, coral reefs and extensive mudflats.

Over a hundred PAs have been created in the Andaman and Nicobar Islands to preserve their very special island ecosystems.

A list of India's National Parks, Sanctuaries, Tiger Reserves, Biosphere Reserves and Elephant Reserves is as follows:

Important National Parks and Sanctuaries in India

Name	*Location*	*Reserve For*
Namdapha Sanctuary	Arunachal Pradesh	Elephant, panther, sambhar, tiger, cheetal, king cobra
Kaziranga National Park	Assam	One-horned rhinoceros, gaur, elephant, leopard, wild buffalo
Sonai Rupa Sanctuary	Assam	Elephant, sambhar, wild boar, one-horned rhinoceros
Gautama Buddha Sanctuary	Bihar	Tiger, leopard, sambhar, cheetal, barking deer
Achanakmar Sanctuary	Chhattisgarh	Tiger, boar, cheetal, sambhar, bison
Valvadore National Park	Gujarat	Wolf, black buck
Wild Ass Sanctuary	Gujarat	Wild ass, wolf, nilgai, chinkara
Gir Forest	Gujarat	India's biggest wildlife sanctuary famous for gir lions
Nagarhole National Park	Karnataka	
Dandeli Sanctuary	Karnataka	Tiger, panther, elephant, cheetal, sambhar, wild boar
Bhadra Sanctuary	Karnataka	Elephant, cheetal, panther, sambhar wild boar
Banerghatta National Park	Karnataka	Elephant, cheetal, deer, grey partridge, green pigeon
Bandipur Sanctuary	Karnataka and Tamil Nadu	Elephant, tiger, panther, sambhar, deer, birds
Ranganthittoo Bird Sanctuary	Karnataka	Important bird sanctuary
Tungabhadra Sanctuary	Karnataka	Panther, cheetal, sloth bear, four-horned antelope
Dachigam Sanctuary	Kashmir	Kashmiri stag
Pachmarhi Sanctuary	Madhya Pradesh	Tiger, panther, boar, sambar, nilgai, barking deer
Gandhi Sagar Sanctuary	Madhya Pradesh	Cheetal, sambhar, chinkara, wild birds
Bandhavgarh National Park	Madhya Pradesh	Tiger, panther, cheetal, nilgai, wild boar
Simlipal Sanctuary	Odisha	Elephant, tiger, leopard, gaur, cheetal
Sunderban Tiger Reserve	West Bengal	Tiger, deer, wild boar, crocodile, Gangetic Dolphin
Jaldapara Sanctuary	West Bengal	Rhinoceros
Ghana Bird Sanctuary	Rajasthan	Water birds, black-buck, cheetal, sambhar
Kanchenjunga National	Sikkim	Snow leopard, musk deer, Himalayan boar park
Vedanthangal Bird Sanctuary	Tamil Nadu	Important bird sanctuary
Dudhwa National Park	Uttar Pradesh	Tiger, panther, sambar, cheetal, nilgai, barking deer
Chandraprabha Sanctuary	Uttar Pradesh	Gir lions, cheetal, sambhar
Corbett National Park	Uttarakhand	Tiger, leopard, elephant, sambhar (named in memory of Jim Corbett)

Tiger Reserves in India

	State	Tiger Reserve	Remarks
1.	Andhra Pradesh	Nagarjunsagar-Srisailam Tiger Reserve	Low tiger density, Left wing extremism (Naxalite)
2.	Andhra Pradesh	Kawal Tiger Reserve	20 tigers (2010)
3.	Arunachal Pradesh	Pakhui Tiger Reserve	Good tiger density
4.	Arunachal Pradesh	Namdapha Tiger Reserve	Low tiger density, Extremism, encroachment by Lisus
5.	Assam	Nameri Tiger Reserve	Encroachment and tree cutting by locals
6.	Assam	Manas Tiger Reserve	Poor tiger density, Recovering from prolonged disturbance due to Bodo unrest
7.	Assam	Kaziranga Tiger Reserve	Good tiger density
8.	Bihar	Valmiki Tiger Reserve	Low tiger density, Extremism
9.	Chhattisgarh	Indravati Tiger Reserve	Severely affected by Naxalites, Out of bounds
10.	Chhattisgarh	Udanti & Sitanadi	Low tiger density Tiger Reserve
11.	Chhattisgarh	Guru Ghasidas National Park in Madhya Pradesh	Separated from Sanjay National Park
12.	Chhattisgarh	Adhanakmar Tiger Reserve	Low tiger density
13.	Goa	Mhadei Tiger Reserve	—
14.	Jharkhand	Palamau Tiger Reserve	Low tiger density, Left wing extremism
15.	Karnataka	Bandipur Tiger Reserve	Good tiger density
16.	Karnataka	Nagarhole (Extension)	Good tiger density Tiger Reserve
17.	Karnataka	Dandeli-Anashi Tiger	Low tiger density
18.	Karnataka	Bhadra Tiger Reserve	Low tiger density
19.	Karnataka	Bannerghatta National Park	Not a Project Tiger reserve – Lion Reserve
20.	Karnataka	Biligiri Rangaswamy	17 tigers (2005) Temple Wildlife Sanctuary
21.	Karnataka	Kudremukh Tiger Reserve	20 tigers (2010)
22.	Kerala	Periyar Tiger Reserve	There are estimated 53 tigers (2010) in the reserve
23.	Kerala	Parambikulam Tiger Reserve	Moderate tiger density
24.	Kerala, Tamil Nadu	Annamalai Tiger Reserve	Moderate tiger density
25.	Madhya Pradesh	Bandhavgarh Tiger Reserve	Good tiger density
26.	Madhya Pradesh	Bori-Satpura Tiger Reserve	Poor tiger density
27.	Madhya Pradesh	Kanha Tiger Reserve	Good tiger density
28.	Madhya Pradesh	Panna Tiger Reserve	Recovering from local extinction of tiger due to poaching, Dacoit infestation
29.	Madhya Pradesh	Pench Tiger Reserve	Good tiger density
30.	Madhya Pradesh	Sanjay Dubri Tiger Reserve	Very low tiger density
31.	Madhya Pradesh	Ratapani Tiger Reserve	—
32.	Maharashtra	Melghat Tiger Reserve	Low tiger density
33.	Maharashtra	Pench Tiger Reserve	Moderate tiger density

Cont'd...

...Cont'd

34.	Maharashtra	Tadoba-Andheri Tiger Reserve	Moderate tiger density; High rate of man-tiger conflicts
35.	Maharashtra	Sahyadri Tiger Reserve	9 tigers (2007)
36.	Maharashtra	Nagzira-Navegaon	Tiger Reserve
37.	Maharashtra	Bor Tiger Reserve	—
38.	Mizoram	Dampa Tiger Reserve	Low tiger density due to ecological reasons
39.	Odisha	Simlipal Tiger Reserve	Low tiger density, Left wing extremism
40.	Odisha	Satkosia Tiger Reserve	Low tiger density
41.	Odisha	Sunabeda Tiger Reserve	—
42.	West Bengal	Buxa Tiger Reserve	Low tiger density offences by jobless tea plantation workers
43.	West Bengal	Sunderbans Tiger Reserve	Good tiger density
44.	Rajasthan	Ranthambore Tiger Reserve	Good tiger density
45.	Rajasthan	Sariska Tiger Reserve	Recovering from local extinction of tiger due to poaching
46.	Rajasthan	Mukundara Hills Tiger Reserve	—
47.	Tamil Nadu	Collect-Mundathurai Tiger Reserve	Low tiger density due to ecological reasons
48.	Tamil Nadu	Mudumalai National Park	Moderate tiger density
49.	Tamil Nadu	Sathyamangalam Tiger	46 tigers (2010) Reserve
50.	Uttar Pradesh	Dudhwa Tiger Reserve	Good tiger density, Poaching along borders
51.	Uttar Pradesh	Pilibhit Tiger Reserve	—
52.	Uttar Pradesh	Suhelwa Tiger Reserve	—
53.	Uttar Pradesh	Corbett Tiger Reserve	Good tiger density

Biosphere Reserves in India

Name	*Location*	*State*	*Type*	*Key Fauna*
Seshachalam hills	Seshachalam hill ranges covering parts of Chittoor and Kadapa districts	Andhra Pradesh	Eastern Ghats	
Great Nicobar Biosphere Reserve	Southernmost Islands of Andaman and Nicobar Islands	Andaman and Nicobar Islands	Islands	Salt water Crocodile
Dehang-dibang	Part of Siang and Dildang Valley	Arunachal Pradesh	Eastern Himalayas	
Manas	Part of Kokrajhar, Bongaigaon, Barpeta, Nalbari, Kamrup and Darrang districts	Assam	Eastern Himalayas	Golden Langur Red Panda
Dibru-Saikhonwa	Part of Dibrugarh and Tinsukia Districts	Assam	Eastern Himalayas	Golden Langur
Rann of Kachchh	Part of Kachchh, Rajkot, Surendra nagar and Patan districts	Gujarat	Desert	Indian Wild Ass
Cold Desert	Pin Valley National-Park and surroundings Chandratal and Sarchu and Kibber Wildlife Sanctuary	Himachal Pradesh	Western Himalayas	Snow Leopard

Cont'd...

...Cont'd

Agasthamalai Biosphere Reserve	Neyyar, Peppara and Shenduruny Wildlife Sanctuary and their adjoining areas	Kerala, Tamil Nadu	Western Ghats	Nilgiri Tahr, Elephants
Panna	Part of Panna and Chattatpur districts	Madhya Pradesh	catchment area of the Ken River	Tiger, Chital, Chinkara, Sambhar and Sloth bear
Pachmarhi Bio sphere Reserve	Parts of Betul district; Hoshangabad district and Chhindwara district	Madhya Pradesh	Semi-Arid	Giant Squirrel Flying Squirrel
Achanakawar Amarkantak	Part of Annupur, Dindori and Bilaspur districts	Madhya Pradesh Chhattisgarh	Maikala hills Range	
Nokrek	Part of Garo	Meghalaya	East Himalayas	Red Panda
Simlipal	Part of Mayurbhanj district	Odisha	Deccan Peninsula	Gaur
Sunderbans	Part of delta of Ganges and Brahmaputra river system	West Bengal	Gangetic Delta	Royal Bengal Tiger
Kanchanjanga	Parts of Kanchenjunga hills	Sikkim	East Himalayas	Snow Leopard Red Panda
Nilgiri Biosphere Reserve	Part of Waynad, Nagarhole, Bandipur and Mudumalai, Nilambur, Silent Valley and Annamalai Hills	Tamil Nadu, Kerala and Karnataka	Western Ghats	Nilgiri Tahr, Lion-tailed macaque
Gulf of Mannar	Indian part of Gulf of Mannar extending from Rameshwaram Islands in the North to Kanyakumari in the South of Tamil Nadu arid Sri Lanka	Tamil Nadu	Coasts	Dugong or Sea Cow
Nanda Devi	Parts of Chamoli district, Pithoragarh district and Bageshwar district	Uttarakhand	Western Himalayas	

List of Project Elephant Reserves

Elephant Range	*Elephant Reserve*	*State*
Eastern India (South-west)	1. Mayurjharna	West Bengal
Bengal-Jharkhand Orrisa	2. Singhbhum	Jharkhand
	3. Mayurbhanj	Odisha
	4. Mahanadi	Odisha
	5. Sambalpur	Odisha
	6. Baitarni	Odisha
	7. South Odisha	Odisha
	8. Lemru	Chhattisgarh
	9. Badalkhol Tamorpingla	Chhattisgarh
North Brahmaputra	10. Kameng	Arunachal Pradesh
Arunachal Pradesh	11. Sonitpur	Assam
South Brahmaputra	12. Dihing-Patkai	Assam
Assam, Arunachal	13. Deomoli	Arunachal Pradesh

Cont'd...

...Cont'd

Kaziranga	14. Kaziranga-Karbi Anglong	Assam
Assam	15. Dhansiri-Lungding	Assam
Nagaland	16. Intanki	Nagaland
Eastern Doors	17. Chirang-Ripou	Assam
Assam-West Bengal	18. Eastern Dooars	West Bengal
East Himalayas	19. Garo Hills	Meghalaya
Meghalaya	20. Khasi Hills	Meghalaya
Nilgiri-eastern ghats	21. Mysore	Karnataka
Karnataka-Kerala	22. Wayanand	Kerala
Tamil Nadu	23. Nilgiri	Tamil Nadu
Andhra Pradesh	24. Rayala	Andhra Pradesh
South Nilgiri	25. Nilabur	Kerala
Kerala-Tamil Nadu	26. Coimbatore	Tamil Nadu
Western Ghats	27. Annamalai	Tamil Nadu
Tamil Nadu-Kerala	28. Anaimudi	Kerala
Periyar	29. Periyar	Kerala
Kerala-Tamil Nadu	30. Srivilliputtur	Tamil Nadu
Northern India	31. Shiwalik	Uttarakhand
Uttarakhand	32. Uttarakhand	Uttarakhand

7

Soils

Soil represents the loose and unconsolidated material derived through the breaking down of rocks. According to Raman, 'soil is formed of rocks that have been more or less changed chemically, together with the remains of plants or animals that live on or in it'. The soil is differentiated into (i) top soil which represents the uppermost layer of the soil and supports all types of plant life, and (ii) sub soil which represents the weathered materials of the parent rocks but does not carry the nutrients and other properties of top soil. The factors affecting soil formation are parent rock materials, climate, topography and organic factors. Each layer of soil is known as 'Soil Profile'.

Major Soil Types of India and their Distribution

India has a large variety of soil group and their spatial distribution is also varied not only from state to state, but also within a state. Several attempts have been made by soil scientists to classify the soils of India according to their structure, texture, colour, pH value, porosity, etc. The earlier studies were done by foreign scholars like Volckar (1893), Leather (1898), Schokalskaya (1932), Champion (1936), etc. Among Indian scholars, Widia (1935), Basu (1937), Vishwanath and Ukil (1944), Chatterjee, Krishnan, Roychaudhary (1954), have classified the soils of India. In 1956, All India Soil and Land Use Survey Organization has done a classification. Similarly in 1957 the National Atlas and Thematic Mapping Organization has published the soil map of India. Indian Council of Agricultural Research (ICAR) set up an All India Soil Survey committee in 1953, which identifies eight soil groups in India, These are (i) Alluvial soil, (ii) Black Soils, (iii) Red Soils, (iv) Laterite Soils, (v) Forest and Mountain soils, (vi) Arid and Desert Soils, (vii) Saline and Alkaline soils and (viii) Peaty and Marshy soils.

Table 7.1 indicates the major soils of India, their extent and their equivalent according to US Department of Agriculture.

Table 7.1 *Distribution of Soil Types of India (with their equivalent according to US Dept. of Agriculture)*

Major Soils (Traditional Name)	*Extent*		*Distribution in States*	*Soil Orders (US Soil Taxonomy)*
	000 ha	*%*		
Alluvial	1,00,006	30.4	J&K, H.P., Punjab, Haryana, Delhi, U.P., Gujarat, Goa, M.P., Maharashtra, Andhra Pradesh, Karnataka, Kerala, Tamil Nadu, Bihar, Odisha, West Bengal, Arunachal Pradesh, Assam, Nagaland, Manipur, Mizoram, Meghalaya, Andaman & Nicobar Islands	Inceptisols, Entisols, Alfisols, Aridisols
Coastal Alluvial	10,049	3.1	Andhra Pradesh, Karnataka, Kerala, Tamil Nadu, Puducherry, Odisha, West Bengal, Gujarat, Lakshadweep, Andaman & Nicobar Islands	Aridisols, Inceptisols, Entisols

Cont'd...

...Cont'd

Red	87,989	26.8	Andhra Pradesh, Karnataka, Kerala, Tamil Nadu, Puducherry, Rajasthan, MP, Maharashtra, Gujarat, Goa, Telangana, Andhra Pradesh, Assam, Manipur, Meghalaya, Nagaland, Mizoram, Tripura, UP, HP, Andaman & Nicobar Islands	Alfisols, Ultisols, Entisols, Inceptisols, Mollisols, Aridisols
Laterites	18,094	5.5	Telangana, Andhra Pradesh, Kerala, Karnataka, Tamil Nadu, Odisha, Maharashtra, West Bengal	Alfisols, Ultisols, Inceptisols
Brown Forest	540	0.2	Karnataka, Maharashtra	Mollisols, Inceptisols
Hill	2,262	0.7	Manipur, Odisha, Tripura, Nagaland, West Bengal, UP, Sikkim	Inceptisols, Endisols
Mountain meadow	60	—	Jammu & Kashmir	Mollisols
Terai	326	0.1	UP, Sikkim	Mollisols, Entisols
Sub-montane	104	—	Jammu & Kashmir	Alfisols
Black	54,682	16.6	MP, Maharashtra, Rajasthan, UP, Bihar, Tamil Nadu, Odisha, Gujarat, Telangana, Puducherry	Vertisols, Mollisols, Entisols, Inceptisols, Aridisols
Desert	26,283	8.0	Rajasthan, Gujarat, Haryana, Punjab	Ardisols, Inceptisoles, Entisols
Others	28,305	—	—	—

Source: www.currentscience.ac.in

Figure 7.1 *Soil Types and Their Distribution in India*

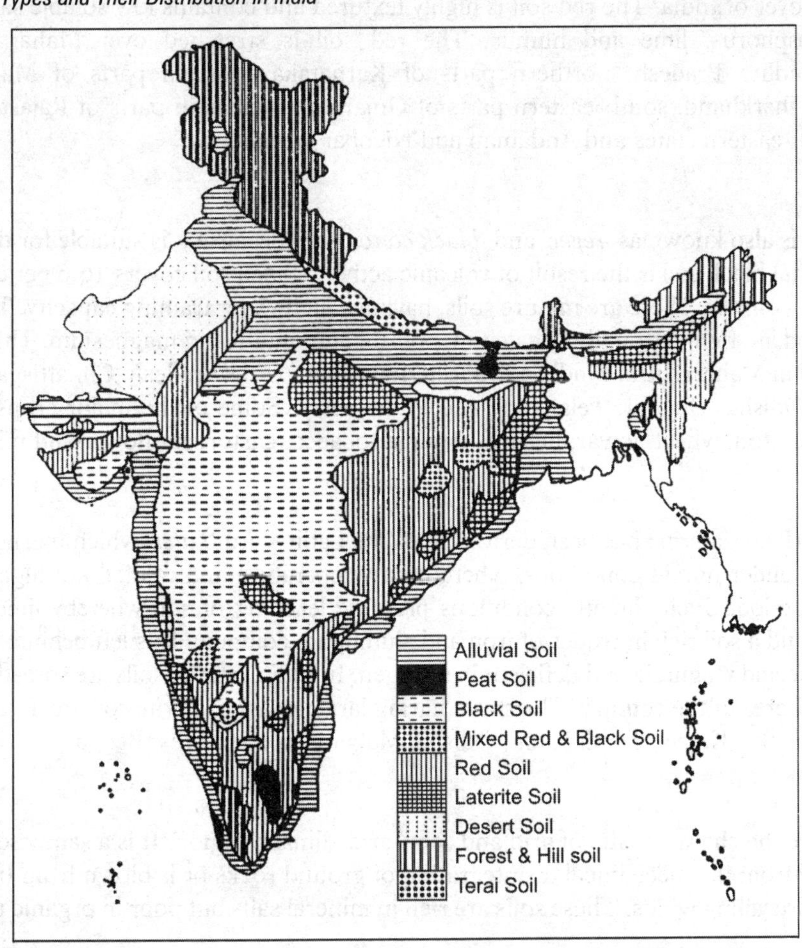

A brief description of the major soil types and their distribution in India is as follows:

Alluvial Soils

Alluvial soils are most fertile soils occupying 30.4 per cent of the total soil area of the country. These soils are formed through the sediments deposited by rivers as in the Indo-Gangetic Plain. Another 3.1 per cent is the coastal alluvial soil, which is found in coastal areas formed by the sea waves. Alluvial soils are classified into newer alluvium (khaddar) and older alluvium (bhangar). The alluvial soils occur mainly in Ganga-Brahmaputra Plains of the northern India. The other rivers like Narmada, Tapi, Mahanadi, Godavari, Krishna and Cauveri have also formed alluvial plains. The states having alluvial soils are: Punjab, Haryana, Uttar Pradesh, Madhya Pradesh, Gujarat, Bihar, West Bengal, Assam, Meghalaya, Nagaland, Arunachal Pradesh, Mizoram, Odisha, Maharashtra, Goa, Andhra Pradesh, Karnataka, Kerala, Tamil Nadu, etc. The coastal alluvial is spread in coastal areas of West Bengal, Odisha, Andhra Pradesh, Tamil Nadu, Puducherry, Kerala, Goa, Gujarat, Lakshadweep and Andaman and Nicobar Islands.

Red Soils

Red soils develop over old crystalline and metamorphic rocks. These are found in hot and humid regions. The colour of the soil is generally red due to high iron content. They occupy 26.8 per cent of the total soil cover of India. The red soil is highly textured and contains low soluble salts, but poor in nitrogen, phosphorus, lime and humus. The red soil is stretched over Maharashtra Plateau, Telangana, Andhra Pradesh, northern parts of Karnataka, western parts of Madhya Pradesh, Chhattisgarh, Jharkhand, south-eastern parts of Gujarat as well as in parts of Rajasthan, Himachal Pradesh, north-eastern states and Andaman and Nicobar islands.

Black Soils

The black soil is also known as *'reger'* and *'black cotton soil'* because it is suitable for the cotton crop. It is believed that black soil is the result of volcanic activity. Black soil covers 16.6 per cent of the total soil area of the country. These are mature soils, having a high water training capacity. These soils have clay texture and are rich in iron, lime, calcium, potash, aluminium and magnesium. The areas of black soil are spread in Maharashtra, Madhya Pradesh, Rajasthan, Uttar Pradesh, Chhattisgarh, Jharkhand, Tamil Nadu, Odisha, Gujarat, Telangana and Puducherry. Some of the major crops grown on the black soil are cotton, wheat, jowar, linseed, virginia tobacco, castor, sunflower and millets.

Laterite Soils

The name laterite or lateritic has been derived from the Latin word *'Later'* which means brick. Laterite soil is formed under humid conditions, where both temperature and rainfall are high and also have alternate dry period. Such climatic conditions promote leaching of soil whereby lime and silica are leached away and a soil rich in oxides of iron and aluminium compounds is left behind. All laterite soils are poor in lime and magnesia and deficient in nitrogen. In India, laterite soils are spread in 5.5 per cent of the total soil area of the country. The states having large areas of laterite soil are Telangana, Andhra Pradesh, Karnataka, Kerala, Tamil Nadu, Odisha, Maharashtra and West Bengal.

Desert Soils

Desert soils are the characteristic of arid and semi-arid climatic region. It is a sandy soil and the sand has originated from the mechanical disintegration of ground rocks or is blown from Indus Basin and the coasts by prevailing winds. These soils are rich in mineral salts but poor in organic matter. The pH value is high. The desert soils consist of aeolian sand (over 90%) and clay (5 to 10%). These soils are spread in 8 per cent of the total soil area of the country and mostly found in arid and semi-arid regions of Rajasthan, Gujarat, Haryana and Punjab.

Forest and Mountain Soils

This type of soil is found on the hill slopes covered by forests. These soils have varied characteristics dependent upon the parent rocks, ground-configuration, climate and humus content. These soils are found in the valleys and hill slopes of the Himalayas on an altitude between 200 to 300 m. Such soil includes loamy podzols and high altitude soils. The podzol soil occupies areas of Assam, Darjeeling, Uttarakhand, Himachal Pradesh and Jammu & Kashmir. Apart from the Himalayan region, the forest soils occur in Western and Eastern Ghats as well as in some parts of the peninsular plateau. These soils are suitable for tea and coffee plantations as well as for spices and tropical fruits. Wheat, maize and barley are also produced in Tamil Nadu, Jammu & Kashmir, Uttaranchal and Himachal Pradesh.

Saline and Alkaline Soils

These soils are known by different names such as *reh, kallar, usar, thur, rakar, karl and chopan.* These soils are mostly found in Karnataka, Telangana, Andhra Pradesh and also in dried parts of Rajasthan, Haryana, Punjab, Uttar Pradesh, Bihar and Maharashtra. In these soils the saline and alkaline efflorescence appears on the surface as a layer of white salt through capillary action. These soils are deficient in nitrogen and calcium and have very low after bearing capacity.

Peaty and Marshy Soils

In area of heavy rainfall where draining system is poor, the peaty and marshy soils have developed. These soils are mostly occur in Sundarban delta as well as in deltas of Mahanadi, Godavari and Kaveri deltas. The Rann of Kachchh is also having such soils. Marshy soils also occur in coastal areas of Tamil Nadu, Kottayan and Allapuzha districts of Kerala and Almora district of Uttarakhand. The peaty soil is black, heavy, highly acidic and deficient in potash and phosphate.

Apart from above main soil groups some other types of soils such as Red and Black Montane are found in India. Karewa soils are the lacustrine deposits found in Kashmir valley and Bhadarwah valley of Jammu Division.

Other Classification of Indian Soils

United States Department of Agriculture (USDA) suggested a detailed classification of soils of the world which is based on soil taxonomy. The same classification has been adopted by Indian Council of Agricultral Research (ICAR) for India also. The type of soils identified accordingly are – *Inceptisols, Entisols, Alfosols, Vertisols, Aridsols, Ultisols, Mollisols* and others. Among these the most widespread in India is the inceptisols, which covers more than 39 per cent of the soil cover of India followed by 28 per cent entisols, 13.5 per cent alfisols, 8.8 per cent vestisols and 4.28 per cent ardisols.

Soil Erosion

Soil erosion is a natural process of removal of top soil by water and wind. It is a process whose rates may be magnified by man. Bare rock surfaces are exposed to various types of physical, chemical and biological processes which led to physical and chemical disruption. Physical process of weathering includes action of water, temperature, glaciers and gravity which cause weathering of rocks through such process as wetting, drying, heating, cooling, freezing, glaciations, solution, etc. On the other hand chemical processes of weathering include hydration, hydrolysis, oxidation, reduction, carbonation, etc. All these processes lead to disintegrate the soil and are the cause of soil erosion. The main types of soil erosion are:

1. Normal or natural erosion
2. Fluvial erosion
3. Aeolian or wind erosion
4. Marine erosion and
5. Accelerated soil erosion by human actions.

Actually, normal soil erosion is a continuous process and its impact on environment is also limited, but when it occurs on a large scale due to water and wind erosion, it not only degrades the environment but is also responsible for the expansion of wasteland due to gullies and ravines, as has been in case of the Chambal region of Madhya Pradesh and Rajasthan. Four types of soil erosion due to water have been identified, these are:

1. sheet erosion,
2. rill erosion,
3. gully erosion, and
4. riparian erosion.

Soil erosion due to wind is a common phenomenon in arid and semi-arid regions and is often a cause of desertification or desert expansion. The process of soil erosion is always accelerated by man either by overgrazing, faulty land practices or by deforestation.

There are certain phenomena associated with the process of soil erosion, such as soil creep, earthflow and mudflow, mass wasting, landslides, etc. The slow downslope movement of soil and regolith is known as *soilcreep*. It is a process in which particles slowly move down the hill slope. In humid climatic regions where slopes are steep, a mass of water-saturated soil may slide downslope during a period of a few hours in the form of an earthflow. When it takes place in the form of mass wasting, it becomes mudflow, a mud stream that pours down canyons in mountainous region and poses a serious environmental hazard. It also occurs in desert regions during a cloud burst. Earthflow and mudflow are the common causes of blockage of highways and railroad lines, supply of irrigation water and also damage property and cause loss of life.

The impact of soil erosion on the environment is both local as well as regional. The following are the general impacts:

1. formation of gullies and ravines and the whole region becomes an undulating area not suitable for human activities
2. washout of the upper layer of the soil, specially along hillside slope
3. uprooting of plants due to soil cutting, thus responsible for deforestation
4. loss of soil fertility
5. loss of pastures
6. expansion of deserts; and
7. increase in frequency of droughts and floods.

In fact, too much soil erosion is one of the causes of eco-imbalance. In India, the problem of soil erosion can be seen in Jammu & Kashmir, Himachal Pradesh, Uttar Pradesh, Bihar, Assam, Arunachal Pradesh, Sikkim as well as in areas of Rajasthan and Madhya Pradesh. It has been estimated that in India about 53 per cent of the area is suffering from the problem of soil erosion. Indian rivers used to carry about 600 million tonnes of soil every year, to the sea. Soil erosion can be checked by adopting the following measures:

1. Biological methods
   (i) Strip-cropping
   (ii) Crop-rotation
   (iii) Application of manure
   (iv) Shelter belt
   (v) Vegetation cover
2. Mechanical methods
   (i) Contouring
   (ii) Terracing
   (iii) Control of gully through retention of run-off, diversion of run-off, new structures
3. By maintaining soil fertility through proper soil management system

8

Social Geography of India

Social geography is the branch of human geography that is most closely related to social theory in general and sociology in particular, dealing with the relation of social phenomena and its spatial components. Very few studies have been done on social geography of India. The notable one is *Social Geography* by Aijazuddin Ahmad, in which various facets of Indian social geography have been discussed.

Indian society is not simply a conglomeration of various ethnic, religious, linguistic, caste and regional collectivities. It is also quite complex in terms of differentiation in each one of the collectivities and their historical and socio-cultural specifications. Indian society is a panorama, to a large extent a unique one, which has absorbed diverse languages, cultures, religions and people of different social organs at different points of time in the past. India is a plural society and is characterized by its unity and diversity. A grand synthesis of cultures, religions and languages of the people belonging to different castes and communities has upheld its unity and cohesiveness despite foreign invasions. In fact, Indian society is the product of its physical environment which also has a characteristic of unity in diversity.

Social Structure

The society is made up of elements drawn from diverse origins. Within the population of India are subsumed tribes – no less than three hundred ethnically differentiated communities – caste groups, language and religious groups, displaying striking differences in social organization and cultural patterns, even material cultures. There are differences in racial strands and ethnic and cultural identities are strongly defined. Almost all religions from tribal forms of animism and totemism to Hinduism, Buddhism, Christianity, Islam and Sikhism have their followers in different proportions. The social diversity is perhaps the most powerful manifestation of Indian identity. The social groups with diverse ethnic origins, representing racial stocks from proto-Australoids and Mongoloids to the different branches of the Mediterraneans, western Brachycephalic groups and the Nordics, found a place for themselves at different points of time adapting themselves to the different ecological niches offered by the physiographic and the climatic setting of the subcontinent. There are strongly defined tribal identities based on ethnic origins. The tribal population includes within its fold hundreds of large and small ethnic groups from the Santhals, Gonds and Bhils to the Mundas, Oraons, Hos, Todas and Apatanis, to mention only a few. The tribal India is characterized by a heterogeneity of a very high order.

Their geographic patterning in the Indian space in itself posed serious problems in the administration of the tribal affairs. The tribes are heavily concentrated in the north-east, in the hills and the plateau surfaces surrounding the Brahmaputra valley as well as in the mid-Indian belt from the Western Ghats to the Rajmahal hills. The tribes have been living in a world of their own in habitats which permitted little interaction within themselves or between them and the rest of the Indian population. The mutually exclusive domains of the tribal and non tribal segments of population appear to have contributed to the regional differentiation of the first order. The tribal identity is thus based on ethnicity, as differentiation took place along the ethno-lingual lines. Tribalism implies allegiance to traditional values of social equality, freedom, communitarian ethos, equality of status, including gender equality, segmentation along clan and kinship structures rather than social stratifications.

The Indian social structure is very complex and differs from region to region. The main components of the Indian social structure are tribes, castes, religions and languages.

Tribes

The tribes are essentially the indigenous people. According to Andre Beteille, 'A tribe is in an ideal state, a self-contained unit. It constitutes a society in itself'. They belong to the different racial, ethnic, linguistic, religious and cultural groups and mostly occupying the areas of isolation. DG Mandelbaum (1972) mentions the following characteristics of Indian tribes:

1. Kinship as an instrument of social bonds;
2. Lack of hierarchy among men and groups (clan and lineages);
3. Absence of strong, complex, formal organizations;
4. Communitarian basis of land holdings;
5. Segmentary character;
6. Little value on surplus accumulation, on the use of capital and on market-trading;
7. Lack of distinction between form and substance of religion; and
8. A distinct psychological make-up for enjoying life.

In India, tribals constitute 8.61 per cent of the total population of the country, numbering 104.28 million (2011 Census) and cover about 15 per cent of the country's area. Table 8.1 shows the tribal population in India, their growth and sex-ratio.

Table 8.1 *Population of Scheduled Tribes in India*

Year	*Total Tribal Population*	*Decadal Growth Rate (%)*	*Sex-ratio*
1961	3,01,30,184	—	987
1971	3,80,15,162	26.2	982
1981	5,61,28,638	35.8	983
1991	6,77,58,380	31.2	972
2001	8,83,26,978	24.5	978
2011	10,42,81,034	23.7	990

It becomes clear that steady growth in total tribal population has been recorded between 1961–2011 although decadal growth rate was highest during 1971–81 (35.8%) and lowest during 2001–2011 (23.7%).

Tribes of India

The major state wise tribes of India are as follows:

Andaman and Nicobar Islands: Andamanese, Chompens, Jarawa, Nicobari, Onges, and Sentineles.

Andhra Pradesh: Andh, Bagata, Balmiki, Bhidhu, Bhil, Chenchu, Dora, Dubla, Gadaba, Gond, Jatapu, Kammara, Konda, Konda-Reddis, Khonds, Kotias/Kotiya, Koya, Kulia, Kutunayakan, Lambadi, Malis, Manne-Dora, Nakha-Dora, Nayaks, Porja, Pradhan, Reddis, Reddi-Doras, Rona/Rena, Sabra, Sugalis, Thoti, Valmiki, Yanadia, Yenadi, Yerukula.

Arunachal Pradesh: Abore, Adi, Aka, Apatani, Dafla, Digaru, Gomba, Khamti, Miji, Mishmi, Monba, Nokte, Singhpo.

Assam: Boro, Chakma, Dimsa, Devari, Hoja, Hozai, Kachari, Lalung, Mikir, Miri, Xaxa

Bihar: Asur, Baiga, Banjara, Bedia, Binjhia, Birhor, Birjia, Chero, Chik, Gond, Ho, Karmali, Kharia, Kherwar, Khond, Kora, Korwa, Lohar, Mahli, Mal-Paharia, Munda, Oraon, Parhaiya, Santhal, Paharia, Sauria, Savar.

Gujarat: Bhil, Dhanka, Dubla, Garsaya, Gond, Kathodia, Rabaria, Siddi, Vaghir, Varli.

Himachal Pradesh: Bhot, Bhotia, Gaddi, Gujjar, Kanaura, Kinnar, Lahauli, Lamba, Pangwal, Sawangi.

Jammu & Kashmir: Bakarwal, Buddhists of Ladakh, Gaddi, Gujjars and Pahadi.

Karnataka: Arandan, Barda, Bamcha, Bavacha, Bhil, Chodhara, Chenchu, Dhanka, Dhodia, Dubla, Ekliga, Godalu, Gamta, Hasadaru, M Hakki, Kadar, Kaikuruba, Kathodi, Kokna, Kolidhor, Koya, Koruba, Kudia, Kaniyan, Kuruman, Kota, Koraga, Konda-Reddi, Konda-Kakush, Kammara, Koraga, Kurumba, Malaikudi, Maleru, Meda, Marti, Mudugar, Malaikandi, Madian, Naikda, Paliyar Pardhi, Patalia, Pomla, Porama, Pulayan, Puniyan, Rathawa, Ragond, Solga, Soligaru, Thoti, Toda, Varava, Varti, Vitolia, Yerawa.

Kerala: Arandan, Erulaj, Eravallan, Kadar, Kannikar, Kochuvelan, Kadiyan, Kammara, Kahunayakan, Kochu, Kondakakus, Koraga, Kota Kudia, Kurambas, Korichachhan, Kuruman, Malai, Mallayan, Mannar, Malai-Kandi, Matri, Pallyan, Palliyar, Paniyan, Palyan, Pulayan, Puniyan, Urali/Uralay Ulladana Vishavan.

Madhya Pradesh: Agariya, Andh Baiga, Bhil, Bhumia, Birhor, Biyar, Chanwar, Dhanwar, Damor, Gadaba, Garasia, Gond, Kamar, Keer, Khairwar, Khond, Kol, Korwa, Meena, Nagasia, Nat, Nihal, Oraon, Panika, Pao, Patika, Pradhan, Saharia, Saonta, Saur, Sawar.

Manipur: Aimol, Hamar, Kuki, Mayon, Ralte, Zou.

Meghalaya: Garo, Hamar, Jaintia, Khasi, Powi.

Mizoram: Lushai, Mizo, Maat, Paki.

Nagaland: Angammi, Ao, Konyak, Kuki, Lotha, Mikir, Mon, Rengma, Sema.

Punjab: Bhot, Bodh, Gaddi, Swangwala.

Odisha: Bagatta, Banjara, Bathudi, Bauga, Bharua, Bhotada, Bhumiya, Bhuyan, Bharua, Binjhal, Birher, Birhor, Chenchu, Dal, Desua, Gondia, Gond, Kora, Ho, Juang, Kammar, Kharia, Khond, Kisan, Kol, Koli, Konda, Kotia, Kuli, Kond, Kolha, Lodh, Lohar, Madia, Mahali, Mirdha, Munda, Oraon, Paitia, Parenga, Pondo, Poraja, Rajor, Shabar, Santhal, Sonti, Soura.

Rajasthan: Andh, Barda, Bavacha/Bamcha, Bhaina, Bhattra, Bhil, Bhila, Bhujiya, Binjhwar, Birhul, Charan, Chodhara, Dhanka, Dhanwar, Dhodia, Dhor Dubla, Gadaba, Gamit, Garsaya, Gond, Halbar or Halbi, Kamar, Kandh, Kathodi or Katkari, Kaudhri, Kharia, Kharwar, Kokna, Kokli, Kol, Koli, Kolam, Korku, Korwa, Koya, Kukka, Kumbi, Kuwar, Malhar, Meena, Naikda, or Naika, nagasia, Nihal, Oraon, Padhar, Parja, Patelia, Paura-Bhils, Pomla, Pradhan, Rathawa, Rawri, Saharia, Saonto, Sawar, Siddi, Thakur/Thakar, Thoti, Varli, Vitolia.

Tamil Nadu: Aranadan, Auidian, Eravalana, Erulia, Irular, Kadar, Kakush, Kammard, Kaniyan, Kochu, Konda, Konda-Reddy, Koraga, Kota, Koya, Kudia, Kulayan, Kochuvelan, Kuruman, Kurumba, Malai-Aryan, Malsar, Malikindi, Malyali, Malyan Mannan, Mudugaar, Muthuran, Paliyar, Paniyan, Palliyan/Pulliyan, Reddi, Sholga, Uralay, Ulladan, Urali, Vishvan.

Uttar Pradesh: Banjara, Bhoksa, Bhotia, Buxa, Hanji, Jaunsari, Kanjar, Raji, Sainsi, Tharu.

West Bengal: Asur, Baiga, Badia, Banjara, Baraik, Bhumij, Birhor, Birjia, Bhoomiji, Bhutia, Cheo, Chic, Chakma, Garo, Gond, Ho, Hajang, Khond, Kora, Karmali, Kharwar, Khund, Kisan, Korwa, Lepcha, Lodha or Kheria, Lepcha, Lohara, Magh, Mahli, Malpharia, Mahali, Malla, Mech, Mru, Munda, Oraon, Rabha, Santhal, Sauria, Savar, Sholaga.

The large tribes of India are the *Gonds* of Chhattisgarh, Maharashtra and Andhra Pradesh; the *Bhils* of Rajasthan, Gujarat, Maharashtra and Madhya Pradesh; *Santhals* of Jharkhand, Odisha and West Bengal. The Gonds and the Bhils are more than four million each. The Santhals are more than three million.

Figure 8.1 *India: Distribution of Bhil Santhal Tribes*

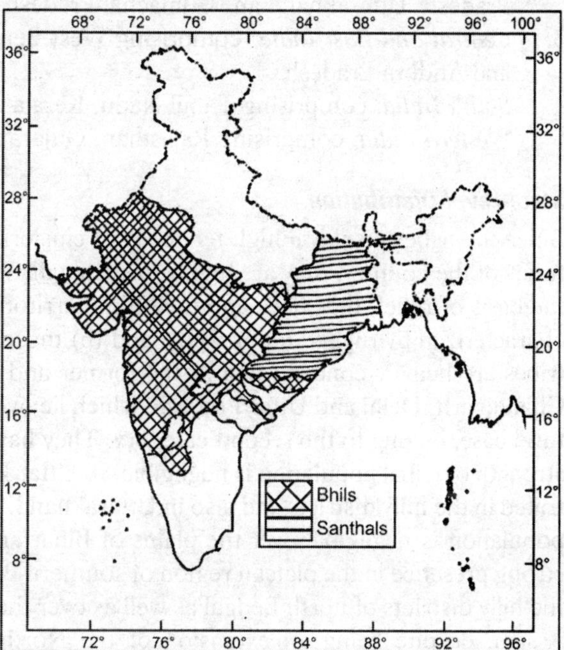

Spatial Distribution of Tribes

The Indian tribes display a very high degree of ethnic diversity both in their racial composition and the dialectic and linguistic affinity. The fact that the scheduled tribes include as many as 285 different communities is an important index of their ethnic diversity. No less impressive is the pattern of their spatial distribution. It has been commonly observed that the tribes reveal strong tendencies of clustering and concentration in the hilly, forested and the geographically inaccessible tracts of the country. This leads to the following two broad generalizations:

1. Either the tribes have generally preferred to live in areas which are by and large unfavourable to sedentary agriculture; or
2. They have been pushed into these enclaves by the peasant groups. Today their way of life, including mode of economy, is intrinsically linked with the ecological base of their present habitat.

B.K. Roy Burmon (1972) divides tribal communities into five territorial groupings, taking into account their historical, ethnic and socio-cultural relations. These are as follows:

1. *North-east India,* comprising Assam, Arunachal Pradesh, Nagaland, Manipur and Tripura.

Figure 8.2 *India: Main Areas of Gond Tribe*

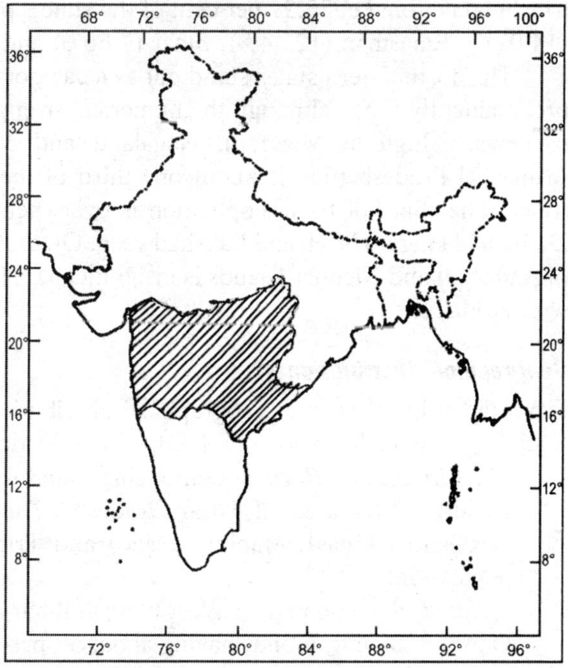

2. *The Sub-Himalayan region* of north and north-west India, comprising hill districts of Uttar Pradesh, Uttarakhand and Himachal Pradesh.
3. *Central and East India,* comprising West Bengal, Jharkhand, Odisha, Chhattisgarh, Telangana and Andhra Pradesh.
4. *South India,* comprising Tamil Nadu, Kerala and Karnataka.
5. *Western India*, comprising Rajasthan, Gujarat and Maharashtra.

State-Level Distribution

Since the tribes display a high tendency of clustering and concentration in the hilly and the forested tracts of the country, they are highly unevenly distributed between the states of the Indian Union. On the basis of relief, the states and the union territories may be grouped into two categories: (a) those characterized by rugged topography, and (b) those with an open relief, (plains and river valleys). The tribes are heavily concentrated in the former and not in the latter. For example, Punjab, Haryana, Chandigarh, Delhi and Uttar Pradesh, which lie over the Indo-Gangetic plain and have a rich agricultural base, belong to the second category. They have either no tribal population or else the numerical strength of tribal population is negligible. In Uttar Pradesh, for example, the tribes are mostly concentrated in the hilly districts and also in Uttarakhand. Likewise, the tribes are either non-existent or their population is negligible over the plains of Bihar and West Bengal. On the other hand, tribes have a strong presence in the plateau region of southern Bihar now Jharkhand. They are also predominant in the hilly districts of north Bengal as well as over the plateau districts of the states bordering on Bihar. Assam, despite being an extension of the North Indian Plain, supports a large chunk of tribal population. In fact, every eighth Assamese is a tribal. The southern states of Kerala, Karnataka, Tamil Nadu and Andhra Pradesh lie mostly in the plateau region and yet the population of tribes in these states is not very large. It varies from 1.03 per cent in Tamil Nadu to 6.31 per cent in Andhra Pradesh. The tribes are, however, numerically strong in the central Indian states from Gujarat and Maharashtra to Orissa and Jharkhand. Among them, Madhya Pradesh has the highest proportion of tribal population, i.e. 23.27 per cent of the state's population, followed by Odisha (22.21%), Gujarat (14.92%), Rajasthan (12.44%), Bihar (7.66%) and West Bengal (5.60%).

The northeastern states stand out as a category in themselves as the population in these states is predominantly tribal, although the numerical strength of tribes is not very large. Tribal proportion is particularly high in Mizoram, Nagaland and Meghalaya (85–95%) and significantly high in Arunachal Pradesh (64%). About one-third of the population in Tripura and Manipur consists of tribes. The share of tribal population is overwhelmingly large in certain union territories, such as Dadra and Nagar Haveli and Lakshadweep. On the contrary, the proportion of tribe in the population of Andaman and Nicobar Islands is insignificant. The percentage given above may vary in accordance with 2011 Census.

Geographical Distribution

From the point of view of geographical distribution, L.P. Vidyathi (ICSSR, Survey of Research in Sociology and Anthropology, Vol. III, 1972) divided the tribal people into four major zones:

1. *The Himalayan Region,* comprising Jammu & Kashmir, Himachal Pradesh (Bhot, Gujjar, Gaddi), Terai area of Uttar Pradesh (Tharus), Assam (Mizo, Garo, Khasi), Meghalaya, Nagaland (Nagas), Manipur (Mao) and Tripura (Tripuri) having 11 per cent of total tribal population.
2. *Middle India,* comprising West Bengal, Bihar and Jharkhand (Santhal, Munda, Oraon and Ho), Odisha (Khonds, Gond) having about 57 per cent of Indian tribal population.

3. *Western India,* comprising Rajasthan (Bhil, Meena, Garasia) Madhya Pradesh-Chhattisgarh (Bhil, etc.), Gujarat (Bhil, Dubla and Dhodia) and Maharashtra (Bhil, Koli, Mahadeo, Kokna) containing about 25 per cent of the Indian tribal population.
4. *Southern India,* comprising Andhra Pradesh-Telangana (Gond, Koya, Konda, Dova), Karnataka (Naikada, Marathi), Tamil Nadu (Irula, Toda), Kerala (Pulayan, Paniyan) and Andaman and Nicobar Islands (Andamanese, Nicobari) containing about 7 per cent of India's tribal population.

Economic/Occupational Classification of Indian Tribes

Majumdhar has given an economic classification of the Indian tribes based on their primary occupation. The following types have been identified:

1. *Tribes hunting in forests:* The main tribes engaged in hunting in India are Chenchu and Chandi of Andhra Pradesh, Kadar, Malapatram and Kurumba of Kerala, Paliyan of Tamil Nadu, and Onge, Jarawa, Senteneles of Andaman and Nicobar Islands. The Raji of Uttar Pradesh, Hill Garo of Meghalaya, Birhor, Korwa, Hill-Kharia of Chotanagpur, and Juang of Odisha.
2. *Tribes engaged in hilly cultivation (Shifting Cultivation of Jhuming):* Most of the tribes living in the hilly areas of Assam, Arunachal Pradesh, Meghalaya, Mizoram, Manipur, Nagaland and Tripura are dependent on shifting cultivation (Jhuming). Some tribes of Odisha, Asur, Gond, Baiga, Munda, of Madhya Pradesh are also engaged in this type of cultivation. Apart from shifting cultivation, these tribes are also engaged in cottage industries.
3. *Tribes engaged in cultivation in plain areas:* Bodo, Miri, Xaxa of Assam, Gonds of Chhattisgarh and Jharkhand.
4. *Simple artisan tribes:* Asurs of Bihar, Agariya of Madhya Pradesh, Kolam of Maharashtra are traditional mat-weavers, and the Irula of Tamil Nadu make beautiful mats and baskets on bamboo.
5. *Pastoral Tribes:* Bakarwals, Gaddis (Jammu & Kashmir and Himachal Pradesh), Maldhan (Gujarat), Todas (Nilgiri Mountains in Kerala and Tamil Nadu), Nageshia of Madhya Pradesh, Maldhan of Gujarat. Some of these tribes, trade milk, while the others rear cattle like sheep, goats for selling them in them in the market.
6. *Tribes living as folk artists:* There are some tribes who earn their livelihood by performing dances, acrobatics, snake charmers, etc. and hence they are termed as 'folk-artists'. The Nuts (Nats), and Saperas of Uttar Pradesh are good examples of this category. Thee Mundupptu of Odisha, are expert acrobats, the Kota of Tamil Nadu are snake charmers.
7. *Agricultural and non-agricultural labour oriented tribes:* Among the tribes engaged in agricultural labour are those which are traditionally agriculturists, but due to their landlessness, they work as agricultural labour on others' land. Non-agricultural tribal labour force includes those tribals who are working in local factories and mines of Bihar, Chhattisgarh, Jharkhand, Madhya Pradesh, Odisha, West Bengal, etc. and tea plantation in Assam and West Bengal.
8. *Tribes engaged in service and trade:* Comparatively, a small percentage of the tribal population is engaged in government and semi-government jobs. Tribes in this category include the Meenas of Rajasthan, Khasi, Mizo and Nagas of North-East India.

Caste System in India

Caste is known as *jati* in common parlance. Caste as a system of social relations has been a central point in Hindu society for several centuries. A caste is an endogamous group, that is, its members marry within the caste. A man is born in a caste and remains in that forever. Caste is a basic attribute of the Indian society. For centuries it has served as a major reference pointing social interactions and continues to do so in some form or the other even today. The origin of caste system is, of course, shrouded in mystery, although there is a general agreement that the beginnings were made in the Vedic Age. It all began as a part of the process of Indo-Aryans society. The Indo-Aryans asserted their superiority by creating a four-fold division of the society which came to be known as *Chatur Varnaya* and placing themselves in the upper strata of the *Brahman* and *Kshatriya* ranks. The people who lived on periphery were assigned a lower status of *Vaisya* and *Sudra*. It is generally known that initially a four-fold *Varna* system emerged in the Vedic age. These four *Varnas* were *Brahman, Kshatriya, Vaisya* and *Sudra*. These groups constituted a hierarchical social division with Brahmans at the top and Sudras at the bottom. Later on, as economy diversified, the Varna divisions ramified in accordance with this process of economic diversification into the *Jati* or caste divisions. The ranking of *Jatis* was relative to one another and in accordance with the *Varna* order.

Castes are often classified into three main groups; (a) functional or occupational castes, (b) racial or ethnic castes, and (c) sectarian castes. Included among the occupational category are the priestly caste of Brahmans, and trading castes, such as the Baniya or Vaishya, Khattri of Punjab, Aggarwal and Oswal of Rajasthan, etc. In fact, there are innumerable caste-groups belonging to the category of occupational castes. They include the whole range of artisans from weavers and carpenters to potters, goldsmiths, barbers, washermen, and so on. Also included in this category are astrologers, oilmen, cattle-breeders, musicians, folk-dancers, acrobats, snake-charmers, and so on.

The racial or ethnic castes vary regionally. There are, for example, Rajbansis and Chandals in Bengal and Bihar, Rajbhar and Chero in Uttar Pradesh and Bihar; Jats, Gujjars and Meos in Punjab, Haryana and Rajasthan; Koli and Mahar in Maharashtra; Nair (Nayar) in Kerala; and Paraiyan in Tamil Nadu and Kerala.

Lingayats appear to be the best example of a sectarian caste. In fact, they originated from a religious sect. In Tamil Nadu, Andhra Pradesh and Karnataka, the different groups of Brahmans, such as Telugu Brahmans, Kanarese Brahmans and Oriya Brahmans, reflect the differences perceived in language, cultural tradition and regional identity. Tamil Visvabrahmans and Telugu Visvabrahmans are other examples of regional variations in caste identity. This confirms the view that caste is relevant within a language domain only. It is in the linguistic context that the origins of a particular lineage group can be traced.

In independent India, the Constitution of India has listed some particular castes as the 'scheduled caste' with a view to provide them protection against discrimination, exploitation and degradation. It provides for their social, economic, political, cultural and educational upliftment. Similarly other backward classes (OBCs) have also been identified at centre and state level to provide certain facilities for their socio-economic upliftment.

Religion

Religion is a matter of faith, a personal affair and a philosophy of life of an individual or a group of individuals. The followers of a certain religion, by virtue of their adherence to a common faith, may develop a community feeling, which leads to solidarity in followers of that religious faith. There is a common code of social conduct based on religious faith such as dressing in a particular way, food habits, festivals and mass assemblies, etc. The main characteristics of religious life are traditionalism, myth and symbols, concept of salvation, sacred places, sacred rituals, sacred writings, sacred community and the sacred experience.

India presents a baffling diversity in religious persuasions and faiths. Although the traditional religion of land is Hinduism, many other faiths and belief systems, from tribal forms of religion to Buddhism, Christianity and Islam, have coexisted for centuries. Table 8.2 indicates the religious population in India during 2001 and 2011.

Table 8.2 *Religious Population in India (2001 and 2011)*

Religion	*Population 2001*	*Percentage of Total Population*	*Population 2011*	*Percentage of Total Population*
Hindus	82.75 cr.	80.4	96.63 cr.	79.8
Muslims	13.8 cr.	13.4	17.22 cr.	14.2
Christians	2.4 cr.	2.3	2.78 cr.	2.3
Sikhs	1.92 cr.	1.8	2.08 cr.	1.7
Buddhists	79 lakh	0.7	84 lakh	0.7
Jains	42.25 lakh	0.4	45 lakh	0.4
Other Religions and Persuasions	66.4 lakh	0.6	79 lakh	0.7
Religion not stated	7.27 lakh	0.07	29 lakh	0.2
Total	102.86 cr.	100	121.09 cr.	100

Spatial Distribution

A demography of religions may be helpful in understanding the dynamics of population of different religious groups. It may also be useful to evaluate the spatial patterns in the distribution of religious groups. As noted above, in terms of numerical strength the followers of Hindu religion enjoy a dominating position. At the all-India level they constitute as much as 79.8 per cent of the total population of the country. Seen in this context, the followers of all other religious persuasions put together have a relatively insignificant share in overall population – just over 19 per cent. Despite this insignificant position, the question of religious minorities continues to be a vexed problem in India. The minority-majority relations have often resulted in social and political problems with far-reaching implications for the civil administration. The Hindu proportion is higher in the rural segment of population – the national average being 84 per cent – thus revealing the predominantly rural character of the Hindu population. As against this, the proportion of Hindus in the urban population is around 76 per cent. Barring the peripheral areas of the country and a few pockets in the interior, Hindus remain a dominant religious group everywhere in the country. However, in the peripheral areas Hindus are outnumbered by the Muslims (e.g. Jammu & Kashmir), Sikhs (e.g. Punjab), Christians (e.g. Mizoram, Nagaland) and Buddhists (e.g. Arunachal Pradesh). In some regions of the country, such as parts of Odisha, Madhya Pradesh and Andhra Pradesh, Hindus account for more than 97 per cent of the total population. The proportion of Hindu population is equally high – above 95 per cent – in the sub-Himalayan districts of Uttar Pradesh and Himachal Pradesh. The proportion of Hindus in population remains well above 90 per cent in eastern Madhya Pradesh, eastern Gujarat, southern Karnataka, Tamil Nadu and coastal Andhra Pradesh. In some districts along the west coast, Hindu percentages are significantly low – below 70 per cent. Similarly, in the tribal areas of the northeast, such as Meghalaya, Nagaland and Mizoram, their proportion is less than 20 per cent, and in some cases even less than 5 per cent. Hindus are also a small minority in the Kashmir valley, so much so that in some districts their proportion is less than 2 per cent. In several districts of the northern states of Uttar Pradesh, Haryana and Bihar which happen to be the areas of Muslim concentration, the Hindu proportion goes down to less than 60 per cent. In some districts their share in population is less than 50 per cent.

Hindus: Hinduism is the most important religion of India. It believes in multiplicity of gods and goddesses. Although, the Hindus are distributed in whole of the country. The proportion of Hindu population is highest in Himachal Pradesh (95%) and lowest in Mizoram (3.6%). It is significant in Andhra Pradesh, Bihar, Gujarat, Himachal Pradesh, Haryana, Karnataka, Madhya Pradesh, Odisha, Rajasthan, Tamil Nadu and Tripura. While it is much lower in Jammu & Kashmir, Mizoram, Meghalaya, Nagaland and Punjab.

Muslims: Islam is the religion of Muslims. The proportion of Muslim population is highest in Jammu & Kashmir (over 68%) and lowest in Mizoram (only 1.1%). Their proportion is higher than national average in Assam, Bihar, Jharkhand, Jammu & Kashmir, Kerala, Uttar Pradesh and West Bengal.

Christianity: Christianity came to India in the first century A.D. when the Syrian Church was established in Kerala. Kerala is having the largest 29 per cent of Christian population. Their proportion is significant in the states of Mizoram, Goa and Nagaland. Andhra Pradesh, Tamil Nadu and Meghalaya also have more than one million population of Christians each.

Sikhs: The Sikh religion or Sikhism was founded by Guru Nanak Sahib in fifteenth century. The sikhs constitute about 2 per cent of the total population of India. The main concentration of sikh population is in Punjab (more than 79%). Haryana, Chandigarh, Delhi, Rajasthan, Uttar Pradesh and Uttarakhand also have sikh population.

Buddhism: Gautam Buddha founded the Buddhism in 563–483 B.C. Their proportion to total population of India is less than one per cent. Their main concentration is in Maharashtra. Other states having Buddhist population are Ladakh area of Jammu & Kashmir, Himachal Pradesh, Sikkim, Tripura and Arunachal Pradesh.

Jainism: Only 0.4 per cent of India's population are the followers of the Jain religion. Jains are mostly inhabited in Maharashtra, Rajasthan, Gujarat, Uttar Pradesh, Uttarakhand and Karnataka.

Zoroastrianism: The Parsis are the followers of Zoroastrianism. The Parsis first reached in India at Diu in 766 A.D. and shifted their colony to Mt. Bharhat and then spread to Navsari, Udvada. Most of the Parsis are now concentrated in Mumbai and some of them in Navsari, Udvada, Surat and Ahmedabad.

Language

India is the home of a very large number of languages. In fact, so many languages and dialects are spoken in India that is often described as a 'museum of languages'. In spite of 'linguistic pluralism', the country is not pluralistic but that of continuum. One dialect merges into the other almost imperceptibly, one language replaces the other gradually. According to the Linguistic Survey of India conducted by Sir George Abraham Grierson towards the end of nineteenth century, there were 179 languages and as many as 544 dialects in the country. However this number varies. The 1961 Census has also recorded 187 languages. But 94 out of 187 languages were spoken by small populations of 10,000 or less. In fact, about 97 per cent of the country's population is affiliated with just 22 languages.

Language Family/Sub-Family

The languages of India belong to several language groups, the major ones being the *Indo-Aryan languages* spoken by 73 per cent of Indians and the *Dravidian languages* spoken by 24 per cent of Indians. Other languages spoken in India belong to the Austro Siatic, *Tibeto-Burman* and few minor language families and isolates. Table 8.3 indicates a broad classification of modern Indian languages.

Table 8.3 *Broad Classification of Modern Indian Languages*

Family	*Sub-family*	*Branch/Group*	*Speech Area*
Austric (Nishada)	Austro-Asiatic	Mon-Khmer	Meghalaya, Nicobar Islands
		Munda	West Bengal, Bihar, Orissa, Assam, Madhya Pradesh, Maharashtra
	Austro-nesian		outside India
Dravidian (Dravida)		South Dravidian	Tamil Nadu, Karnataka, Kerala
		Central Dravidian	Andhra Pradesh, Madhya Pradesh, Orissa, Maharashtra
		North Dravidian	Bihar, Orissa, West Bengal, Madhya Pradesh
Sino-Tibetan (Kirata)	Tibeto-Myanmari	Tibeto-Himalayan	Jammu & Kashmir, Himachal Pradesh, Sikkim,
		North Assam	Arunachal Pradesh
		Assam-Myanmari	Assam, Nagaland, Manipur, Mizoram, Tripura, Meghalaya
	Siamese-Chinese		outside India
Indo-European (Arya)	Indo-Aryan	Iranian	Outside India
		Dradic	Jammu & Kashmir
		Indo-Aryan	Jammu & Kashmir, Punjab, Himachal Pradesh, Uttar Pradesh, Rajashtan, Haryana, Madhya Pradesh, Bihar, Orissa, West Bengal, Assam, Gujarat, Maharashtra, Goa.

Source: Social Geography, Aijazuddin Ahmad, Rawat Publications, Jaipur, 2011, p. 282.

Distribution of Major Languages

The official language of the Union Government of India, according to the Article 343 of the Constitution of India are Hindi in Devanagri script and English. But individual states can legislate their own official languages, depending upon their own demographics. For example the official language of Andhra Pradesh is Telugu, in Karnataka – Kannada, in Gujarat – Gujarati, in Maharashtra – Marathi, in Punjab – Punjabi, in Tamil Nadu – Tamil, in Jammu & Kashmir – Urdu, in Kerala – Malayalam and English, etc.

At present there are 22 official languages in India. The Table 8.4 lists the 22 languages set out in the Eighth Schedule as of May 2008, together with regions where they are used (English language is not included in Eighth Schedule, it is one of the official languages of Union of India).

Table 8.4 *Official Languages of India, their Family, Number of Speakers and Areas*

Language	*Family*	*Speakers (in million 2001)*	*State(s)*
Assamese	Indo-Aryan North Eastern	13	Assam, Arunachal Pradesh
Bengali	Indo-Aryan Eastern	83	West Bengal, Tripura, Assam, Jharkhand, Andaman & Nicobar Islands
Bodo	Tibeto-Burman	1.4	Assam
Dogari	Indo-Aryan North-Western	2.3	Jammu & Kashmir
English	West Germanic	10.35	All over India
Gujarati	Indo-Aryan Western	46	Gujarat, Dadra and Nagar Haveli, Daman & Diu
Hindi	Indo-Aryan Central	258–422	Bihar, Chhattisgarh, Delhi, Himachal Pradesh, Jharkhand, Madhya Pradesh, Rajasthan, Uttar Pradesh, Uttarakhand

Cont'd...

...Cont'd

Kannada	Dravidian	40	Karnataka
Kashmiri	Indo-Aryan Southern	5.5	Jammu & Kashmir
Konkani	Indo-Aryan Southern	2.5–7.6	Goa
Maithili	Indo-Aryan Eastern	12–32	Bihar
Malyalam	Dravidian	33	Kerala, Lakshadweep, Puducherry
Manipuri (also Meital or Meithal)	Tibeto-Burman	1.5	Manipur
Marathi	Indo-Aryan Eastern	72	Maharashtra, Goa, Dadra & Nagar Haveli, Daman & Diu
Nepali	Indo-Aryan Northern	2.9	Sikkim, West Bengal
Oriya	Indo-Aryan Eastern	33	Odisha
Punjabi	Indo-Aryan North Western	34	Punjab, Haryana, Delhi, Chhattisgarh, Rajasthan, Uttarakhand, Jammu
Sanskrit	Indo-Aryan	0.01	Uttarakhand
Santhali	Munda	6.5	Santhal tribes of Chhotanagpur Plateau (In Bihar, Chhattisgarh, Jharkhand & Odisha)
Sindhi	Indo-Aryan North-Western	2.5	Non regional
Tamil	Dravidian	61	Tamil Nadu, Puducherry, Andaman & Nicobar Islands
Telugu	Dravidian	74	Andhra Pradesh, Telangana, Puducherry, Adaman & Nicobar Islands
Urdu	Indo-Aryan Central	70	Jammu & Kashmir, Telangana, Andhra Pradesh, Delhi, Bihar, Uttar Pradesh

'Classical' Languages

In 2004, the Government of India declared that languages that met certain requirements could be accorded to the status of a 'Classical Language' in India. The languages thus far declared to be classical are *Tamil* (in 2004), *Sanskrit* (in 2005), *Telugu* (in 2008), *Kannada* (in 2008), *Malayalam* (in 2013) and *Oriya* (in 2014).

Other Regional Languages

In India there are several regional languages. The regional languages moving more than one million speakers are: Bhojpuri, Rajasthani, Magadh/Magahi, Chattisgari, Haryanvi, Marwari, Malvi, Mewari, Khorth/Khotta, Bundeli/Bundelkhan, Pahari, Bagheli/Baghel Khan, Laman/Lambadi, Awadhi, Harauti, Garhwali, Nimadi, Sadon Sadri, Kumauni, Tulu, Dhundhri, Surgujia, Bagri Rajasthani, Banjari, Nagpuria (Varhadi), Surjapuri and Kangri.

9

Population

India is the second largest country in the world after China in terms of its total population. According to 2011 Census the total population of India is 1,21,01,93,422. India has about 2.4 per cent of the total area of the world but its share in world's population is over 17.64 per cent.

The fist population Census in India was carried out in 1872 in the regime of Lord Mayo. But the complete Census was actually conducted in 1881 and since then the Census in India has been conducted regularly after every ten years. The latest Census 2011 is the 15th National Census of the country. The population of India is a great human resource and has a direct impact on socio-economic development and also determines the quality of life of the people.

A brief review of the various facets of population is given here, which helps in proper understanding of the demographic structure of the country.

Growth of Population

The decadal growth of population in India from 1901 to 2011 has been shown in Table 9.1

Table 9.1 *India: Decennial Growth of Population, 1901–2011*

Census Year	*Population*	*Absolute*	*Decadal Growth Rate (%)*
1901	23,83,96,327	—	—
1911	25,20,93,390	1,36,97,063	5.75
1921	25,13,21,213	–7,72,177	–0.31 (Year of Demographic Divide)
1931	27,89,77,238	2,76,56,025	11.00
1941	31,86,60,580	3,96,83,324	14.22
1951	36,10,88,090	4,24,27,510	13.31
1961	43,92,34,771	7,81,46,681	21.64
1971	54,81,59,652	10,89,24,881	24.80
1981	68,33,29,097	13,51,69,455	24.66
1991	84,64,21,039	16,30,91,942	23.87
2001	1,02,87,37,436	18,23,16,397	21.54
2011	1,21,01,93,422	18,14,55,986	17.64

Source: censusindia.gov.in/2011.

It becomes clear from the Table 9.1 that a steady growth of population has been recorded in India since 1901, except for a decrease during the decade of 1911–21. This was due to high mortality during this period because of epidemics like cholera, influenza, malaria, plague, small-pox and also due to severe droughts in the year 1911, 1913, 1915, 1918 and 1920. Census year 1921 is considered as a 'Year of Demographic Divide'.

The total population of India in 1901 was 23,83,96,327 which became 1,21,01,93,422 in 2011. The decadal growth rate increased from 5.75 per cent in 1911 to 17.64 per cent in 2011. Although decadal growth rate was recorded highest in 1971, when it was 24.80 per cent but, in 1981 the decadal growth rate was 24.66 per cent and in 1991 it was 23.54 per cent; in fact from 1931 to 1981, there had been a steady increase in decadal growth rate from 11.00 per cent to 24.66 per cent. Thereafter a decline in growth rate was recorded, which was 23.87 per cent in 1991, 21.54 per cent in 2001 and finally 17.64 per cent in 2011. Demographers have identified four periods of population growth, these are:

I	Period of stagnant population	1901–1921
II	Period of steady growth	1921–1951
III	Period of rapid growth	1951–1981
IV	Period of high growth with declining trend	1981–2011

The most significant decade in the history of population growth of India is the 2001–2011, when the percentage of decadal growth declined as compared to the previous census decade. This is because of family planning, lower infant mortality, changing status of women and increase in standard of living.

The states which are having higher decadal growth rate than country's growth rate are Arunachal Pradesh, Bihar, Chhattisgarh, Gujarat, Haryana, Jammu & Kashmir, Himachal Pradesh, Madhya Pradesh, Maharashtra, Manipur, Meghalaya, Rajasthan, Uttar Pradesh and Uttarakhand. In Union Territories, Dadra and Nagar Haveli and Daman and Diu have recorded the highest decadal growth rate of 55.50 and 53.54 respectively. Both Puducherry and Delhi have also recorded higher growth rate than country's average.

Distribution of Population

The distribution of population in states of India and Union Territories is given in Table 9.2.

Table 9.2 *Population Distribution in India–2011*

Rank	State/Union Territory	Type (Status)	Population	% of India's Population
1.	Uttar Pradesh	State	19,98,12,341	16.5
2.	Maharashtra	State	11,23,74,333	9.28
3.	Bihar	State	10,40,99,452	8.6
4.	West Bengal	State	9,12,76,115	7.54
5.	Andhra Pradesh*	State	8,45,80,777	6.99
6.	Madhya Pradesh	State	7,26,26,809	6.00
7.	Tamil Nadu	State	7,21,47,030	5.96

Cont'd...

...Cont'd

8.	Rajasthan	State	6,85,48,437	5.66
9.	Karnataka	State	6,10,95,297	5.05
10.	Gujarat	State	6,04,39,692	4.99
11.	Odisha	State	4,19,74,218	3.47
12.	Kerala	State	3,34,06,061	2.76
13.	Jharkhand	State	3,29,88,134	2.72
14.	Assam	State	3,12,05,576	2.58
15.	Punjab	State	2,77,43,338	2.29
16.	Chhattisgarh	State	2,55,45,198	2.11
17.	Haryana	State	2,53,51,462	2.09
18.	Delhi	UT	1,67,87,941	1.39
19.	Jammu & Kashmir	State	1,25,41,302	1.04
20.	Uttarakhand	State	1,00,86,292	0.83
21.	Himachal Pradesh	State	68,64,602	0.57
22.	Tripura	State	36,73,917	0.30
23.	Meghalaya	State	29,66,889	0.25
24.	Manipur	State	25,70,390	0.21
25.	Nagaland	State	19,78,502	0.16
26.	Goa	State	14,58,545	0.12
27.	Arunachal Pradesh	State	13,83,727	0.11
28.	Puducherry	UT	12,47,953	0.10
29.	Mizoram	State	10,97,206	0.09
30.	Chandigarh	UT	10,55,450	0.09
31.	Sikkim	State	6,10,577	0.05
32.	Andaman and Nicobar Islands	UT	38,058	0.03
33.	Dadra and Nagar Haveli	UT	3,43,709	0.03
34.	Daman and Diu	UT	2,43,247	0.02
35.	Lakshadweep	UT	64,473	0.01
Total	India	28+7	1,21,01,93,422	100

* Andhra Pradesh including Telangana.

The uneven distribution of population in India is evident from the fact that Uttar Pradesh with an area of 7.25 per cent of the country has a population of 16.5 per cent country's population in 2011. The top five states in India according to population in 2011 are Uttar Pradesh (16.5%) Maharashtra (9.28%), Bihar (8.6%), West Bengal (7.54%) and Andhra Pradesh (6.99%), Andhra Pradesh now has been divided into two states viz. Telangana and Andhra Pradesh. Next to Andhra Pradesh is the Madhya Pradesh having 6.00 per cent population.

Among the lowest five, one state and four Union Territories are included. These are Sikkim (0.05%), Andaman and Nicobar Islands (0.03%), Dadra and Nagar Haveli (0.03%), Daman and Diu (0.02%) and Lakshadweep (0.01%).

The distribution of population in the country is influenced by physical factors such as topography, climate, natural vegetation, and soils; economic factors like agricultural productivity, mineral resources, industrial and transportation as well as by socio-cultural and political factors.

Density of Population

Density of population shows the spatial variation in the distribution of population. According to Census 2011, the average density of population in India is 382 per sq km. In 2001 the density of population in India was 325 per sq km, thus, there is a net increase of 57 persons per sq km. The striking feature of India's density is that it is consistently increasing since 1901 as evident from Table 9.3.

Table 9.3 *Density of Population of India (1901–2011)*

Census Year	*Density of Population (per sq km)*	*Census Year*	*Density of Population (per sq km)*
1901	77	1961	142
1911	82	1971	177
1921	81	1981	216
1931	90	1991	267
1941	103	2001	325
1951	117	2011	382

There is a great variation in state wise density of population. According to 2011 Census, the state wise density of population varies from 1,102 persons per sq km in Bihar to 17 persons per sq km in Arunachal Pradesh. Among Union Territories, Delhi is having density of 11,297 persons per sq km, followed by Chandigarh (9,252), Pondicherry (2,598), Lakshadweep (2,013) and Daman and Diu (2,169). The remaining two UTs, i.e. Dadra and Nagar Haveli and Andaman and Nicobar Islands are having density of 698 and 46 persons per sq km in 2011. Table 9.4 indicates the state wise density of population in 2001 and 2011.

Table 9.4 *State wise Density of Population*

Rank in 2001	*State*	*Density of Population per sq km in*
	2001	*2011*
1. West Bengal	903	1,030
2. Bihar	881	1,102
3. Kerala	819	859
4. Uttar Pradesh	690	828
5. Punjab	484	550
6. Tamil Nadu	480	555
7. Haryana	478	573
8. Goa	364	394
9. Assam	340	397
10. Jharkhand	338	414
11. Maharashtra	315	365
12. Tripura	305	350
13. Andhra Pradesh	277	308

Cont'd...

...Cont'd

14. Karnataka	276	319
15. Gujarat	258	308
16. Odisha	236	269
17. Madhya Pradesh	196	236
18. Rajasthan	165	201
19. Uttarakhand	159	189
20. Chhattisgarh	154	189
21. Nagaland	120	119
22. Himachal Pradesh	109	123
23. Manipur	111	122
24. Meghalaya	103	132
25. Jammu & Kashmir	100	124
26. Sikkim	76	86
27. Mizoram	42	52
28. Arunachal Pradesh	13	17
Union Territories		
1. Delhi	9,340	11,297
2. Chandigarh	7,900	9,252
3. Pondicherry	2,034	2,598
4. Lakshadweep	1,895	2,013
5. Daman and Diu	1,413	2,169
6. Dadra and Nagar Haveli	449	698
7. Andaman and Nicobar Islands	043	46
India	**324**	**382**

Top Five High Population Density States and UTs

States	*Density in 2011*	*UTs*	*Density in 2011*
1. Bihar	1,102	1. Delhi	11,297
2. West Bengal	1,030	2. Chandigarh	9,252
3. Kerala	859	3. Puducherry	22,598
4. Uttar Pradesh	828	4. Daman and Diu	2,169
5. Punjab	550	5. Lakshadweep	2,013

Top Five Lowest Population Density States

States	*Density of Population in 2011*
1. Arunachal Pradesh	17
2. Mizoram	52
3. Sikkim	86
4. Nagaland	119
5. Meghalaya	122

Among UTs, Andaman and Nicobar Islands have recorded lowest density of population in 2011 which is 46 per sq km.

Sex Ratio

Sex ratio is expressed as the number of females per thousand males. It is a very important demographic attribute which determines the marital status, workforce, growth of population as well as gender equality. In 1901 the sex ratio in India was 972, which is highest during 1901 to 2011. While the lowest sex ratio was recorded in 1971 as 930. The temporal variation in sex ratio pattern of India is shown in Table 9.5.

Table 9.5 *Sex Ratio in India (1901–2011)*

Census Year	*Sex Ratio (Females per 1000 of Males)*	*Census Year*	*Sex Ratio (Females per 1000 of Males)*
1901	972	1961	941
1911	964	1971	930
1921	955	1981	934
1931	950	1991	927
1941	945	2001	933
1951	946	2011	940

The State wise and UTs wise sex ratio in 2011 is as follows:

Table 9.6 *State wise Sex Ratio in India (2011)*

States/Union Territories	*Sex Ratio 2011*	*States/UTs*	*Sex Ratio 2011*
States			
Andhra Pradesh	992	Nagaland	931
Arunachal Pradesh	920	Odisha	978
Assam	954	Punjab	893
Bihar	916	Rajasthan	926
Chhattisgarh	991	Sikkim	889
Goa	968	Tamil Nadu	995
Gujarat	918	Tripura	961
Haryana	877	Uttarakhand	963
Himachal Pradesh	974	Uttar Pradesh	908
Jammu & Kashmir	883	West Bengal	947
Jharkhand	947	**Union Territories**	
Karnataka	968	Andaman and Nicobar Islands	878
Kerala	1,084	Chandigarh	818
Madhya Pradesh	930	Dadra & Nagar Haveli	775
Maharashtra	925	Daman and Diu	618
Manipur	987	Delhi	866
Meghalaya	986	Lakshadweep	946
Mizoram	975	Puducherry	1,038

Among states, Kerala is having sex ratio of 1,084 in 2011, which is highest in Indian states. Kerala is the only state in India having more than one thousand females per one thousand males. While Haryana with 877 sex ratio is the lowest, Union Territories including Daman & Diu is having sex ratio of 618 which is lowest not only in UTs but in states also.

Highest Sex Ratio in States and UTs in 2011

	States/UTs	*Sex Ratio*
1.	Kerala	1,084
2.	Puducherry	1,038
3.	Tamil Nadu	995
4.	Andhra Pradesh	992
5.	Manipur	987
6.	Meghalaya	986

States/UTs having Lowest Sex Ratio in 2011

	States	*Sex Ratio*		*Union Territories*	*Sex Ratio*
1.	Haryana	877	1.	Daman & Diu	618
2.	Jammu & Kashmir	883	2.	Dadra and Nagar Haveli	775
3.	Sikkim	889	3.	Chandigarh	818
4.	Uttar Pradesh	908	4.	Delhi	866
5.	Bihar	916	5.	Andaman & Nicobar Islands	878
6.	Gujarat	918			
7.	Arunachal Pradesh	920			

Another aspect of sex ratio is the child sex ratio, which is very important for the future of family and society. The top three states in 2011 regarding the highest child sex ratio are Mizoram (971), Meghalaya (970) and Chhattisgarh (964). Among UTs, the top three positions are held by Andaman and Nicobar Islands (966), Puducherry (965) and Dadra and Nagar Haveli (921).

The states having lowest child sex ratio are Haryana (830), Punjab (846) and Jammu & Kashmir (859); whereas among UTs, Delhi (866), Chandigarh (867) and Lakshadweep (908) are at the lowest. The child sex ratio in India has continuously declined from 976 in 1961 to 914 in 2011. It is a cause of concern both for government and public, because this gender inequality will create several problems in future.

Literacy

Literacy rate is an indicator of the level of educational development of a country and also a key variable which determines the quality of life. The Indian literacy rate has grown to 74.04 per cent (2011) from 12 per cent at the end of British rule in 1947. There is a wide gender disparity in the literacy rate in India; effective literacy rate (age 7 and above) in 2011 was 82.14 per cent for men and 65.46 per cent for women. Table 9.7 shows progress of literacy in India during 1951–2011.

Table 9.7 *India – Literacy Rate (1951–2011)*

Census Year	Total (%)	Male (%)	Female (%)
1951	18.33	27.16	8.86
1961	28.30	34.40	15.34
1971	34.45	45.96	21.97
1981	43.57	54.38	29.76
1991	52.21	64.13	39.29
2001	64.84	75.26	53.67
2011	74.04	82.14	65.46

Although literacy rate has been improved but is still below the 85 per cent, the target set by Planning Commission for the year 2011–12.

There is a great variation in literacy rate of States and Union Territories. This becomes clear from the Table 9.8.

Table 9.8 *Indian States Union and Territories – Ranking by Literacy Rate*

Rank	State/UTs	Literacy Rate % 2011	Literacy Rate % 2001	Decadal Difference (%)
	INDIA	74.04	64.83	9.21
1.	Kerala	93.91	90.86	3.05
2.	Lakshadweep	92.28	86.66	5.62
3.	Mizoram	91.58	88.80	2.78
4.	Tripura	87.75	73.19	14.56
5.	Goa	87.40	82.01	5.39
6.	Daman & Diu	87.07	78.18	8.89
7.	Puducherry	86.55	81.24	5.31
8.	Chandigarh	86.43	81.94	4.49
9.	Delhi	86.34	81.67	4.67
10.	Andaman & Nicobar Islands	86.27	81.30	4.97
11.	Himachal Pradesh	83.78	76.48	7.30
12.	Maharashtra	82.91	76.88	6.03
13.	Sikkim	82.20	68.81	13.39
14.	Tamil Nadu	80.33	73.45	6.88
15.	Nagaland	80.11	66.59	13.52
16.	Manipur	79.85	69.93	9.92
17.	Uttarakhand	79.63	71.62	8.01
18.	Gujarat	79.31	69.14	10.17
19.	Dadra & Nagar Haveli	77.65	57.63	20.02
20.	West Bengal	77.08	68.64	8.44

Cont'd...

...Cont'd

21. Punjab	76.68	69.65	7.03
22. Haryana	74.64	67.91	8.73
23. Karnataka	75.60	66.64	8.96
24. Meghalaya	75.48	62.56	12.92
25. Odisha	73.45	63.08	10.37
26. Assam	73.18	63.25	9.93
27. Chhattisgarh	71.04	64.66	6.38
28. Madhya Pradesh	70.63	63.74	6.89
29. Uttar Pradesh	69.72	56.27	13.45
30. Jammu & Kashmir	68.74	52.52	13.22
31. Andhra Pradesh	67.66	60.47	7.19
32. Jharkhand	67.63	53.56	14.07
33. Rajasthan	67.06	60.41	6.65
34. Arunachal Pradesh	66.96	54.34	12.61
35. Bihar	63.82	47.00	16.82

The facts emerged from the above table are:

- Kerala ranks first in country with a literacy rate of 93.91 per cent, followed by Lakshadweep, 92.28 per cent and Mizoram, 91.58 per cent.
- The top five states UTs in literacy rate are: Kerala, Lakshadweep, Mizoram, Tripura and Goa.
- Bihar with literacy rate of 63.82 per cent ranks lowest in the country. The other two states having low literacy rate are Arunachal Pradesh (66.96%) and Rajasthan (67.06%).
- The States and Union Territories which have achieved Planning Commission's target of 85 per cent are Kerala, Lakshadweep, Mizoram, Tripura, Goa, Daman and Diu, Puducherry, Chandigarh, NCT of Delhi and Andaman and Nicobar Islands.

Another important fact regarding literacy in India is that it is 84.9 per cent in urban areas and 68.9 per cent in rural areas.

Rural and Urban Population

According to the Census of India 2011, the rural-urban population figures are as follows:

Population (in Crore)

	2001	*2011*	*Difference*
India	102.9	121.0	18.1
Rural	74.3	83.3	9.0
Urban	28.6	37.7	9.1

It becomes clear from the above mentioned figures that:

- For the first time since independence, the absolute increase in population is more in urban areas that in rural area.
- The percentage of rural population is 68.84 per cent, whereas that of urban population is 31.16 per cent.

- The proportion of rural population has declined from 72.19 per cent in 2001 to 68.84 per cent in 2011.
- The growth rate of urban population during 2001–2011 was 31.8 per cent, while percentage growth rate of rural population in the same period was 12.2 per cent.

The other aspects of rural-urban population are:

- In 2011 the sex ratio in rural areas was 947, while in urban areas it was 926.
- The sex ratio between 0–6 years children in rural areas was 919 and in urban areas, 902.
- There is a decline of 8.9 million children in rural areas, while the urban areas have shown increase of 3.9 million.
- The literacy rate in rural areas in 2011 was 68.9 per cent, while in urban areas it was 85.0 per cent.
- The male literacy rate in rural areas in 2011 was 78.6 per cent and in urban areas it was 89.7 per cent.
- The female literacy rate in rural areas was 58.8 per cent and in urban areas it was 79.9 per cent.

The number of urban units in India are as follows:

Towns:

Census 2001	5,161	
Census 2011	7,935	Increase: 2,774

Statutory Towns:

Census 2001	3,799	
Census 2011	4,041	Increase: 242

Census Towns:

Census 2001	1,362	
Census 2011	3,894	Increase: 2,532

The number of rural units in 2011 was:

Villages:

Census 2001	6,38,588	
Census 2011	6,40,867	Increase: 2,279

Rural Urban Population in India 2011:

Persons:

Total	1,21,01,93,422
Rural	83,30,87,662
Urban	37,71,05,760

Highlights

- During 2001–11 the growth of rural population has been 12.18 per cent.
- Growth in rural population in India is steadily declining since 1991.
- Meghalaya (27%) and Bihar (24%) witnessed largest growth of rural population among states.

- Four states recorded decline in rural population during 2001–2011. These are Kerala (by 26%), Goa (19%), Nagaland (15%) and Sikkim (5%).
- More than 30 per cent growth in urban population has been recorded in states of Sikkim (153%), Kerala (93%) and Tripura (76%).

The Most Populous Metropolitan Areas in India

Rank	*Metropolitan Area*	*State/Territory*	*Population (2011)*
1.	Mumbai	Maharashtra	1,84,14,288
2.	Delhi	National Capital Region	1,63,14,838
3.	Kolkata	West Bengal	1,41,12,536
4.	Chennai	Tamil Nadu	86,96,010
5.	Bangalore	Karnataka	84,99,399
6.	Hyderabad	Telangana	77,49,334
7.	Ahmedabad	Gujarat	63,52,254
8.	Pune	Maharashtra	50,49,968
9.	Surat	Gujarat	45,85,367
10.	Jaipur	Rajasthan	30,73,350
11.	Kochi	Kerala	22,77,630
12.	Patna	Bihar	22,31,554

10

Agriculture, Livestock and Fisheries

Agriculture

Today, India ranks second worldwide in farm output. Agriculture and allied sectors like forestry and fisheries accounted for 13.7 per cent of the GDP in 2013 and about 50 per cent of the total workforce. As per the 2010 FAO world agriculture statistics, India is the world's largest producer of many fresh fruits and vegetables, milk, major spices, select fresh meat, select fibre crops such as jute, several staples such as millets and castor oil seed. India is the second largest producer of wheat and rice. India is also the world's second and third largest producer of several dry fruits, agriculture-based textile, raw materials, roots and tubers, crops, pulses, framed fish, eggs, coconut, sugarcane and numerous vegetables. India ranks within the world's five largest producers of over 80 per cent agricultural produce items, including many cash crops such as coffee and cotton in 2010. Since independence, India has made immense progress towards food security. In fact, agriculture plays a very vital role in Indian economy.

Characteristics of Indian Agriculture

The characteristic features of the Indian agriculture are as follows:

1. Emphasis on subsistence agriculture
2. Dependence on monsoon
3. Predominance of food crops
4. Seasonal pattern
5. Variety of crops and mixed cropping
6. Pressure of population on agriculture
7. Insignificant place given to fodder crops
8. Importance of animals
9. Limited mechanization
10. Small size holdings and fragmentation of fields
11. Labour intensive agriculture
12. Low productivity
13. Rapid growth of commercial agriculture
14. Plantation agriculture in limited areas
15. Lack of definite agricultural land use policy
16. Impact of 'Green Revolution', etc.

Types of Agriculture

In India, various types of agriculture have developed due to variation in the physical and socio-economic conditions. The variety of farm practices ranges from the traditional to most modern commercial farming. The types of agriculture prevalent in India are as follows:

Subsistence Agriculture: Majority of farmers in many parts of India are doing subsistence agriculture. The farmers used to cultivate small and scattered holdings with the help of draught animals and family members. The techniques used in such farming are still primitive and the main objective of such agriculture is to produce food for their own consumption. Generally food crops are grown along with limited production of vegetables, pulses and oilseeds.

Shifting Cultivation: This is a very primitive method of cultivation and largely practiced by the tribals of Assam, Meghalaya, Nagaland, Manipur, Tripura, Mizoram, Arunachal Pradesh and Kerala. In this type of cultivation a piece of forest land is cleared by tribal people by felling or burning of trees and crops are grown. After 2–3 years when the fertility of the soil decreases, it is abandoned and the same practice is performed in another piece of land. Dry paddy, wheat, maize, millets, tobacco and sugarcane are the main crops grown in this type of agriculture.

Shifting cultivation is known by different names in various parts of India, such as *Jhoom* in Assam, *Podu* in Andhra Pradesh and Odisha, *Onam* or *Ponam* in Kerala, *Bewar, Masha, Penda, Bera* in Madhya Pradesh and Chhattisgarh.

Terrace Farming: In hilly and mountainous areas, terrace farming is done on the slopes. To avoid steep slopes, terraces have been formed and on these terraces farming is done.

Plantation Agriculture: Plantation agriculture was introduced in India by Britishers in the 19th century. It involves growing and processing of a single crop purely for commercial purpose. Crops like tea, coffee, rubber, cocoa, banana, spices, coconut, etc. are grown in plantation agriculture. The plantation agriculture is prevalent in North-Eastern State of India, Sub-Himalayan region of West Bengal, Nilgiri, Anaimalai and Cardamom hills of south India. Fruit plantations are common in Himachal Pradesh and Jammu & Kashmir.

The other types of agriculture practised in India are dry farming, irrigated farming, commercial farming, intensive and extensive farming.

Cropping Seasons

In India, two main cropping seasons are *rabi* and *kharif*, the third season is known a *zaid*.

Rabi Crops are sown just before or at the beginning of winter season and harvested before the beginning of summer season. The crops grown in rabi are wheat, mustard, barley, gram, rape seed, etc. These crops are also known as winter crops.

Kharif Crops are sown in the beginning of the rainy season and harvested before the winter season. Kharif crops include rice, maize, bajra, jowar, cotton, groundnut, jute, tobacco, sugarcane and pulses.

Zaid Crops: A brief cropping season is practiced in irrigated areas. The crops are sown in February-March and harvested in June. Crops like urad, moog, melons, water melons, etc. are grown as zaid crops. Green vegetables and oil seeds sown in August-September and harvested by November-December constitute the group of zaid kharif crops.

Crops

A variety of crops are grown in different parts of India due to varied physical conditions. These crops can be grouped into following categories:

1. *Food Crops:* Rice, Wheat, Maize, Jowar, Bajra, Ragi and Pulses.
2. *Cash Crops:* Cotton, Jute, Sugarcane, Oil seeds, Groundnut, Linseed, Sesamum, Castor seed, Rape seed, Mustard, etc.
3. *Plantation Crops:* Tea, Coffee, Spices, Banana, Coconut, Cardamom, etc.
4. *Horticulture:* Fruits – Apple, Peach, Pear, Apricot, Almond, Walnut, Strawberry, Mango, Citrus fruits, etc. and vegetables.

The important crops, their geographical conditions of growth, producing areas and production are briefly mentioned in Table 10.1.

Table 10.1 *INDIA: Major Crops: Conditions of Growth, Areas and Production*

Major Crops	*Conditions of Growth*			*Distribution/ States*	*Production 2012–13 (Million Tonnes)*
	Temperature (in °C)	*Rainfall (in cm)*	*Soil*		
Rice	10°–20°C at the time of sowing 30°–35°C during harvesting	100 to 200 cm	Alluvial, Clayed and Loamy soil	West Bengal, Uttar Pradesh, Punjab, Andhra Pradesh, Tamil Nadu, Odisha, Madhya Pradesh, Bihar, Assam, Karnataka and Kerala	104.40
Wheat	10°–15°C (winter) 21°–26°C (Summers)	15 to 75 cm	Clayed Loamy soil	Uttar Pradesh, Punjab, Madhya Pradesh, Haryana, Rajasthan, Bihar, Gujarat and Maharashtra	92.46
Maize	18°–20°C	25 to 50 cm	Alluvial soil	Andhra Pradesh, Telangana, Karnataka, Bihar, Gujarat, Maharashtra, Rajasthan, Madhya Pradesh, Himachal Pradesh	22.23
Bajra	20°–35°C	40 to 50 cm	Loamy soil	Rajasthan, Uttar Pradesh, Haryana, Maharashtra, Gujarat, Punjab, Tamil Nadu	
Barley	15°–18°C	35 to 60 cm	Loamy and Clayed soil	Rajasthan, Madhya Pradesh, Haryana, Punjab	
Cotton	21°–30°C	50 to 75 cm	Black soil	Gujarat, Maharashtra, Andhra Pradesh, Punjab, Haryana, Maharashtra, Karnataka, Rajasthan	34.00 Million Bales
Sugarcane	20°–26°C	150 cm	Loamy soil	Uttar Pradesh, Maharashtra, Karnataka, Tamil Nadu, Gujarat, Andhra Pradesh, Haryana, Punjab, Bihar	388.96
Tea	25°–30°C	150 to 250 cm	Loamy Forest soil	Assam, West Bengal, Karnataka, Tamil Nadu, Tripura, Himachal Pradesh	01.00
Coffee	16°–28°C	150 to 250 cm	Friable Loamy soil	Karnataka, Kerala, Tamil Nadu	0.30
Rubber	25°–35°C	250 to 300 cm	Loamy soil	Kerala, Tamil Nadu, Karnataka	0.80
Tobacco	15°–38°C	50 to 80 cm	Friable Sandy soil	Andhra Pradesh, Karnataka, Gujarat, Uttar Pradesh, Tamil Nadu, Bihar, West Bengal	0.82
Jute and Mesta	20°–30°C	150 to 200 cm	Alluvial soil	West Bengal, Bihar, Assam, Odisha, Uttar Pradesh, Andhra Pradesh	11.30

The largest producing states of important crops in India during 2012–13 are as follows:

Table 10.2 *Largest Producing States of Important Crops During 2012–13*

Crops	*States*	*Production (in Million Tones)*	*Percentage Share of all India Production*
Rice	West Bengal	14.96	14.33
	Uttar Pradesh	14.41	13.86
	Punjab	11.37	10.89
Wheat	Andhra Pradesh	30.36	32.77
	Punjab	16.11	17.42
	Madhya Pradesh	13.13	14.2
Maize	Andhra Pradesh/Telangana	4.81	21.64
	Karnataka	3.43	15.43
	Bihar	2.41	10.48
Total Pulses	Madhya Pradesh	4.98	26.99
	Uttar Pradesh	2.43	13.17
	Maharashtra	2.41	13.06
Groundnut	Andhra Pradesh	1.11	3.37
	Tamil Nadu	0.86.	18.11
	Gujarat	0.76	16
Rapeseed and Mustard	Rajasthan	3.65	46.68
	Haryana	0.96	12.28
	Madhya Pradesh	0.92	11.76
Soyabean	Madhya Pradesh	7.8	53.13
	Maharashtra	4.67	31.81
	Rajasthan	1.47	10.01
Sugarcane	Uttar Pradesh	134.85	39.78
	Maharashtra	62.17	18.34
	Karnataka	35.73	10.54
Cotton*	Gujarat	8.73	25.68
	Maharashtra	7.65	22.5
	Andhra Pradesh	7.35	21.62
Jute and Mesta	West Bengal	8.31	73.54
	Bihar	2.01	18.5
	Assam	0.58	5.13

*Production in million bales, each bale 170 kg.

*Production in million bales, each bale 180 kg.

Source: Pocket Book on Agricultural Statistics–2013, Department of Agriculture and Cooperation, Directorate of Economics and Statistics, New Delhi.

A number of other crops are also grown in India. Among the production of fruits, dry fruits and spices are also important. The states producing these crops are as follows.

Table 10.3 *Fruit crops, Spices and their Distribution*

Crops	*Distribution in India*
Apple	Himachal Pradesh (Kullu and Shimla Region), Jammu & Kashmir (Kashmir valley), Hilly areas of Uttarakhand
Banana	Kerala, Tamil Nadu, Maharashtra, Gujarat
Mango	Uttar Pradesh, Bihar, Andhra Pradesh, Maharashtra, Odisha, West Bengal, Tamil Nadu
Grapes	Uttarakhand, Himachal Pradesh, Maharashtra, Punjab, Andhra Pradesh
Strawberry	Jammu & Kashmir, Himachal Pradesh, Uttarakhand
Oranges	Maharashtra, Karnataka, Tamil Nadu, Meghalaya, Sikkim
Pineapples	Assam, Meghalayas, West Bengal, Tripura
Cardamom	Karnataka, Sikkim, Kerala, Tamil Nadu
Cashew nut	Kerala, Andhra Pradesh
Chillies	Maharashtra, Andhra Pradesh, Odisha
Pepper	Kerala, Karnataka, Tamil Nadu
Turmeric	Andhra Pradesh, Odisha
Saffron	Jammu & Kashmir

Figure 10.1 *India: Rice Producing Areas*

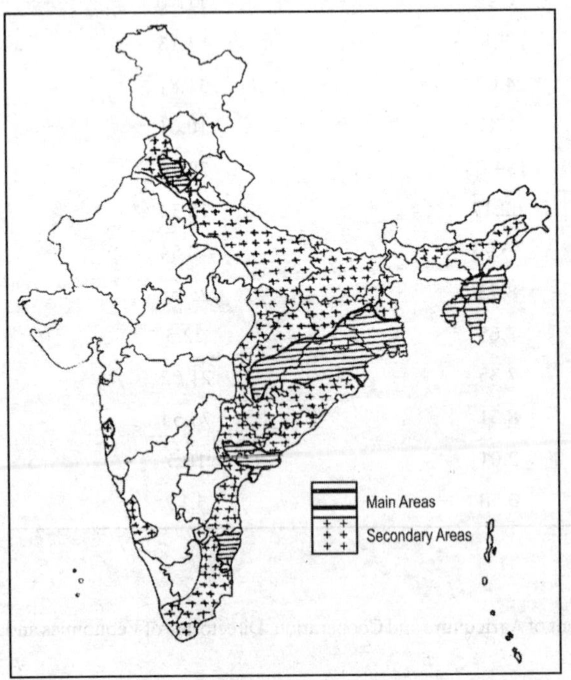

Figure 10.2 *India: Wheat Producing Areas*

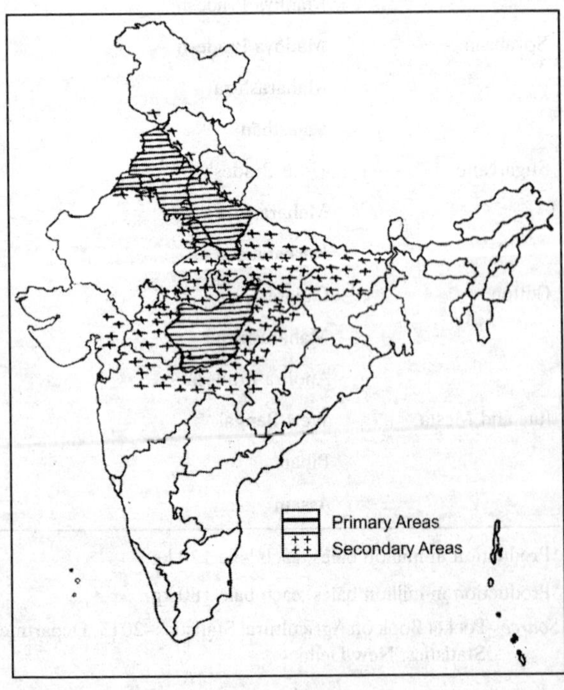

Figure 10.3 *India: Maize and Barley Producing Areas*

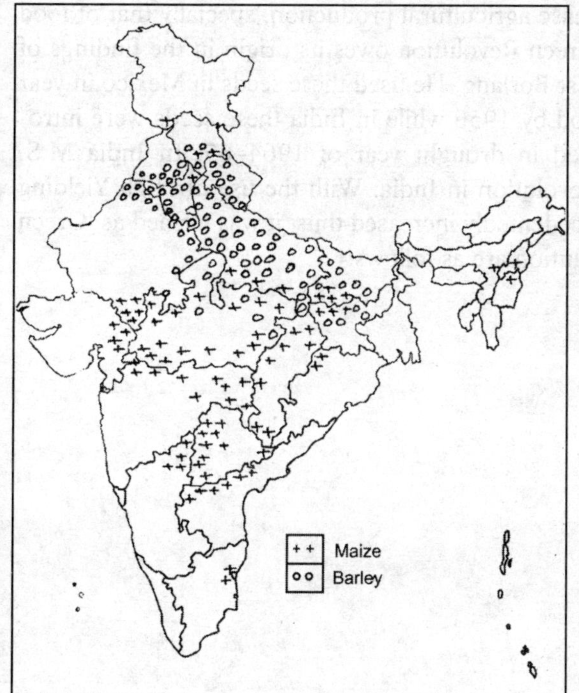

Figure 10.4 *India: Cotton and Jute Producing Areas*

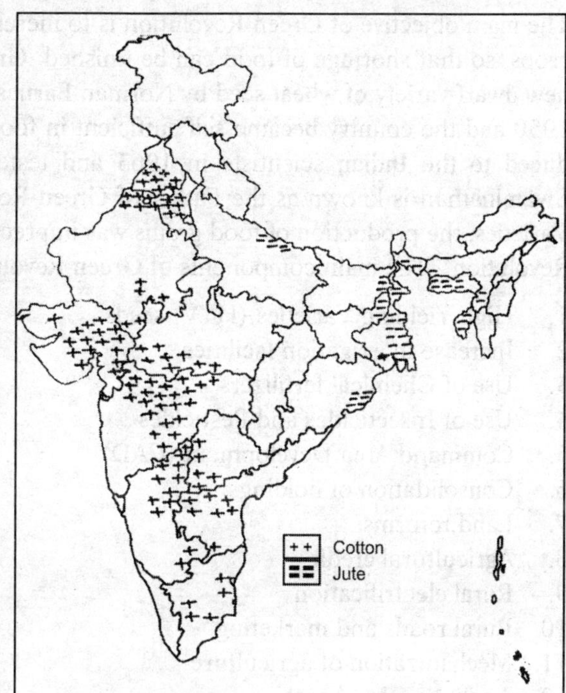

Figure 10.5 *India: Sugarcane Producing Areas*

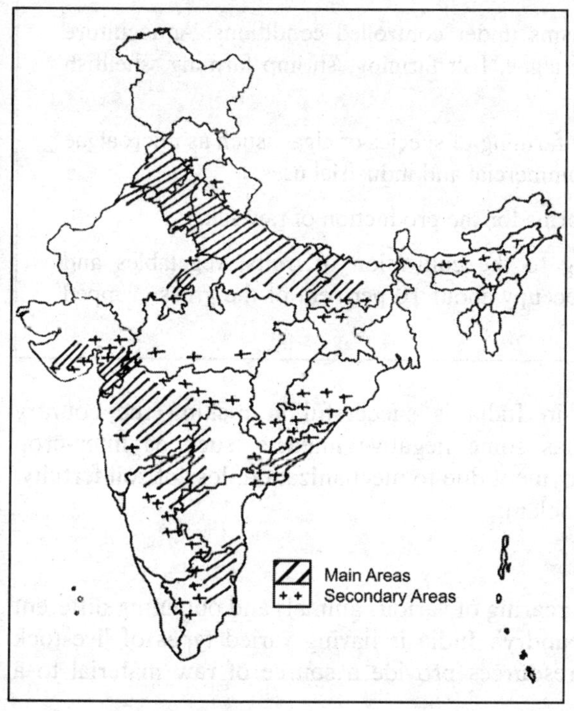

Figure 10.6 *India: Tea and Coffee Producing Areas*

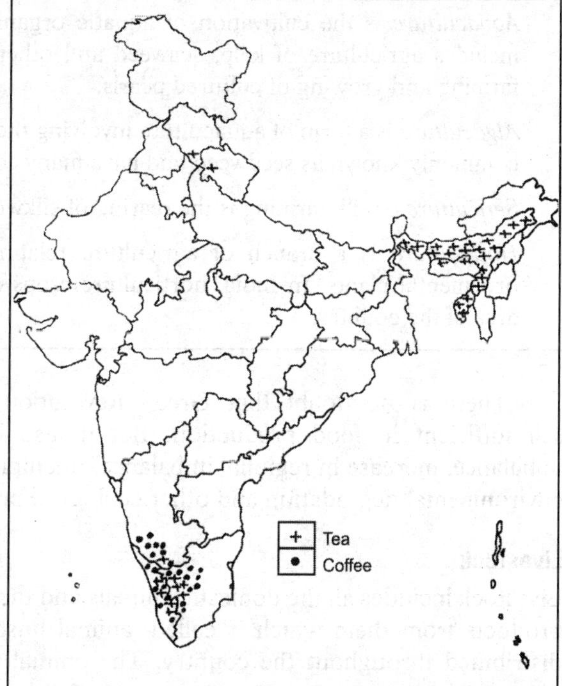

Green Revolution in India

The main objective of Green Revolution is to increase agricultural production, specially that of food crops, so that shortage of food can be finished. Green Revolution owes its origin in the findings of new dwarf variety of wheat seed by Norman Earnest Borlang. He used these seeds in Mexico in year 1950 and the country became self sufficient in food by 1956 while in India these seeds were introduced to the Indian scientists in 1963 and tested in drought year of 1964–65. In India M.S. Swaminathan is known as the Father of Green Revolution in India. With the use of High Yielding Varieties, the production of food grains was unprecedentedly increased thus, it was termed as 'Green Revolution'. The main components of Green Revolution are as follows:

1. High Yielding Varieties (HYV) seeds
2. Increase in irrigation facilities
3. Use of Chemical fertilizers
4. Use of Insecticides and Pesticides
5. Command Area Development (CAD)
6. Consolidation of holdings
7. Land reforms
8. Agricultural credit
9. Rural electrification
10. Rural roads and marketing
11. Mechanization of agriculture
12. Agricultural research

Developing Forms of Agriculture in India

Aquaculture is the cultivation of aquatic organisms under controlled conditions. Aquaculture includes agriculture of kelp, seaweed and other algae, fish farming, shrimp farming, shellfish farming and growing of cultured pearls.

Algaculture is a form of aquaculture involving the farming of species of algae, such as microalgae commonly known as seedweed and have many commercial and industrial uses.

Sericulture or silk farming is the rearing of silkworms for the production of raw silk.

Horticulture is a branch of agriculture relating to the cultivation of fruits, vegetables and ornamental plants. In India, horticulture crops occupy about 10 per cent of the gross cropped area of the country.

There is no doubt that Green Revolution in India is successful in making the country self-sufficient in food production. But it also has some negative impacts, such as inter-crop imbalance, increase in regional imbalance, unemployment due to mechanization, loss of soil fertility, environmental degradation and other ecological problems.

Livestock

Livestock includes all the domestic animals and the rearing of various animals and obtaining different products from them which is called 'animal husbandry'. India is having varied types of livestock distributed throughout the country. The animal resources provide a source of raw material to a number of industries and animals are also used as a mode of transport and also an important input in Indian agriculture. Animals products such as milk, cheese, paneer, butter, meat, eggs, etc. are in great demand in the country. In India, domestic animals include cattle, buffaloes, sheep, goats, horses, ponies, donkeys, camels, pigs, etc. Their number according to 2007 animal Census is as follows:

Table 10.4 *Livestock Population in India–2007*

Species	*Number (Millions)*	*Species*	*Number (Millions)*
Cattle	199.08	Pigs	11.13
Buffalo	105.34	Mullers	0.14
Sheep	71.56	Donkeys	0.44
Goats	140.54	Yaks	0.08
Horses and Ponies	0.61	Total Livestock	529.70
Camels	0.52		

India has the largest population of cattle and buffaloes. Their total number in 2007 was 199.8 and 105.34 million respectively. The largest number of cattle is in Madhya Pradesh followed by Uttar Pradesh, Maharashtra, Bihar and Rajasthan. There are 27 breeds of cattle and a large percentage of them do not belong to any defined breed. The major breed of cattle and their main areas are as follows:

Table 10.5 *Important Cattle Breed and their Distribution*

Breed	*States*	*Breed*	*State*
Alambadi	Tamil Nadu	Kenwaria	Uttar Pradesh and Madhya Pradesh
Amrit mahal	Karnataka	Kherigarh	Uttar Pradesh
Bachaur	Bihar	Khillari	Maharashtra
Bargur	Tamil Nadu	Krishna	Maharashtra and Andhra Pradesh
Dangi	Maharashtra	Malvi	Madhya Pradesh and Andhra Pradesh
Deoni	Andhra Pradesh	Mewati	Rajasthan
Gaolao	Madhya Pradesh	Nagori	Rajasthan
Gir	Gujarat, Maharashtra and Rajasthan	Rathi	Rajasthan
Hallikar	Karnataka	Ongole	Andhra Pradesh and Telangana
Haryana	Haryana	Tharparkar	Gujarat and Rajasthan
Hissar	Haryana		
Kanjivaram	Tamil Nadu		
Kankrej	Gujarat		

Buffaloes are mainly kept for milk but also used as draught animals. India accounts for more than half the buffaloes of the world and their numbers are increasing. The main breeds of buffaloes found in India are as follows:

Table 10.6 *Important Breeds of Buffaloes and their Distribution*

Breed	*States*
Bhadwari	Uttar Pradesh and Madhya Pradesh
Jafarabadi	Gujarat
Mehsana	Gujarat
Murrah	Punjab, Haryana and Rajasthan
Nagpuri	Maharashtra and Madhya Pradesh
Nili-Ravi	Haryana and Punjab
Surti	Gujarat

Sheep are reared for a variety of purposes, including mutton and wool. The total number of sheep in 2007 was 71.56 million. The four major regions of sheep in India are:

1. *Temperate Himalayan region* extending in Jammu & Kashmir, Himachal Pradesh and Uttarakhand.
2. *Dry western region* comprises Rajasthan, Punjab, Haryana, Western Uttar Pradesh, Madhya Pradesh and Gujarat.
3. *The semi-arid southern region* comprising Maharashtra, Karnataka, Telangana and Tamil Nadu.
4. *The Eastern region* covering drier parts of Bihar, West Bengal, Assam and Odisha.

In order to improve the quality and productivity, the government has established sheep breeding farms at Bhaisor (Uttar Pradesh), Fatehpur (Rajasthan), Mamidipally (Andhra Pradesh), Daksum (Jammu & Kashmir) and Challekeri (Karnataka). A Central Sheep Breeding Farm is located at Hissar (Haryana).

Goats: There are 140.54 million goats in India in 2007, which are about 15 per cent of the total number of goats in the world. Although goats are distributed throughout the India, but their main concentration is more in Rajasthan, Uttar Pradesh, Bihar, Jharkhand, West Bengal, Madhya Pradesh, Chhattisgarh and Maharashtra. The important breeds of goat are:

Table 10.7 *Important Breeds of Goats and their Distribution*

Angore is also known as:		
Chamba, Gaddi, Chegu	—	Jammu & Kashmir and Himachal Pradesh
Pashmina	—	Jammu & Kashmir
Jamunapuri	—	Uttar Pradesh, Rajasthan and Madhya Pradesh
Barabari	—	Western Uttar Pradesh and Haryana
Beetal	—	Punjab
Marwari, Mehsana, Kathiawar, Zalwadi	—	Rajasthan, Gujarat and Madhya Pradesh
Barari, Surti, Deccani	—	In Peninsular India

Dairy Development in India is related with livestock industry. After independence much emphasis has been given on dairy development programmes in Five Year Plans. With the result, the production of milk in 2012–13 was 132.4 million tonnes. In recent years the important scheme that has taken for dairy development is the *Operation Flood*, which is an integrated dairy development programme, specially in backward and hilly areas. The dairy industry in India is flourishing both in public and private sectors.

Fisheries

India is the world's third largest producer of fish contributing over 4 per cent of the total global output. The country ranks next only to China in the production of inland fish. The fisheries sector in the country is a source of livelihood for over 11 million people engaged fully, partially, or in subsidiary activities pertaining to this sector. It contributes over 4 per cent of the gross domestic product in agriculture or about 1.2 per cent of the total GDP and accounts for 20 per cent of the total value of agricultural exports.

The fish catch in India is of two types:

1. *Marine Fisheries:* Includes coastal, off-shore and deep sea fisheries.
2. *Inland or Fresh water Fisheries:* When fish production is done from rivers, lakes, ponds, canals, reservoirs, tanks, etc. It also includes fisheries done at estuaries, delta channels, back-water, lagoons, etc.

India's fish production has increased since 1950–51, when the total production was only 7.52 lakh tonnes to 90.19 lakh tonnes in 2012–13. Table 10.8 shows fish production in India during 2010–11 to 2012–13.

Table 10.8 *Fish Production in India*

Years	*Production in Lakh Tonnes*		
	Marine	*Inland*	*Total*
2010–11	32.50	49.81	82.31
2011–12	33.72	52.94	86.66
2012–13	32.75	57.44	90.19

Marine Fisheries: India has a coastline of over 7,517 km including that of Andaman and Nicobar Islands and Lakshadweep Islands and its continental shelf spreads over 3,11,680 sq km. This entire area is suitable for marine fisheries. The import fish caught along the coast are *Shark, Sardine, Herring, Anchovies, Mumbai Duck, Fly Fish, Ribbon Fish, Mackerel* and *Indian Salmon*. Marine fish contribute 50 per cent of the total fish production. Kerala is the leading producer of marine fish followed by Maharashtra, Karnataka, Gujarat and Goa. The higher fish production in the Arabian Sea is due to the broader continental shelf. The East coast of India contributes about 28 per cent of the total marine fish in the country. The total production of marine fish in 2012–13 was 32.75 lakh tonnes.

Inland Fisheries: More than 50 per cent of the country's fish production comes from the inland fisheries including freshwater fisheries like lakes, ponds, tanks, rivers, reservoirs, as well as in brackish water like delta channels, blackwater, lagoons, estuaries, etc. The total production of inland fisheries in 2012–13 was 57.44 lakh tonnes. West Bengal is the highest producer of inland fisheries followed by Andhra Pradesh, Bihar, Uttar Pradesh, Assam, Odisha, Karnataka, Maharashtra, Tamil Nadu, etc. Inland fresh water fisheries are more common in eastern part of the country, specially lakes, ponds and rivers of West Bengal, Bihar, Jharkhand and Odisha. Delta channels of Sunderbans, Mahanadi and Subarnarekha and Chilika lake in Odisha, delta channel of Krishna and Godawari in Andhra Pradesh lake Pulicat and the delta channel of the Cauveri in Tamil Nadu are all major areas of brackish water fisheries.

For the development of fisheries in India, government has taken several steps and also started programmes. The '*Blue Revolution*' in India was started in 1970 during Fifth Five Year Plan, when the central government sponsored the Fish Farmers Development Agency (FFDA). Under this programme, there has been tremendous increase in the production of shrimp. Andhra Pradesh and Tamil Nadu have developed shrimp in a big way. The Nellore District of Andhra Pradesh is known as the 'Shrimp Capital of India'. A new scheme for development of reservoir fisheries was initiated in 1989–90 for development of 27,000 hectare reservoir area in Gujarat, Maharashtra and Karnataka.

The development of fishing harbours, viz. Kochi, Chennai, Vishakhapatnam, Roychowk, and Paradwip, 28 minor fishing harbours and 113 fish landing centres helped in growth of marine fishing. The Central Institute of Fisheries Education, Mumbai and its units and Inland and Fisheries Training Centre, Barrackpur and Regional Training Centre, Agra provide facilities for fisheries education and training. The other institutes like Central Marine Fisheries Institute, Kochi, the Central Institute of Fisheries Technology, Kochi, and the Central Inland Fisheries Research Institute, Barrackpur are providing training and research in the field of fisheries. In 2004, a Marine Fishing Policy has also been launched with the objective to develop a sustainable fishing activity in the country.

11

Irrigation

In India the monsoon rainfall is uncertain, irregular, variable, seasonal and unevenly distributed, therefore, need for irrigation for agricultural development is indispensable. The process of supplying water to crops by artificial means such as canals, wells, tubewells, tanks, etc. from the sources of water such as rivers, ponds or underground water is called irrigation. The three sources of irrigation in India are canal, wells including tubewells and tanks. About 60 per cent of total irrigation is done by wells and tubewells, 29 per cent by canals and 11 per cent by tanks.

Irrigation has been practiced in India in one way or other since ancient times. The Grand Anicut across the Cauvary built in the first century AD, the Viranarayana and Gangai Konda-Cholapuram tanks in Tamil Nadu and the Anantharaja Sagara tank in Andhra Pradesh built in the 10th and 14th centuries AD respectively and the Bojpur Lake (Bhopal) constructed during the 11th century shows that Indians had acquired skill in the development of irrigation works in ancient times.

During Medieval period, the then rulers have also given much emphasis on construction of wells and tanks both for drinking water as well as for irrigation. The Hindu rulers and, the Mughal emperors and British rulers have developed irrigation at different times in the history of India. Feroze Shah Tughlak (1351–88) built the present Western Yamuna canal to carry water to Safidon in Hissar (Haryana) towards the middle of 14th century. There are several instances in Indian history regarding use of water for irrigation. The Britishers have also built a network of canals. There is no need for detailed description. After independence, India has started planned economic development in 1951 to resolve and rebuild the country's agricultural economy through large scale development of irrigation and power. Accordingly, much emphasis has been given in Five-year Plans for the development of a large number of major and medium irrigation projects. Similar efforts have also been done by state administration to develop irrigation facilities in their respective states.

Sources of Irrigation

Canals

Canals are one of the oldest and most important sources of irrigation in India. A high priority was given to irrigation in every Five Year Plan with the result of which a network of canals has been developed. Canals are effective source of irrigation in areas of low level relief, deep fertile soils, perennial source of water and extensive command area. The main canal irrigated areas are located in the northern plains of

India. Uttar Pradesh, Punjab, Haryana, Rajasthan and Bihar accounts for about 60 per cent of the canal irrigated areas of the country. In south and central India, Andhra Pradesh, Maharashtra, Karnataka, Madhya Pradesh, Odisha, Chhattisgarh and Tamil Nadu, canal irrigation has been developed. In Peninsular India, canal irrigation has been least developed due to rocky topography.

The important canals of the country are listed below:

Table 11.1 *Important Canals of India*

State	*Name of the Canals*
Punjab	Upper Bari Doab Canal, Sirhind Canal, Bhakra Canal, Bist Doab Canal
Uttar Pradesh	Upper Ganga Canal, Lower Ganga Canal, Sharda Canal, Eastern Yamuna Canal, Agra Canal, Betwa Canal, Western Gandak Canal, Matatilla Canal, Tehri Dam Canal
Haryana	Western Yamuna Canal, Bhakra Canal, Jui Canal, Gurgaon Canal
Bihar	Son Canal, Kosi Canal, Gandak Canal
Rajasthan	Indira Gandhi Canal, Chambal Project, Mahi Project, Gang Canal
Madhya Pradesh and Chhattisgarh	Chambal Project, Tawa Project, Mahanadi Canal, Tandula Canal, Weinganga Canal, Barana Project.
Andhra Pradesh	Godawari Delta Project, Krishra Delta Project, Nagarjunasagar Project
West Bengal	Mayurakshi Project, Midanpur Canal
Odisha	Canals from Hirakund Dam, Mahanadi Delta Canal
Karnataka	Tungbhadra Project, Malpurabha Project, Bhadra Project, Visveswaraya Canal
Tamil Nadu	Mettur Canal, Periyar Project, Lower Bhawani Project
Maharashtra	Mutha Project, Tapi Canal system, Pravara Canals
Gujarat	Mahi Project, Ukai Project

There are number of medium and minor irrigation projects that exists in every state. Apart from these, multi-purpose projects have been developed throughout the country to provide irrigation, hydro-electricity as well as to control floods, etc.

Table 11.2 *Important Multi-Purpose Projects of India*

1. Nagarjuna Sagar Project in Andhra Pradesh
2. Tungbhadra Project in Andhra Pradesh and Karnataka
3. Pochampad Project in Andhra Pradesh
4. Kakrapara Project in Gujarat
5. Ukai Project in Gujarat
6. Mahi Project in Gujarat
7. Bhadra Project in Gujarat
8. Upper Krishna Project in Karnataka
9. Ghataprabha Valley Development Project in Karnataka
10. Malaprabha Project in Karnataka
11. Tawa Project in Hoshingabad district (Madhya Pradesh)
12. C.V.D. Project on the Chambal River

Cont'd...

...Cont'd

13. D.V.C. (Damodar Valley Corporation)
14. Mayurakshi Project in Jharkhand and West Bengal
15. Kangsabati Project in Bankura district of West Bengal
16. Tista Valley Project in West Bengal.
17. Sabarmati Project in Gujarat
18. Panam Project in Panchmahal District of Gujarat
19. Kanjan Project in Gujarat
20. Mahanadi Reservoir in Madhya Pradesh
21. Hasdeo Bango and Bargi Projects in Madhya Pradesh
22. Sarda-Sahayak and Ramganga Project in Uttarakhand
23. Ukadi, Krishna and Upper Penganga Project in Maharashtra
24. Thein Project in Punjab
25. Bhima Project on Pavna River in Pune and Sholapur districts of Maharashtra
26. Hirakund Project in Odisha
27. Tawakwadi Project in Maharashtra
28. Bhakra Nangal Project in Himachal Pradesh/Punjab
29. Beas Project in Himachal
30. Parambikulam Alliyar Project in Tamil Nadu
31. Indira Gandhi Canal in Rajasthan
32. Gandak Project on the Gandak River, Bihar
33. Rihand Project in Uttar Pradesh
34. Kosi Project in Bihar

Merits of Canal Irrigation

1. Most of the canals provide perennial irrigation and supply of water as and when required.
2. After construction it is a cheap source of irrigation.
3. The sediments brought through canals enrich the fertility of the soil.
4. Canals are helpful in food control.
5. Apart from irrigation, canal also helps in development of fisheries.
6. Canal water is also used for drinking purposes as well as a source of industrial water supply.

Demerits of Canal Irrigation

1. Misuse of canal water.
2. Loss of water due to seepage, percolation, and evaporation.
3. Excessive flow of water in the files raises the ground water level and through capillary action alkaline salts come over surface and makes large areas unfit for agriculture as seen in Punjab, Haryana, Rajasthan and Uttar Pradesh.
4. Problem of water logging.
5. Overflow of canal water during rainy season, sometimes a cause of flood in nearby areas.
6. Water-logged areas become the breeding grounds for mosquitoes.
7. Regular maintenance of canals is a costly affair.

Tanks

Tank irrigation is the most feasible and widely practiced method of irrigation all over the peninsula, where most of the tanks are small in size and build by individuals or groups of farmers by raising bunds across seasonal streams. Tank irrigation is more popular in Peninsular India because:

1. Hard rocks and undulating relief restrict the construction of canals.
2. Due to hard rock structure, there is little percolation of rain water.
3. Most of the rivers are seasonal.
4. During the rainy season, several streams become torrential, therefore, the only way to use their water is to impound it by construction of dams.
5. The scattered nature of population and agricultural fields also favours tank irrigation.

About 14 to 16 per cent of the net irrigated area in the country was under tank irrigation till 1960s. Since then, it has declined to about 3-4 per cent through tanks. But tanks are still important source of irrigation in several states, especially, Andhra Pradesh, Telangana, Karnataka, Tamil Nadu, Odisha, Kerala and West Bengal.

Tank irrigation accounts for 14 per cent of the net irrigated area of Andhra Pradesh and Telangana. The Vizianagram district of North Andhra Coast has about 68 per cent of its net irrigated area through tanks. Several other districts are having 20 to 40 per cent tank irrigated area. Tamil Nadu has about a fifth of the net irrigated area dependent on tanks. Tanks are the most popular source of irrigation in the south-eastern districts of Tamil Nadu. In Karnataka, the tank irrigated area has declined from 30 per cent in 1966–69 to 10 per cent in 1995–96 and to 6.5 per cent in 2006–07.

In Kerala about 10 per cent of net irrigated area is tank served, while in Odisha, tank irrigation is more popular in Ganjam and interior districts. In West Bengal, tank irrigation is important in the western districts of Murshidabad to Maldah and Purulia. Tanks are a source of irrigation in several districts of Rajasthan (Pali, Udaipur, Rajsamand, Ajmer and Sirohi), Maharashtra (Bhir, Bhandara and Chandrapur) and in the Balaghat district of Madhya Pradesh and Madhubani in Bihar.

Tanks are not only a dependable source of irrigation but they also contribute towards the control of floods and droughts and help in maintenance of ecological balance in the catchment area through control of soil erosion and in their ayacuts through recharge of ground water. Their role in the regeneration and preservation of rural and agricultural economy is now being realized especially after the introduction of the National Watershed Development Project for Rain-fed Areas (NWDPRA) initiated during the Sixth Plan and their use in pisciculture.

Wells and Tubewells

Wells and tubewells irrigate the largest cropped area of the country. About 54 per cent of the total irrigated area is under tubewell and well irrigation. Tubewell irrigation is extensively developed in northern plains of India. About 95 per cent of the tubewells are located in Uttar Pradesh, Punjab, Haryana, Gujarat, Bihar, Madhya Pradesh, Maharashtra, Andhra Pradesh, Karnataka, West Bengal and Odisha.

The well irrigation has been used in India from time immemorial. Various methods are used to lift the ground water from the well for irrigation, drinking and for other purposes. Some of the widely used methods are the *Persian wheel*, *reht*, *charas* or *mot*, and *dhinghly* (*lever*). Well irrigation is popular in areas where sufficient sweet water is available. These areas include a large part of the Great Northern Plain, the deltaic regions of Mahanadi, Godavari, Krishna and Cauvery. The greater part of Peninsular India is not suitable for well irrigation due to rocky structure, uneven relief and lack of underground water. Similarly many parts of Rajasthan, Punjab, Haryana and Gujarat and some parts of Uttar Pradesh have brackish ground water, which is not suitable for irrigation as well as for drinking purpose.

Tubewells were first constructed in India in the state of Uttar Pradesh in 1931, to augment the supply in the Upper Ganga Canal and later on in Haryana in 1954 along Western Yamuna Canal. They have since grown into extremely popular source of irrigation. Tubewell-irrigated area is currently the largest among all the sources in all the states of the Great Plains from Punjab to West Bengal. In Punjab, except Bhatinda and Faridkot districts, where salinity of groundwater comes in a way, every district has at least 70 per cent of its net irrigated area by tubewells. The northern districts of Haryana (Ambala and Karnal) and Rewari and Gurgaon in South have more than 80 per cent of their net irrigated area served by tubewells.

In Uttar Pradesh, except Bundelkhand, almost every district has more than 50 per cent of its net irrigated area served by tubewells and the proportion rises to over 75 per cent in the western and eastern regions. The Haridwar and Udhamsinghnagar districts of Uttarakhand have over 75 per cent of net irrigated area served by tubewells. In north plains of Bihar, two-third area is irrigated by tubewells. In West Bengal, tubewell irrigation is popular in central region. In Rajasthan, Gujarat and Madhya Pradesh also tubewell irrigation is popular in some districts.

In areas where tubewells are not possible due to physical nature of structure, borewells have been used. Deep borewells are sunk several hundred metres deep and water is lifted with pumps. Borewells have become popular in Odisha, Maharashtra, Karnataka, Rajasthan, Gujarat, Madhya Pradesh, Chhattisgarh, Tamil Nadu, Jharkhand, Kerala and Assam.

Up to 1951, the total number of tubewells was 2,500, it became more than two lakhs in 1961 and in 2010 the electric operated tubewells were 40 lakhs, and 35 lakhs tubewells and borewells were diesel-operated. The area irrigated by tubewells rose to 44.61 lakh hectare in 1970–71 to 142.6 lakh hectare in 1990–91, 337.4 lakh hectare (including other wells) in 2001 and 240.56 lakh hectare in 2006–07.

Merits of Tubewell Irrigation

1. Tubewells can be installed in a short period of time at convenient places by the government and farmers.
2. The cost of construction is low.
3. It is an independent source of irrigation.
4. Tubewell water can be used as and when required.
5. Several chemicals such as nitrate, chloride, sulphate, etc. are generally found in tubewell water which add fertility to soil.
6. Once installed, the farmers are free from the regular charges for water as in case of canal.
7. The problem of salinity and reh is not in case of tubewell irrigation.

Problems/Demerits of Tubewell Irrigation

Although tubewells have gained wide acceptance as the most efficient form of irrigation and a key factor in the Green Revolution but it is turning out to be a threat to the agriculture based on it in many areas because of too many tubewells extracting too much groundwater in the recent years. The Expert Group on Groundwater Management and Ownerships (Planning Commission, 2007) notes that in 2004, an alarming 28 per cent of the blocks in the country were in the category of semi-critical, critical or over exploited compared to 4 per cent in 1995. In six major states (Gujarat, Haryana, Maharashtra, Punjab, Rajasthan and Tamil Nadu) the proportion of blocks in these categories was as high as 54 per cent. This figure is increasing year after year. The Central Ground-Water Authority (CGWA) constituted under the Environmental (Protection) Act of 1986, is empowered to control the withdrawal of groundwater by new industries/projects in 'over-exploited', 'critical' and 'semi-critical' areas.

The other disadvantages of tubewell irrigation are:

1. The irrigated area is very limited.
2. The well may dry up and be rendered useless for irrigation if excessive water is taken out of it.
3. During drought, the groundwater level falls and enough water is not available.
4. Tubewells can draw a lot of groundwater from its neighbouring areas, thus making these areas unfit for agriculture.
5. The operation cost (electricity or diesel) is quite high.

A master plan for artificial recharge of groundwater has been prepared by the CGWA. It has identified a total 4.5 lakh km^2 area in the country needing artificial recharge of groundwater.

12

Mineral and Power Resources

Mineral Resources

India's mineral resources are sufficiently rich and varied to provide the country with necessary base for industrial development. Due to varied geological structure, India is endowed with variety of minerals. Most of the minerals are products of pre-palaeozoic age and are mainly associated with metamorphic and igneous rocks of the Peninsular India. The famous geologist D.N. Wadia has observed, 'The mineral wealth of India, though by no means inexhaustible, it is varied enough to provide for sound economic and industrial development of the country but has at the same time, certain important deficiencies. It has been estimated that nearly 100 minerals are known to be produced or worked in India, of which nearly 30 may be considered more important including several which although comparatively unimportant in quantity today are capable of material development in future with expansion of industries. The country has fairly abundant reserves of coal, iron and mica, adequate supplies of manganese ore, titanium and aluminum, raw materials for refractory and limestone; there is a deficiency in ores of copper, lead and zinc. There are workable deposits of tin and nickel'. Since Wadia's estimation of mineral resources in India, several new minerals and areas have not only been identified but production of minerals has also been going on. The new scientific researches based on latest technology has given a new vision to the mining sector and increased mineral production in the country.

Distribution of Minerals

India is rich in mineral resources but their distribution is highly uneven. Most of the minerals in India are largely confined in Peninsular India where metamorphic and igneous rocks are found. The northern plains of India and Himalayan region are almost devoid of the metallic minerals. The states which are rich in minerals are Jharkhand, Chhattisgarh, Odisha, Bihar, West Bengal, Madhya Pradesh, Rajasthan, Karnataka, Tamil Nadu, Gujarat, Telangana, Andhra Pradesh, and Assam. In fact, the mineral resources of India are concentrated in a few belts, these are:

1. The North-Eastern Peninsular Belt
2. The Central Belt
3. The Southern Belt
4. The South-Western Belt
5. The North-Western Belt

1. The *North-Eastern Peninsular Belt:* This belt covers Chotanagpur Plateau (Jharkhand) and Odisha Plateau, West Bengal and Parts of Chhattisgarh. This is the richest mineral belt in India which contains large quantity of coal, iron ore, manganese, mica, bauxite, copper, kyanite, chromite, beryl, apatite and many other minerals. The Chotanagpur Plateau is known as the 'Mineral Heartland of the India'. This region contains about 93 per cent of iron ore, 80 per cent coal, 70 per cent chromite and 100 per cent kyanite of the country.
2. The *Central Belt:* This belt covers parts of Chhattisgarh, Madhya Pradesh, Telangana, Andhra Pradesh and Maharashtra. In this region large deposits of manganese, bauxite, limestone, marble, coal, gems, mica, copper, graphite, etc. are found.

Figure 12.1 *India: Major Mineral Belts*

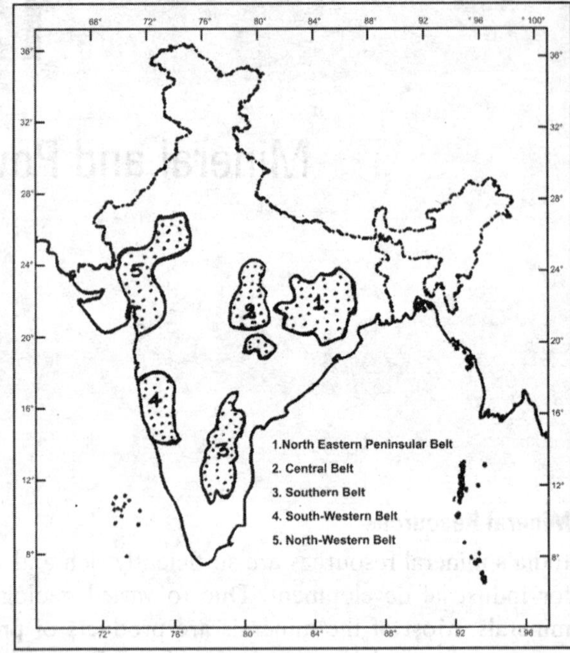

3. *The Southern Belt:* This belt covers mostly the Karnataka Plateau region stretching over the states of Karnataka, Telengana, and Tamil Nadu. This belt is rich in gold, iron ore, chromite, manganese, lignite, mica, bauxite, gypsum, asbestos, dolomite, china clay and limestone.
4. *The South-Western Belt:* This belt stretches over Goa, Karnataka and Kerala. It has deposits of iron ore, garnet, clay, ilmenite, zircon, bauxite, limestone and soapstone.
5. *The North-Western Belt:* This belt extends along the Aravallis in Rajasthan, Gujarat and Maharashtra. This belt is having rich deposits of copper, lead, zinc, uranium, mica, stealite, beryllium and precious stones, etc.

Apart from the above mineral belts the Assam has petroleum and lignite reserves. The Himalayan region, although poor in mineral deposits but possibility of mineral deposits like copper, lead, zinc, bauxite, antinomy, nickel, cobalt, etc. is there. The Indian Ocean or more specific continental shelf of Arabian sea and the Bay of Bengal are rich in petroleum and natural gas. The sea bed also contains minerals like manganese, phosphate, barium, aluminium, silicon, iron, titanium, sodium, potassium, chromium, monazite, ilmenite, magnetite and garnet.

Types of Minerals

Minerals are classified as:

1. *Metallic Minerals*
   (i) *Ferrous Minerals:* Iron, Manganese, Chromite Nickel, etc.
   (ii) *Non-Ferrous Minerals:* Copper, Lead, Zinc, Tungsten, Gold, Silver, Bauxite, Tin, Vanadium, Magnesium, etc. which do not have iron content.
2. *Non-Metallic Minerals:* Mica, Gypsum, Nitrate, Dolomite, Limestone, Phosphite, Pyrites, etc.
3. *Mineral Fuels:* Coal, Petroleum and Natural gas and nuclear minerals.

Important Minerals of India

Metallic Minerals

The production of major metallic minerals in India in year 2012–13 is as follows:

Table 12.1 *Production of Metallic Minerals in India*

Minerals	*Production in Thousand Tonnes*	*Minerals*	*Production in Thousand Tonnes*
Iron Ore	1,36,019	Lead and Zinc Ore	8,582
Manganese Ore	2,322	Lead Conc.	184
Copper Ore	3,639	Zinc Conc.	1,492
Copper Conc.	12	Tin Conc.	47,776 kg
Bauxite	15,360	Silver	3,73,901 kg
Chromite	2,950	Gold	1,588 kg

Source: Bulletin of Indian Bureau of Mines

Iron Ore

Iron ore is the most important mineral and one of the major raw-materials of iron and steel industry which provide a base for many other industries. The four types of iron ore found in India are-*Haematite, Magnetite, Limonite* and *Siderite*. Haematite and Magnetite are mostly available in the country. Both these types are superior types of iron ores having 60 to 70 and 60 to 65 per cent metal content respectively. The major iron ore producing regions of the country are Singhbhum district of Jharkhand, Gurumahisani, Sulepat and Badam Pahar in Mayurbhanj district of Odisha, Keonijhar and Sundargarh districts of Odisha, Bailadilla of Bastar district in Chhattisgarh in Karnataka Sandur-Hospet of Bellory district, Baba Budan hills and Kundermukh in Chikmagalur district and parts of Shimoga, Chitradurg and Tumkur districts. The districts of Chandrapur, Bhandara and Ratnagiri, Telangana, Andhra Pradesh, Tamil Nadu and Goa are also major iron ore producing states. The leading iron ore producing states in India are Karnataka (24.5%), Odisha (21.98%), Chhattisgarh (19.92%), Goa (17.92%) and Jharkhand (14.32%).

India is having rich iron ore deposits. More than 90 per cent of iron ore reserves exist in Karnataka (40% 85%), Odisha (17.14%), Jharkhand (13.81%), Chhattisgarh (9.59%), Telangana/Andhra Pradesh (7.25%) and Goa (3.88%).

Manganese

Manganese is an important raw material for smelting iron ore and making ferro alloys. India is the world's 8th largest producer of manganese. Manganese deposits are found in almost all geological formations, however, it is mainly associated with Dharwar System.

Odisha is the leading producer of manganese. The other major producers are Maharashtra, Madhya Pradesh, Karnataka and Andhra Pradesh. Goa, Gujarat, Jharkhand and Rajasthan also produce Manganese. Table 12.2 indicate the leading manganese producing states and their mining areas.

According to the Indian Bureau of Mines (IBM), the total reserves of manganese in India are about 295 million tones, while Geological Survey of India (GSI) estimates total reserves as 406 million tonnes, out of which 104 million tonnes are proved reserves.

Table 12.2 *Major Manganese Producing States of India*

State	*Percentage of the Total Production of the Country*	*Major Mining Areas*
1. Odisha	38.55	Bolangir, Kalahandi, Koraput, Sambalpur, Sundargarh
2. Maharashtra	22.91	Bhandara, Nagpur, Ratnagiri
3. Madhya Pradesh	19.68	Balaghat and Chhindwara
4. Karnataka	13.20	Bellary, Chitradurga, Shimoga, Tumkur, Uttari-Kannad
5. Andhra Pradesh	4.58	Cuddapah, Guntur, Srikakulam, Vijainagram, Vishakhapatnam

Copper

Copper is a useful metal in electrical industry for making wires, electric motors, transformers, generators, etc. It is alloyable, malleable and ductile. The major copper producing regions are Balaghat and Betul districts in Madhya Pradesh, Khetri (Jhunjhunu district), Ajmer, Bhilwara, Chittorgarh, Jaipur and Sirohi districts of Rajasthan, Singhbhum in Jharkhand; Madhya Pradesh produces 55.97 per cent copper in the country, followed by Rajasthan 40.80 per cent and Jharkhand 3.14 per cent.

Bauxite

Bauxite ore is the source of aluminium. Bauxite is found mainly in tertiary deposits and is associated with laterite rocks occurring extensively either on the plateau or hill ranges of Peninsular India and also in the coastal tracts of the country.

India is the sixth largest bauxite producer and fourth largest bauxite reserve country of the world. Odisha is the leading producer of the bauxite in India and produces nearly fifty per cent of bauxite. Table 12.3 indicates leading bauxite producing states and their major areas.

Figure 12.2 *India: Iron Ore and Manganese*

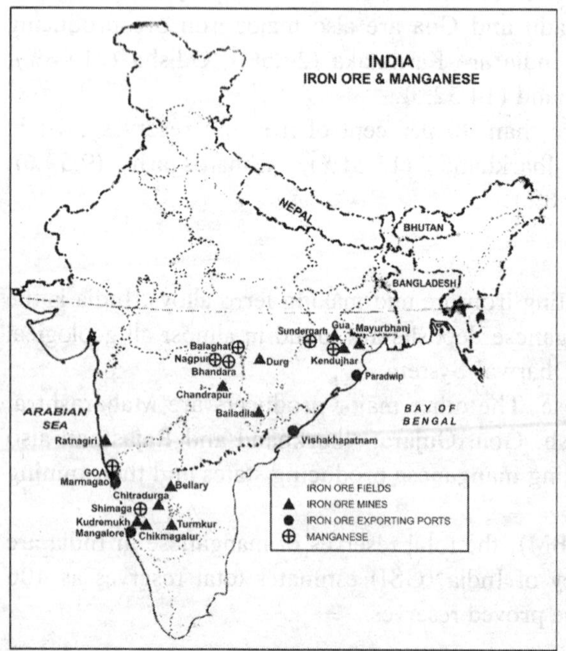

Figure 12.3 *India: Copper and Bauxite*

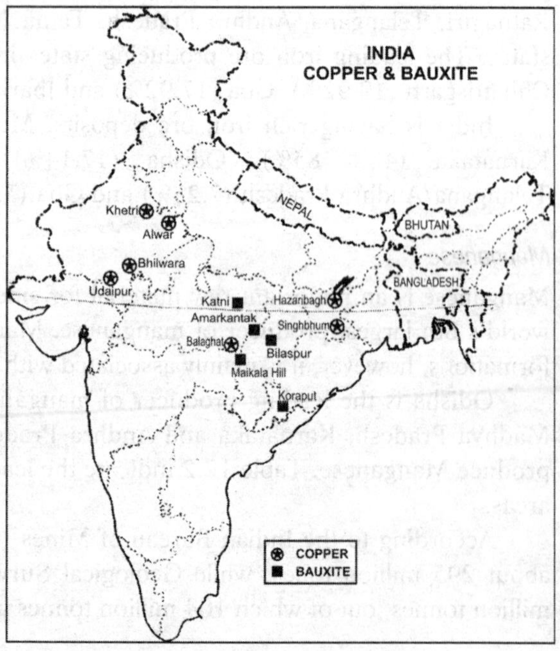

Table 12.3 *Major Bauxite Producing States of India*

State	*Percentage of the Total Production of India*	*Major Mining Areas*
Odisha	50.16	Bolangir, Kalahandi, Koraput, Sundergarh, Sambalpur
Gujarat	15.82	Amreli, Bhavnagar, Jamnagar, Junagarh, Kachchh
Jharkhand	11.87	Dumka, Lohardaga, Palamu, Ranchi
Maharashtra	9.87	Kolhapur, Pune, Ratnagiri, Satara, Thane
Chhattisgarh	6.18	Amarkantak-Pathar, Bilaspur, Durg, Raigarh and Sarguja
Tamil Nadu	2.74	Madurai, Nilgiri and Salem
Madhya Pradesh	2.35	Balaghat, Katni, Jabalpur, Maikal range, Mandla, Shahdol

The other bauxite producing states of the country are Goa, Gujarat, Jharkhand and Rajasthan. India is also having about 4 per cent of the total bauxite reserves of the world. Nearly 51 per cent of these reserves lie in Odisha and the remaining are in Andhra Pradesh, Gujarat, Chhattisgarh, Maharashtra, Madhya Pradesh and Jharkhand.

Lead

Lead is a widely used metal obtained from galena which is found associated with limestone, sandstones and calcareous slates. Rajasthan is the leading producer of lead. The main mines are located at Zavar, Rajpura-Dariba (Udaipur district), Auchaga-Gulabpura (Bhilwara district) as well as in Rajsamand, Dungarpur, Banswara, Sawai Madhopur and Dholpur districts. In Andhra Pradesh, lead in found in Cuddapah, Guntur, Khamman, Kurmman, Kurmool, and North Arcot areas. In Uttarakhand it is found in Tehri and Pithoragarh and in Jharkhand, Hazaribagh, Palamn, Ranchi and Singhbhum are the main areas. Madhya Pradesh, Himachal Pradesh, Jammu & Kashmir and West Bengal also produce limited amount of lead. India also has lead reserves of about 525 million tonnes.

Zinc

Zinc is also found along with lead and its producing areas are also same as lead. Rajasthan state produces about 98 per cent of the total zinc production. Sikkim, Jammu & Kashmir, Madhya Pradesh, Maharashtra, Tamil Nadu and Meghalaya also produce small quantity of zinc.

Tungsten

Tungsten is obtained from the wolfram ore. In India tungsten is mined at Degana in Rajasthan, Bankura district of West Bengal, Sakoli basin in Bhandra and Nagpur districts of Maharashtra, Kolar mines in Karnataka, Chittor and East Godawari districts in Andhra Pradesh, Ahmedabad district of Gujarat and Singhbhum district in Jharkhand.

Chromite

India is the third largest producer of chromite in the world. About 90 per cent chromite is produced in Odisha. Karnataka, Maharashtra, Jharkhand, Telangana and Andhra Pradesh are other chromite producing states in India.

Silver

Silver usually occurs with lead and zinc and also in the chalcopyrite and pyrrhotite ores of copper and in the ores of gold. Rajasthan, Jharkhand, Madhya Pradesh, Karnataka, Telangana, Sikkim, Bihar,

Uttarakhand, Jammu & Kashmir, Meghalaya, Chhattisgarh, Tamil Nadu are the silver producing states. India's production of silver in 2012–13 was 3,73,901 kg.

Gold

Gold is a precious mineral and is in great demand for jewellery and ornament industry. In 2012–13, India produced 1,588 kg gold. About 90 per cent gold is produced in Karnataka. The Kolar and Hutti Gold Files are notable for gold production. Small quantity of gold is also produced in West Bengal, Jharkhand, Andhra Pradesh and Rajasthan.

Non-Metallic Minerals

A large number of non-metallic minerals are found in India. However, at present only a few of these are important for the commercial production. These minerals are used in a variety of industries among which cement, fertilizer, iron and steel, ceramics, refractories, electricals and chemicals are important. The production of non-metallic minerals is given in Table 12.4.

Table 12.4 *Production of Important Non-Metallic Minerals in India (2012–13)*

Non-Metallic Minerals	*Production in '000 Tonnes*
Baryter	1,378
Ball Clay	1,855
Calcite	7.2
Clay	1987
Diamond	31,989 (Carat)
Felspar	1,291
Fire Clay	816
Garnet (Abrasive)	1,277
Graphite R.O.M.	132
Gypsum	3,537
Kaolin	7,679
Laterite	3,372
Limestone	2,79,736
Magnesite	213
Mica	1,254
Phosphorite	2,124
Pyrophylline	245
Quartz	1,041
Quartizite	365
Steatite	939
Wollastonite	141

Source: Bulletin Indian Bureau of Mines

Distribution of Non-Metallic Minerals

Mica

Mica represents a group of minerals characterized by perfect cleavage in a single direction due to which it can be split into extremely thin, flexible and tough sheets or limnae. Although mica is widely distributed mineral in all crystalline rocks of India, marketable occurrence of it is restricted to a few pegmatite veins.

Andhra Pradesh is the largest producer of mica in India and produces more than 71 per cent of the total country's mica production followed by Rajasthan, Jharkhand and Bihar.

Table 12.5 *Leading Mica Producing States of India*

State	*Percentage of the Total Production of India*	*Major Mining Areas*
Andhra Pradesh	71.15	Khammam, Krishna, Nellore, Vishakhapatnam, and West Godavari
Rajasthan	16.03	Ajmer, Bhilwara, Dungarpur, Sikar, Tonk, Udaipur districts
Jharkhand	11.57	Dhanbad, Giridih, Hazaribagh, Kodarma, Ranchi, Singhbhum
Bihar	01.25	Bhagalpur, Gaya, Munger

Occurrences of mica have been reported from Punalur and Neyyar in Kerala, Tiruchirapalli, Salem, Coimbatore and Nilgiri districts of Tamil Nadu, Balaghat, Narsingpur, Chhindwara districts of Madhya Pradesh and the Surguja district of Chhattisgarh. In 2012–13 the total mica production of India was 12,54,914 kg.

Limestone

Limestone ($CaCO_3$) is used in a wide range of industries, specially in cement, iron and steel, chemicals, fertilizers, sugar, paper, aluminium and ferro-manganese, etc. Andhra Pradesh, Rajasthan, Madhya Pradesh, Gujarat, Karnataka, Tamil Nadu and Chhattisgarh are the leading states both in reserves and production of limestone.

Rajasthan produces 16.15 per cent of the total production of limestone, followed by Madhya Pradesh (16.01%), Andhra Pradesh (15.81%) and Gujarat (10.77%). The other limestone producing states are Chhattisgarh (9.42%), Tamil Nadu (9.29%), Karnataka (8.41%), Maharashtra (6.06%), Himachal Pradesh (4.58%) and Odisha (1.68%). Small quantity of limestone is also produced in Jammu & Kashmir, Jharkhand, Himachal Pradesh, Uttarakhand, Kerala and Uttar Pradesh.

Dolomite

Dolomite is also a carbonate rock and is formed directly by means of chemical sedimentation as well as the action of magnesium solutions on the lime sedimentation and rocks. It is used in the metallurgical industry, especially in the iron and steel industry. The estimated total reserves of all grades of dolomites in India are 7,533 million tonnes (India, 2012). The total production of dolomite in the country in 2012–13 was 6.71 million tonnes.

The leading dolomite state is the Odisha which produces 28.71 per cent of the country's production, followed by Chhattisgarh (27.38%), Andhra Pradesh (18.35%). Jharkhand (7.88%), Rajasthan (5.74) and Karnataka (5.66%). Other dolomite producing states are, Arunachal Pradesh, Haryana, Himachal Pradesh, Maharashtra, Tamil Nadu, Uttarakhand, Uttar Pradesh and West Bengal.

Gypsum

Gypsum is mainly used in making ammonia sulphate, fertilizers and in cement industry. Rajasthan is the leading producer of gypsum in the country and produces about 99 per cent of the total production of country. The main gypsum producing districts of Rajasthan are Bikaner, Barmer, Churu,

Hanumangarh, Jaisalmer, Jodhpur and Pali. The small quantity of gypsum is also produced in Tamil Nadu, Jammu & Kashmir, Gujarat, Uttarakhand, Andhra Pradesh, Himachal Pradesh, Karnataka and Madhya Pradesh. The total production of gypsum in India in 2012–13 was 35,37,755 tonnes.

Asbestos

Asbestos is of great commercial value due to its fibrous structure and great resistance to fire. It is used for making fire-proof cloth, paper, millboard, sheeting, paint, etc. and also for manufacturing insulators, felts, etc. The two leading producers of asbestos in India are Rajasthan and Andhra Pradesh including Telangana. Rajasthan only produces over 94 per cent of country's total asbestos production. The other asbestos producing states are Karnataka, Jharkhand, Madhya Pradesh, Chhattisgarh, Tamil Nadu, Gujarat, Uttarakhand and Nagaland.

Magnesite

Magnesite is primarily used for the manufacturing of refractory bricks, as well as for files, fire-proof floorings and for extraction of the metal magnesium. The total *in situ* reserves of magnetite are 415 million tonnes of which 76 million tonnes are in proved category. Major deposits are found in Uttarakhand, Tamil Nadu, and Rajasthan, while minor occurrences are in Jammu & Kashmir, Karnataka, Himachal Pradesh and Kerala. Tamil Nadu is the largest producer and produces about 74 per cent magnetite of the country, followed by Uttarakhand (19.4%), Karnataka (6.29%) and Rajasthan (0.10%).

Kyanite

Kyanite occurs in metamorphic aluminous rocks and is primarily used in metallurgical, ceramic, refractory, electrical, glass, cement and other industries due to its ability to stand high temperatures. Jharkhand is the largest producer of Kyanite and produces about four-fifth of the total production of the country. Maharashtra and Karnataka produces 14.5 and 5.6 per cent respectively. Some deposits of Kyanite are reported from Odisha, Rajasthan, Tamil Nadu and Andhra Pradesh.

Sillimanite

The occurrence and uses of sillimanite are almost the same as Kyanite. The total *in situ* reserves of sillimanite are estimated to be 58.8 million tonnes. These are found in Tamil Nadu, Odisha, Kerala, Andhra Pradesh and West Bengal. Odisha is the largest producer of sillimanite and produces about 55 per cent of the total country's production. Kerala, Maharashtra, Rajasthan, Karnataka and Meghalaya are other producing states of sillimanite.

Apatites and Rock Phosphate

Apatites and rock phosphates occur naturally as native phosphates, apatites or rock phosphates on in the form of phosphorite.

Apatite deposits occur in fewer locations, the chief ones being in the Singhbhum district in Jharkhand and in Vishakhapatnam district in Andhra Pradesh. Deposits of phosphorite are also found in Udaipur district of Rajasthan. Rajasthan has the largest reserves and it is by far also the largest producer of phosphorite in India and produces 66 per cent of the country's total production. Madhya Pradesh produces 22.55 per cent, Uttarakhand 6.36 per cent and Uttar Pradesh 4.71 per cent phosphorite of the country.

West Bengal is the largest producer of apatite and produces about 93 per cent of the total country's production, followed by Andhra Pradesh which produces about 7.74 per cent.

Salt

There are three sources of salt production in India: (i) the sea-water along the peninsular coasts; (ii) the salt lakes, brine springs and wells of Rajasthan and Kutch; and (iii) the rock salt deposits of Himachal Pradesh.

Sea-water is the most important source of salt production in India. The main producers of sea-water salt are Gujarat, Maharashtra, Kerala, Tamil Nadu, Andhra Pradesh and West Bengal. In Rajasthan salt is produced from salt lakes. The important salt lakes of Rajasthan are Sambhar Lake, Didwana, Lunkaransar, Pachbhadra, and Phalodi.

Rock salt occurs in India in only Mandi district of Himachal Pradesh.

Diamonds

Diamond is mainly found in the Vindhyan formations of Bundelkhand (Madhya Pradesh), Andhra Pradesh, Telangana, and Karnataka. Panna district of Madhya Pradesh is the main diamond producing area in India.

Atomic Minerals

Uranium

The most important region for uranium mining is the Jadugoda region of Singhbhum district in Jharkhand. Gaya district of Bihar and the sedimentary rocks of Saharanpur district in Uttar Pradesh are also having uranium deposits. Monazite sand is another source of uranium. Monazite sand rich in uranium is found in Kerala. Some uranium is also found in the copper and zinc mines of Rajasthan. India produces about 2 per cent of world's uranium. The newly found uranium reserves are in Jummalapalle in Andhra Pradesh which is one of the biggest reserves in India.

Thorium

Thorium is also derived from monazite which contains 10 per cent thoria and 0.3 per cent urania. It is extracted from monazite sand mainly found in coastal region of Kerala. Kerala, Jharkhand, Bihar, Tamil Nadu and Rajasthan are the thorium producing states in India.

The other atomic minerals are Beryllium and Lithium found mainly in Jharkhand, Madhya Pradesh and Rajasthan.

Power Resources

Coal

Coal is one of the important minerals which is mainly used in the generation of thermal power and smelting of iron ore. Coal occurs in rock sequences mainly of two geological ages, namely Gondwana and tertiary deposits. There are four types of coal-*Anthracite, Bituminous, Lignite* and *Peat*. Anthracite coal is the highest quality coal containing 80 to 90 per cent carbon. Bituminous coal is having 60 to 80 per cent of carbon and is the most popular coal in commercial use. Lignite contains 40 to 60 per cent of coal; while peat is having less than 40 per cent carbon, therefore, it is the lowest type of coal. About 80 per cent coal in India is of bituminous type and is of non-coking grade.

The total coal reserves of India are estimated as 2,50,795 million tonnes. The major states having large proportion of coal reserves in the country are:

Jharkhand	28.97%
Odisha	24.62%
Chhattisgarh	16.24%
West Bengal	10.98%
Andhra Pradesh	7.06%
Maharashtra	3.45%

The other states having less than one per cent coal reserves are – Uttar Pradesh, Meghalaya, Assam, Bihar, Arunachal Pradesh and Nagaland.

Table 12.6 shows the production of coal in India:

Table 12.6 *Coal Production in India*

Year	*Production (lakh tonnes)*
1960–61	360.65
1970–71	756.50
1980–81	1,290.00
1990–91	2,240.00
2000–01	3,510.00
2010–11	5,010.00
2012–13	5,571.58

It becomes clear from the Table 12.6 that India's total coal production in 1960–61 was 756 lakh tonnes which rose to 3,510 lank tones in 2000–01 and 5,571 lakh tonnes in 2012–13.

Jharkhand is the leading coal producing state of India and produces about 29 per cent of the country's total coal production. The major coal fields of Jharkhand are Jharia, Bokaro, Giridih or Karharbari, Karanpur, Ramgarh, Hutar, Dalteuganj and Deogarh. Odisha produces about 15 per cent of the total coal production of the country. Most of the coal fields of Odisha are found in Dhenkanal, Sambalpur and Sundargarh districts. The Talcher coal field located in these districts covers an area of 500 sq km and is the second largest coal reserves in the country.

Chhattisgarh is the second largest producer of coal and third in coal reserves. Coal deposits occur in Chhattisgarh in the Surguja, Korba and Raigarh districts. Madhya Pradesh is the fourth largest producer of coal in India. The main coal fields of Madhya Pradesh are Sone river valley, Pench and Kanhan river valley, Kanhan river valley and the Satpuras.

West Bengal, Andhra Pradesh, Maharashtra, Assam, Meghalaya, Arunachal Pradesh, Jammu & Kashmir, Tamil Nadu, Rajasthan and Gujarat are also having coal deposits and also produce coal, mostly of poor quality.

Petroleum or Mineral Oil

Crude petrol consists of hydrocarbons of liquid and gaseous states varying in chemical composition, colour and specific gravity. It is an essential source of energy for all internal combustion engines in automobiles, railways and aircrafts. Its numerous by-products are processed in petrochemical industries. Crude oil is obtained mainly from the sedimentary rocks of marine origin. In India, crude oil is found in the sedimentary rocks of tertiary period. Being lighter than water, crude oil overlain with gas, gets accumulated in the anticlines above the water surface.

Oil in India was discovered in upper Assam near Margherita for the first time in 1860 by Assam Railway and Trading Company. Subsequently in 1889, oil was discovered at Digloi. In 1917 oil was also discovered at Badarpur (Assam). In 1954, production of oil was started in Naharkatiya region. Oil explorations and production was systematically taken up after the Oil and Natural Gas Commission (ONGC) was set up in 1956 with the efforts of ONGC. Oil was discovered in the Gulf of Cambay in 1961 and in Bombay High in 1976.

The following basins have been identified for the petroleum and natural gas:

1. The upper Assam Basin
2. The western Bengal Basin
3. The western Himalayan Basin
4. The Rajasthan – Saurashtra – Kachchh Basin
5. The Northern Gujarat Basin
6. The Ganga Valley Basin
7. The coastal Tamil Nadu, Andhra and Kerala Basin
8. The Andaman and Nicobar Basin
9. Offshore of the Khambat, Bombay and Bassein

The oil producing areas of India are as follows:

Oil fields of North-East India

The major oil fields of north-east India are located in Brahmaputra valley in Assam and its neighbouring states.

In Assam Digloi, Naharkatiya, Morceri-Hugrijan are the main oil fields. Other discovered oilfields are at Rudrasagar, Sibsagar, Lakawa, Gabki, Badarpur, Barholta and Anguri.

In Arunachal Pradesh, Tripura and Nagaland also, oil reserves have been found.

On-shore Oil Fields of Western India

Gujarat produces over 18 per cent of the total oil production of India. ONGC discovered the oil bearing rock strata in an area of about 15,360 sq km around Gulf of Khambhat. Ankleshwar was the first major oil-field discovered in 1958. The other oilfields are Khambhat or Lunej, Ahmedabad and Kalol.

Figure 12.4 *India: Coal and Petroleum*

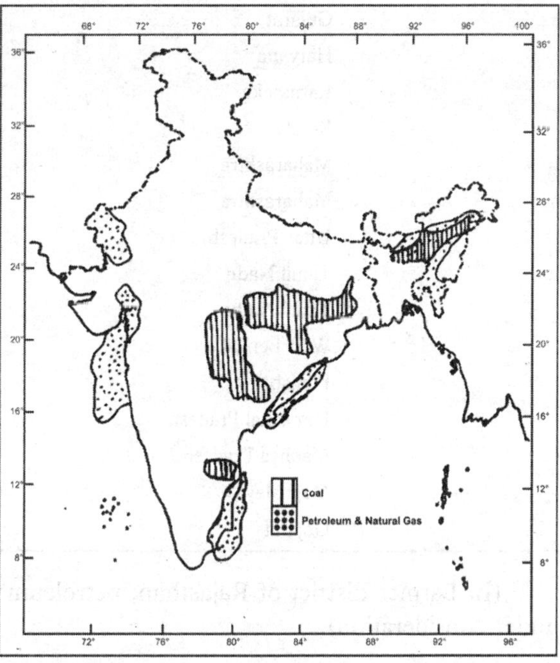

Western Coast Off-shore Oil Fields

Extensive surveys have been conduced by ONGC in the offshore areas of Kachchh, Khambhat, Konkan, Malabar and Coromandal coast, Krishna-Godawari delta and Sundarbans. Success or commercial scale has been achieved at Mumbai High, Bassein and Aliabet.

Mumbai High oil fields was discovered in 1974 and production began in 1976. Oil is taken from a depth of over 1,400 metres with the help of a specially designed platform known as '*Sagar Samrat*'. It produces about 65 per cent of crude oil of India.

Bassein and Aliabet are other oilfields.

East Coast Oilfields: In East coast region Godawari, Krishna and Kavery Basins have great potential for oil and gas production. The Rawa oilfield in Krishna-Godawari off-shore basin is expected to produce 1 to 3 million tones of crude oil annually.

Rajasthan Oilfields are the recent discovery. In Barmer district of Rajasthan and its adjoining area, oil block covers an area of approximately 5,000 sq km. Crude oil has been found in Bayatu, Nagar, Koslu, Guda. Till now more than 3.66 barrel oil producing fields have been discovered in Barmer-Sanchore Basin and in Mangla Oil field, commercial production has also been started in 2009 and it will start in other oil fields also.

In 1950–51 the production of crude oil was 0.27 million tonnes only, it raised to 0.50 million tonnes in 1960–61. The production of crude oil became 7.20 million tonnes in 1970–71 and 32.5 million tones in 2000–01. In year 2012–13 the total production was recorded as 39.73 million tonnes.

Petroleum Refining

Oil extracted from oil wells is in its crude form and contains many impurities. It is refined in oil refineries before use. After refining, various products such as diesel, kerosene, petrol, lubricants, etc. are obtained. The first oil refinery in India was started in 1901 at Digboi in Assam. The state wise oil refineries in India are as follows:

State	*Location Refineries/Companies*
Andhra Pradesh	Vishakhapatnam, HPCL
Andhra Pradesh	Tatipaka, ONGC
Assam	Digboi, IOC
Assam	Namaligarh, RPL
Bihar	Barauni, IOD
Gujarat	Jam Nagar, RPL
Gujarat	Koyali, IOC
Haryana	Panipat, IOCL
Karnataka	Mangalore, MRPL
Kerala	Kochi, KRL
Maharashtra	Trombay, HPCL
Maharashtra	Trombay, IOCL
Uttar Pradesh	Mathura, IOCL
Tamil Nadu	Manali, CPCL
Tamil Nadu	Nagapattinam, CPCL
West Bengal	Haldia, IOC
Punjab	Bhatinda
Himachal Pradesh	Manali
Madhya Pradesh	Bina
Meghalaya	Umium
Gujarat	Uadinor

(In Barmer district of Rajasthan, petroleum refinery is proposed to be established, site is still under consideration).

Natural Gas

Natural gas has emerged as the most sought after fuel in India in the recent years. The demand for gas has been rising in the country at 20 per cent per annum now for over two decades. Natural gas has two distinct advantages, first it is eco-friendly and second relatively cheap. Natural gas is a clean fuel with numerous uses. Power industry is the most important customer. Besides this, natural gas is also used in domestic and commercial purposes. In India, its three major consumers are power utilities (37%), fertilizers (24%) and captive users and LPG (20%).

The total natural gas reserve of India is estimated as about 450 billion cubic metres. Out of the total reserves, about 75 per cent lies in the Bombay High and Bassein Oilfields, 12 per cent in Gujarat, 7 per cent in Andhra Pradesh and 6 per cent in Assam. There are rich deposits of natural gas in Godavari and Krishna basins, Ankleshwar and Gulf of Khambat in Gujarat and Thanjavur and Shingleput districts of Tamil Nadu. New reserves of natural gas have also been found in Barmer district of Rajasthan, Kangra district in Himachal Pradesh, and Firozpur district of Punjab.

The production of natural gas has increased from 17 million cubic metres in 1960–61 to 20,920 million cubic metres in 2000–01. In year 2012–13 the production of natural gas in India was 39,733 million cubic metres.

Electricity

The electricity sector in India had an installed capacity of 2,50,256 GW at the end of July 2014. India became the world's third largest producer of electricity in year 2013 with 4.8 per cent global share in electricity generation. India generated around 967 TWH (9,67,150.32 GWh) of electricity. The total annual generation of electricity from all types of sources was 1,102.9 Terawatt-hours (TWH) in 2013.

In India the sources of electricity are thermal, hydro, nuclear and other renewable sources. Their installed capacity is given in Table 12.7

Table 12.7 *Total Installed Power Generation Capacity in India (2011)*

Full	*MW*	*Percentage of Total*
Total Thermal	1,19,040.98	65.16
Coal	1,00,09.38	54.79
Gas	17,742.85	9.71
Oil	1,199.75	0.65
Hydro (Renewable)	38,706.40	21.18
Nuclear	4,780.00	2.61
Other renewable sources	20,162.24	11.03
Total	1,82,162.62	100

The all India percentage installed generating capacity on 31 March 2013 was coal–58 per cent, Gas–9 per cent, Diesel–1 per cent, Hydro–18 per cent, Nuclear–2 per cent and RES – 12 per cent. The total installed capacity of India in 2013 was 22,33,44 MW.

Thermal Power

Thermal electricity is generated by using coal, oil (diesel) and natural gas. Thermal electricity has recorded a much faster growth in India both with respect to installed and actual generation. In 1950–51 the installed capacity of thermal electricity was 1.1 thousand MW. It became 45.8 thousand MW in 1990–91 and 76.3 thousand MW in 2000–01. The installed power generation capacity in

2010–11 was 119.04 thousand MW, which was 65.16 per cent of the total installed power generation capacity of India. In thermal power, coal contributes 58 per cent, gas 9 per cent and diesel only 1 per cent. The National Thermal Power Co-operation (NTPC) plays an important role in the development of thermal power in India. Maharashtra ranks first in production of thermal power, followed by Uttar Pradesh, Gujarat and West Bengal. A list of important stations in India is as follows:

Table 12.8 *Important Thermal Power Stations in India*

States	*Thermal Power Stations*
Andhra Pradesh	Bhadrachalam, Kothagudam, Manuguru, Nellore, Ramagudam, Vijayawada.
Assam	Bongaigaon Chandarpur, Namrup
Bihar	Barauni, Kahalgaon, DVC, Durgapur
Chhattisgarh	Korba
Delhi	Badarpur, Indraprastha, Rajghat
Gujarat	Ahmedabad, Banas, Dhuvaram, Gandhinagar, Kachchh, Kandla, Mahuva, Porbandar, Sabarmati, Shahpur, Sikka, Ukai Utaran, Wankbori
Haryana	Faridabad, Panipat, Yamuna Nagar
Jammu & Kashmir	Kalakot
Jharkhand	Bokaro, Chandrapur, Subarnrekha
Madhya Pradesh	Amarkantak, Satpura, Singrauli, Vindhyachal
Maharashtra	Ballarshah, Bhusaval, Chandrapur, Chola, Dhobal, Khapar-Kheda, Koradi, Nasik, Paras, Parli, Trombay, Ujjaini, Uran.
Manipur	Loktak (Imphal)
Punjab	Bhatinda, Rupnagar
Rajasthan	Kota, Suratgarh, Chhabra, Anta
Odisha	Balimela, Talcher
Tamil Nadu	Ennore, Mettur, Neyveli,
Uttar Pradesh	Bahraich, Dorighat, Gorakhpur, Harduagani, Jawaharpur, Kanpur, Mau, Moradabad, Obra Panki, Parichha, Tundla
West Bengal	Birbhum, Bundel, Durgapur, Farakka, Gauripur, Kalaghat, Kolkata, Murshidabad, Titagarh, Santaldih.

Hydro Power

Hydroelectricity is generated using the force of running water falling from a height or 'head'. Depending on the head, there are four types of hydropower schemes in India: (i) the run-of the-river schemes, (i) the storage reservoir schemes, (iii) the trans-basin diversion schemes, and (iv) the pumped storage reservoir schemes. The most important advantage of the hydropower schemes is their low cost power generation, i.e. cost of water is nil and it is renewable source.

The first hydro-electric plant in India was established at Darjeeling in 1898. It was followed by the Mettur (Tamil Nadu) and Srivasamudram (Karnataka). The real progress in hydro-power production took place only after independence. Several multipurpose projects were constructed, such as Bhakra-Nangal, Rihand, Hirakund, Nagarjun Sagar, Damodar Valley Corporation, Chambal, Tungbhadra, Koyana, etc. The National Hydroelectric Power Corporation (NHPC) was set up in 1975. In 1950–51 the installed capacity of hydro electricity was only 0.6 thousand MW which rose to 11.8 thousand MW in 1980–81, 18.8 thousand MW in 1990–91, 25.1 thousand MW in 2000–01 and 38.70 thousand MW in 2010–11. Andhra Pradesh (including Telangana) ranks first, followed by Karnataka, Tamil Nadu, Punjab, Maharashtra, Kerala and Madhya Pradesh in hydro-power generation. A list of important hydro-electric power plants located in states is given in Table 12.9.

Table 12.9 *Important Hydro-Power Plants in India*

States	*Hydro-Power Stations*
Karnataka	Mahatma-Gandhi (Jog Falls), Sivasamudram (Kaveri), Bhadra, Munirabad, Saravati, Tungbhadra, Krishnaraja-Sagar.
Kerala	Idikki (Priyar), Kallada, Kuttiaddy, Pallivasal, Parambikulam, Poringal, Panniar, Sabarigiri, Periyar
Madhya Pradesh	Jawahar Sagar and Pratap Sagar on Chambal, Twa
Maharashtra	Bhola, Bhatnagar-Beed, Girma, Khopali, Koyna, Purna, Paithon, Vaiterna
Andhra Pradesh	Machkund, Nagarjun-Sagar, Nizam-Sagar, Sileru, Srisalem
Bihar	Kosi
Gujarat	Akrimota, Sardar-Sarovar, Ukai (Tapi), Hathmati (Sabarmati), Bhadra (Kathiawad)
Jammu & Kashmir	Dool-Hasti, Lower Jhelum, Salal, Baghliar
Jharkhand	Maithon, Panchet, Tilaiya, Mayurakshi
Odisha	Hirakud (Mahanadi), Balimela, Rengali (Brahmani), Indiravati
Punjab and Himachal Pradesh	Bhakra-Nangal on Sultej, Dehar on Beas, Giri-Bata, Harike Binwa, Andhra, Chamera, Pong, Siul, Bassi
Rajasthan	Ranapratap Sagar and Jawahar Sagar on Chambal river
Tamil Nadu	Bhavanisagar, Mettur, Periyar, Aliayar, Kodayar, Moyar, Suruliyar, Papnasam
Uttarakhand	Tehri-dam and Koteshwar-dam on Bhagirathi
Uttar Pradesh	Rihand, Ramganga, Chibro on Tons
West Bengal	Panchet
North-eastern States	Dikhu, Doyan (both in Nagaland), Gomuti (Tripura), Loktak (Manipur), Kopali (Assam), Khandong and Kyrdemkulai (Meghalaya), Sirlui and Barabi (Mizoram), Ranganadi (Arunachal Pradesh)

There are several other medium and small hydro-power plants in the country located in various states.

India has vast potentialities of hydro-power development. The basin-wise estimated potential of hydropower is shown in Table 12.10

Table 12.10 *Basin-wise Estimated Potential of Hydropower in India (Potential in Thousand MW at 60 per cent load factor)*

Basin	*Potential*	*Percentage of Total*
Indus	20.0	23.8
Brahmaputra	35.0	41.7
Ganga	11.0	13.1
Central Indian Basin	3.0	3.6
West flowing rivers	9.0	10.7
Total	84.0	100

Nuclear or Atomic Power

Nuclear power is the fourth largest source of electricity in India after thermal, hydroelectric and renewable sources of electricity. As of 2013, India has 21 nuclear reactors in operation in 7 nuclear power plants, having an installed capacity of 53.8 MW and producing a total 30,292.91 GWH of electricity. Seven other reactors are under construction. At present nuclear power constitutes only less than 4 per cent of the total energy production in India.

The Atomic Energy Institution at Trombay was established in 1954 and renamed as the 'Bhabha Atomic Research Centre (BARC)' in 1967. The first nuclear power station with a capacity of 320 MW was set up at Tarapur near Mumbai in 1969, followed by Rawat Bhata (near Kota) in 1972.

India's operating nuclear power reactors are as follows:

Table 12.11 *India's Operating Nuclear Power Reactors*

Reactors	*State*
Tarapur Units – 1, 2, 3 and 4	Maharashtra
Rawatbhata Units – 1, 2, 3, 4, 5 and 6	Rajasthan
Kaiga Units – 1, 2, 3 and 4	Karnataka
Narora Units – 1 and 2	Uttar Pradesh
Madras Units – 1 and 2	Tamil Nadu
Kakrapur Units – 1 and 2	Gujarat
Kadankulam	Tamil Nadu

The projects under construction are Madras (Tamil Nadu), Kakrapur, Unit 3 and 4 (Gujarat), Rawatbhata, Unit 7 and 8 (Rajasthan), Kudan Kulam, Unit 2 (Tamil Nadu).

The planned nuclear power projects are – Gorakhpur (Haryana), Chutka (Madhya Pradesh), Mahi Banswara (Rajasthan), Kaiga Unit 5 and 6 (Karnataka) and Madras, Unit 4 and 5 (Tamil Nadu), Jaitapura (Maharashtra).

Renewable/Non-Conventional Energy Sources

India is endowed with an abundance of renewable energy sources among which solar energy, wind energy, bio-mass energy, fuel-cells, tidal energy, hydrogen energy and geo-thermal energy are notable. The development of these sources was started in 1970s, but more concerted efforts in this direction were initiated in 1981, when Commission for Additional Sources of Energy (CASE) was set up. A separate Department of Non-conventional Energy Sources (DNES) was created in the Ministry of Energy in 1982. It was upgraded into Ministry of Non-conventional Energy Sources (MNES) in 1992.

Solar Energy

Solar radiation is an important source of energy. India receives 5,000 trillion KWH of solar radiation during the year. Most parts of the country have 300 clear sunny days in a year. Solar energy can be tapped through thermal and photovolic (PV) routes for heat or electricity applications respectively. In many parts of the country, solar energy has been used in water heating, for lightning as well as in solar air heaters and geysers. Solar photovoltaic technology enables the conversion of solar radiation into electricity. More than 10 lakh solar power units have been installed in the country. Gujarat state produced 654.8 MWp of solar energy, followed by Rajasthan 510.25 MWp and Madhya Pradesh 132.0 MWp. Other States using solar energy are Andhra Pradesh, Chhattisgarh, Haryana, Jharkhand, Karnataka, Maharashtra, Odisha, Punjab, Tamil Nadu, Uttar Pradesh, Uttarakhand and West Bengal. In 2010, India started Jawaharlal Nehru National Solar Mission. India's largest ultra mega solar power plant is planned to set up in Madhya Pradesh's Rewa district.

Wind Energy

Among all the renewable energy sources, wind is perhaps the most important. It is the most viable and cost effective and the fastest growing energy source. Wind power mechanism is simple, pollution free and eco-friendly and with virtually no running and maintenance costs. The development of wind power in India began in 1990's and has significantly increased in last few years.

As of 31 March 2014 the installed capacity of wind power in India was 21,136.3 MW which mainly spread across Tamil Nadu, Gujarat, Maharashtra, Karnataka, Rajasthan, Madhya Pradesh, Andhra Pradesh, Kerala, Odisha and West Bengal. Wind Power accounts for 8.5 per cent of India's total installed capacity and it generates 1.6 per cent of the country's power.

There is a growing number of wind energy installations in states across India. In January 2014 the states of India had a cumulative installed capacity of 21,264 MW. The state wise situation of wind power is given in Table 12.12

Table 12.12 *State-wise Wind Power Capacity (2013–14)*

State	*Capacity as on 31–3–2014 (MW)*
Tamil Nadu	7,253
Gujarat	3,414
Maharashtra	2,976
Rajasthan	2,820
Karnataka	2,409
Andhra Pradesh	753
Madhya Pradesh	439
Kerala	55
Others	4.30
Total	21,264

India has a vast potential of wind energy development. The India's total potential of wind energy is estimated as 48,561 MW. The maximum potential of wind energy lies in Karnataka (11,531 MW) followed by Gujarat (10,645 MW) and Andhra Pradesh (8,968 MW). Madhya Pradesh, Rajasthan, Maharashtra, Tamil Nadu, Kerala and Odisha are the other states having wind energy development potentialities.

Ocean Energy (Tidal Energy)

The tidal waves and sea waves are the main sources of ocean energy. India has a long coast line of about 6,100 km, but the ocean energy production has not yet developed. The potential sites for ocean energy development are the Gulf of Kutchh (estimated potential 1,200 MW), Gulf of Cambay (7,000 MW) and the Durgaduari Creek in the Sunderbans Delta (100 MW). The first Indian plant of 3.65 MW capacity is being installed in the Durgaduari Creek in the Sunderbans. Most recently India joined the tidal power wave with the approval of a commercial-sale tidal power approval of a Gulf of Kutchh.

Ocean Thermal Energy Conversions (OTEC) extracts solar energy through a heat engine operating across the temperature difference between warm surface water and cold deep water. India OTEC programme started in 1980 to install a 20 MW plant off the Tamil Nadu Coast and in 1982 an OTEC cell was formed in National Institute of Ocean Technology. But till now India has not been able to extract ocean thermal energy for commercial use.

Bio-Energy

Biomass comprises all forms of combustible matter of plant and animal origin having a major composition of cellulose, hemi-cellulose and lignin. Biomass materials like wood agricultural residues and animal dung cakes have been traditionally used in India as a fuel in rural India. Under Biomass Programme launched by Ministry of Non-Conventional Energy Sources (MNES) in 1987, efforts are being made to make use of biomass in a efficient manner.

The current availability of biomass in India is estimated at about 500 million metric tonnes per year. The Government of India has installed a total 288 biomass power generation projects aggregating to 2,665 MW capacity. The states which have taken up biomass power generation are Andhra Pradesh, Tamil Nadu, Karnataka, Maharashtra and Uttar Pradesh. Table 12.13 shows biomass potential installed capacity in key Indian states.

Table 12.13 *Biomass Potential and Installed Capacity in Key Indian States*

State	*Power Potential (MW)*	*Installed Capacity (by 2011)*
Punjab	2,413.2	74.5
Uttar Pradesh	1,594.3	592.5
Haryana	1,120.8	35.8
Rajasthan	1,093.5	73.3
Maharashtra	1,014.2	403
Madhya Pradesh	841.7	1.0
Karnataka	631.9	1.0
Andhra Pradesh	625	365.18
Gujarat	457.7	363.25
Chhattisgarh	248.5	0.5
Kerala	195.9	231.9

Source: Biomass Atlas by IISC, Bangalore.

13

Industries

India ranks among the world's top ten countries in industrial output. Manufacturing industries in India provide more employment than any other sector, except agriculture. Manufactured goods make the bulk of India's exports and the country is now in a position to sell abroad plant and machinery for a wide range of industries on a turnkey basis. According to the World Bank, India's industrial manufacturing GDP output in 2012 was the 10th largest in the world. The Indian industrial sector underwent significant changes as a result of the economic liberalization in Indian economic reforms of 1991, which removed import restrictions, brought in foreign competition, led to the privatization of certain government-owned public sector industries, liberalized the FDI regime, improved infrastructures and led to an expansion in the production of fast moving consumer goods. In India, the process of industrialization started with the launching of the First Five Year Plan and it has continued through successive plan periods.

Factors Affecting Location of Industry

The locational factors are the most important in the growth of industries in a particular area. Each type of industry have had its specific locational requirement, but there are certain common factors which have influenced the industrial locations in India, these are as follows:

Geographical Factors

1. Raw Materials
2. Power
3. Skilled Labour
4. Transportation
5. Market
6. Water Supply
7. Site (Proper land)
8. Climate

Non-Geographical Factors

1. Capital
2. Technology
3. Government Policies
4. Efficient Organization (Management)
5. Financial Facilities such as Banking, Insurance, etc.)

Textile Industry

The Indian textile industry has an overwhelming presence in the economic life of the country. Apart from providing one of the basic necessities of life, the textile industry also plays a vital role through its contribution to industrial output, employment generation and the export earning of the country. The textile sector contributes about 14 per cent to industrial production; four per cent to the GDP and 17 per cent to country's export earning. Textile industry includes cotton, silk, jute and man-made fibres (synthetic fibres). At present, India is the third largest producer of silk and the fifth largest producer of synthetic fibre. India is the largest producer and exporter of cotton yarn and jute goods.

Cotton Textile

The cotton and blended textile industries are the largest group of industries in the organized sector in India. The cotton textile industry was most firmly rooted in India before machine-made cloth from England entered the market in the 19th century. The Indian cotton textile industry is composed of four main sectors: (i) the mills including spinning, weaving and composite units (ii) the handlooms and power looms, (iii) the hosiery sector, and (iv) the garment sector.

In India, modern cotton textile industry made a beginning in 1818, when the first cotton mill was set up on the Bowreah Estate at Ghoosury near Kolkata, but it was soon closed. A real start in the industry came with the setting up of the Bombay Spinning and Weaving Company in Mumbai in 1851. The new location proved so favourable for the industry that several mills came up here in quick succession. The number of mills increased from only 8 in 1876 to 79 in 1885, 170 in 1890 and 252 in 1903. By 1939, the number of cotton mills was increased to 415. Mumbai, Ahmedabad, Coimbatore, Chennai, Kanpur, Ludhiana, Nagpur, Solapur became the centres of the textile industry. After independence the growth of textile industry has gained new pattern. The total number of textile mills in the country is 1,719, out of which 188 mills are in public sector.

The production of cotton textile has increased from 8,368 million sq metres in 1980–81 to 6,181 million sq metres in 2010–11. Maharashtra is the leading producer of cotton textile, which produces about 39 per cent of the country's total production followed by Gujarat 34.5 per cent and Tamil Nadu 6.4 per cent. The other cotton textile producing states are Punjab, Madhya Pradesh, Uttar Pradesh, Rajasthan, Puducherry, Karnataka and Kerala. The important producing centres of cotton textile are as follows:

Maharashtra: Mumbai, Solapur, Pune, Nagpur, Jalgaon, Akola, Sangli, Miraj, Kolhapur, Kalyan and Thane

Gujarat: Ahmedabad, Bhavnagar, Bharuch, Kalol, Kadi, Kelot, Khambat, Nanded, Porbandar, Rajkot, Vadodara and Viramgaon

Tamil Nadu: Coimbatore, Chennai, Madurai, Perambur, Salem, Tiruchirapalli, Tirnuveli and Tuticorin

Uttar Pradesh: Agra, Aligarh, Bareilly, Etawah, Lucknow, Mirzapur, Modinagar, Modipuram, Moradabad, Saharanur and Varanasi

Madhya Pradesh: Bhopal, Dewas, Gwalior, Indore, Jabalpur, Ratlam and Ujjain

Punjab: Amritsar, Ludhiana and Phagwara.

Rajasthan: Bhilwara, Siganganagar, Kishangarh, Ajmer, Beawar, Jaipur, Pali and Udaipur

West Bengal: Kolkata, Haora, Hugli, Panihar, Murshidabad, Sirampur and Shiampur

Kerala: Alleppey, Alwaye, Kollam, Thiruvananthapuram and Trichur

Karnataka: Bangalore, Belgaum, Gulbarga, Chennapatnam, Chitradurga, Mangalore and Mysore

Andhra Pradesh/Telangana: East Godawari, Guntur, Hyderabad and Udaygiri

Bihar: Gaya, Bhagalpur and Patna

Haryana: Bhiwani, Hissar and Rohtak

Silk Textile

India's silk industry is quite old and since ancient times India has been well known for the production of silk. All the five commercial varieties of silk, viz. Mulberry, Tropical Tasar, Oak Tasar, Eri and Munga are produced in India. India is the second largest producer of silk in the world after China, contributing 18 per cent of the total silk production of the world.

There are about 90 silk mills in India. In addition, there are several medium and small sized units and it is also popular in the form of handloom industry. About 98 per cent of total silk production comes from Karnataka, West Bengal and Jammu & Kashmir. Other silk producing states are Bihar, Jharkhand, Chhattisgarh, Uttar Pradesh, Punjab, Assam, Tamil Nadu, Gujarat, Andhra Pradesh, Madhya Pradesh, Meghalaya, Manipur, Tripura and Himachal Pradesh.

Figure 13.1 *India: Centres of Cotton Textile Industry*

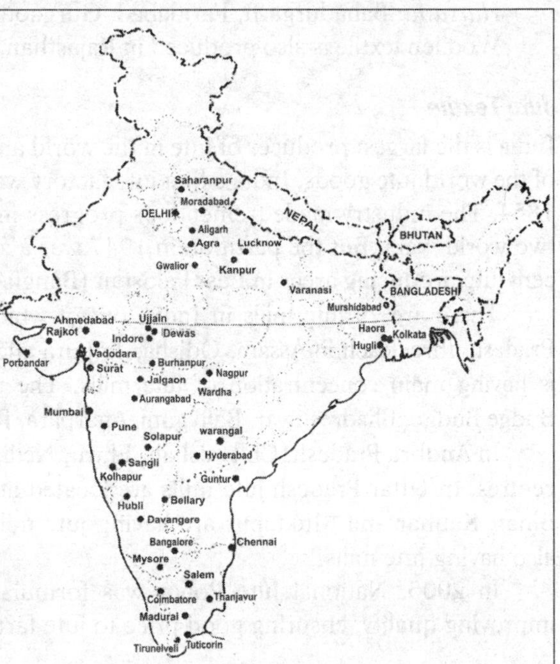

Karnataka is the largest producer of silk textile. The main centres of the silk industry are Bangalore, Belgaum, Gokak, Kolar, Mandya, Mysore and Tumkar. In West Bengal silk textile units are located at Basawa, Bishenpur, Chak-Islampur, Kolkata, Madhu-Ghat. In Jammu & Kashmir Anantnag, Baramulla, Doda, Jammu, Riasi, Srinagar and Udhampur are main centres. Coimbatore, Dharampuri, Kancheepuram, Nilgiri, Selem and Tirunelveli are the main silk producing centres of Tamil Nadu.

India is one of the main exporters of silk textile. It's silk is exported to USA, UK, Kuwait, Russia, Saudi Arabia, Oman, UAE, etc.

Woollen Textile

The modern woollen industry made a beginning in India in 1876, when the first woollen mill was started by the British India Corporation at Kanpur. It was later renamed as 'Kanpur Woollen Mills'. Dhariwal in Punjab (1881) had the second woollen mill soon. Other woollen mills were set up at Mumbai (1982) Bangalore (1884), Raymond Mills at Thane, Jalgaon, Ratnagiri and Boregaon (1925). Now there are more than 150 units in organized sector.

The main concentration of woollen textile industry is in Punjab, Maharashtra and Uttar Pradesh. Other states producing woollen goods are Gujarat, Jammu & Kashmir, Karnataka and West Bengal.

Punjab: The leading centres of woollen textile industry are Dhariwal, Amritsar, Ludhiana, and Kharar. Worsted, tweeds, serges, fiannels, and blankets are produced in these centres. Ludhiana specializes in hosiery and knitwear. Ferozepur is one of the country's largest trading centres of raw wool.

Maharashtra: Mumbai is the leading centre of woollen textile having 45 woollen mills. The other mills are located at Thane, Jalgaon, Ratnagiri and Boregaon.

Uttar Pradesh: Kanpur, Agra, Mirzapur, Modinagar, Shahjahpur, and Varanasi are the main centres of woollen textile.

Gujarat: Ahmedabad, Jamnagar, Kalol and Vadodara are having woollen mills.

Haryana: Bahadurgarh, Faridabad, Gurgaon, and Panipat are the centres of woollen textile.

Woollen textile is also produced in Rajasthan, Himachal Pradesh, West Bengal and Tamil Nadu.

Jute Textile

India is the largest producer of jute in the world and contributing about 35 per cent of the total output of the world jute goods. India's first jute factory was established at Rishra (20 km north of Kolkata) in 1854. The industry made tremendous progress in later part of the 19th century and was boosted by two world wars. But the partition in 1947 was a setback. The partition resulted in transfer of 80 per cent jute producing areas in East Pakistan (Bangladesh), whereas most of the mills remained in India.

There are 79 jute mills in India, out of which 62 are located in West Bengal, Bihar and Uttar Pradesh, three each in Assam, Odisha, Tripura and Chhattisgarh. The Hooghly region of West Bengal is having main concentration of jute mills. The important centres are Titagarh, Jagatdal, Howrah, Budge Budge, Bhadreshwar, Ballygunj, Agarpara, Rishra, Srirampur, Bansberia, Uluberia and Naihati.

In Andhra Pradesh, Chitvalshah, Eluru, Nellimaralla, Guntur and Ongole are the jute producing centres. In Uttar Pradesh jute mills are located at Kanpur and Shahjawan (near Gorakhpur) and in Bihar, Katihar and Muktapur are having jute mills. Gauripur (Assam), Raigarh (Chhattisgarh) are also having jute mills.

In 2005, National Jute Policy was formulated with the objective of increasing productivity, improving quality, ensuring good price to jute farmers and enhancing the yield per hectare.

Iron and Steel Industry

The iron and steel industry is the basic to the industrial development. The development of iron and steel industry opened the doors to rapid industrial development in India. Almost all sectors of the Indian industry depend heavily on the iron and steel industry for their basic infrastructure. The history of iron and steel industry in India is nearly 4,000 years old. The famous iron pillar near Qutub Minar in Delhi is dating back to 350 AD. The first attempt of modern iron and steel industry was made in 1830 at Porto Nova (Near Chennai, Tamil Nadu), but it was not successful. Pig iron was produced successfully in India first by Barakar Works of the Bengal Iron Works Co. at Kulti (West Bengal) in 1877. In fact, the growth of iron and steel in India was started in 1907 with the establishment of Tata Iron and Steel Company (TISCO) at Jamshedpur (Sakchi at that time). The Indian Iron and steel Company (IISCO) was setup in 1919 at Burnpur followed by setting up of Mysore Steelworks at Bhadravati (now Visveswaraya Iron and Steel Works) in 1923. The iron and steel industry witnessed rapid growth after independence, specially during second five year plan period when integrated steel projects were started at Bhilai, Rourkela and Durgapur. India is now the fourth largest crude steel producer in the world and the largest producer of sponge iron. Presently steel contributes to nearly two per cent of the gross domestic product (GDP) and employs over five lakh people.

For iron and steel industry the basic raw material required is iron ore and cooking coal. Besides these, limestone, dolomite, manganese and fire clay are also essential. All these raw materials are gross (weight losing), therefore, the best location for the iron and steel plants is near the source of raw material. In India, there is a crescent shaped region comprising parts of Chhattisgarh, Northern Odisha, Jharkhand and West Bengal, which is extremely rich in high grade iron ore, good quality cooking coal and other supplementing raw materials is most suitable for the development of iron and steel industry. The important iron and steel plants of India are as follows:

Tata Iron and Steel Corporation Limited (TISCO)

- Established in 1907, production of pig iron started in 1908 and steel production in 1911.
- Location-at Sakchi (renamed Jamshedpur) in Singhbhum district of Bihar.
- Iron ore is available from Noamundi of Singhbhum in Jharkhand and Mayurbhanj in Odisha.
- Coal from Jharia and Raniganj mines
- Dolomite, limestone and fireclay from Sundergarh district of Odisha.
- Water from river Subarnarekha.
- Good transport facilities are available, well connected with Kolkata, Mumbai and Chennai.

A second plant established by TISCO is the *Dubari Steelworks* located at Gopalpur in Odisha.

Figure 13.2 *India: Major Iron and Steel Industry Centres*

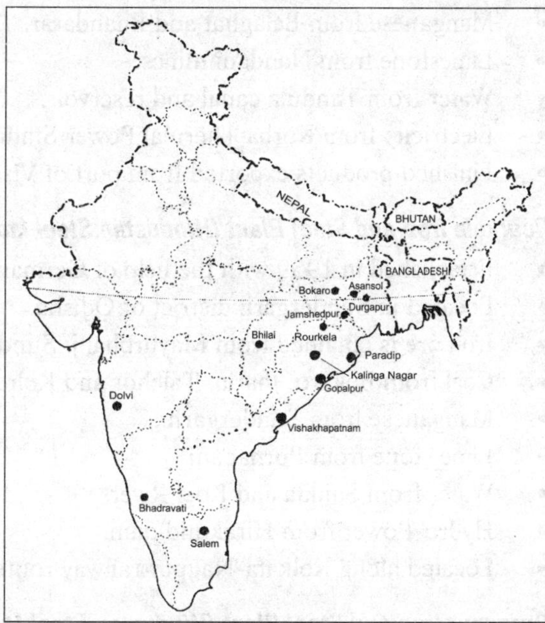

Indian Iron and Steel Corporation (IISCO)

- Three plants established at Kulti in 1864, Hirapur in 1908 and Burnpur in 1937 and merged together as Indian Iron and steel company (IISCO).
- Brought under government control in 1972.
- Iron ore is available from Guna mines of Singhbhum district of Jharkhand and Mayurbhanj in Odisha.
- Coal is obtained from Jharia mines.
- Electricity from Damodar valley corporation.
- Dolomite and limestone from Sundergarh district of Odisha.
- Water from Damodar river.
- Good transport facilities are available, well connected with Kolkata.

Visweswarya Iron and Steel Plant (VISL) or Bhadravati Iron and Steel Plant

- Established in 1923, taken over by government in 1962.
- Located at Bhadrawati (Karnataka).
- Iron ore is obtained from Kundermuck and Baba Budan Mills (Chikmagalur district, Karnataka).
- Electricity from Shoravati Power Project.
- Limestone from Bhundiguda just 25 km away.
- Manganese from Shimoga and Chittradurga mine.
- Water from Bhadravati river.

Bhilai Iron and Steel Plant (Hindustan Steel Ltd.)

- Established with technical co-operation with Russia in Durg district of Chhatisgarh in 1959.
- Iron ore is obtained from Dhalli-Rajhara mines

- Coal from Jharia.
- Manganese from Balaghat and Bhandarar.
- Limestone from Nandani mines.
- Water from Tandula canal and reservoir.
- Electricity from Korba Thermal Power Station.
- Finished products exported from port of Vishakhapatnam.

Rourkela Iron and Steel Plant (Hindustan Steel Ltd.)

- Established in 1959 with the help of Germany.
- Located in Sundergarh district of Odisha.
- Iron ore is obtained from Mayurbhanj, Sundergarh and Keonjhar.
- Coal from Bokaro, Jharia, Talchor and Kolra.
- Manganese from Sundergarh.
- Lime stone from Purnapani
- Water from Sankla and Koel Rivers.
- Hydro-Power from Hirakund dam.
- Located along Kolkata-Nagpur railway route.

Durgapur Iron and Steel Plant (Hindustan Steel Ltd.)

- Established in 1956 with the help of Great Britain at Durgapur in West Bengal.
- Iron ore from Singhbhum (Jharkhand) and Kendujhar (Odisha).
- Coal from Raniganj and Jharia.
- Manganese from Keonjhar.
- Limestone from Birmitrapur
- Electricity from Damodar Valley Corporation.
- Water from Damodar river.

Bokaro Steel Plant (BSL)

- Established with Soviet collaboration in 1964 at Bokaro (Jharkhand)
- Iron ore from Kiriburn of Keonjhar district.
- Coal from Jharia, Bokaro and Kargali mines.
- Limestone from Daltonganj in Palamu district.
- Dolomite from Bilaspur district
- Water from Tenu Dam across Damodar river.
- Electricity Damodar Valley Corporation.

Other Steel Plants

1. Salem Steel Plant (1982) at Salem (Tamil Nadu)
2. Vijaynagar Steel Plant near Hosepet in Bellary district of Karnataka.
3. Vishakhapatnam Steel Plant (1971).
4. Daitari Steel Plant at Daitari near Pardwip in Odisha.
5. Dolvi Steel Plant at Dolvi in Ratnagari district of Maharashtra.
6. Pasco Steel, Paradwip with the help of Pohang Steel Company of South Korea.
7. Tata Steel Kalinganagar in Odisha.

Mini Steel Plants: In addition to the above mentioned iron and steel plants, more than 225 mini-steel plants are also producing steel.

The production of Indian Steel Industry during the last five years is given in Table 13.1.

Table 13.1 *Indian Steel Industry (Production in Million Tonnes)*

Category	*2009–10*	*2010–11*	*2011–12*	*2012–13*	*2013–14*
Pig Iron	5.88	5.68	5.371	6.870	7.950
Sponge Iron	24.33	25.08	19.63	14.33	18.20
Total Finished steel (alloy + Non-alloy)	60.62	68.62	75.70	81.63	87.67

Source: Joint Plant Committee, Ministry of Steel, Government of India.

Aluminium Industry

Aluminium smelting is the second important metallurgical industry of India, next only to iron and steel industry. Because of its flexibility, good conductivity of electricity and heat and its capacity to be moulded into desired shape, aluminium is used for various purposes, such as generation and distribution of electricity, manufacturing of airplanes, railway coaches, bus bodies, household utensils, defence and nuclear accessories, etc. The source of aluminium is the bauxite.

The aluminium smelting industry came into being in India during the second world war at Alupuram (Alwaye) by the Aluminium company in 1938. It was later on converted into a public sector company in 1944. Aluminium Corporation of India started production of alumina in 1942 at Jaykaynagar (West Bengal). Thus, at the time of independence, there were only two aluminium smelting plants in India. During the second five-year plan, two plants have been established at Hirakud in Odisha and Renukoot in Uttar Pradesh by the India Aluminium Company Ltd. (INDAL) and Hindustan Aluminium Corporation respectively. The Bharat Aluminium Company (BALCO) came into being in 1965 and its first plant was established at Korba in Chhattisgarh. The production of alumina started in 1973 and aluminium in 1975. The second plant of this company was established at Ratnagiri in Maharashtra. In 1965 another company, viz. Madras Aluminium (MALCO) established its plant at Mettur. At present, India is having following major aluminium producing plants:

Bauxite Reserves of India	:	3.1 per cent of world reserve
Alumina Production	:	3.7 per cent of world production
Aluminium Production	:	3.39 per cent of world production

Top Leading Aluminium Producing Companies in India (2009–10)

Company	*Capacity in Thousand Tonnes*	*Production in Thousand Tonnes*
HINDALCO	445	523.4
NALCO (Angul)	345	360
BALCO (Korba)	345	367
MALCO (Metter)	40	23
VAL (Jharsuguda)	250	82

1. The India Aluminium Company Ltd. (INDAL) having three smelting units located at Alupuram (Alwaye in Kerala), Hirakund (Odisha) and Belgaum (Karnataka)
2. The Aluminium Corporation of India, Jaykaynagar (near Asansol)

3. The Hindustan Aluminium Corporation Ltd. (HINDALCO), Renukoot
4. The Madras Aluminium Company Ltd. (MALCO), Mettur
5. The Bharat Aluminium Company Ltd. (BALCO) Korba
6. The National Aluminium Company Ltd. (NALCO), Koraput

Chemical and Allied Industries

Chemical Industry is an integral component of the Indian economy, which contributes around 7 per cent of the Indian GDP. The industry is integral to the development of agricultural and industrial development in India and has key linkages with various other industries, such as automobiles, consumer durables, engineering, food processing and more.

The chemical industry is one of the earliest domestic industries in India, contributing considerably to both the industrial as well as economic growth of the country since it achieved independence in 1947. The industry presently produces around 70,000 commercial products, which range from toiletries and cosmetics to plastics and pesticides.

Indian Chemical Sector Classification	
Organic Chemicals	15%
Inorganic Chemicals	9%
Drugs and Pharmaceuticals	15%
Man-Made Fabrics	10%
Fertilizers	10%
Soap and Toiletries	9%
Polymers	6%
Paints and Dyes	5%
Pesticides	3%
Other Chemicals	3%

Heavy Inorganic Chemicals

Soda ash, caustic soda, calcium carbide and sulphuric acid are included in heavy inorganic chemicals.

Soda Ash

The first soda ash (Na_2CO_3) plant in India was setup at Dhrangadhra (Gujarat) in 1932. As sodium chloride is derived from marine salt and limestone, the principle raw material used in soda ash are available in Gujarat-Saurashtra region. The 30 per cent of India's soda ash production comes from Tata Chemicals Ltd., Mithapur. The other centres are Porbandar, Dhrangadhra, Veraval (all in Gujarat), Tuticorin (Tamil Nadu), Varanasi (UP) and Naya Nagal (Punjab).

Caustic Soda

The first plant of caustic soda was set up in 1936 at Mettur (Tamil Nadu). There are more than 40 units producing caustic

Figure 13.3 *India: Centres of Aluminum Industry*

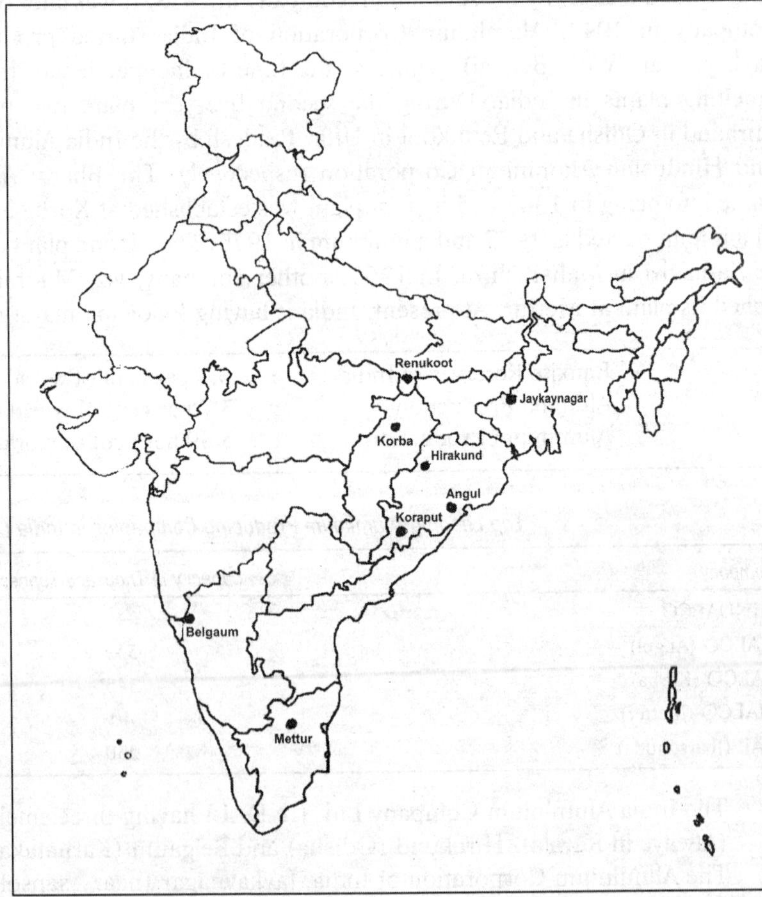

soda in India. The important centres of caustic soda production are Vadodara, Nagada, Thane, Naya Nagal, Alwar, Kurnool, Dhrangadhra, Renukoot, Ankleshwar and Ahmedabad.

Calcium Carbide

There are nine units producing calcium carbide in India. These are located at Hagari (Bellary, Karnataka), Chingawanam (Kerala), Shakarnagar (Tamil Nadu), Birla Nagar (West Bengal), Chembur (Mumbai), Kota (Rajasthan), Dinnedevarapada, Kurnool (Andhra Pradesh) and Panna, (Madhya Pradesh).

Sulphuric Acid

In India, sulphuric acid is produced in more than 100 units. The leading sulphuric acid producing units are located at Mumbai, Debari (Udaipur, Rajasthan), Khetri (Jhunjhunu, Rajasthan), Vadodara, Ahmedabad and Kolkata.

The production of major chemicals in India is given in Table 13.2

Table 13.2 *Production of Major Chemicals in India (000' MT)*

Year	*Alkali Chemicals*	*Inorganic Chemicals*	*Organic Chemicals*	*Pesticides and Insecticides*	*Dyes and Dyestuffs*	*Total Major Chemicals*
2008–09	5,427	513	1,254	85	32	7,311
2009–10	5,602	518	1,281	82	42	7,525
2010–11	5,981	572	1,342	85	47	8,027

Source: Department of Chemicals and Petrochemicals. Note: MT = Metric Tonne.

Heavy Organic Chemicals

The major heavy organic chemicals are acetone, phenol, methanol, acetic acid and acetic anhydride. These are having wide applications in manufacture of several other chemical products. The Hindustan Organic Chemicals Ltd. (HOCL) is a major producer of organic chemicals. It was established in 1960 at Rasayani (Raigad, Maharashtra). Another unit of HOCL was established at Cochin (Kerala). A subsidiary company called 'Hindustan Fluorocarbons Ltd.' has been established in Medak district of Andhra Pradesh.

Petrochemicals

The petrochemicals industry forms an important part of chemical industry. It is one of the fast growing sectors of Indian economy. Petrochemicals include the production of synthetic fibres, polymers, elastomers, synthetic detergent, plastics, etc. Petrochemicals are organic chemicals derived from thc polymerization of naphtha and natural gas. The Indian Petrochemical Corporation Ltd. (IPCL) and Petro File Corporation Ltd. (PCL) are two important organizations in this field. The important centres of petrochemicals industry are: Mumbai, Barauni, Mathura, Sahupuram, Patalganga, Durgapur, Pimpri, Rishra, Bongaigoon, Chennai, Haldia, Visakhapatnam and Mathura.

Plastic industry has two separate branches of production; one engaged in production of raw materials and another in conversion and fabrication of various goods. Their production is also associated with petrochemicals industry. Vadodara, Bharuch, Jamnagar,(Gujarat), Vishakhapatnam (AP), Haldia (West Bengal), Tenughat (Bihar), Payal (Ludhina, Punjab), Mangalore (Karnataka), Thane, Belapur (Maharashtra) are the important centres of the production of raw materials. While Delhi, Kolkata, Bangalore, Vadodara, Vapi, Surat, Kanpur, Amritsar, Coimbatore, Mumbai and Chennai are important centres of plastic industry. The production of major petrochemicals is given in Table 13.3.

Table 13.3 *Production of Major Petrochemicals in India (000 Metric Tonnes)*

Year	Synthetic Fibres	Polymers	Elastomers	Synthetic Detergent Intermediates	Performance Plastics	Total Major Petrochemicals
2008–09	2,343	5,060	96	552	141	8,193
2009–10	2,610	4,791	106	618	172	8,287
2010–11	2,719	5,292	95	638	191	9,007

Source: Department of Chemicals and Petrochemicals.

Pharmaceuticals

The pharmaceuticals industry is one of the fastest growing industries in India. The public sector units played a key role in the development of pharmaceuticals industry in India. The Indian Drugs and Pharmaceuticals Ltd. (IDPL) was incorporated in 1961. The company has five plants at Rishikesh (Uttarakhand), Gurgaon (Haryana), Hyderabad (Tamil Nadu), Chennai (Tamil Nadu) and Muzaffarpur (Bihar).

The Rajasthan, Uttar Pradesh and Odisha have their 'Drugs' and Pharmaceutical units located at Jaipur, Lucknow and Bhubaneshwar.

The Hindustan Antibiotics Ltd. (HAL) was incorporated in 1954 and has three subsidiaries viz. KAPL–Bangalore (Karnataka), MAPL–Nagpur (Maharashtra) and MSDPL–Imphal (Manipur).

There other companies are located at Kolkata, which were nationalized in 1951. They include Bengal Immunity Ltd (BIL), Bengal Chemicals and Pharmaceuticals (BCPL) and the Smith Stainistreet Pharmaceuticals Ltd. (SSPL). The BCPL is the largest producer of anti-snake venom in India.

Apart from above mentioned public sector companies, there are more than 15,000 private drug manufacturing units in the country located in Mumbai, Vadodara, Chennai, Delhi, Kolkata, Kanpur, Patna, Bhopal, Gurgaon, Hyderabad, Bangalore, etc. Several foreign companies are also involved in this industry.

Fertilizer Industry

One of the vital industries for the Indian economy is the Indian fertilizer industry as it manufactures a very critical raw material for agriculture which is the major occupation of the country. The Indian fertilizer industry has grown in size and stature as it ranks third in the world.

The first super phosphate factory in India was setup at Ranipet near Chennai (Tamil Nadu) in 1906 and production of synthetic ammonia and ammonium sulphate began in 1944 and 1947 respectively. India's fertilizer industry is essentially a post-Independence development. The setting up of the Sindri plant in Jharkhand by the central government in 1951 can be taken as the turning point in the history of this industry. At present, there are 57 large fertilizer plants and 64 medium and small-sized plants in India producing urea, DAP (Diammonium Phosphate), complex fertilizer, ammonium sulphate (AS) and Calcium-Ammonium Nitrate (CAN).

Table 13.4 indicates the installed capacity of various fertilizer units in India.

Table 13.4 *Installed Capacity of Various Fertilizer Units in India*

Produce	No. of Units	Total Installed Capacity (LMT per Annum)
Urea	31	210.84
Diammonium Phosphate (DAP)	12	72.99
Complexes	19	52.22
Single Super Phosphate (SSP)	80	77.33

The following are the public sector fertilizer units in India:

1. The Fertilizer Corporation of India Ltd. (FCI) having four units at Sindri (Jharkhand), Nagal (Punjab), Tailcher (Odisha), Ramagundam (Andhra Pradesh).
2. The National Fertilizers, Ltd. (NFL) – Units at Bhatinda, Nagal, Panipat and Bijaipur.
3. The Hindustan Fertilizer Corporation Ltd. (HFC) – Units-Barauni (Bihar), Durgapur (West Bengal) and three at Namrup (Assam).
4. The Rashtriya Chemicals and Fertilizers Ltd. (RCF) – Five Units at Trombay and one at Thal (Maharashtra)
5. The Fertilizer and Chemicals Travancore: Two units at Kochi and one at Udhyoga Mandalam.
6. Madras Fertilizers Limited, established in 1966.
7. The Paradwip Phosphate Ltd. (PPL)
8. Brahmaputra Valley Fertilizer Corporation Ltd. (BVFCL)

In co-operative sector also, a number of fertilizer plants have developed. The Indian Farmers Fertilizer Co-operative Ltd. (IFFCO) has six units at Aonla, (two units), Kalal and Kandla (Gujarat) and Phulpur two units (Allahabad).

Gujarat and Uttar Pradesh produce about 42 per cent of the total national output of fertilizers. Andhra Pradesh, Rajasthan, Odisha, Maharashtra, Madhya Pradesh, Punjab, Goa, West Bengal, Karnataka, Haryana and Assam are other producing states. Table 13.5 shows the production and imports of fertilizers in India.

Table 13.5 *Production and Imports of Fertilizers in India*

	Production			*Imports*		
Year	*2009–10*	*2010–11*	*2011–12*	*2009–10*	*2010–11*	*2011–12*
Urea	211.12	218.80	222.88	52.09	66.09	56.43
Diammonium Phosphate (DAP)	42.46	35.37	39.41	58.89	74.09	53.00
Complex Fertilizers	80.38	87.27	90.69	—	—	—
Muriate of Potash (MOP)	Nil	Nil	Nil	52.86	63.57	24.91

Source: Department of Chemicals and Petrochemicals.

Paper and Pulp Industry

The paper industry in India is more than a century old. The first paper mill was set up in 1832 at Serampore (West Bengal), although it was a failure. Subsequently paper mills were established at Ballygunj (Kolkata) in 1870, Lucknow in 1879, Titagarh in 1881, Pune in 1887 and Raniganj in 1891. At present there are over 850 paper mills manufacturing a wide variety of items required by the consumers.

The paper mills are manufacturing industrial grades, cultural grades and other special papers. The paper industry in India could be classified into three categories according to the raw material consumption: (1) Wood based (2) Agro-based, and (3) Waste paper. The number of wood-based industries are around 14 and remaining 836 are based on non-conventional raw materials (agro residues and recycled fibre and waste paper). Paper industry is a raw material based industry. The raw material used in paper making is soft wood, bamboo, grasses, bagasse, rags and waste paper. In India bamboo alone contributes about 70 per cent of raw material for paper industry. The Indian paper industry is among the top 12 global players today, with an output of more than 13.5 million tonnes.

The distribution of paper mills in India is as follows:

States	Centres of Paper Mills
Andhra Pradesh/Telangana	Ballarshah, Bhadrachalam, Bodhan, Khammam, Kurnool, Rajamundry, Sirpur and Tirupati.
Maharashtra	Chinchwad, Goregaon, Jagaon, Kalyan, Kamptee, Khandala, Khopoli, Paravarnagar, Sangli, Vikhroli and Bellarpur. (Sangli and Bellarpur are famous for news prints)
West Bengal	Alambazar, Chandrahati, Dum-Dum, Ganganagar, Howrah, Kon Kinara, Kolkota, Naihati, Raniganj, Sheoraphali, Titagarh and Tribeni
Odisha	Brajranagar, Chowdar and Rayagada
Karnataka	Bangalore, Bhadrawati, Belagnla, Dandeli, Mandya, Najanguda and Ramangaram.
Gujarat	Barejadi, Bevla, Gondal, Kalol, Khadki, Songarh, Utran, Utvada, Valsal and Vapi.
Madhya Pradesh	Shadol, Bhopal, Dhemka, Indore, Nepanagar, Ratlam, and Vidisha.
Tamil Nadu	Chennai, Charanmahadevi, Selam, Pallipalayam, Tambaran and Udamalpet.
Uttar Pradesh	Budaun, Jagadhari, Lal-Kuan, Lucknow, Mainpuri, Meerut, Modinagar, Naini and Saharanpur.
Bihar	Barauni, Dalmianagar, Dumka, Patna, Rameshwarnagar and Samistipur.
Kerala	Ernakulam, Kozhikhode, Punalur, and Rajnpuram
Assam	Cachar, Guwahati, Jorhat and Nayagaon
Nagaland	Mokokchung

Security paper mill is located at Hoshangabad and National news print and paper mills (NEPA) is located at Nepanagar.

Sugar Industry

Sugar industry in India is the second largest organized agro-based industry, next to cotton textile. India ranks second in sugar production in the world following Brazil. The Indian sugar industry uses sugarcane in the production of sugar and hence maximum number of sugar mills are located in sugarcane producing states of India including Uttar Pradesh, Maharashtra, Gujarat, Tamil Nadu, Karnataka and Andhra Pradesh.

There are 453 sugar mills in India. Co-operative sector has 252 mills, private sector has 134 mills and 67 mills are in public sector. Total annual production of sugar is 18.5 million tonnes. Maharashtra has produced 7.7 million tonnes of sugar, followed by Uttar Pradesh 6.7 million tonnes, Karnataka 3.29 million tonnes in year 2012–13. The important districts having sugar mills are as follows:

State	Districts having Sugar Mills
Maharashtra	Nasik, Ahmednagar, Pune, Satara, Solapur, Sangli and Kolhapur
Uttar Pradesh	Meerut, Muzaffarnagar, Bijnor, Saharanpur, Moradabad, Deoria, Basti, Gonda, Gorakhpur and Kheri districts.
Tamil Nadu	Coimbatore, North and South Arcot and Tiruchirapalli districts
Karnataka	Belgaum, Mandya, Bijapur, Bellary, Shimoga and Chitradurga districts
Andhra Pradesh	East and West Godavari, Krishna, Visakhapatnam, Nizamabad, Medak and Chittor districts.
Gujarat	Sur t, Valsad, Bhavnagar, Junagarh, Rajkot and Jamnagar districts.
Bihar	Champaran, Sarn, Muzaffarnagar and Darbhanga districts.

Cement Industry

Indian cement industry is the 2nd largest after China accounting about 8 per cent of the total global production. It had a total capacity of about 347 million metric tonnes in 2012–13. The production capacity is projected to reach 550 million metric tonnes by year 2020.

The first attempt to manufacture cement on modern lines was made in Chennai in 1904, but failed. The real start was made later in 1915, when the Indian Cement Co. Ltd. set up a plant at Porbandar (Gujarat). Two more factories one each at Katni (MP) and Lakeri (Rajasthan) commenced production. Six new factories were erected during 1922–23 at Dwarka (Gujarat), Japla, Malegaon, Banmori, and Kaimur (MP) and Shahabad (Karnataka). In 1936, 10 out of the 11 existing cement companies merged into the Associated Cement Cos. Ltd. (ACC). After independence emphasis has been given on the growth of cement factories with the result India has more than 125 plants and over 300 mini cement plants; the main cement producing states are Andhra Pradesh, Telangana, Rajasthan, Madhya Pradesh, Tamil Nadu, Gujarat, Maharashtra, Chhattisgarh, Jharkhand, Uttar Pradesh, Himachal Pradesh, West Bengal and Odisha.

Engineering Industry

Engineering is the largest segment in the Indian Industry. It is a diverse industry with a number of segments, and can be broadly categorized into two segments namely, heavy engineering and light engineering. Powered by significant investments in power projects and infrastructure development the sector has witnessed tremendous growth in recent years. This sector accounts for about 20 per cent of India's total export and is the largest foreign exchange earner for the country in terms of merchandised goods.

Engineering industry includes transport equipment, machine tools, transformers, switchgears, furnaces, cast, and forged simple and precision parts for turbines, automobiles and railways; it also includes car, motorcycle, scooter, tractors and other types of vehicles.

Heavy Machinery

Manufacturing of heavy machinery made a beginning in 1958, when Heavy Engineering Corporation Ltd. was set up at Ranchi in Jharkhand. The plants produce castings, forging, and rolls to meet the requirements of heavy machine tools and heavy machine building plants. The Mining and Allied Machinery Corporation Ltd. Durgapura produces various items of mining equipments.

India today produces a wide range of industrial machinery including textile, jute, sugar, paper, cement, chemical, agricultural, fertilizer, pharmaceuticals, dairy, metallurgical, etc. The main producing centres of industrial machinery are Mumbai, Kolkata, Coimbatore, Ahmedabad, Ludhiana, Gwalior, Chennai, Allahabad, Yamunanagar, Pune, Delhi, Shahibabad, Jamshedpur, Durgapur, Rourkela, Vadodara, Hyderabad, Sonepat, Faridabad, etc.

Machine Tools

The manufacturing of machine tools commenced in 1930s, but its real progress started with the establishment of the Hindustan Machine Tools (HMT) factory in public sector at Bangalore in 1953. Its two units are located at Bangalore and others are at Pinjore (Haryana), Kalamassery (Kerala), Hyderabad, Srinagar and Ajmer. Other notable factories are – The Heavy Machine Tools at Ranchi, The Praga Tools Ltd. Secunderabad, The National Instrument Factory at Kolkata with its Jadavpur unit and The Instrumentation Ltd. Plant at Kota and Palghat (Kerala). All these units are in public sector. Over 100 private sector units have developed in different parts of the country.

Railway Locomotives, Coaches and Wagons

Railway locomotive factories are:

1. Chittaranjan Locomotive Works, Chittaranjan (CLW)
2. The Diesel Locomotive Works, Varanasi
3. Tata Engineering and Locomotive Co., Jamshedpur (TELCO).

Other railway equipments are manufactured at Bhilai, Jamshedpur, Durgapur and Rourkela. Railway coaches are manufactured at:

1. The Integral Coach Factory, Perambur near Chennai.
2. The Bharat Earth Movers Ltd. (BEML) at Bangalore.
3. Rail Coach Factory at Kapurthala (Punjab).

Some private companies also produce railway coaches. Rail wagons are produced in 13 units in private sector and three railway workshops are located in Maharashtra, Uttar Pradesh and Delhi.

Automobile Industry

India's passenger car and commercial vehicle manufacturing industry is the sixth largest in the world with an annual production of more than 3.9 million units in 2011. In 2009, India emerged as Asia's fourth largest exporter of passenger cars.

The majority of India's car manufacturing industry is based around three clusters in the south, west and north. The southern cluster consisting of Chennai is the biggest with 35 per cent of revenue share. The western hub near Mumbai and Pune contributes to 33 per cent of market and the northern cluster around the National Capital Region contributes 32 per cent. Chennai houses the Indian operations of Ford, Hundai, Renault, Mitsubishi, Nissan, BMW, Hindustan Motors, Daimier, Caparo, Mini and Dastan. Gurgaon and Maneser in Haryana from the northern cluster where the country's largest car manufacture Maruti Suzuki is based. The Chakam Corridor near Pune, Maharashtra is the Western cluster with companies like General Motors, Volkswagen, Skoda, Mahindra and Mahindra, Tata Motors, Mercendes Benz, Land Rover, Jaguar Cars, Fiat and Force Motors having assembly plants in the area. Another emerging cluster is in the state of Gujarat with manufacturing facility of General Motors in Halol and Tata Nano at Sanad. Kolkota, Noida and Bangalore are also centres of automobile industry.

In India, buses and trucks are manufactured at Chennai and Mumbai, Jeeps and four and three-wheelers tempos and small trucks are manufactured at Mumbai, Pune and Gurgaon. Two-wheelers (scooters and motor-cycles) are manufactured at Pune, Mumbai, Faridabad, Chennai, Mysore, Ludhiana and Tirupati.

Tractors are manufactured at Faridabad, Pinjore, Delhi, Mumbai and Chennai.

Ship Building

In India, Ship building is still mostly a central public sector industry. The two PSUs, the 'Hindustan Shipyard Ltd' (HSL), Visakhapatnam and the 'Cochin Shipyard Ltd. (CSL) accounting for most of the production and repairs with minor contributions from other 4 PSUs, i.e. 'Hooghly Dock and Port Engineer's Ltd.' (HDPEL), 'Central Inland Water Transport Corporation' (CIWTC) at Kolkata, and two PSUs, the 'Alcok Ashdown' and the 'Shalimar works' owned by the Gujarat and West Bengal state governments respectively. Of remaining 26 yards, 3 are defence undertaking and 23 are small/medium private shipyards catering to domestic requirement of medium and small crafts.

Aircraft Industry

The aircraft industry in India is restricted to the Hindustan Aeronautics Ltd. (HAL). A defence undertaking was set up in 1964 with its corporate office at Bangalore. The company has 12 divisions located in 6 states. Its range of production includes aircrafts, engines, accessories, avionics, structures for satellites, industrial and marine gas turbine engines.

Light Mechanical Goods

India has become a major producer of a large variety of light mechanical engineering goods, particularly sewing machines, office equipments, bicycles, watches, etc. The sewing machine units are located at Mumbai, Delhi, Secunderabad and Jalandhar. India has emerged as the second largest producer of bicycles after China in the world. Bicycles production units are located at Sonepat (Haryana), Ludhiana (Punjab), Ambattur (Near Chennai), Chandigarh, etc.

The wrist watches industry has grown in last few decades. The Titan Industries, Hosur (Tamil Nadu) controls about 48 per cent of the market. Hindustan Machine Tools, Bangalore is the second largest producer. Timex Watches, Sakura Seimitsu India and Kamala Dials and Devices are other important producers.

Heavy Electrical Engineering

The heavy electrical industry produces electrical equipment like boilers, turbines, generators, transformer, switchgears, etc. which are required for generation, transmission and distribution of power. The Bharat Heavy Electricals Ltd. (BHEL) is the major supplier of such items. BHEL is a central PSU set up in 1956 in technical collaboration with UK, former USSR, Czechoslovakia, Japan and USA and also receive technical assistance from France and Germany. It is a multi unit company having branches at Bangalore, Bhopal, Goindwal (Jammu), Hyderabad, Tiruchirapalli, Ranipat (Tamil Nadu), Jhansi, Jagdishpur (UP), Haridwar and Rudrapur (Uttarakhand).

Apart from BHEL, there are several other companies producing heavy power equipments.

Electronic Industry

In last two and a half decades, the electronic industry in India is growing at a brisk pace. This industry in India constitutes just 0.7 per cent of the global electronic industry. However the demand in the Indian market is growing rapidly and India is also an exporter of a vast range of electronic components and products. The electronics sector industries are grouped into 7 categories, viz. (1) Semiconductors, (2) Consumer electronics, (3) Industrial electronics, (4) Computers, (5) Communication and broadcasting equipment, (6) Strategic electronics, and (7) Electronics components.

The main centres of electronic industry in India are – Mohali (Chandigarh, Punjab), Hyderabad, Bangalore, Bharuch (Gujarat), Palakkad (Kerala), Mumbai, Kolkata, Delhi, Chennai, Mysore, Noida, Pune, Gurgaon, Jaipur, Ahmedabad, Coimbatore, Madurai, Tiruchi, Bhubaneshwar, Indore, Kochi, Visakhapatnam, Chandigarh, Thiruvananthapuram, etc. Most of the strategic electronic production comes from the 'Bharat Electronics Ltd.' (BEL) and 'Electronics Corporation of India Ltd. (ECL).

Industrial Regions in India

In the whole world, there is a tendency of concentration of industries in certain regions, because of the favourable geo-economic conditions. The same is true regarding India also. In India there are certain regions in which industrial growth has been taken and they have become the industrial regions. Several geographers have delineated industrial regions but the parameters used by them, however differ from each other. The major and minor industrial regions identified in India are as follows:

Major Industrial Regions

1. Mumbai-Pune Region
2. Hugli Region
3. Bangalore-Tamil Nadu Region,
4. Gujarat Region,

5. Chotanagpur Region,
6. Vishakhapatnam-Guntur Region,
7. Gurgaon-Delhi-Meerut Region, and
8. Kollam-Thiruvananthapuram Region.

Major Industrial Regions of India

Major Industrial Regions	Industries Developed	Industrial Centres
1. Mumbai-Pune Industrial Region	Cotton Textile, engineering goods, petroleum refining, synthetic and plastic goods, drugs, fertilizers, electrical equipments, electronics, software, fertilizers, shipbuilding, transport equipments, food industries.	Mumbai, Pune, Kolaba, Kalyan, Thane, Trombay, Pimpri, Nasik, Manmad, Solapour, Kolhapur, Ahmednagar, Satara and Sangli
2. Hugli Industrial Region	Jute industry, Cotton textile, paper, engineering, textile machinery electrical, chemical, pharmaceuticals, fertilizers	Kolkata, Haora, Haldia, Serampur, Rishra, Sodepur, Budge-Budge, Birlanagar, Titagarh, Shibpur, Naihati, Kakinara, Hugli, Birlanagar, Bansbera, Belgurriah, Triveni, Belur, etc.
3. Bangalore–Chennai Industrial Region	Heavy engineering, cotton textile, machine tool, telephone, aircraft engineering goods, diesel engines, light engineering goods, aluminium, sugar, cement, glass, paper, chemicals, film, cigarette, leather goods, fertilizer, iron and steel, software, electronics.	Bangalore, Salem, Chennai, Madurai.
4. Gujarat Industrial Region	Cotton textile, petrol-chemicals, refinery, silk and synthetic fabrics, motor, tractor, diesel engine, textile machinery, engineering, dyes, pharmaceuticals, sugar, dairy products.	Jamnagar, Vadodara, Bharuch, Koyali, Ahmedabad, Khera, Surendranagar, Rajkot, Surat, Valsad, Ankleshwar.
5. Chotanagpur Industrial Region	Iron and steel, engineering, cement, machine tools, fertilizers, paper, locomotives, heavy electrical, heavy engineering.	Ranchi, Dhanbad, Chaibasa, Sindri, Hazaribag, Bokaro, Jamshedpur, Rourkela, Durgapur, Asansol and Dalmianagar.
6. Vishakhapatnam – Guntur Industrial Region	Shipbuilding, petro-chemical refinery, sugar, textile, jute, paper, aluminium, cement, fertilizer, light engineering.	Vishakhapatnam, Vijayawada, Guntur, Vijaynagar, Eluru, Rajahmundry and Kurnool.
7. Gurgaon–Delhi–Meerut Industrial Region	Electronics, Cotton, Woollen and synthetic fabrics, hosiery, sugar, cement, machine tools, tractor, cycle, chemical and Vanaspati, agricultural implements, software, etc.	Gurgaon, Delhi, Noida, Shahadara, Faridabad, Meerut, Modinagar, Ghaziabad, Ambala and Mathura.
8. Kollam-Thiruvananthapuram Industrial Region	Cotton textile, sugar, rubber, glass, matchbox, fertilizer, paper food processing, coconut coir products, aluminium, cement, petroleum refinery, etc.	Kollum, Kocchi, Thiruvananthapuram, Alppuzha and Punalur

Minor Industrial Regions

1. Ambala-Amritsar
2. Saharanpur-Muzaffarnagar-Bijnor
3. Indore – Dewas – Ujjain

4. Jaipur – Ajmer
5. Kolhapur – South Kannada
6. Northern Malabar
7. Central Malabar
8. Adilabad – Nizamabad
9. Allahabad – Varanasi – Mirzapur
10. Bhopal – Munger
11. Durg – Raipur
12. Bilaspur – Korba
13. Brahmaputra Valley

Industrial Districts

Apart from above mentioned major and minor industrial regions, there are several districts/centres which are distinct in industrial development. These are: Kanpur, Agra, Lucknow, Gorakhpur, Aligarh, Bhopal, Gwalior, Hyderabad, Alwar, Kota, Nagpur, Cuttack, Jalpaiguri, Purnia, Jabalpur, Bareilly, etc.

Figure 13.4 *India: Major Industrial Regions*

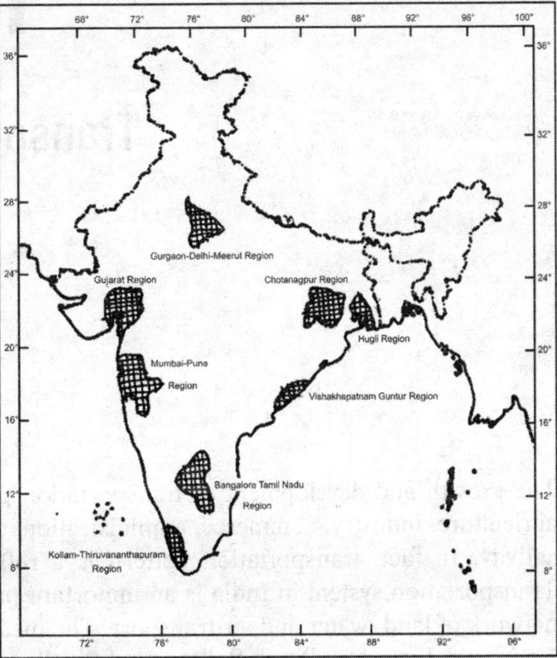

14

Transportation

The growth and development of transportation provide a medium, contributing to the progress of agriculture, industry, commerce, administration, defence, education, health or any other community activity. In fact, transportation pattern is a reflection of regional and/or national development. Transportation system in India is an important part of the national economy. India has widespread network of land, water and air transport. The means of transport in India include:

1. *Land Transport:* Road, Railway and Pipeline.
2. *Water Transport:* Inland and Oceanic Waterways.
3. *Air Transport:* National and International.

A brief account of transport in India is given here.

Roads

India has a road network of over 46,89,842 km in 2013, which is the second largest road network in the world. At 0.66 km of roads per square kilometre of land the density of India's road network is similar to that of the USA (0.65) and far higher than China (0.16) and Brazil (0.20). In terms of 1,000 people, India has less than 3.8 km of roads.

The first evidence of road development in the Indian Subcontinent can be traced back to approximately 4,000 BC from the ancient cities of Harappa and Mohenjodaro of the Indus Valley civilization.

In ancient India, ruling emperors and kings had constructed numerous brick roads in the cities and also have few connecting roads mostly within their territory and sometimes inter-territory also. One of the most famous highways of medieval India was the *Grand Trunk Road*, built by Sher Shah Suri between 1540–1545 which began in Sonargaon near Dhaka in Bangladesh and ended at Peshawar in modern Pakistan. In India it linked several important cities like Kolkata, Patna, Varanasi, Kanpur, Agra, Delhi, Ambala, Ludhiana, Jalandhar and Amritsar.

Britishers have also developed roads in India and most of the important towns were linked with road transport. But in vast country like India the road network was too less and the conditions of the roads were also poor. India inherited a poor road network infrastructure at the time of its independence in 1947. But afterwards planned development has started. The boost in road development has occurred with the establishment of the National Highway Authority of India on 15 June 1989.

Road transport is vital to India's economy. It enables the country's transportation sector to contribute 4.7 per cent towards India's Gross Domestic Product. India's road network carries over 65 per cent of its freight and about 80 per cent of passenger traffic.

The classification of Indian roads according to their administration is as follows:

Road Classification	*Administrative Authority*	*Total Kilometres (as in 2011)*
National Highways	Ministry of Road Transport and Highways (Central Government)	92,851
State Highways	State Governments	1,63,896
Major and other District Roads	Local governments, district and municipal authorities	17,63,898
Rural Roads	Panchayats and other local authorities	27,49,805

National Highways

National highways are the main highways running through the length and breadth of the country connecting state capitals, ports, large industrial and tourist centres, etc. The National Authority of India is the authority responsible for the development, maintenance and management of National Highways. In May 2014, the total length of the National Highways in India was over 92,861 km of which 22,757 km are 4 lane or 6 lane modern highways. A complete list of National Highways in India is given at the end of this chapter.

The National Highways Development Project (NHDP) has been undertaken in 1999 for the rapid development of national highways under special programmes. The NHDP has two components:

Golden Quadrilateral (GQ): Comprising National Highways connecting four metro cities of Delhi, Mumbai, Chennai and Kolkata having total length of 5,846 km.

North-South and East-West Corridor: Comprising National Highways connecting Srinagar to Kanyakumari including Kochi-Salempur and Silchar to Porbandar. The total length of corridor is about 7,300 km.

Other aspects of the project include:

- Port connectivity programme: Four laning of roads measuring about 400 km for providing connectivity to the ports.
- Four laning of about 10,000 km of existing National Highways to be taken under build-operate-transfer (BOT) basis.
- Improving highway upgrading of about 2,100 km of single lane roads to two lane roads throughout the country.

Express Highways

These are multi-lane, well-paved highways used for traffic. Express ways are the highest class of roads having approximately 942 km length. The following are the Express ways in India:

Express Way Name	*State*	*Distance (Km)*
Ahmedabad-Vadodara	Gujarat	95
Mumbai-Pune	Maharashtra	93
Jaipur-Kishangarh	Rajasthan	90
Allahabad bypass	Uttar Pradesh	86
Durgapur	West Bengal	65
Chennai bypass	Tamil Nadu	32

Cont'd...

Cont'd...

Delhi-Gurgaon	Delhi/Haryana	28
Noida – Greater Noida	Delhi/UP	24.53
Delhi-Noida Direct Flyway	Delhi/UP	9.2
Hyderabad Elevated	Telangana	11.6
Hosur Road Elevated	Karnataka	9.985
Kona Exp. Way	West Bengal	8
Raipur-Bhilai-Durg	Chhattisgarh	26
Yamuna Exp. Way	Uttar Pradesh	165
Bangalore-Mysore Infrastructure corridor	Karnataka	41

State Highways

State Highways are developed and maintained by the various agencies of the state and union territories. The state highways are the important state roads which link district headquarters, cities and towns within the state and also connect them with National Highways or highways of the neighbouring states. The total length of State Highways in India is 1,63,896 km in 2011. Maharashtra is having 32,232 km State Highway, which is the highest among the Indian states. Gujarat is at the second place with a length of 19,796 km, followed by Madhya Pradesh 11,789 km, Rajasthan 10,047 km and Karnataka 10,021 km.

District Roads

These roads connect the towns and large villages with the district headquarters and also with cities of another districts. These are developed and maintained by district PWD or by Zila Parishad. The length of district road in the country is 17,63,898 km.

Village Roads/Rural Roads

These roads are constructed and maintained by the Village Panchayats. The total length of rural roads in the country is 27,49,805 km.

International Highways

Under the agreement with the Economic and Social Commission on Asia and Pacific (ESCAP). Some of the country's highways linking neighbouring countries have been declared international highways. There are two categories of such highways:

1. The main arterial routes linking the capitals of neighbouring countries, such as (i) Lahore-Mandalay (Myanmar) route passing through Amritsar-Delhi-Agra-Kolkata-Golaghat-Imphal, (ii) Agra-Gwalior-Hyderabad-Bangalore-Dhanushkodi road, and (iii) Barhi-Kathmandu road.
2. Routes joining major cities, ports, etc. with arterial network such as (i) Agra-Mumbai road (ii) Delhi-Multan road, (iii) Bangalore-Chennai Road, and (iv) Golaghat-Ledo Road.
   The World Bank provides funds for the maintenance of these roads.

Border Roads Organization (BRO)

The Border Roads Organization was established in 1960 for the development of roads of strategic importance in the northern and north-eastern borders of the country. Apart from construction and maintenance of roads in border areas, the BRO has also developed the road infrastructure in

Maharashtra, Karnataka, Andaman and Nicobar islands and Chandigarh. The Zojila-Kargil and Mani-Leh roads were completed by it in 1998. The Pathankot-Jammu-Srinagar national highway is also maintained by BRO. It has also constructed roads and air field in Afghanistan, Bhutan, Tajikistan and Myanmar. So far BRO has constructed more than 32,000 km of roads and surfaced about 40,500 km of roads.

Special Accelerated Road Development Programme in the North-Eastern Region (SARDP-NE)

The SARDP-NE formed for improving road connectivity to the remote places of the NE region of India envisages 2/4 laning of 3,251 km of NH, two laning of about 2,500 km of state roads and 1,888 km of roads of strategic importance. This will ensure connectivity of 34 district headquarters out of the total of 85, which are still not connected to the NH in the 8 NE States.

Special Features of Road Transport

- As per an estimate, the road network carries 65 per cent of freight and 80 per cent of passenger traffic.
- The Jawahar tunnel is located at Banihal pass connecting Jammu and Srinagar. It is the longest tunnel road in India.
- The longest National Highway is New No. NH–44 which runs from Srinagar to Kanyakumari, covering a distance of 3,745 km.
- The shortest National Highway is the NH–47-A (5.9 km), which connects Kundanoor of Maradu in Kochi city to the Kochi port at Willingdon Island.
- The highest roadway of the world is the Manali-Leh Highway.
- Jhansi is the junction of North-South and East-West corridors.
- The National Highway Authority of India was constituted by an Act of Parliament, the National Highways Authority of India Act, 1988. The authority was actually operationalized in February, 1995.

Railways

Indian Railways is a state owned enterprise, owned and operated by the Government of India. It is one of the largest networks in the world, ranks fourth after the USA, Russia and China. The India's railway networks comprises 1,15,000 km (71,000 mile) of track over a route of 65,436 km (40,660 mile) and 7,172 stations. In 2013–14 Indian Railways carried 8,425 million passengers annually or more than 23 million passengers daily. Indian Railways is the world's ninth largest commercial or utility employer by number of employees with over 1,307 million employees.

The railway system started in India on 16 April 1853, when the first railway line between Bombay (Mumbai) and Thane (34 km) was inaugurated. This linc was extended to Kalyan on 1 May 1854 and to Khopali on 12 May 1856. The Khandala-Pune section was opened to traffic on 14 June 1858.

Meanwhile, the construction of railway lines was going on in eastern part of the country and first section of Eastern Indian Railway, from Haora to Hugli, a distance of 37 km was inaugurated on 15 August 1854. This was extended to Pundooah on 1 September 1854 and to Raniganj coal mines on 3rd February 1855. The line from Kanpur to Allahabad was opened in 1859. Mughal Sarai also appeared on the railway map of India in 1862. In 1860, the Kanpur-Etawah section was opened to traffic between 1862 and 1866; all the gaps between Howrah and Delhi were filled. In South India, first 105 km long railway line from Royapurani to Arcot was opened in 1856. In 1870, all the routes between Kolkata and Mumbai were started and in 1871, the Mumbai-Chennai route was also started.

The railway development was quite fast after 1900 and the rail network has been spread over many parts of the country. At the time of India's independence (15 August 1947) the total length of Railways was 65,217 km, out of which 10,523 km went to Pakistan and 54,694 km to India.

Originally, railways were operated by private companies owned by the British, but in 1950, the entire railway management was taken over by the Central Government. Now the entire system of railway is managed by the Railway Board, under the Ministry of Railways. The railway network has been divided into 17 railway zones which are subdivided into divisions. These divisions are the basic operating units of the Indian railways. The following are the railway zones of India:

S. No.	*Railway Zones*	*Route (km)*	*Headquarter*
1.	Central	3,905	Mumbai
2.	Western	6,182	Mumbai
3.	Southern	5,098	Chennai
4.	Eastern	2,414	Kolkata
5.	Northern	6,968	Delhi
6.	North-Eastern	3,667	Gorakhpur
7.	South-Eastern	2,631	Kolkata
8.	South-Central	5,951	Secunderabad
9.	Northeast Frontier	3,907	Guwahati
10.	East-Central	3,628	Hajipur
11.	North-Western	5,459	Jaipur
12.	East coast	2,677	Bhubaneshwar
13.	North-Central	3,151	Allahabad
14.	South-East-Central	2,447	Bilaspur
15.	South-Western	3,177	Hubli
16.	West-Central	2,965	Jabalpur
17.	Konkan Railway	741	Belapur, Mumbai

Indian railway uses four gauges: Broad Gauge (1.676 mm or 5ft 6 in) which is wider than the Standard Gauge (1.435 mm or 4ft 8½ in). The Metre Gauge is 1.000 mm (3ft 3⅜ in) and two Narrow Gauges, 762 mm (2ft 6 in) and 610 mm (2ft).

Broad Gauge is predominant gauge used by Indian Railways. The total length of broad gauge track is 1,05,000 km, which is 91 per cent of the entire track length.

Metre Gauge has about 8,000 km of track length, which is 7 per cent of the track length of all gauges.

Narrow Gauge has 2,000 km track length. The Kalka-Shimla Railway, the Kangra Valley Railway and the Darjeeling Himalayan Railway are three notable hill lines that use narrow gauge.

The total length of track used by Indian railways is about 1,15,000 km, while the total route length of network is 65,000 km. About 23,541 km or 36 per cent of the route was electrified on 31 March 2013.

UNESCO World Heritage Sites

There are two UNESCO World Heritage Sites on Indian Railways:

1. The Chatrapati Shivaji Terminus
2. Mountain Railways of India
   (i) *Kalka-Shimla Railway* in Himachal Pradesh (also featured in Guinness Book of World Records in 2003 for offering the steepest rise in altitude in the space of 96 km.
   (ii) *Darjeeling Himalayan Railway* in West Bengal.
   (iii) *Nilgiri Mountain Railway* (Metre gauge) in Tamil Nadu.

Notable Trains

- *Palace on Wheels* is a specially designed luxury tourist train service. It departs from New Delhi and covers Jaipur, Sawai Madhopur, Chittorgarh, Udaipur, Jaisalmer, Jodhpur, Bharatpur and Agra; returns to Delhi.
- *Royal Rajasthan on Wheels,* a luxury tourist train starts from New Delhi and covers Jodhpur, Udaipur, Ranthambore National Park, Khajuraho, Varanasi, Sarnath and Agra; back to Delhi
- *Maharaja Express,* a luxury train operated by IRCTC on the five circuits covering more than 12 destinations across North-West and Central India.
- *Deccan Odyssey,* luxury tourist train covers tourist centres in Maharashtra and Goa.
- *The Golden Chariot* runs on two circuits, Pride of the South and Splendour of the South
- *Mahaparinivam Express* to cover Buddhist Circuit.

Other Trains

- *Thar Express* connecting Khokhrapar (Pakistan) and Munabao (India), restarted on 18 February 2006.
- *Life Line Express,* a special train also known as 'Hospital on Wheel', which provides health care to the rural areas.
- *Vivek Express* between Dibrugarh and Kanyakumari has the longest run in terms of distance on Indian Railways network. It covers 4,286 km in about 82 hours and 30 minutes.
- *Bhopal Shatabdi Express* is the fastest train in India having a minimum speed of 150 km per hour on the Faridabad-Agra section.
- *Double Decker AC trains* have been introduced in India between Pune-Mumbai, Dhanbad and Howrah and Delhi-Jaipur route.

Pipeline Transport

Pipelines are used for transportation of liquids and gas, because these are convenient, efficient and economical. Pipelines are mostly used to transport crude oil from oilfields and port terminals and petroleum products from refineries to markets. These are also used to transport petroleum products, natural gas, milk, etc. and even solids can also be transported through pipe lines after converting them into slurry. In India, pipeline transportation is growing at a fast rate. The important pipelines of India are as follows:

1. Naharkatia-Nunmati-Barauni Pipeline
2. Mumbai High-Mumbai-Ankleshwar-Kayoli Pipeline
3. Salaya-Koyali-Mathura Pipeline
4. Hajira-Bijapur-Jagdishpur (HBJ) Gas Pipeline
5. Jamnagar-Loni LPG Pipeline
6. Kandla-Bhatinda Pipeline
7. Haldia-Kolkata Pipeline

8. Dehej-Hazira-Uran Pipeline
9. Kuchi-Mangalore-Bangalore Pipeline
10. Vishakhapatnam-Secunderabad Pipeline
11. Mundra-Delhi Pipeline

There are several other pipelines that have also been laid in different parts of the country. The Gas Authority of India Ltd. (GAIL) has drawn up an amlaitions plan to supply gas through pipeline in various parts of the country. Table 14.1 shows pipeline network in India.

Table 14.1 *Pipeline Network in India*

Pipelines	*Area (km)*	*Capacity*
16 crude pipelines	8,558	106.45 MMT
LPG pipelines	2,313	3.94 MMT
Gas pipelines	13,508	334 MMSCMD
Gas pipelines-capacity	9,300	264 MMSCMD
Gas pipeline authorized by petroleum and Natural Gas Regulatory Board	4,300	184 MMSCMD

Source: Ministry of Petroleum and Natural Gas

Water Transport

Inland Water Transport

India has about 14,500 km of navigable and potentially navigable waterways, which consist of Ganga-Bhagirathi-Hooghly rivers, the Brahmaputra, the Barak river, the rivers in Goa, the backwaters in Kerala, inland waters in Mumbai and the deltaic region of the Godwari-Krishna rivers. About 44 million tonnes of cargo is moved annually through these water ways using mechanized vessels and country boats.

India is having the following National Waterways:

National Waterway 1: Allahabad – Haldia stretch of the Ganga – Bhagirathi-Hooghly river of total 1,620 km. It was declared National Waterway–1 in year 1986.

National Waterway 2: Sadiya – Dhubri stretch of the Brahmaputra river of total length 891 km. Declared National Waterway–2 in year 1988.

National Waterway 3: Kollam – Kottapuram stretch of West coast canal and Champakara and Udyog Mandal of total length 205 km. Declared as National Waterway – 3 in year 1993.

National Waterway 4: Kakinada – Pondicherry stretch of canals and Kaluvelly tank. Bhadrachalan – Rajahmundrry stretch of River Godavari and Vazirabad-Vijayawada stretch of river Krishna of total length 1,095 km. Declared as National Waterway – 4 in year 2008.

National Waterway 5: Talcher – Dharma stretch of rivers, Geokali – Charbatia stretch of East coast canal, Charbatia-Dhamara stretch of Matai river and Mahanadi delta rivers of total length 620 km. Declared National Waterway in 2008.

National Waterway 6: Lakhipur-Bhanga stretch of 121 km of the Barak River. It was accepted as National Waterway in January 2013.

The Central Inland Water Transport Corporation (CIWTC): Constituted in 1967, it has its headquarter at Kolkata. It is engaged in transportation of goods by inland waterways in the Ganga, Brahmaputra, Hugli and Sunderban regions. Regular cargo service is operating between Kolkata and Karimganj (Assam), Kolkata and Bangladesh and Haldia and Patna.

Inland Waterways Authority of India (IWAI): Came into existence on 27 October 1986 for the development and regulation of inland waterways for shipping and navigation.

Shipping

India has a long history of shipping. In ancient times also Indian maritime trade was flourished. Since independence, India has made a significant progress in shipping; at present nearly 95 per cent of the total trade volume is moved by shipping. Today, India has the largest merchant shipping flat among the developing countries and ranks 17th amongst the countries with largest cargo carrying flat.

Coastal Shipping is an important aspect of water transport, which involves movement of goods and passengers from one part to another part within the country. In India, coastal shipping is prevalent between Kandla-Mumbai, Mumbai-Marmagao, Mumbai-Kochi, Kochi-Nagappattinam, Nagappattinam-Chennai, Chennai-Vishakhapatnam, Chennai-Kolkata, Vishakhapatnam-Kolkata, etc. The 12 major (now 13, with inclusion of Port Blair in the Andaman and Nicobar island) ports are having coastal shipping facility between them. There are 200 minor ports in India.

The major sea-ports of India are located on Eastern and Western coasts of India. These are:

Ports of Eastern Coast: Kolkata, Paradip, Vishakhapatnam, Chennai, Tuticorin and Ennore.

Ports of Western Coast: Kandla, Mumbai, Marmagaon, New Mangalore, Cochin and Jawaharlal Nehru (renamed as a Nhava Sheva Port) (14 km South of Mumbai).

Port Blair Port is located in the Andaman and Nicobar Island. The largest port of India is the Jawaharlal Nehru (Nhava Sheva) Port, while the largest natural port is Vishakhapatnam. Kandla port is considered as a tidal port.

Air Transport

Air transport is the fastest means of transport not only for international travel but also for movement within a vast country like India. In India, air transport was started in 1911, between Allahabad and Naini over a distance of 10 km. The British, French and Dutch introduced air transport in 1929–30. India National Airways was formed in 1933 and it introduced air service between Karachi and Lahore (Now in Pakistan). By the end of World War-II, major cities Mumbai, Delhi, Kolkata and some other places were provided with air service. At the time of independence of India, four companies were in operation and in 1951 four more companies joined the service. In 1953, Air Transport was nationalized and two corporations Air India International and Indian Airlines were formed. In 1992, government had introduced on 'Open Sky Policy' for cargo. On 1 March 2007, the Government of India approved the merger of Air India and Indian Airlines.

Airports Authority of India (AAI) was constituted by an Act of Parliament and came into being on 1 April 1995 by merging erstwhile National Airports Authority and International Airports Authority of India. Now Airports Authority of India is a single organization entrusted with the responsibility of creating, upgrading, maintaining and managing civil aviation infrastructure both on ground and air space in the country.

At present following are the international airports in India:

- Chatrapati Shivaji International Airport, Mumbai
- Indira Gandhi International Airport, Delhi
- Anna International Airport, Chennai
- Subhash Chandra Bose International Airport, Kolkata
- Goa Airport, Vasco De Gama
- Thiruvananthapuram International Airport

- Sardar Vallabhbhai Patel International Airport, Ahmedabad
- Sri Guru Ram Dass Jee International Airport, Amritsar
- Bengaluru International Airport, Bengaluru
- Guwahati, Lokpriya Gopinath Bordolio International Airport
- Cochin International Airport
- Rajiv Gandhi International Airport, Hyderabad
- Calicut Airport, Calicut
- Srinagar Airport, Srinagar
- Port Blair Airport (VirSavarkar International Airport)
- Nagpur Airport (Dr Babasaheb Ambedkar International Airport)

Besides these international airports, there are 92 domestic airports and 23 civil enclaves at defence airfields.

Air India is the India's national flag carrier after merging with Indian Airlines in 2011 and plays a major role in connecting India with rest of the world. Indigo, Jet Airways, Air India, Spice Jet, Go Air are the major carriers. These airlines connect more than 80 cities across India and also operate overseas routes after the liberalization of Indian aviation. Several other foreign airlines connect Indian cities with other major cities across the globe.

Pawan Hans Helicopters Limited is a public sector company that provides helicopter services to ONGC to its off-shore locations and also to various state governments in India particularly in North-East India.

There are 45 heliports in India (as on 2013). India also has the world's highest helipad at the Siachen Glacier, a height of 6,400 metre above the mean sea level.

The subsidiaries of Air India are – Air India Cargo, Air India Regional, Air India Express, Air India Charters Ltd. and Air India Air Transport Services.

The Air India's short-haul routes mainly include domestic cities and cities in South-East Asia and South-West Asia while the long-haul routes destinations in East Asia, North America and Europe. India had bilateral air service agreements with 104 countries.

There are several private airlines in operation providing air service to various parts of India. A few important airlines are Blue Dart Aviation, Deccan 360, Deccan Aviation, Go Air, Indigo, Invasion Air, Jagson Airlines, Jet Airways, Jet Lite, Jet Konnect, Kingfisher Airlines, Spice Jet, Quic Jet Cargo Airlines Private Ltd., etc.

List of National Highways of India (with New Numbers)

The List of National Highways in India by Highway Number is a listing of National Highways of India sorted by highway number. On 28 April 2010 the Ministry of Road Transport and Highways officially published a new numbering system for the national highway network in the Gazette of the Government of India. It is a systematic numbering scheme based on the orientation and the geographic location of the highway. This was adopted to ensure more flexibility and consistency in the numbering of existing and new national highways.

According to the new system, all north-south oriented highways will have even numbers increasing from the east to the west. Similarly, all east-west oriented highways will have odd numbers increasing from the north to the south of the country. In the new scheme, the single longest national highway will be NH 44 which will run from Srinagar to Kanyakumari.

- The *NH No.* is the new highway number (e.g. 44 for NH 44).
- The *States* column shows the states through which the highway runs.
- The *Length* column is the total length of the highway in kilometres.
- The *Route* column shows major villages/towns/municipalities along the route. The previous highway numbers are shown in parentheses as well as at List of National Highways in India by old highway number.

NH No.	*States*	*Length (km)*	*Route*
1	J&K	535	(NH1A) Uri, Baramula, (NH1D) Srinagar, Kargil, Leh
2	Assam, Nagaland, Manipur, Mizoram	1,214	(NH37) N 15 near Dibrugrah, Sibsagar, (NH61) Jhanji, Amguri, Mokokchung, Wohka, (NH39) Kohima, (NH150) Imphal, Churachandpur, (NH54) Seling, Serchhip, Lawngtla, Tuipang
102		107	(NH39) NH2 near Imphal, Moreh
202		460	(NH155) NH2 near Mokokchung, Tuensang, Sampurre, Meluri, (NH150) Jessami, Ukhrul, NH2 near Imphal
302		10	(NH54A) NH2 near Therait, Lunglei
502		23	(NH54B) NH2 near Venus Saddle, Saiha
3	Punjab, Himachal Pradesh	427	(NH1) Atari, Amritsar, (NH70) Jalandhar, Hoshiarpur, Naduan, Hamirpur, Toni Devi, Awa Devi, (NH21) Mandi, Kullu, Manali
103		58	(NH88) NH3 near Hamirpur, Bhota, Ghumarwain, N 154 near Ghaghas
303		60	(NH20A) N 154 near Nagrota, (NH88) Daulatur, Ranital, Jawalamukhi, NH3 near Nadaun
503		70	(NH88) N 154 near Mataur, Kangra, (NH20A) Ranital, Dera Gopipur, NH3 near Mubarakpur
703		144	(NH71) N 44 near Jaland, Nakodar, Mahatpur, Moga, Badhni, NH7 near Barnala
4	Andaman & Nicobar	230	(NH223) Mayabandar Port Blair, Chiriyatapu
5	Punjab, Chandigarh, Haryana, Himachal Pradesh	637	(NH95) Firozepur, Moga, Jagraon, Ludhiana, (NH21) Kharar, (NH22) Chandigrah, Kalka, Solan, Shimla, Theog, Narkanda, Rampur, Chini, Shipkila
105		67	(NH21A) NH5 near Pinjore, Baddi, Nalagarh, N 205 near Swarghat
205		183	(NH21) Kharar Ropar, Swarghat, (NH88) Nauni, Darlaghat, NH5 near Shimla
6	Meghalaya, Mizoram	507	(NH40) N 37 near Jorabat, (NH44) Shillong, (NH53) Badarpur, (NH154) Panchgram, (NH54) Kolasib, Kanpui, Aizawl, NH2 near Selling
106		82	(NH44E) Shillong, Nongstoin
206		126	(NH40) Jowai, Dauki, Mylliem
306			(NH54) Kolasib, Silchar
7	Punjab, Chandigarh, Haryana, Himachal Pradesh, Uttarakhand	770	(NH10) Fazilka, (NH15) Abohar, Malaut, (NH64) Bathinda, Barnala, Sangrur, Patiala, Rajpura, (NH73) Panchkula, Raipur Rani, (NH72) Narayangarh, Dhanana, Ponta-Sahib, Dehradun, (NH58) Rishikesh, Devprayag, Rudraprayag, Karnaprayag, Chamoli, Badrinath, Mana

Cont'd...

...Cont'd

107		74	(NH109) Rudraprayag, Guptkashi Phata, Gaurikund
307		46	(NH72A) Dehradun, Mohand, Biharigarh, Chhutmalpur
507		109	(NH123) Harbatpur, Vikasnagar, Kalsi, Barkot,
707		151	(NH72B) Paonta Sahib, Rajban, Shillai, Minas, Minus-Tiuni, Hatkoti
907		60	(NH73A) Paonta Sahib, Darpur, Ledi, Mustafabad, Jagadhri, Yamuna Nagar
8	Assam, Tripura	371	(NH44) Karimganj, Patharkandi, Churaibari, Ambasa, Teliamura, Agartala, Udaipur, Sabrum Indo
108			(NH44A) Namu, Mamit, Lengpui, Sairang, Aizawl
9	Punjab, Haryana, Delhi, Uttar Pradesh, Uttarakhand	811	(NH10) Malaut, Dabwali, Sirsa, Fatehabad, Hisar, Hansi, Rohtak, Bahadurgarh, (NH24) Delhi, Ghaziabad, Moradabad, (NH87) Rampur, Bilaspur, (NH74) Rudrapur, (NH125) Sitarganj-Khatima, Tanakpur, Pithoragarh
109		284	(NH87) Rudrapur, Haldwani, Nainital, Bhowali, Almora, Ranikhet, Dwarahat, Chaukhutiya, Gairsaing, Aobadri, Karnaprayag
309		287	(NH74) Rudrapur, (NH121) Kashipur, Ramnagar, Dhumakot, Thalisain, Tripalisan, (NH119) Bubakhal, Pauri, Srinagar
509	UP	239	(NH93) Moradabad, Chandausi, Babrala, Aligarh, Agra
709	Haryana	68	(NH71A) Rohtak, Gohana, Panipat
10		174	(NH31) Siliguri, (NH31A) Sivok, Kalimpang, Gangtok, (NH310) border
110		76	(NH55) Siliguri, Karsiyang, Darjeeling
11	Rajasthan	495	(NH15) Jaisalmer, Pokaran, (NH11) Bikaner, Sri Dungargarh, Ratangarh, Fatehpur
12	WB	612	(NH34) Dalkola, Raiganj, Gajol, Maldah, Farakka, Morgram, Baharampur, Krishananagar, Ranaghat, Barasat, (NH117) Kolkata, Kakdwip, Bok-Khali
112		59	(NH35) Barasat, Gaighata, Bangaon, border
13		1,150	(NH229) Tawang, Jang, Sela Lake, Baisakhi, Senge, Mohan Camp, Dirang, Dangsing, Bomdila, Tenga Valley, Kimi, Palizi, Seppa, Sagalee, Midpu, Hoz, Yazali, Ziro Town, Daporijo, Bam, Along, Biru, Pangin, Pasighat (NH–52).
14	WB	306	(NH60) Morgram, Rampur Hat, Siuri, Raniganj, (NH60A) Bankura, Garhbeta, Salbani, Kharagpur
114	WB	109	(NH2B Ext) Mallarpur, Mayureswar, Prantik, (NH2B) Bolpur, Bhedia, Guskhara, Talit, Barddhaman
314	WB	87	(NH60A) Bankura, Puruliya
15	Assam	664	(NH52) Baihata-Charali, Mangaldai, Dhekiajui, Tezpur, Banderdeva, North Lakhimpur, (NH52B) Kulajan, (NH37) Dibrugarh, Tinsukia, (NH52) Rupai, Mahadevpur, Wakro
115		64	(NH37) Dum Duma, Saikhoaghat, Kundil Bazar, Roing
215	Assam		(NH52B) Mahadevpur, Namchik, Changlang, Khonsa, Kanubari, Dibrugarh
315		111	(NH153) Makum, Ledo, Lekhapani, border
415		59	(NH52A) Ghopur, Itanagar, Daimukh, Banderdeva
515		111	(NH52) Kulajan, Jonai, Pasighat
715	Assam	197	(NH37A) Tejpur, (NH37) Kaliabor, Jakhalabandha, Bokakhat, Jorhat, Jhanji

Cont'd...

...Cont'd

16	WB, Orrisa, AP, Tamil Nadu	1,659	(NH6) Kolkata, (NH60) Kharagpur, (NH5) Baleshwar, Bhubaneshar, Vishakhapatnam, Vijayawada, Nellore, Chennai
116		52	(NH41) Kolaghat, Haldia Port
216		397	(NH214) Kathipudi, Kakinada, (NH214A) Machilipatriam, Ongole
316	Odisha		(NH203) NH16 near Bhubaneshwar–Puri–Konark; (NH203A) Puri–Satpada
516			(NH217) NH16 near Narendrapur–Gopalpur
716	AP, Tamil Nadu	147	(NH205) Chennai, Tiruttani, Reangunta
17	Assam	477	(NH31) N 10 near Sivok, Bagrakot, (NH31C) Chalsa, Nagarkata, (NH31) Goyerkata, Bispara, Falakata, Sonarpur, Koch-Bihar, Tufanganj, Golakganj, Bilasipara, (NH31B) North Salmana, (NH37) Goalpara, Boko, N 27 near Guwahati
117		14	(NH31) NH17 near North Salmara, N 27 near Bijni
217		307	(NH51) NH17 near Paikan, Tura, (NH62) Dalu, Baghmara, Rongjeng Damra, NH17 near Dudhnai
317		67	(NH31C) NH17 near Birpara, Madari Hat, Rajabaht Khawa, N 27 near Salsabari
517			NH17 near Goyerkata, N 27 near Dhupgari
717		41	NH17 near Chalsa, N 27 near Mainaguri
18	Jharkhand, WB	359	(NH32) NH19 near Govindpur, Dhanbad Cas, Puruliya, Balarampur, (NH33) Chandil, Ghatshila, (NH6) Baharagora, (NH5) junction with N 49, Baripada, Betnoti, NH16 near Baleshwar
118	Jharkhand	5	NH18 near Asanbani, Jamshedpur
19	Delhi, UP, Bihar, WB	1,435	(NH2) Delhi, Mathura, Agra, Kanpur, Allahabad, Varanasi, Mohania, Aurangabad, Bobhi, Barhi, Bagadar, Gobindpur, Asansol, Palsit, Kolkata
119	Bihar	75	(NH2C) NH19 near Dehri, Jadunathpur, border Bihar/UP
319	Bihar	117	(NH30) N 19 near Mohania, Dinara, Charpokhari, N 922 near Ara
519		28	(NH2A) N 19 near Sikandara, N 27 near Bhognipur
719	UP, MP	124	(NH92) N 19 near Etawah, Bhind, N 44 near Gwalior
919	Haryana	74	(NH71B) N 19 near Palwal, Sohna, Dharuhera, N 352 near Rewari
20	Bihar, Jharkhand	658	(NH31) N 31 near Bakhtiyarpur, Bihar Sharif, Nawada, Rajauli, Kodarma, (NH33) Barhi, Hazaribag, (NH75) Ranchi, Khunti, Murhu, Chakradharpur, Chaibasa, Jaintgarh, Parsora, (NH215) Kendujhargarh, NH16 near Panikholi
120	Bihar	92	(NH82) N 20 near Bihar Sharif, Nalanda, Rajgir, Hisua, N 22 near Gaya
320		86	(NH23) N 20 near Ramgarh, Gola, N 18 near Chas
520		105	(NH125) Parsora, N 143 near Rajamundra
21	Rajasthan, UP	262	(NH11) Jaipur, Dausa, Bharatpur, Agra
22	Bihar	416	(NH77) border near Sonbarsa, Sitamarhi, Muzaffarpur, (NH19) Hajipur, (NH83) Patna, Punpun, Gaya, Bodh Gaya, (NH99) Dobhi, Hunterganj, Chatra, N 39 near Chandwa
122	Bihar	101	(NH28) N 22 near Muzaffarpur, Dholi, Mushrigharari, N 31 near Barauni
322	Bihar	58	(NH103) N 22 near Hazipur, N 122 near Mushrigharari
522	Bihar	119	(NH100) N 22 near Chatra, Hazaribagh, N 19 near Bagodar
722	Bihar	74	(NH102) N 22 near Muzaffarpur, Rewaghat, N 31 near Chhapra

Cont'd...

...Cont'd

922	Bihar	140	(NH30) N 22 near Patna, (NH84) Ara, Bhojpur, Buxar
23		222	(NH11A) Kothum, (NH11B) Lalsot, Karauli, Bari, Dhaulpur
24	UP	293	(NH29) Sonauli (border), Pharenda, Gorakhpur, (NH97) Ghazipur, Zamania, N 19 near Saiyad Raja
25		353	(NH112) Barmer, Kawas, Madhasar, Dhudhwa, Bagundi, Tilwara, Balotra, Pachpadra, Kalyanpur, Jodhpur, Kaparda, Bilara, Jaitaran, (NH14) Bar, Beawar
125	Rajasthan	168	(NH114) Jodhpur, Balesar, Dechu, Pokaran
26	AP	526	(NH201) N 53 near Bargarh, Barapali, Balangir, Bhawanipatna, (NH43) Boriguma, Koraput, Salur, Vizianagaram, NH16 near Natavalsa
27	MP, UP, Bihar, Assam	3,507	(NH112) Porbandar, (NH8A) Bamanbore, Morvi, (NH15) Samakhiali, (NH14) Radhanpur, Palanpur, (NH76) Pindwara, Udaipur, Mangarwar, Chittaurgarh, Kota, Baran, Shivpuri, Ganj, (NH25) Jhansi, (NH128) Kanpur, Lucknow, (NH28) Faizabad, Gorakhpur, Gopalganj, Pipra, Kothi, (NH57) Muzaffarpur, Darbhanga, Forbesganj, Araria, (NH31) Purnia, Dalkola, Islampur, (NH31D) Shiliguri, Jalaiguri, (NH31) Mainaguri, Dhupgar, Falakata, (NH31D) Sonapur, (NH31C) Salsaguri, (NH31) Bongaigaon, Bijni, Patacharkuchi, Nalbari, (NH37) Guwahati, Dishpur, (NH36) Nagaon, (NH54) Dobaka, Lumding, Haflong, Silchar
127		42	(NH37) N 27 near Nagaon, Samaguri, N 715 near Jakhalabandha
127A		38	(NH152) Pathsala–Bhutan border
227	Bihar	214	(NH104) N 27 near Chakia, Narhar, Pakri Bridge, Madhuban, Shivhar, Sitamarhi, Harlakhi, Umgaon, Jaynagar, Laukaha, Laukahi, N 27 near Narahia
327		32	(NH31C) N 27 near Bagdogra, Naksal Bari, Galgalia
527	Bihar	14	(NH57A) N 27 near Forbesganj, Jogbani
527B	Bihar	54	(NH105) N 27 near Darbhanga, Aunsi, N 227 near Jaynagar
527D	Bihar	69	(NH28A) N 27 near Piprakothi, Sagauli, Raxaul, Nepal border
727	Bihar	169	(NH28B) N 27 near Kushinagar, Chhitanuni Rail-cum-Road bridge, Bagaha, Lauriya, Bettiah, N 527D near Chhapwa
927	Bihar	152	(NH28C) N 27 near Barabanki, Baharaich, Nepalganj (border)
28	UP	305	(NH233) Nepal border, Naugarh Nagar, Bansi, Basti Tanda, (SH73) Azamgarh, N 31 near Varanasi
28A	Bihar	68	(NH28) Pipra, Motihari, Raxaul
28B	Bihar, UP	150	Chapwa, Kushinagar
128	UP	164	(NH232) N 28 near Tanda Ambedkar Nagar, Sultanpur, Amethi, N 30 near Rae Bareli
29		320	(NH36) N 27 near Dabaka (Sutargaon), Amlakhi, (NH39) Dimapur, (NH150) Kohima Chizam, NH202 near Jessami
129		105	(NH39) N 29 near Dimapur, Bokajan, Golaghat, N 715 near Numaligarh
30	UP, MP, CG, AP	2,010	(NH74) NH9 near Sitarganj, Pilibhit, (NH24) Bareilly, Shahjahanpur, Sitapur, (NH24B) Lucknow, (NH231) Raebareli, (NH96) Pratapgarh, (NH27) Allahabad, Mangawan, (NH7) Rewa, Katni, (NH12A) Jabalpur, Mandla, Chilpi, (NH200) Simga, (NH43) Raipur, Dhamtari, Keskal, (NH221) Jagdalpur, Konta, Nellipaka, Bhadrachalam, Paloncha, Kottagudam, Tiravuru, Mailavaram, N 65 near Kondapalle
130	CG	291	(NH200) N 30 near Simga, (NH111) Bilaspur, Kathgora, N 43 near Ambikapur
330	UP	103	(NH96) N 30 near Allahabad, Pratapgarh, Sultanpur, N 27 near Faizabad

Cont'd...

...Cont'd

530		66	(NH24) N 30 near Bareily, NH9 near Rampur
230		17	(NH24A) N 30 near Bakshi-ka-Talab, Chenhat, junction with N 27/N 731/N 30/N 27, N 30 near Bakshi-ka-Talab
31	UP, Bihar	968	(NH232A) N 27 near Unnao, (NH232) Lalganj, (NH231) Raebareli, Salon, Pratapgarh, Machhlishahr, (NH56) Jaunpur, (NH29) Varanasi, (NH19) Ghazipur, Ballia, Chhapra, (NH30) Hajipur, (NH31) Bakhtiyarpur, Mokama, Begusarai, Khagari, Bihpur, Kora, Katihar, Harishchanderpur, N 12 near Pandua
131		132	(NH106) N 31 near Bihpur, Kishanganj, Madhepura, Birpur (border)
231	Bihar	196	(NH107) N 31 near Maheshkund, Sonbarsa Raj, Simri Bakhtiyarpur, Saharsa, Madhepura, Sarsi, (NH31) Purnia, N 31 near Kora
331	Bihar	65	(NH101) N 31 near Chhapra, Baniapur, N 27 near Muhumadpur
431	Bihar	70	(NH30A) N 31 near Phatuha, Chandi, Harnaut, N 31 near Barh
531	Bihar	93	(NH85) N 31 near Chhapra, Siwan, N 27 near Gopalganj
731	UP	221	(NH56) N 31 near Jaunpur, Sultanpur, N 27 near Lucknow
32	TN	314	(NH45) N 48 near Chennai, Chengalpattu, (NH66) Tindivanam, (NH45A) Puducherry, Cuddalore, Chidambaram, Karaikal, (Puducherry), N 83 near Nagapatinam
132	TN	37	(NH45) N 32 near Tindivanam, N 38 near Viluppuram
332		38	(NH45A) N 32 near Puducherry, N 38 near Viluppuram
33	Bihar	443	(NH110) N 139 near Arwal, Jahanabad, Bandhuganj, Ekangarsarai, (NH82) Biharsharif, (NH80) Mokama, Luckeesarai, Munger, Bhagalpur, Kahalgaon, Sahibganj, Rajmahal, Barharwa, N 12 near Farakka
34	UK, UP, MP	1,426	(NH108) Gangotri Dham, Bhatwari, Uttarkashi, (NH94) Dharasu, NearTerhi, Ampata, (NH58) Rishikesh, (NH74) Haridwar, (NH119) Najibabad, Bijnore, (NH58) Meerut, (NH91) Ghaziabad, Bulandshahr, Aligarh, Etah, Kannauj, (NH86) Kanpur, Hamirpur, Mahoba, Chattarpur, (NH12A) Hirapur, Damoh, (NH7) Jabalpur, N 44 near Lakhnadon
134		94	(NH94) N 34 near Dharasu, Kuthnaur, Yamnotri
334	CG, UP	207	(NH58) N 34 near Haridwar, Roorkee, Muzaffarnagar, (NH235) Meerut, Hapur, N 34 near Bulandshar
534		151	(NH119) N 34 near Najibabad, Kotdwar, Satpauli, Bubakhal
734		108	(NH74) N 34 near Najibabad, Nagina, N 309 near Kashipur
934	MP	156	(NH86) N 34 near Hirapur, Banda, (NH26A) Sagar, Jeruwakhera, Khurai, Bina
234	UP	156	(NH91A) N 34 near Kannauj, Bela, (NH92) Etawah, Kishni, N 34 near Bhongaon
35	UP		(NH76) N 34 near KaBrai, Banda, Karwi, Mau, Allahabad, (NH7) Mirzapur, N 19 near Varanasi
135		130	(NH7) N 35 near Mirzapur, Lalganj, Drummondganj, Mauganj, N 30 near Mangawan
335		110	(NH232) N 35 near Banda, Fathepur, N 31 near Lalganj
36	TN	334	(NH45C) N 132 near Vikravandi, Panruti Vadulur, Neyveli-Township, Sethiathop, Kumbakonam, (NH226) Thanjavur, Gandarvakottai, (NH210) Pudukottai, (NH226) Tirumayam, Kilasevalpatti, Tirupattur, Madagupatti, Shivganga, N 87 near Manamadurai
136		68	(NH226) N 36 near Thanjavur, Thiruvaiyaru, Kunnam, Perali, N 38 near Perambalur

Cont'd...

...Cont'd

336	Kerala	52	(NH210) N 36 near Pudukkottai, Kiranur, N 83 near Tiruchirappalli
536		109	(NH210) N 36 near Tirumayam, Devakottai, Tiruvadanai, N 87 near Ramanathapuram
37		313	(NH53) NH2 near Imphal, Jiribam, Jirighat, Lakhipur, Silchar, (NH44) Badarpur, Karimganj, border
38	TN	604	(NH234) N 75 near Vellore, Polur, Tiruvannamalai, (NH45) Viluppuram, Ulundurpettai, Perambalur, (NH45B) Tiruchchirappali, Tovarankurichchi, Melur, Madurai, Aruppukkottai, N 138 near Tuticorin Port
138		54	(NH7A) N 38 near Tuticorin Port, N 44 near Palayankottai
39	MP, CG, Jharkhand	869	(NH75) N 44 near Jhansi, Chhatarpur, Khajuraho, Panna, Satna, Rewa, Sidhi, Dudhinagar, Garhwa, Daltenganj, Latehar, Chandwa, N 20 near Ranchi
139	Bihar	236	(NH98) N 39 near Rajhara, Chhatarpur, Hariharganj, Aurangabad, Daudnagar, Arwal, Naubatpur, N 31 near Patna
339	J&K	34	N 39 near Nowgong, N 34 near Srinagar
539		173	(NH12A) N 39 near Jhansi, Pirthipur, Tikamgarh, N 934 near Shahgarh
40		420	(NH18) N 44 near Kurnool, Nandyal, Cuddapah, Pileru, Putalapattu, Chittoor, N 48 near Ranipettai
140	AP		N 40 near Putalpattu, N 71 near Tirupati
41	Gujarat	290	(NH8A) N 41 near Samakhiyali, Gandhi Dham, Mandvi, Naliya, Narayan Sarovar
141	Gujarat	31	(NH8A) N 41 near Gandhi Dam, Kandla Port
42		337	(NH205) N 44 near Anantapur, Kadiri, (NH219) Madanapalli, Kuppam, N 44 near Krishnagiri
43	MP, CG, Jharkhand	806	(NH78) N 30 near Katni, Umaria, Shahdol, Nagar, Ambikapur, Pathalgaon, Jashpurnagar, Chhattisgarh, (NH23) Gumla, (NH33) Ranchi, N 18 near Chandil
143		247	(NH23) N 43 near Gumla, Palkot, Kolebira, Thethaitanagar, Jharkhand, Panposh, Rajamundra, N 49 near Barakot
44	J&K, Himachal Pradesh, Punjab, Haryana, Delhi, Uttar Pradesh, Madhya Pradesh, Maharashtra, Telangana, Andhra Pradesh, Karnataka, Tamil Nadu	3,745	(NH1A) Srinagar, Banikul, Jammu, Pathankot, Mukerian, (NH1) Jalandar, Ludhiana, Pajpura, Ambala, Karnal, Panipat, (NH2) Delhi, Faridabad, Palwal, Mathura, (NH3) Agra, Dhaulpur, Morena, (NH75) Gwalior, Datia, (NH26) Jhansi, Lalitpur, Sagar, Narsinghpur, (NH7) Lakhnadon, Seoni, Nagpur, Jamb, Pandha-kawada, Adilabad, Nirmal, Ramayampet, Hyderabad, Kurnool, Gooty, Anantapur, Penukonda, Devanahalli, Bengaluru (Bangalore), Hosur, Krishnagiri, Dharmapuri, Salem, Kanniyakumari
244		246	(NH1B) N 44 near Khanabal, Symthanpass, Kishtwar, Doda, N 44 near Batote
144	J&K	15	(NH1C) N 44 near Domel, Katra
344	Punjab	173	(NH72) N 44 near Ambala, (NH73) Dhanana, Saha, Yamunanagar, Saharanpur, N 334 near Roorkee
544	TN	332	(NH47) N 44 near Salem, Coimbatore, Palakkad, Thrissur, N 66 near Ernakulum
744		232	(NH208) N 44 near Tirumangalum, Srivilliputtur, Rajapalaiyam, Tenkasi, Puliyur, Punalur, Kottarakara, N 66 near Kollam
944		23	(NH47B) N 44 near Kavalkinaru, Aralvaymozhi, N 66 near Nagerkoil
45	MP	280	(NH12) N 46 near Obaidullaganj, Bareli, Tendukheda, N 30 near Jabalpur

Cont'd...

...Cont'd

46	MP	604	(NH3) N 44 near Gwalior, Shivpuri, Guna, (NH12) Biora, Bhopal, (NH69) Obeddullaganj, Hoshangabad, N 47 near Betul
146	MP	188	(NH86) N 46 near Bhopal, Vidisha, N 44 near Sagar
47	Gujarat, MP	1,080	(NH8A) N 27 near Bamanbore, Limbdi, (NH59) Ahmedabad, Godhra, Dahod, (NH59A) Indore, (NH69) Betul, Saoner, N 44 near Nagpur
147	Gujarat	50	(NH8C) N 47 near Sakhej, Gandhinagar, N 48 near Chilloda
48	Delhi, Haryana, Rajasthan, Gujarat, Maharashtra, Karnataka, Tamil Nadu	2,807	(NH8) Delhi, Bawal, Kotputli, Jaipur, (NH79A) Kishangarh, (NH79) Nasirabad, Bhiwara, (NH76) Chittaurgarh, (NH8) Udaipur, Ahmedabad, Vadodara, Ankleshwar, (NH3) Mumbai, (NH4) Thane, Pune, Satara, Kolhapur, Belgam, Chitradurga, (NH7) Bengaluru (Bangalore), (NH46) Krishnagiri, Vellore, (NH4) Arcot, Chennai
148		105	(NH11A) N 48 near Manoharpur, Dausa, N 23 near Lalsot
248	Rajasthan	53	(NH11C) old NH8 through Jaipur from km220 to km 273.50
348		28	(NH4B) N 48 near Palspe, Jawaharlal Nehru Port
448	Rajasthan	57	(NH8) N 48 near Kishangarh, (NH79) Ajmer, N 48 near Nasirabad
548		5	N 66 near Kalamboi, N 348 at km 16.687
648	Karnataka, Tamil Nadu	143	(NH207) N 48 near Nelamangla, Dodaballapur, Devenhalli (NH44), Sarajpur, Bagalur, N 48 near Hosur
748	Karnataka, Goa	160	(NH4A) N 48 near Belgaum, Anmod, Panda, N 66 near Panaji
948	Karnataka, Tamil Nadu		(NH209) N 48 near Bengaluru, Kanakapura, Malvalli, Kollagal, Chamarajnagar, N 544 near Coimbatore
49	CG, Jharkhand, WB	706	(NH200) N 130 near Bilaspur, Pamgarh, Raigarh, Kanaktora, Jharsuguda, Kuchinda, Pravasuni, (NH6) Deogarh, Barakot, Palalaharba, KenduJhargarh, Bangriposchi, Baharagora, NH16 near Kharagpur
149			(NH23) N 49 near Palalaharha–Talcher, N 55 near Nuahata
50	Karnataka	544	(NH218) N 65 near Homanabad, Gulbarga, Jevargi, (NH13) Bijapur, Hospet, N 48 near Lakshraisagara
150	Karnataka	137	N 50 near Gulbarga, Yadagiri, N 167 near Devasuguru
150A	Karnataka	618	N50 near Jewargi, Siruguppa, Bellary, Hiriyur, Chikkanayakanahalii, Nagamangala, Srirangapatna, Mysore, Nanjangud, NH–948 near Chamarajnagar
150E	Karnataka, Maharashtra	105	NH–50 near Gulbarga, Chowdapur, Afzalpur, NH–52 near Solapur
51	Gujarat	522	(NH8E) Dwarka, Bhogat, Porbandar, Navibander, Shil, Mangrol, Somnath, Kodinar, Una, Mahuva, Talaja, Bhavnagar
151		99	(NH8D) N 51 near Gadu, Vanthali Junagadh, N 27 near Jetpur
52	Punjab, Haryana, Rajasthan, MP, Maharashtra	2,317	(NH71) Sangrur, (NH65) Narwana, Hisar, (NH11) Fatehpur, (NH12) Jaipur, Tonk, Kota, Aklera, Rajgarh, (NH3) Biora, Dewas, Indore, Sendhwa, (NH211) Dhule, Aurangabad, Beed, Osmanabad, (NH13) Solapur, (NH218) Bijapur, (NH63) Hubli, Ankola
152	Haryana	155	(NH65) N 52 near Narwana, Kaithal, (NH22) Ambala, NH7 near Panchkula
352	Haryana	218	(NH71) N 52 near Narwana, Jind, Rohtak, Jhajjar, Rewari, N 48 near Bawal
552	Rajasthan	75	(NH116) N 52 near Tonk, Uniara, Sawai Madhopur
752		92	(NH90) Aklera, Atru, Baran

Cont'd...

...Cont'd

53	Gujarat, Maharashtra, CG	1,781	(NH6) Hajira, Surat, Uchchhal, Dhule, Jalgaon, Akola, Amravati, Nagpur, Bhandara, Deori, Rajnandgaon, Durg, Raipur, Arang, Saraipali, Bargarh, Sambalpur, (NH200) Deograh, Dubri, Chandhikhol, (NH5A) Haridaspur, Paradip Port
153		39	(NH216) N 53 near Saraipali, N 49 near Sarangarh
353		146	(NH217) N 53 near Ghorai, Mahasamund, Bagbahra, Nauparha, N 59 near Khariar
54	Punjab	340	(NH15) Panthankot, Gurdaspur, Amritsar, Zira, Faridkot, (NH64) Bathinda, NH9 near Dabwali
154		279	(NH20) N 54 near Pathankot, Nurpur, Palampur, Jogindarnagar, (NH21) Mandi, Sundar Nagar, Ghaghas, Bilaspur, N 205 near Nauni
55	CG	264	(NH42) N 53 near Sambalpur, Redhakhol, Angul, Nuahata, Dhenkanal, NH16 near Cuttack
56		290	(NH79) N 27 near Chittaurgarh, (NH113) Nimbahera, Pratapgarh, Banswara, Jhalod, Umbi, N 47 near Dahod
156	MP	10	(NH79) N 56 near Nimbahera, Border to MP
57		297	(NH224) N 26 near Balangir, Sonapur, Bauda, Dashapalla, Nayagarh, NH16 near Khordha
58	Rajasthan	620	(NH65) Fatehpur, Ladnun, (NH89) Nagaur, Merta City, (NH8) Ajmer, Beawar, Devgarh, Udaipur
59		352	(NH217) N 353 near Khariar, Titlagrah, Lankagarh, Baligurha, Sorada, Asika, NH16 near Brahmapur
60	Maharashtra	395	(NH3) N 53 near Dhule, (NH50) Nashik, Sangammer, Ale, N 48 near Pune
160	Maharashtra	149	(NH3) N 60 near Nasik, N 48 near Thane
61	Maharashtra	652	(NH222) N 160 near Bhiwandi, Kalyan, Murbad, Ale, Ahmadnagar, Shevgaon, Gevrai, Majalgaon, Parbhani, Nanded, Bhokar, N 44 near Nirmal
62	Rajasthan	748	(NH15) Abohar, Ganganagar, Suratgarh, Lunkaransar, (NH89) Bikaner, (NH65) Nagaur, Jodhpur, (NH14) Pali, Sirohi, Pindwara
162		89	(NH14) Pali, Bar
63	AP	560	(NH16) Nizamabad, Metpalli, Mancheral, Chinnur, Kotapad, N 26 near Boriguma
163	AP	344	(NH202) N 63 near Bhopalpatnam, Venkatapuram, Warangal, N 44 near Hyderabad
64	Gujarat	332	(NH8) N 48 near Ahmedabad, (NH228) Anand, Dandi
65	Maharashtra, AP	926	(NH9) N 48 near Pune, Indapur, Solapur, Umarga, Homnabad, Hyderabad, Suryapet, Vijayawada, Vuyyuru, Pamarru, N 216 near Machllipatnam
165		107	(NH214) N 65 near Pamarru, Mandavalli, Pallevada, Digamarru, N 216 near Narsapur
66	Maharashtra, Goa, Karnataka, Kerala, Tamil Nadu	1,593	(NH17) N 48 near Panvel, Indapur, Mahad, Rajapur, Kudal, Panaji, Margao, Karwar, Honavar, Udupi, Mangalore, Kasaragod, Kannur, Ponnani, Edappalli, (NH47) Ernakulam, Thiruvananthapuram, N 44 near Kanniyakumari
166	Maharashtra	149	(NH204) Ratnagiri, Tink, Pali, N 48 near Kolhapur
366		10	(NH17A) N 66 near Cortalim, Vasco, Monnugao Port
566			(NH17B) N 748 near Ponda, Verna, N 366 near Vasco
766	Kerala, Karnataka	268	(NH212) N 66 near Kozikode, Kalpetta, Gundlulpet, Mysore, N 948 near Kollegal

Cont'd...

...Cont'd

966		122	(NH213) N 66 near Ferokh, N 544 near Palakkad
966A		15	(NH47C) N 544 near Kalamassery, N 66, Vallarpadam
966B		8	(NH47A) N 66, Willington Island
67	Karnataka, Andhra Pradesh	300	(NH63) N 48 near Hubli, Gadag, Koppal, Hosepet, Bellary, N 44 near Gooty
167	Karnataka, Telangana	264	N 67 near Bellary, Raichur, Makhtal, N 44 near Badepalli
367	Karnataka	157	NH67 near Bhanapur, Kukunur, Yelburga, Gajendragad, Badami, Guledagudda, Bagalkot, NH52 near Gadankeri
68	Rajasthan	429	(NH15) Jaisalmer, Barmer, Sanchor, Tharad, Bhabhar, Radhanpur
69	Karnataka, Andhra Pradesh	625	(NH206) N 66 near Honnavar, Shimoga, (NH234) Banavar, Hulyar, Sira, Madhugiri, Chintamani, (NH4) Mulbagal, Palmaner, N 40 near Chittor
169	Karnataka	215	(NH13) N 69 near Shimoga, Tirthahalli, Kopa, Karkal, N 66 near Mangalore
169A	Karnataka	87	NH–169 near Thirthahali, Agumbe, Hebri, NH–66 near Udupi
369	Karnataka	105	(NH13) N 69 near Shimoga (Bhadravati), Channagiri, Holalkere, N 48 near Chitradurga
71	AP	108	(NH205) N 42 near Madanpalle, Pileru-Tirupati, N 716 near Renigunta
73	Karnataka	316	(NH234) N 66 near Mangaluru, (NH206) Banavara, Arsikere, N 48 near Tumkur
173	Karnataka	72	NH73 near Mudigere, Chikmagalur, NH69 near Kadur
75	Karnataka, Tamil Nadu	533	(NH48) N 73 near Bantval, Hassan, (NH4) Nelamangala, Bengaluru (Bangalore), Kolar, (NH234) Mulbagal, Venkatagirikota, Pernampet, Gudiyattam, Katpadi, N 48 near Vellore
275	Karnataka	367	NH–75 near Bantwal, Puttur, Sullia, Madikeri, Pertiyapatna, Mysore, Srirangapatna, Mandya, Ramanagaram, NH75 near Bengaluru (Bangalore)
77	TN	177	(NH66) N 48 near Krishnagiri, Uttangarai, Tiruvannamalai, Gingee, N 32 near Tindivanam
79	TN	137	(NH68) N 44 near Salem, Attur, N 38 near Ulundurpettai
81	TN	340	(NH67) N 544 near Coimbatore, Palladam, Karur, Krishnarayapuram, (NH227) Tiruchirappalli, Lalgudi, Kallakudi, Kizhapalur, Udaiyarpalaiyam, Jayamkondacholapuram, Gangaikondacholapuram, Kattumannarkoil, Komarakshi, N 32 near Chidambaram
181	TN	211	(NH67) N 81 near Coimbatore, Mettuppalaiyam, Udagamandalam (ooty), Gudalur, N 766 near Gundulpet
83	TN	394	(NH209) N 544 near Coimbatore, Pollachi, (NH45) Dinigul (NH67) Tiruchchirappalli, Thanjavur, N 32 near Nagappattinam
183		322	(NH45) N 83 near Dindigul, (NH220) Teni, Uttamapalayam, Nadu, Kumily, Pirmed, Kanjirappali, Kottayam, Adur, N 744 near Kottarakara
85	TN	413	(NH49) N 66 near Kochi, Ernakulam, Muvattupula, Devikulam, Teni, Madurai, (NH230) Tiruppuvanam, Povandhi, Sivaganga, Kalaiyarkoil, Tiruvadanai, Todni point
86	MP		Bhopal, Sehore, Dewas
87	TN	154	(NH49) N 38 near Tiruppuvanam, Manamadurai, Ramanathapuram, Rameswaran, Dhanushkodi

15

International Trade

Trade means exchange of goods and services and when it involves more than one country it is known as 'international trade' or 'foreign trade'. International trade arises because of (i) differences in available resources (ii) differences in stage of industrial development, (iii) differences in people's culture and skill, (iv) differences in transportation facilities, and (v) differences in geographical environment. International trade is mutually beneficial as no country is self-sufficient. India has a long tradition of international trade with other South-east Asian, West Asian and East Asian countries since ancient times. With the establishment of British rule in India, its trade link with Britain became strong. India used to export raw materials such as cotton, jute, tea, iron ore, etc. to Britain and import manufactured goods, machinery, medicines, etc. from that country, in other words, it was fully controlled by Britishers. But after independence, the entire scenario has been changed and the international trade is done according to the country's need and policy of trade. India's international trade has undergone a sea change during the last six decades in terms of volume, composition as well as direction. Although India's contribution in world trade in quite low, yet it plays a significant role in the world economy. A brief review of India's international trade and its changing pattern is given here.

India: Foreign Trade Policy

Although, India has steadily opened up its economy, its traffic continues to be high when compared with other countries and its investments norms are still restrictive. This leads some to see India as a 'rapid globalizer', while others still see it as a highly 'protectionist'.

Till early 1990s, India was a closed economy and restrictions on imports were extensive. The country began to cautiously reform in 1990s, liberalizing only under conditions of extreme necessity. Since that time trade reforms have produced remarkable results. India's trade to GDP ratio has increased from 15 per cent to 35 per cent of GDP between 1990 and 2005.

In recent years, the government's stand on trade and investment policy has displayed a marked shift from protecting 'producers' to benefiting 'consumers'. This is reflected in its Foreign Trade Policy for 2004–09 which states that 'For India to become a major player in world trade … . We have also to facilitate those imports which are required to stimulate our economy'. For every year foreign trade policy has been declared by the Ministry of Commerce and Industry. A detailed framework has been given in such trade policies for import and export.

In all the foreign trade policies emphasis has been given for export promotion. Specific steps have been taken in this direction and achievements made in recent years. Trade policy reforms like specification and streamlining the procedures and addition of items in the Special Import License (SIL) to improve premium and incentive to exports, have created a free environment of trade, strengthened export production base and removed procedural irritants; imports are being gradually liberalized to facilitate flow of raw materials and inputs to the export.

India has become a founder member of World Trade Organization (WTO) which helps in the exports of agricultural and labour-intensive products and services. The regional and bilateral initiatives include comprehensive agreement with the European Commission, Indo-US Commercial Alliance, Operationalization of SAARC, Preferential Trading Agreements (SAPTA) and efforts to promote an India Ocean Rim Block. Country-specific Initiatives have been taken to promote trade with several countries and adjoining countries like Bangladesh, Nepal, Bhutan, Sri Lanka, Myanmar, etc.

India has recently signed trade agreements with its neighbours and in seeking new ones with the East Asian countries and the United States. Its regional and bilateral trade agreements or variants of them are:

- Indo-Sri Lanka Free Trade Agreement
- Trade Agreements with Bangladesh, Bhutan, Sri Lanka, Maldives, China and South Korea
- Indo-Nepal Trade Treaty
- Comprehensive Economic Cooperation Agreement (CECA) with Singapore
- Framework Agreements with the Association of South East Asian Nations (ASEAN), Thailand and Chile.
- Preferential Trade Agreements with Afghanistan, Chile, and Mercour (the latter is a trading zone between Brazil, Argentina, Uruguay and Paraguay). In fact, India is having bilateral agreements with many of the European, American, African countries besides countries of Asia, Australia, Russia, New Zealand, etc.

During the last twenty years, several incentives/schemes have all been given to promote international trade of India. Some notables are:

- The Export Promotion Capital Goods Scheme
- The Duty Exemption Scheme
- Export Oriented Unit Scheme (EOU)
- Export Processing Zones (EPZ)
- Electronic Hardware Technology Parks (EHTP)
- Software Technology Parks (STP)
- The Special Economic Zones (SEZ)
- Agri-Export Zones, etc.

India's Export, Import and Balance of Trade

The nature of India's foreign trade has changed over the years. Though there has been an increase in the total volume of export and import, the value of import continued to be higher than that of export, thus the trade balance is always negative. Table 15.1 shows the total export, import and balance of trade.

Table 15.1 *India's Exports Import and Balance of Trade*

Year	*Values in Rs '00 Crores*			*Percentage Growth*		
	Export	*Import*	*Balance of Trade*	*Exports*	*Imports*	*Balance of Trade*
2008–09	8,407.55	13,744.36	–5,336.81	28.19	35.77	49.72
2009–10	8,455.34	13,737.36	–5,182.02	0.57	–0.78	–2.90
2010–11	11,429.22	16,834.67	–5,405.45	35.17	23.45	4.31

Not only of the years shown in table but during the last ten years or even more, there has been an increase in trade deficit. This deficit is attributed to the price rise.

Export Merchandize/Commodities

India's merchandize trade during 2010–11 and 2011–12 is given in Table 15.2.

Table 15.2 *India's Merchandize Trade (US dollar Billion)*

Items	*2010–11*	*2011–12*
Export	251.1	304.6
Oil Exports	41.5	55.6
Non-oil Exports	209.6	249.0
Import	369.8	489.4
Oil Import	106.0	154.9
Non-oil Imports	263.8	334.5
Non-oil non-gold Imports	223.3	278.3
Total Balance	–118.7	–184.8

The principal commodities' percentage share of export is given in Table 15.3.

Table 15.3 *India's Export of Principal Commodities*

Commodity Group	*Percentage Share*	
	2010–11	*2011–12*
1. **Primary Products**	13.1	15.0
Agricultural and Allied	9.6	12.3
Products, Ores and Minerals	3.4	2.7
2. **Manufactured Goods**	62.9	61.3
Leather and Manufactures	1.6	1.6
Chemical and Related Products	11.5	12.2
Textile and Textile Products	23.1	22.0
Gems and Jewellery	16.1	15.4
3. **Petroleum Products**	16.5	18.3
4. **Others**	7.5	5.4
Total Export	100	100

The major commodities exported from India are:

1. *Agricultural Products:* Tea, coffee, rice, wheat, cotton, tobacco, cashew, species, oil meal, sugar, molasses, Marine products.
2. *Ores and Minerals:* Iron ore, processed minerals
3. *Manufactured Goods*
   (i) Leather and manufactures
   (ii) Chemicals and Related Products – Basic chemicals, pharmaceuticals, cosmetics, plastic, rubber, glass, paints and enamels, residual chemicals and allied products.
   (iii) Engineering Goods – Manufactures of metals, machinery and instruments, transport equipments, iron and steel, electronic goods.
   (iv) Textile and Textile Products – Cotton yarn, fabrics, natural silk yarn, man-made yarn, fabrics, man-made staple fibre, woollen yarn, fabrics, readymade garments, jute and jute manufactures, coir and coir manufactures, carpets.
   (v) Gems and Jewellery
   (vi) Handicrafts
   (v) Petroleum Products
   (vi) Others

Exports Destination

Destination-wise, the UAE continues to remain the biggest destination for India's exports during 2011–12 with a share of 11.8 per cent, followed by the US (11.3%), China (5.9%), Singapore (5.5%) and Hong Kong (4.2%). These five countries together accounted for around 39 per cent of India's total exports during 2011–12. Reflecting the uncertainties prevailing in Europe, growth of India's exports to Germany, Netherlands, Italy and Belgium decelerated during 2011–12, while the same to France showed a decline of 11.5 per cent. Growth to exports to Japan, SAARC region, Africa and Latin American countries also showed moderation. Table 15.4 shows the India's export to principal regions of the world.

Table 15.4 *India's Exports to Principal Regions*

Region/Country	*Percentage Shares*	
	2010–11	*2011–12*
1. Organisation for Economic Co-operation and Development (OECD) Countries.	13.2	33.8
European Union (EU)	18.3	17.2
North America	10.6	11.9
US	10.1	11.3
Asia and Oceania	2.8	3
Other OECD Countries	1.5	1.6
2. Organisation of the Petroleum Exporting Countries (OPEC)	21.3	19
3. Eastern Europe	1.1	1.1
4. Developing Countries	38.2	40.7
Asia	27.9	29.6
South Asian Association for Regional Cooperation (SAARC)	4.6	4.3

Cont'd...

...Cont'd

Other Asian Developing Countries	23.3	25.3
People's Republic of China	6.2	5.9
Africa	6.3	6.7
Latin America	4	4.4
5. Others/Unspecified	6.2	5.4
Total	100	100

Composition of Import Trade

At the time of Independence, nearly 70 per cent import of India consisted of either manufactured consumer goods or inputs of manufacturing industries. But after that, the import of manufactured goods decreased gradually and import of variety of goods increased. Specially a tremendous increase has been seen in petroleum and petroleum products. The number of commodities India imports is very large. These are classified by DGCI & S into six categories as: (i) Food and allied products, (ii) Fuel, (iii) Fertilizer (iv) Paper board, manufactures and newsprint (v) capital goods, and (vi) others.

As per the available data on commodity-wise imports for 2011–12, petroleum and petroleum products continued to be a major item of India's imports, followed by capital goods and gold and silver. Petroleum and petro-products account for nearly 37.7 per cent of India's total merchandize imports.

Import of capital goods was 21.2 per cent in 2010–11, while it registered 20.3 per cent in 2011–12. On the other hand gold and silver percentage of total import was 10.5 per cent. Table 15.5 shows the import of principal commodities in year 2010–11 and 2011–12.

Table 15.5 *Imports of Principal Commodities*

Commodity/Group	*Percentage Shares*	
	2010–11	*2011–12*
1. Petroleum, Crude and Other products	28.7	31.7
2. Capital Goods	21.2	20.3
3. Gold and Silver	11.5	12.5
4. Organic and Inorganic Chemicals	4.1	3.9
5. Coal, Coke and Briquettes, etc.	2.7	3.6
6. Fertilizers	1.9	2.4
7. Metalliferrous Ores, Metal Scrap, etc	2.6	2.7
8. Iron and Steel	2.8	2.4
9. Pearls, Precious and Semi-Precious stones	9.4	6.2
10. Others	15.1	14.3
Total	100	100

China is the largest source of imports with a share of 11.8 per cent of total imports in 2011–12, followed by UAE, Switzerland, Saudi Arabia and the USA. These five countries together constituted around 37 per cent of India's imports. Among the OPEC countries, share of Indonesia, Iraq, Kuwait and Saudi Arabia in India's imports increased significantly. In contrast, share of Iran in India's total imports declined from 3.0 per cent in 2010–11 to 2.8 per cent in 2011–2012. Table 15.6 indicates the shares of groups/countries in India's imports.

Table 15.6 *Share of Groups/Countries in India's Imports*

	Commodity/Group	Percentage Shares	
		2010–11	2011–12
1.	OECD Countries	30.6	29.7
	EU	12.0	11.7
	France	1.0	0.8
	Germany	3.2	3.2
	UK	1.5	1.5
	North America	6.0	5.3
	US	5.4	4.8
	Asia and Oceania	5.4	5.7
	Other OECD Countries	7.2	7.0
2.	OPEC	33.6	35.4
3.	Eastern Europe	1.5	1.7
4.	Developing Countries	33.0	32.3
	Asia	27.1	25.8
	SAARC	0.6	0.6
	Other Asian Developing Countries	26.5	25.3
	People's Republic of China	11.8	11.8
	Africa	3.6	4.0
	Latin America	2.4	2.4
5.	Other/Unspecified	1.3	0.9
	Total	100	100

Foreign Trade Trends

The following trends are apparent in India's foreign trade during 2010–11 and 2012–13.

- Export during 2013 were valued at US 26,259 million dollar which was 4.23 per cent higher in Dollar terms during 2012.
- Commodity-wise export data shows that share of petroleum products in total merchandize exports declined from 20.1 per cent of 2011–12 to 16.6 per cent in 2012–13.
- Export of primary products showed a sharp rise in 2012–13 whereas the growth in exports of engineering goods, textile products, gem and jewellery and leather and manufactures and chemical products was either lower or negative in 2012–13.
- Country-wise pattern of exports shows that USA has become the biggest destination for India's exports with a share of 14.4 per cent replacing UAE (13.4%), followed by China (5.1%), Singapore (4.5%) and Hong Kong (4.2%) in 2012–13.

Imports

- The total imports were valued at US 41,181 million dollars.
- Oil imports during 2013 were 15.45 per cent higher than oil imports in 2012.

- Petroleum and petroleum products continued to be a major item of India's imports, followed by capital goods and gold and silver.
- During 2012–13, the share of European Union in India's imports declined to 11.2 per cent as compared to 12.3 per cent during 2011–12, the same of OPEC group of countries and Latin American countries increased significantly. China continued to be the largest source of imports, followed by UAE, Saudi Arabia, USA and Switzerland.

Salient Features of the International Trade of India

The following are the salient features of international trade of India:

1. Unfavourable balance of trade.
2. Diversity of export and more export of manufactured goods.
3. Petroleum, crude oil and petroleum products having highest share of import.
4. Maritime trade.
5. Trade through selected ports.
6. India's share in international trade is less than one per cent.
7. Most of the overseas trade is done in public sector.
8. Increasing import of raw material and capital goods, etc.

Multiple Choice Questions

SET – 1

INDIA: Introduction, Geology and Physiography

1. In respect of area, the rank of India in the world is:
   (a) Eighth (b) Seventh
   (c) Sixth (d) Fifth

2. Which one of the following country is having larger area than India?
   (a) Brazil (b) South Africa
   (c) France (d) Kazakhstan

3. Which of the following country is having the longest border with India?
   (a) Pakistan (b) Nepal
   (c) Bangladesh (d) China

4. The country having the shortest border with India is:
   (a) Myanmar (b) Nepal
   (c) Bhutan (d) Afghanistan

5. In which of the following state of India, the Tropic of Cancer does not pass through?
   (a) Rajasthan
   (b) Chhattisgarh
   (c) Odisha
   (d) Jharkhand

6. Which is the northernmost point of India?
   (a) Siachen (b) Indra Col
   (c) Kargil (d) Khardungla

7. Which one of the following state is not having common border with Bangladesh?
   (a) Meghalaya (b) Tripura
   (c) West Bengal (d) Manipur

8. What is the percentage of the land area of India in the total land area of the world?
   (a) 2.3% (b) 2.4%
   (c) 2.5% (d) 2.6%

9. Indra Point is the southernmost point of India. Where it is located?
   (a) Kerala
   (b) Lakshadweep
   (c) Nicobar Island
   (d) Tamil Nadu

10. The total length of the coastline of India including coastlines of islands is:
    (a) 7,516 km (b) 6,100 km
    (c) 4,096 km (d) 2,933 km

11. The length of Indo-Pakistan boundary is:
    (a) 4,096 km (b) 3,917 km
    (c) 3,310 km (d) 2,933 km

12. With which country does Radcliff Line form India's boundary?
    (a) China (b) Bangladesh
    (c) Nepal (d) Pakistan

13. Which one of the following is not an Indian state adjoining Pakistan?
(a) Rajasthan (b) Haryana
(c) Uttarakhand (d) Gujarat

14. One of the states through which the Tropic of Cancer passes is:
(a) Haryana (b) Jharkhand
(c) Bihar (d) Maharashtra

15. How many states are there through which the Tropic of cancer passes in India?
(a) 8 (b) 7
(c) 6 (d) 5

16. If a latitude passes through Sikkim it will also pass through:
(a) Rajasthan
(b) Jammu & Kashmir
(c) Gujarat
(d) Himachal Pradesh

17. Which one of the following is not a coastal state of India?
(a) Odisha (b) Telangana
(c) Tamil Nadu (d) Karnataka

18. The correct sequence of the four Indian states in descending order according to their area is:
(a) Rajasthan, Uttar Pradesh, Madhya Pradesh, Maharashtra
(b) Rajasthan, Madhya Pradesh, Maharashtra. Uttar Pradesh
(c) Uttar Pradesh, Rajasthan, Madhya Pradesh, Maharashtra
(d) Madhya Pradesh, Uttar Pradesh, Rajasthan, Maharashtra

19. The smallest state of India according to area is:
(a) Sikkim (b) Nagaland
(c) Goa (d) Mizoram

20. In area, which is the smallest Union Territory of India
(a) Chandigarh (b) Daman & Diu
(c) Pondicherry (d) Lakshadweep

21. Which one of the following states of India has the longest coastline?
(a) Gujarat (b) Maharashtra
(c) Tamil Nadu (d) Andhra Pradesh

22. Arrange the following states of India according to their area in a descending order and select the correct answer using the code given below:
1. Uttar Pradesh 2. Madhya Pradesh
3. Maharashtra 4. Rajasthan

(a)	2	4	1	3
(b)	4	2	1	3
(c)	4	2	3	1
(d)	2	4	3	1

23. The largest Union Territory of India according to area is:
(a) Delhi
(b) Andaman and Nicobar Island
(c) Dadra and Nagar Haveli
(d) Chandigarh

24. The three smallest states of India according to their area in ascending order are:
(a) Goa, Sikkim, Tripura
(b) Goa, Nagaland, Mizoram
(c) Sikkim, Goa, Tripura
(d) Goa, Mizoram, Tripura

25. Arrange the following states of India in respect of area in a descending order and select the correct answer using the code given below:
(1) Assam (2) Jharkhand
(3) Himachal Pradesh (4) Uttarakhand

Code:

(a)	1	2	3	4
(b)	1	2	4	3
(c)	1	3	2	4
(d)	2	1	4	3

26. The total number of districts in India is:
(a) 620 (b) 630
(c) 640 (d) 650

27. The latitudes that pass through Kachchh also pass through:
(a) Odisha (b) Haryana
(c) Arunachal Pradesh (d) Mizoram

28. Which is not a correct pair?

	State	Capital
(a)	Manipur	Imphal
(b)	Assam	Shillong
(c)	Nagaland	Kohima
(d)	Arunachal Pradesh	Itanagar

29. Which is not a correct pair?

Union Territory	Capital
(a) Andaman and Nicobar Island	Port Blair
(b) Daman and Diu	Diu
(c) Dadra and Nagar Haveli	Silvassa
(d) Pondicherry	Pondicherry

30. Kavaratti is the capital of:
(a) Dadra and Nagar Haveli
(b) Manipur
(c) Lakshadweep
(d) Nagaland

31. Which region of India was a part of the oldest landmass Gondwana land?
(a) Penisular India
(b) North-East India
(c) Thar Desert
(d) Himalayan region

32. Which is the oldest system of rock formation in India?
(a) The Purana rock system
(b) The Archaean rock system
(c) The Dravidian rock system
(d) The Aryan rock system

33. Which of the following regions of India is not having rocks of Archaean system?
(a) Chotanagpur Plateau
(b) Madhya Pradesh
(c) Karnataka
(d) Chattisgarh

34. According to Wadia, which system of rock is the most ancient metamorphosed sedimentary rock-system?
(a) The Dravidian rock system
(b) The Vindhyan system
(c) The Cuddaph system
(d) The Dharwar system

35. Arrange the following geological formations in a chronological order using the code given below:
1. Cuddapah Formation
2. Decan Lava Trap
3. Gondwana Formation
4. Vindhyan Formation

Code:
(a) 3,1,4 and 2 (b) 1,2,4 and 3
(c) 1,4,3 and 2 (d) 3,1,2 and 4

36. Which of the following areas is not having rocks of Cuddaph system?
(a) Kurnool district of Andhra Pradesh
(b) Southern part of Western Ghats
(c) Chhattisgarh
(d) Lesser Himalayas

37. Which river valley has the Gondwana rocks?
(a) Brahmaputra (b) Ganga
(c) Damodar (d) Sutlej

38. Which region of India can be considered a shield?
(a) Himalayas (b) Deccan
(c) Aravallis (d) North-East India

39. The Great Boundary Fault lies in the state of:
(a) Uttarakhand and Himachal Pradesh
(b) Jammu & Kashmir and Himachal Pradesh
(c) Maharashtra and Gujarat
(d) Gujarat and Rajasthan

40. Which one of the following systems has the most mineralized rocks in India?
(a) Dharwar (b) Vindhyan
(c) Cuddapah (d) Gondwana

41. Which is the correct chronological sequence of the Indian Geological Time Scale?
(a) Archaean, Aryan, Purana, Dravidian
(b) Archaean, Purana, Dravidian, Aryan
(c) Purana, Dravidian, Archaean, Aryan
(d) Dravidian, Purana, Aryan, Archaean

42. Examine the following statements and select the correct answer using the code given below:
1. The Gondwana land drifted northward during the carboniferous period.
2. The Penisular India was a part of the Gondwana land.
3. The Gondwana land consists of South Africa, Australia, Antarctica and India.

Code:
(a) 1 and 2 are correct
(b) 1 and 3 are correct
(c) 2 and 3 are correct
(d) 1, 2 and 3 are correct

43. The Delhi series belongs to:
(a) Vindhyan system (b) Dravidian system
(c) Cuddaph system (d) Dharwar system

44. Which one of the following statements is not correct?
    (a) The Central ranges of Himalayas are made of Archaean rocks.
    (b) The entire Peninsula of India is made of Archaean rocks.
    (c) The Peninsular gneiss is known as Bengal gneiss.
    (d) Archaean rocks are found in Rajasthan and Chotanagpur.

45. Which one of the following series is not corrlctly matched?

Series	State
(a) Champion Series	Andhra Pradesh/ Telangana
(b) Raico Series	Delhi
(c) Bijwar Series	Madhya Pradesh
(d) Damuda Series	Jharkhand

46. **Assertion (A)** Vindhyan rocks are found in Meghalaya.
    Reason (R) Vindhyan are not subjected to metamorphic activity.
    Code:
    (a) Both A and R are true and R is the correct explanation of A.
    (b) Both A and R are true, but R is not the correct explanation of A.
    (c) A is true, but R is false.
    (d) A is false, but R is true.

47. In which system of rocks are most of the India's coal deposits found?
    (a) Vindhyan system
    (b) Gondwana system
    (c) Cuddaph system
    (d) Dharwar system

48. In which areas are middle deccan trap rocks found?
    (a) Maharashtra
    (b) Tamil Nadu
    (c) Coastal regions of India
    (d) Central India and Malwa

49. **Assestion (A)** The Bhagar land belongs to the pleistocine period.
    Reason (R) The Karewas of Kashmir are Pleistocine deposits.
    Select your answer using following codes:
    (a) Both A and R are true, and R is the correct explanation of A.
    (b) Both A and R true, but R is not the correct explanation of A.
    (c) A is true but R is false
    (d) A is false but R is true

50. Tertiary Period is the most significant period in India's geological history because of which of the following reasons:
    (a) Upliftment of Himalayas
    (b) Origin to Deccan Trap
    (c) Coal Formation Occurred
    (d) Glaciation Started

51. Match List I with List II and select the correct answer using the code given below:

List I (Region)	List II (Geological Period)
A. Deccan Traps	1. Late Cenozoic
B. Western Ghats	2. Pre-Cambrian
C. Araval	3. Cretaceous Eocene
D. Narmada-Tapi Alluvial deposits	4. Cambrian
	5. Pleistocene

Code:

	A	B	C	D
(a)	3	5	1	4
(b)	3	1	2	5
(c)	2	1	3	4
(d)	1	4	2	5

52. Which one of the following statement is not correct?
    (a) During the Carboniferous Period the entire Gondwana land was glaciated.
    (b) The Rajmahal hills do not exhibit glaciated land forms.
    (c) Aravalli was the axis of glaciations during the Carboniferous Period.
    (d) The Nilgiri and Cardomam Hills represent evidence for south-north gradient.

53. Which one of the following is the oldest formation?
    (a) Satpura geosyncliene
    (b) Eastern Ghat geosynclines
    (c) Dharwar geosyncline
    (d) Delhi geosyncline

54. In which geological period the Greater Himalayas originated?
(a) Miocene (b) Pleistocine
(c) Post-Pliocene (d) Eocene upto Oligocine

55. In which geological period Nothern Plains of India came into existence?
(a) Quarternary Period
(b) Post-Pliocene Period
(c) Miocene Period
(d) Eocene Period

56. Himalayas are also known as:
(a) Himadri (b) Himavan
(c) Himachal (d) All the above

57. The name given to the foothill zone of Himalayas is:
(a) Middle Himalayas (b) Shiwalik
(c) Inner Himalayas (d) Greater Himalayas

58. Which of the following hills are found, where the Western Ghats and Eastern Ghats meet?
(a) Anaimalai hills (b) Cardamom hills
(c) Nilgiri hills (d) Shevaroy hills

59. Where are the Shevaroy hills located?
(a) Kerala (b) Tamil Nadu
(c) Karnataka (d) Andhra Pradesh

60. Arrange the following mountain ranges from north to south and select the correct answer using the code given below:
(1) Karakoram range
(2) Ladakh range
(3) Pir-Panjal range
(4) Zaskar range
Code:
(a) 1, 2, 3 and 4 (b) 2, 1, 4 and 3
(c) 2, 1, 3 and 4 (d) 1, 2, 4 and 3

61. Arrange the following peaks in descending order of height and select the correct answer using the code given below:
(a) Nanda Devi (b) Kamet
(c) K2 (d) Nanga Parbat
Code:
(a) 3, 4, 1 and 2 (b) 3, 1, 2 and 4
(c) 4, 3, 2 and 1 (d) 1, 4, 3 and 2

62. Which one of the following is the highest peak of the Peninsular India?
(a) Doda-Beta (b) Anai-Mudi
(c) Cardamom (d) Kodaikanal

63. Which one of the following is the highest peak in the Eastern Himalayas?
(a) Saramati (Naga Hills)
(b) Abor Hills (Arunachal Pradesh)
(c) Dapha Bum (Mishmi Hills)
(d) Brail Range (Cachar Hills)

64. Which one of the following is the third highest peak in the world?
(a) Annapurna (b) Kanchanjunja
(c) K2 (d) Nanda Devi

65. What is the Nepalese name of the Mt. Everest?
(a) Chomlungma (b) Himavan
(c) Sagarmatha (d) Himadri

66. Which among the following is the highest peak of the Great Himalayas?
(a) Namcha Barwa (b) Trisul
(c) Gurla Mandhata (d) Annapurna

67. Which one of the following is not a peak of middle Himalayas?
(a) Gosainath (b) Pir Panjal
(c) Dhaola Dhar (d) Nag Tiba

68. The longest range of middle Himalayas is:
(a) Nanda Devi (b) Pir Panjal
(c) Mussoorie (d) Trisul

69. The famous valley which is located in between Pir Panjal and Zaskar Range is:
(a) Kangra Valley (b) Kulu Valley
(c) Kashmir Valley (d) Jhelem Valley

70. Which one of the following is not a hill resort of middle Himalayas?
(a) Mussoorrie (b) Nainital
(c) Almora (d) Dehra Dum

71. Which one of the following is not correctly matched?
(a) Gurushikhar-Rajasthan
(b) Mahabaleshwar-Madhya Pradesh
(c) Ayodhya Hills-West Bengal
(d) Harish chandragarh-Western Ghats

72. Pensi La connects:
(a) Srinagar with Leh
(b) Sikkim with Tibet
(c) Himachal Pradesh with Tibet
(d) Himachal Pradesh with Kargil

73. In which part of Himalayas are 'Dums' located?
(a) Greater Himalayas
(b) Middle Himalayas
(c) Outer Himalayas
(d) Trans-Himalayas

74. Which is not a Trans-Himalayan range?
(a) Ladakh (b) Badrinath
(c) Kailash (d) Zaskar

75. Which one of the following is not a regional division of Himalaya as suggested by Sir S. Burrard?
(a) The Himachal Himalaya
(b) The Punjab Himalaya
(c) The Kumaon Himalaya
(d) The Assam Himalaya

76. Jawahar Tunnel passed through which pass?
(a) Bomid La (b) Dihang Pass
(c) Banihal Pass (d) Bara Lancha

77. Which is not a correct match?

	Pass	Location/State
(a)	Nathu La	Sikkam
(b)	Thang La	Ladakh
(c)	Debasa	Himachal Pradesh
(d)	Diphu Pass	Jammu & Kashmir

78. In which State is Rohtang Pass located?
(a) Jammu & Kashmir
(b) Himachal Pradesh
(c) Uttarakhand
(d) Arunachal Pradesh

79. Which pass joins Srinagar with Kargil and Leh?
(a) Bara Lancha
(b) Zoji-La
(c) Jelepa-La
(d) Chang-La

80. Through which pass does Arunachal Pradesh have shortest access to Myanmar?
(a) Diphu Pass (b) Bomdi La
(c) Dihang Pass (d) Likhapani

81. Which one of the following is the highest pass in the Himalayas?
(a) Banihal pass
(b) Khardung La
(c) Balcha Dhura
(d) Pir Panjal Pass

82. Which one of the following is not correctly matched?

	Pass	State
(a)	Bomdila	Sikkim
(b)	Shipkila	Himachal Pradesh
(c)	Aghil	Jammu & Kashmir
(d)	Nathu La	Sikkim

83. Nanda Devi Peak forms a part of:
(a) Punjab Himalaya (b) Nepal Himalaya
(c) Kumaon Himalaya (d) Assam Himalaya

84. Which one of the following is the correct sequence of the given hills starting from the north and going towards the south?
(a) Nallamalai Hills – Nilgiri Hills – Javadi Hills – Anaimalai Hills
(b) Anaimalai Hills – Javadi Hills – Nilgiri Hills – Nallamalai Hills
(c) Nallamalai Hills – Javad Hills – Nilgiri Hills – Anaimalai Hills
(d) Anaimalai Hills – Nilgiri Hills – Javadi Hills – Nallamalai Hills

85. Which of the following hills are found where the Eastern Ghats and Western Ghats meet?
(a) Cardamom Hills (b) Aravalli Hills
(c) Shevoroy (d) Nilgiri Hills

86. Consider the following statements and select the correct answer using the code given below?
1. Himalaya is the controller of Indian monsoon as well as climate of India.
2. Himalaya region is rich in mineral resources.
3. Several pilgrimage centres are located in Himalaya region

Code:
(a) Only 1 is correct
(b) 1 and 2 are correct
(c) 2 and 3 are correct
(d) 1 and 3 are correct

87. **Assertion (A)** Himalayas have come out of the geosyncline known as 'Tethys Sea'.
Reason (R) Himalayas are the result of northward movement of Indian plate and southward movement of Asian plate.
Code:
(a) Both A and R are true and R is the correct explanation of A
(b) Both A and R are true, but R is not a correct explanation of A
(c) A is true but R is false
(d) A is false but R is true

88. The Himalayan Ranges located north of Great Himalaya is known as:
(a) Shiwalik (b) Trans-Himalayas
(c) Outer Himalayas (d) Inner Himalayas

89. Examine the following statements and select the correct answer using the code given below:
1. The Main Central Thrust separates the Zasker series from the Greater Himalayas.
2. The Main Boundary Thrust separates the Greater Himalayas from the Lesser Himalayas
3. The Himalayas Front Fault separates the Northern Plains of Indian form the Siwaliks
4. The Main Central Thrust like between the Greater Himalayas from the Lesser Himalayas
Code:
(a) 1, 2 and 3 are correct
(b) 2, 3 and 4 are correct
(c) 1, 2 and 4 are correct
(d) 1, 2,3 and 4 are correct

90. Where are the Shevaroy hills located?
(a) Karnataka (b) Andhra Pradesh
(c) Tamil Nadu (d) Kerala

91. Which of the following in not correct?
(a) Western Ghats are relatively higher in their north.
(b) Anai Mudi is the highest peak of Western Ghats.
(c) Tapi River lies to the south of Satpura.
(d) Nadmada and Tapi rivers are said to be old rift valleys.

92. If there were no Himalayan ranges, what would have been the most likely geographical impact on India?
1. Most of the country would experience the cold waves from Siberia.
2. Indo-Gangetic plain would be devoid to such executive alluvial soils.
3. The pattern of monsoon would be different from what it is at present.
Which of the statement given above is/are correct?
(a) Only 1 (b) 1 and 3
(c) 2 and 3 (d) All of these

93. When you travel in Himalayas you will see the following:
1. Deep gorges
2. U-turn river courses
3. Parallel Mountain ranges
4. Steep gradients causing land-sliding

Which of the above can be said to be the evidence for the Himalayas being young fold mountains?
(a) 1 and 2 (b) 1, 2 and 4
(c) 3 and 4 (d) All of these

94. The highest peak of Zaskar Range is:
(a) Kanchenjunga
(b) Kamet
(c) Namcha Barwa
(d) Dhaulagiri

95. Match List I with List II and select the correct answer using code given below:

List I (Peak)	List II (Range)
A. Gurushikhar	1. Annamula
B. Dodabeta	2. Aravalli
C. Annuaimudi	3. Nilgiri
D. Dhupgarh	4. Satpura

Code:

	A	B	C	D
(a)	2	1	3	4
(b)	2	3	1	4
(c)	2	4	3	1
(d)	3	2	1	4

96. Which of the following ranges is the youngest?
(a) Satpura (b) Aravalli
(c) Himalaya (d) Sahyadri

97. Examine the following statement and select the correct answer using the code given below:
   1. The Bhabar lies along the foot of Shiwaliks.
   2. The Bhabar extends from Indus to Tista.
   3. The Bhabar is generally 8 to 16 km wide.
   4. All the rivers disappear in a Bhabar tract.

   Code:
   (a) 1, 2 and 3 are correct
   (b) 1, 2 and 4 are correct
   (c) 1, 3 and 4 are correct
   (d) 1, 2, 3 and 4 are correct

98. Which of the following statements is not correct?
   (a) The Khadar has the fossils of mammals.
   (b) The Khadar alluvium is of dark colour.
   (c) The Khadar is called as 'bet' in Punjab.
   (d) The Khadar land consists of new alluvium.

99. Which one of the following statements is not correct?
   (a) The Bhabar tract is wider in the east.
   (b) The Bhabar tract is not suitable for cultivation.
   (c) The trees with long roots are there in Bhabar tract.
   (d) In the Bhabar tract the possible beds are parallel to slope of the river bed.

100. The upland of the Ganga Delta regions is called:
   (a) Doab (b) Chars
   (c) Bhur (d) Thur

101. The marshy lands in Sundarban Delta are known as:
   (a) Bhur (b) Chars
   (c) Reh (d) Bils

102. In Tarai region, area of which state/states is/are included
   (a) Punjab (b) Uttarakhand
   (c) Uttar Pradesh (d) All of these

103. Examine the following statements and select the correct answer using the code given below:
   1. The Eastern Coast of India is an emergent and indented coast.
   2. The Western Coast of India is a submerged and indented coast.
   3. The backwaters and lagoon in Kerala are known as kayals
   4. The coastal sand dunes in Kerala are known as Teris.

   Code:
   (a) 1, 2 and 3 are correct
   (b) 1, 3 and 4 are correct
   (c) 2, 3 and 4 are correct
   (d) 1, 2, 3 and 4 are correct

104. Which one of the following statements is not correct?
   (a) Panchan Island lies near Gujarat.
   (b) Osam and Barda are volcanic hills in Kathiawar.
   (c) Kori Creek lies in Maharashtra.
   (d) Moor Islands lies off Sunderban Delta.

105. Rajmahal hills are located in:
   (a) Chatonagpur Plateau
   (b) Malwa Plateau
   (c) Bundelkhand
   (d) Vindhyachal Plateau

106. Which of the following regions is included in Bundelknand?
   (a) Satna Plateau (b) Rewa Plateau
   (c) Mirzapur Plateau (d) All of these

107. Which of the following is not a sub division of the Deccan Plateau?
   (a) Karnataka Plateau
   (b) Malwa Palteau
   (c) Chhatisgarh Palteau
   (d) Dandakaranya

108. Bharner and Kaimur hills are the part of:
   (a) Sitapura Range (b) Aravali Range
   (c) Vindhyan Range (d) Western Ghats

109. Which one of the following is not a pass of Western Ghats?
   (a) Palghat (b) Rajghat
   (c) Borghat (d) Thalghat

110. The highest peak of Eastern Ghats is:
   (a) Aroya-Konda (b) Nimalgiri
   (c) Singa-Raju (d) Mahendragiri

111. The Southernmost part of west coastal plain is known as:
   (a) Malabar Coast (b) Konkan Coast
   (c) Kerala Coast (d) Tamil Nadu Coast

112. Where is Utkal Plain located?
    (a) Along Andhra Pradesh Coast
    (b) Along Odisha Coast
    (c) Along Tamil Nadu Coast
    (d) Along Kerala Coast

113. 'Saddle Peak', the highest peak of Andaman and Nicobar Islands is located in:
    (a) Great Nicobar (b) Middle Andaman
    (c) Little Andaman (d) North Andaman

114. Which of the following is the main reason for greater breadth of east coastal plain of India?
    (a) Eastern Coastline is sub emergent.
    (b) Eastern Coastline is emergent.
    (c) Rivers flowing into Bay of Bengal are perennial.
    (d) Many of the rivers in this region form estuaries.

115. The number of Islands included in Andaman group is:
    (a) 280 (b) 305
    (c) 325 (d) 352

116. The Andaman and Nicobar Islands are separated by:
    (a) Mountain Chain
    (b) Ten Degree Channel
    (c) Submerged ridge
    (d) Nine Degree Channel

117. The Southernmost island of Lakshadweep is known as:
    (a) Amindivi Island (b) Wheeler
    (c) Minicoy Island (d) Moor Island

118. Sri Hari Kota is located at:
    (a) The mouth of Pulicat Lake
    (b) On Mangalore Coast
    (c) Krishna River Delta
    (d) Gulf of Mannar

119. Duncan Pass is located between:
    (a) North and South Andaman
    (b) South and Little Andaman
    (c) Andaman and Nicobar
    (d) North and Middle Andaman

120. The most common characteristic of the hidden islands located in Arabian Sea is:
    (a) They are very small in size
    (b) They are extended parts of the main land
    (c) They are all of coral origin
    (d) They have a very dry climate.

Answer Key – SET 1

Que. No	Answer	Que. No	Answer	Que. No	Answer	Que. No	Answer
1	b	31	a	61	a	91	a
2	a	32	b	62	b	92	d
3	c	33	c	63	c	93	d
4	d	34	d	64	b	94	b
5	c	35	c	65	c	95	b
6	b	36	b	66	d	96	c
7	d	37	c	67	a	97	a
8	b	38	b	68	b	98	b
9	c	39	d	69	c	99	a
10	a	40	a	70	d	100	b
11	c	41	b	71	b	101	b
12	d	42	d	72	d	102	d
13	c	43	c	73	c	103	d
14	b	44	b	74	b	104	c
15	a	45	a	75	a	105	a
16	a	46	c	76	c	106	d
17	b	47	b	77	d	107	b
18	b	48	d	78	b	108	c
19	c	49	d	79	b	109	b
20	d	50	a	80	a	110	a
21	a	51	b	81	b	111	c
22	c	52	b	82	a	112	b
23	b	53	c	83	c	113	d
24	a	54	d	84	c	114	b
25	b	55	a	85	d	115	c
26	c	56	d	86	d	116	b
27	d	57	b	87	a	117	c
28	b	58	c	88	b	118	a
29	b	59	b	89	b	119	b
30	c	60	d	90	c	120	c

SET – 2
INDIA: Drainage, Lakes and Waterfalls

1. Examine the following statements and select the correct answer using the code given below:
   1. The Bhagirathi River originates from the Gaumukh Glacier.
   2. The Alaknanda River originates from the Satopanth Glacier.
   3. The Mana pass lies along th course of Alaknanda.

   Code:
   (a) 1 and 2 are correct
   (b) 1 and 3 are correct
   (c) 2 and 3 are correct
   (d) 1, 2 and 3 are correct

2. The total number of river basins in India is:
   (a) 113 (b) 110
   (c) 115 (d) 124

3. The river basins having catchment area of more than 20,000 Sq km are known as:
   (a) Marco basin (b) Major basin
   (c) Large basin (d) Medium basin

4. The total number of large basins in India is:
   (a) 10 (b) 12
   (c) 13 (d) 14

5. The total number of minor basins in India is:
   (a) 53 (b) 55
   (c) 57 (d) 60

6 Which of the following river basins in India is having the largest area?
   (a) Brahmaputra (b) Ganga
   (c) Godavari (d) Mahanadi

7. Arrange the following river basins in descending order according to their area and select the correct answer using code given below:
   1. Brahmaputra 2. Godavari
   3. Mahanadi 4. Narmada

   Code:
   (a) 2, 1, 3 and 4 (b) 1, 2, 3 and 4
   (c) 3, 2, 4 and 1 (d) 1, 3, 4 and 2

8. Which one of the following is the largest river basin in Peninsular India?
   (a) Mahanadi (b) Tapi
   (c) Godavari (d) Krishna

9. Which one of the following river basins is totally in India?
   (a) Ravi (b) Beas
   (c) Sutlej (d) Chenab

10. Sarda and Sarju are the tributaries of:
    (a) Kosi (b) Gomati
    (c) Manas (d) Ghagra

11. Which one of the following tributaries is not correctly matched?

	Main River	Tributary
(a)	Krishna	Koyna
(b)	Mahanadi	Jawai
(c)	Godavari	Sabari
(d)	Kaveri	Akravati

12. Which one of the following is the largest tributary of river Ganga?
    (a) Ghagra (b) Gandak
    (c) Yamuna (d) Kosi

13. Match List I with List II and select correct answer using the code given below:

List I (River)	List II (Tributary)
A. Ganga	1. Pen-Ganga
B. Godavari	2. Bhima
C. Yamuna	3. Bagmat
D. Krishna	4. Dhansan

    Code:

	A	B	C	D
(a)	3	1	2	4
(b)	3	1	4	2
(c)	4	3	2	1
(d)	4	2	3	1

14. Which one of the following is not correctly matched?

	Tributary	River
(a)	Tista	Ganga
(b)	Banas	Chambal
(c)	Rihand	Son
(d)	Chandrabhaga	Chenab

15. Which one of the following east flowing rivers of India has rift valley due to down warping?
(a) Damodar (b) Mahanadi
(c) Son (d) Yamuna

16. The correct sequence of the eastward flowing rivers of the peninsular India from north to south is:
(a) Subernrekha, Mahanadi, Godaveri, Krishna, Kaveri, Pennar, and Vaigai
(b) Subernrekha, Mahanadi, Krishna, Godaveri, Kaveri, Vaigai and Pennar
(c) Mahanadi, Subernrekha, Godaveri, Krishna, Kaveri, Pennar, and Vaigai
(d) Mahanadi, Subernrekha, Krishna, Godaveri, Kaveri, Vaigai and Pennar

17. What is the correct sequence of the rivers Godavari, Mahanadi, Narmada and Tapti in descending order of their length?
(a) Godavari-Mahanadi-Narmada-Tapti
(b) Godavari-Narmada-Mahanadi-Tapti
(c) Narmada-Godavari-Tapti-Mahanadi
(d) Narmada-Tapti-Godavari-Mahanadi

18. Consider the following pairs:

	Tributary River	**Main River**
1.	Chambal	Narmada
2.	Son	Yamuna
3.	Manas	Brahmaputra

Which of the pairs given above is/are correctly matched?
(a) 1, 2 and 3 (b) 1 and 2
(c) 2 and 3 (d) Only 3

19. Which one of the following rivers does not originate in India?
(a) Beas (b) Chenab
(c) Ravi (d) Sutlej

20. Rivers that pass through Himachal Pradesh are:
(a) Beas and Chenab only
(b) Beas and Ravi only
(c) Chenab, Ravi and Yamuna
(d) Beas, Chenab, Ravi and Yamuna

21. At which one of the following places do two important rivers of India originate-while one of them flows towards north and merges with another important river flowing towards Bay of Bengal, the other one flows towards the Arabian Sea?
(a) Amarkantak (b) Badrinath
(c) Mahabalashwar (d) Nasik

22. Which river forms its delta in the Cuttack district of Odisha?
(a) Damodar (b) Mahanadi
(c) Godavari (d) Krishna

23. Teesta is a tributary of:
(a) Ravi (b) Ganga
(c) Brahmaputra (d) Mahanadi

24. Which river is known of its changing course behavior?
(a) Damodar (b) Kosi
(c) Ghaghara (d) Son

25. Which river has its source in Maharashtra?
(a) Godavari (b) Mahanadi
(c) Narmada (d) Vaigai

26. Which of the following rivers is believed to be older than the Himalayas?
(a) Indus (b) Ganga
(c) Brahmaputra (d) Sutlej
Code:
(a) 1, 2 and 3
(b) 2, 3, and 4
(c) 1, 2, and 4
(d) 1, 3 and 4

27. Which river flows between the Satpuras and the Vindhyas?
(a) Tapi (b) Narmada
(c) Gandak (d) Godwari

28. 'The Tributary of Ganga', which flows northward is:
(a) Ganghara (b) Gandak
(c) Son (d) Kosi

29. Which one of the following rivers flows towards the west?
(a) Kaveri (b) Narmada
(c) Krishna (d) Godavari

30. Most of the rivers flowing west from the Western Ghats do not from delta because of:
(a) Lack of eroded material
(b) The steep gradient
(c) Lack of vegetation free area
(d) Low velocity

31. Which one of the following is not correctly matched?
(a) Purna-Tapi
(b) Parbati-Chambal
(c) Jodh Ganga-Alaknanda
(d) Penganga-Godavari

32. Which one of the following rivers is shortest in length?
(a) Ravi (b) Jhelum
(c) Chenab (d) Beas

33. Which one of the following rivers is shortest in length?
(a) Kosi (b) Ramganga
(c) Gandak (d) Ghaghra

34. Ganga is also known as:
(a) Mandakini (b) Jahnavi
(c) Bhagirathi (d) All of these

35. Sarda, Surju and Rapti are the tributaries of:
(a) Ghaghara (b) Gantak
(c) Kosi (d) Ramganga

36. Which is not a correct match?

	River	**Tributary**
(a)	Kosi	Arin
(b)	Ghaghara	Trishul
(c)	Chambal	Kali Sindh
(d)	Damodar	Barakar

37. River Brahmaputra has originated from:
(a) Nepal
(b) Kong Shu Lake
(c) Kailash Mts
(d) Namcha Barua

38. The combined stream of Ganga and Brahmaputra is known as:
(a) Meghana (b) Teesta
(c) Padama (d) Manas

39. Which one of the following is not a major Tributary of Brahmaputra?
(a) Subaushri
(b) Kemang
(c) Manas
(d) Suvernareka

40. Which is not a correct match?

	River	**Origin Place**
(a)	Barak	Manipur Mountains
(b)	Teesta	Himalayan range, North Sikkim
(c)	Manas	Southern Sikkim
(d)	Kopali	Malangha

41. Godavari is the longest river of Peninsular India, what is its length?
(a) 1,465 km (b) 1,400 km
(c) 1,310 km (d) 1,420 km

42. Hemavati and Arkavati are the tributaries of:
(a) Mahanadi (b) Kaveri
(c) Krishna (d) Godavari

43. Krishna River has originated from:
(a) Amarkantak (b) Trimlak plateau
(c) Mahabaleshwer (d) Karnataka plateau

44. Bhima and Tungabhadra are the tributaries of river:
(a) Krishna (b) Narmada
(c) Sabarmati (d) Pennar

45. Dhualadhar fall is on river:
(a) Tapi (b) Mandvi
(c) Rupen (d) Narmada

46. Through which of the following states does the river Pennar flow?
(a) Madhya Pradesh, Chhattisgarh and Odisha
(b) Karnataka, Telengana and Andhra Pradesh
(c) Karnataka, Kerala and Tamil Nadu
(d) Telangana, Karnataka and Tamil Nadu

47. Which one of the following is not a west flowing river of Central India?
(a) Rupen (b) Mahi
(c) Varaha (d) Sabarmati

48. Consider the following statements and select the correct answer using code given below:
1. In India, the drainage basin of Brahmaputra is bigger than that of the Indus Basin
2. The Mahanadi drainage basin is bigger than that of the Krishna river basin.

Code:
(a) Only 1 is correct
(b) Only 2 is correct
(c) Both 1 and 2 are correct
(d) Neither 1 nor 2 is correct

49. Which is not a correct match:

	Present Name	Ancient Name
(a)	Ravi	Ravati
(b)	Chenab	Asikni
(c)	Beas	Vipasa
(d)	Sutlej	Burli

50. Which one of the following rivers does not join Ganga from the southern side?
(a) Son (b) Karmanasa
(c) Gomati (d) Punupun

51. Which is not a tectonic lake in India?
(a) Tso Moriri (b) Bhimtal
(c) Kolleru (d) Nainital

52. In which states of India are glacial lakes found?
(a) Jammu & Kashmir (b) Uttar Predesh
(c) Himachal Pradesh (d) All of these

53. In which state is crater lake Loner located?
(a) Maharashtra (b) Karnataka
(c) Odisha (d) Gujarat

54. Which of the following is/are the example(s) of lagoon lakes?
(a) Chilka (b) Pulicat
(c) Asthamudi (d) All of these

55. Gohna lake of garhwal is an example of-
(a) Aeolion lake (b) Glacial lake
(c) Landslide lake (d) Dissolution lake

56. Match List I with List II and select the correct answer using the code given below:

	List I (Lake)	List II (State)
a.	Bhimtal	Andhra Pradesh
b.	Kolleru	Uttarakhand
c.	Nakki lake	Ladakh
d.	Tso Morari	Rajasthan

Code:

	A	B	C	D
(a)	2	1	4	3
(b)	2	4	1	3
(c)	4	3	2	1
(d)	4	3	1	2

57. Which is not a correct match?

	Lake	State
a.	Koliveli Lake	Tamil Nadu
b.	Roop kund	Himachal Pradesh
c.	Vembauattu	Kerala
d.	Loktak Lake	Manipur

58. Which is the largest lake in India?
(a) Wular Lake (b) Hussain Sagar
(c) Chandra Tal (d) Chilka Lake

59. The largest artificial lake in India is:
(a) Chilka Lake
(b) Jaisamand (Dhaber Lake)
(c) Wular Lake
(d) Dal Lake

60. Which is the largest freshwater lake in India?
(a) Wular Lake (b) Pushksr Lake
(c) Bojwet land (d) Dal Lake

61. **Assertion (A)** West flowing rivers of Peninsular India have no deltas.
Reason (R) These rivers do not carry any alluvial sediments.
Code:
(a) Both A and R are true and R is the correct explanation of A
(b) Both A and R are true, but R is not the correct explanation of A
(c) A is true but R is False
(d) A is false but R is true

62. Which of the following pairs are correctly matched?

	Water falls	River
1.	Kapildhara falls	Godavari
2.	Jog Falls	Sharavati
3.	Sivasamudran falls	Kaveri

Select the correct answer using the codes given below:
(a) 1 and 2 (b) 2 and 3
(c) 1 and 3 (d) All of these

63. Which one of the following water falls is not correctly matched?

	Waterfalls	States
(a)	Vantawung	Mizoram
(b)	Barehipani	Odisha
(c)	Bundla	Himachal Pradesh
(d)	Teeragarh	Jharkhand

64. Which one of the following waterfalls is not in Karnataka?
(a) Hundru (b) Jog
(c) Gokak (d) Irupu

65. Which one of the following waterfalls is not correctly matched?

Waterfalls	River
(a) Gokak	Ghataprabha
(b) Budha Ghag	Mahanadi
(c) Dudhsagar	Mandvi
(d) Dhunwadhar	Narmada

66. In which state is Chulia waterfalls located?
(a) Madhya Pradesh
(b) Uttar Pradesh
(c) Rajasthan
(d) Bihar

67. Which one of the following is not a correct match?

Waterfalls	States
(a) Catherine	Tamil Nadu
(b) Elephant	Manipur
(c) Kune	Maharashtra
(d) Vasundhara	Uttarakhand

68. In which state are Gokak, Megod and Barkana waterfalls located?
(a) Karnataka
(b) Tamil Nadu
(c) Himachal Pradesh
(d) Jharkhand

69. Which one of the following is not a waterfall of Maharashtra?
(a) Palaruvi
(b) Kune
(c) Pandavgat
(d) Palani

70. Match List I with List II and select the correct answer using the code given below:

List I (Waterfalls)	List II (State)
A. Ayyaar	1. Himachal Pradesh
B. Barakara	2. Karnataka
C. Bundla	3. Odisha
D. Duduma	4. Tamil Nadu

Code:

	A	B	C	D
(a)	4	2	3	1
(b)	4	2	1	3
(c)	3	4	1	2
(d)	3	4	2	1

71. Which of the following waterfalls is not correctly matched?

State	Waterfalls
(a) Odisha	Barehipani
(b) Himachal Pradesh	Bundla
(c) Jharkhand	Teerathgarh
(d) Mizoram	Vantawung

72. Match List I with List II and select the correct answer using the code given below:

List I (Waterfalls)	List II (River)
(a) Dudhsagar	1. Ghataprabha
(b) Duduma	2. Machkund
(c) Gokak	3. Sharavati
(d) Jog	4. Mandovi

Code:

	A	B	C	D
(a)	2	3	4	1
(b)	2	4	1	3
(c)	4	2	3	1
(d)	4	2	1	3

73. Which river form the famous Marble Falls?
(a) Tapi (b) Cauvery
(c) Narmada (d) Sharavati

74. Chitrokote Falls is formed by which river and in which district is it located?
(a) Indravati in Bastar
(b) Tapi in Nasik
(c) Krishna in Vijaywara
(d) Narmada in Jabalpur

75. In which state is Tirathgarh waterfall located?
(a) Jharkhand (b) Chhattisgarh
(c) Madhya Pradesh (d) Karnataka

Answer Key – SET 2

Que. No	Answer	Que. No	Answer	Que. No	Answer
1.	a	26.	d	51.	c
2.	a	27.	b	52.	d
3.	c	28.	c	53.	a
4.	d	29.	b	54.	d
5.	b	30.	b	55.	c
6.	b	31.	c	56.	a
7.	a	32.	d	57.	b
8.	c	33.	b	58.	d
9.	b	34.	d	59.	b
10.	d	35.	a	60.	a
11.	b	36.	b	61.	c
12.	c	37.	b	62.	b
13.	b	38.	c	63.	d
14.	a	39.	d	64.	a
15.	b	40.	c	65.	b
16.	a	41.	a	66.	c
17.	b	42.	b	67.	b
18.	d	43.	c	68.	a
19.	d	44.	a	69.	d
20.	d	45.	d	70.	b
21.	a	46.	b	71.	c
22.	b	47.	c	72.	d
23.	c	48.	b	73.	c
24.	b	49.	d	74.	a
25.	a	50.	c	75.	b

SET – 3
INDIA: Climate, Natural Vegetation, Biodiversity and Soils

1. 'We may speak of the climates but not the climate of India'. Who has given this statement?
   (a) Flohn (b) Blandford
   (c) O.H.K. Spate (d) Halley

2. Consider the following statements and select the correct answer using the code given below:
   1. The western disturbances originate from the Mediterranean sea.
   2. The origin of the western disturbances is associated with the jet streams.

   Code:
   (a) Only 1 is correct
   (b) Only 2 is correct
   (c) Both 1 and 2 are correct
   (d) Neither 1 nor 2 is correct

3. Which one of the following records the lowest temperature in January?
   (a) Bangalore (b) Chennai
   (c) Hyderabad (d) Thiruvananthapuram

4. The highest mean daily temperature in the month of July is recorded at:
   (a) Amritsar (b) Bikaner
   (c) Jaisalmer (d) Jodhpur

5. How is La-Nina different from El-Nino?
   1. La-Nino is characterized by an unusually cold temperature in equatorial Indian Ocean, while El-Nino is characterized by an unusually cold temperature in equatorial Pacific Ocean.
   2. El-Nino has negative effects on Indian monsoon, but La-Nino has no effect on Indian monsoon.

   (a) Only 1 (b) Only 2
   (c) Both 1 and 2 (d) None of these

6. Western coast of India receives more rainfall than the Eastern coast because:
   (a) It is narrower than Eastern coast.
   (b) Western Ghats are higher and more continuous to cause obstruction to winds.
   (c) Eastern Ghats are extended parallel to wind direction.
   (d) Arabian Sea monsoonal winds exist for a longer time duration.

7. Which one of the following is the correct sequence of the given Indian cities in the decreasing order of their normal annual rainfall?
   (a) Kochi-Kolkata-Delhi-Patna
   (b) Kolkata-Kochi-Patna-Delhi
   (c) Kochi-Kolkata-Patna-Delhi
   (d) Kolkata-Kochi-Delhi-Patna

8. Consider the following statements about the flow of south-western monsoon in India:
   1. The Arabian Sea branch merges with Bay of Bengal branch over the Punjab and joins Himalayas.
   2. Breaks or spells of dry weather are not the characteristic features of south west monsoon.

   Which of the statements given above is/are correct?
   (a) Only 1 (b) Only 2
   (c) Both 1 and 2 (d) neither 1 nor 2

9. Consider the following statements:
   1. Monsoon rains occur in periodic intervals of 10 days.
   2. Malabar Coast usually receives the first monsoonal rainfall of the season.

   Which of the statements given above is/are correct?
   (a) Only 1 (b) Only 2
   (c) Both 1 and 2 (d) Neither 1 nor 2

10. Which of the following pairs of cities have the largest annual average rainfall difference as compared to the other three pairs?
    (a) Kolkata and Bhubaneswar
    (b) Jabalpur and Nagpur
    (c) Guwahati and Shillong
    (d) Mumbai and Pune

11. Which among the following is/are the major factor(s) responsible for the monsoon type of climate in India?
    1. Location
    2. Thermal Contract
    3. Upper air circulation
    4. Inter tropical convergence zone

Select the correct answer using the code given below:

(a) Only 1 (b) 2 and 3
(c) 2, 3 and 4 (d) all the above

12. The heaviest rainfall in India is recorded at:
(a) Mawsyuran (b) Cherrapunji
(c) Shillong (d) Tura

13. The normal date of the burst of monsoon in Mumbai is:
(a) 10th June (b) 5th June
(c) 1st June (d) 25th June

14. The winter rains in northern plains is due to:
(a) South-West Monsoon
(b) Retreating Monsoon
(c) Easterly Depressions
(d) Westerly Depressions

15. The period of south west monsoon in India is:
(a) June to September
(b) May to August
(c) June to October
(d) May to July

16. Who has given the Thermal concept regarding the origin of monsoon?
(a) Flohn (b) Hulley
(c) Koteswaram (d) Ramswani

17. In which areas do Jet Streams blow during summers?
(a) 20°N to 35°N latitudes
(b) 35°N to 45°N latitudes
(c) 10°N to 25°N latitudes
(d) 10°N to 40°N latitudes

18. Who emphasized the role of Tibetan Plateau in the origin of Indian monsoon?
(a) P. Koteswaraw (b) Flohn
(c) Krishnamurti (d) Rama Rattan

19. **Assertion (A)** The occurrence of El-Nino results into weak monsoon.
Reason (R) Shifting of Inter-Tropical Convergence Zone is the main cause of Indian monsoon
(a) Both A and R are true and R is the correct explanations of A
(b) Both A and R are true but R is not the correct explanation of A
(c) A is true but R is False
(d) A is False but R is True

20. What is the date of the arrival of Indian monsoon in Andaman and Nicobar Island?
(a) 10th May (b) 15th May
(c) 20th May (d) 25th May

21. Which of the following is the correct ascending order of annual rainfall received by given Indian cities?
(a) Lucknow, Cuttack, Jaisalmer, Ahmedabad
(b) Jaisalmer, Ahmedabad, Lucknow, Cuttack
(c) Ahmedabad, Jaisalmer, Cuttack, Lucknow
(d) Cuttack, Lucknow, Ahmedabad, Jaisalmer

22. The average annual temperature of a metrological station is 25°C, its average annual rainfall is 63 cm and annual range of temperature is 9°C. The station in question is:
(a) Chennai (b) Cherrapunji
(c) Allahabad (d) Kolkata

23. **Assertion (A)** Bengaluru receives much highest average rainfall than Mangalore.
Reason (R) Bengaluru has the benefit of receiving rainfall both from South-West and North-East monsoon
(a) Both A and R are true and R is the correct explanation of A
(b) Both A and R are true but R is not the correct explanation of A
(c) A is true but R is False
(d) A is False but R is True

24. Which one of the following states of India receives the lowest average annual rainfall?
(a) Jammu & Kashmir (b) Odisha
(c) Himachal Pradesh (d) Jharkhand

25. Which is not the region of very high rainfall in India?
(a) Meghalaya (b) Odisha
(c) Nagaland (d) Mizoram

26. Norwesters is West Bengal are known as:
(a) Loo (b) Cherry Blossoms
(c) Kalabaisakhi (d) Mango showers

27. In which state are the rains received from thunderstorms at the end of summer known as Mango showers?
(a) Tamil Nadu (b) Kerala
(c) Telangana (d) Goa

28. The first attempt to divide India into climate regions was made by:
(a) B.L.C Jhonson (b) K.L. Rao
(c) W. Koppen (d) H.E. Blandford

29. How many climate regions of India have been identified by Koppen?
(a) Nine (b) Ten
(c) Seven (d) Twelve

30. Which one of the following state is not having tropical Savannah type (AW) climate?
(a) Odisha (b) Haryana
(c) Telengana (d) Jharkand

31. In which one among the following categories of protected areas in India are local people not allowed to collect and use the biomass?
(a) Biosphere reserves
(b) National Parks
(c) Wetlands
(d) Wildlife Sanctuaries

32. Amongst the following Indian states, which one has the minimum total forest cover?
(a) Sikkim (b) Goa
(c) Haryana (d) Kerala

33. The percentage of total forest cover of India in 2013 was:
(a) 21.23 (b) 20.56
(c) 19.83 (d) 23.18

34. The top three states of India in descending order according to their forest area are:
(a) Madhya Pradesh, Maharashtra, Odisha
(b) Arunachal Pradesh, Odisha, Maharashtra
(c) Madhya Pradesh, Arunachal Pradesh, Chhattisgarh
(d) Chhattisgarh, Odisha, Madhya Pradesh

35. Which state is having the highest percentage of forest cover in total area of the state?
(a) Arunachal Pradesh (b) Nagaland
(c) Madhya Pradesh (d) Mizoram

36. Which one of the following states is having the least forest cover percentage in India?
(a) Haryana (b) Punjab
(c) Gujarat (d) Bihar

37. Match List I with List II and select the correct answer using the code given below:

List I **(Type of vegetation)**	**List II** **(Region)**
A. Mangrove	1. Odisha
B. Tropical wet evergreen	2. Western Ghats
C. Tropical wet deciduous	3. West Bengal
D. Dry Savannah and Steppe	4. Gujarat

Code:

	A	B	C	D
(a)	3	2	1	4
(b)	1	4	3	2
(c)	3	4	1	2
(d)	1	2	3	4

38. In which state/union territory are tropical wet evergreen forests found?
(a) Madhya Pradesh
(b) Odisha
(c) Andaman and Nicobar Islands
(d) Telangana

39. Which one of the following is not a main area of mangroves?
(a) Sunderbans (b) Mahanadi Delta
(c) Gulf of Khambat (d) Tarai

40. What is the approximate percentage of reserved forest area in India?
(a) 52% (b) 53%
(c) 54% (d) 51%

41. Consider the following statements and select the correct answer using the code given below:
1. Teak forest is deciduous in nature
2. The largest area of teak is in the state of Assam.

Code:
(a) Only 1 is correct
(b) Only 2 is correct
(c) Both 1 and 2 are correct
(d) Neither 1 nor 2 is correct

42. Consider the following statements and select the correct answer using the code given below:
    1. *Sabai* grass is found in the sub-Himalayas Tarai regions.
    2. The *khus khus* grass is grown in Rajasthan.
    3. The *elephant grass* is grown in Andhra Pradesh.

    Code:
    (a) 1 and 2 are correct
    (b) 2 and 3 are correct
    (c) 1 and 3 are correct
    (d) 1, 2 and 3 are correct

43. In which part of India are thorn forests commonly found?
    (a) Assam and Meghalaya
    (b) Tarai Region
    (c) Rajasthan and Gujarat
    (d) Malabar Coast

44. What type of vegetation is usually found in river estuaries?
    (a) Deciduous forests
    (b) Littoral forests
    (c) Thorn forests
    (d) Grass

45. Where is sandal-wood commonly found?
    (a) Topical deciduous forests
    (b) Topical evergreen forests
    (c) Alpine forests
    (d) Tropical scrubland areas

46. Examine the following statements and select the correct answer using the code given below:
    1. Haldu is a tree of monsoon climate.
    2. Samuel is found in Assam, Bihar and Tamil Nadu.
    3. Shisham occurs on the Chhotanagar Plateau
    4. Rosewood grows along the slopes of Western Ghats.

    Code:
    (a) 1, 2 and 3 are correct
    (b) 1, 2 and 4 are correct
    (c) 1, 3 and 4 are correct
    (d) 1, 2, 3 and 4 are correct

47. Which one of the following statements is correct?
    (a) White willow grows in the valley of Kashmir.
    (b) Blue pine grows along the entire length of Himalayas.
    (c) Deodar grows in North-East Himalayas.
    (d) Ebony prefers well drained rocky soil.

48. Match the following lists:

	List I (Biosphere Reserve)		List II (State)
A.	Similipal	1.	Sikkim
B.	Dihang Diband	2.	Uttarakhand
C.	Nokrek	3.	Arunachal Pradesh
D.	Kanchenjunga	4.	Odisha
		5.	Meghalaya

Select the correct answer using code given below:

Code:

	A	B	C	D
(a)	1	3	4	5
(b)	4	5	2	1
(c)	1	5	2	4
(d)	4	3	5	1

49. Which one among the following has the maximum number of National Parks?
    (a) Andaman and Nickobar Islands
    (b) Arunachal Pradesh
    (c) Assam
    (d) Meghalaya

50. Which one of the following is not a Biosphere reserve?
    (a) Agasthyamalai
    (b) Nallamalai
    (c) Nilgiri
    (d) Pachmarhi

51. Which one of the following is located in Baster region?
    (a) Baundhavgarh National Park
    (b) Randeli Sanctuary
    (c) Rajaji National Park
    (d) Indrawati National Park

52. Consider the following statements:
    1. Silent Valley National Park is in the Nallamalal range.
    2. Pathrakavedu Hydro-electric plant project is proposed to be built near the Silent Valley National Park

3. The Kunthi river originates in Silent Valley rain forests.

Which of the statements given above is/are correct?

(a) 1 and 3 (b) Only 2
(c) 2 and 3 (d) All of these

53. Consider the following regions
1. Eastern Himalayas
2. Eastern Mediterranean regions
3. North-West Australia

Which of the above is/are biodiversity hotspot(s)?

(a) Only 1 (b) 1 and 2
(c) 2 and 3 (d) All of these

54. Consider the following areas:
1. Bandipur
2. Bhitarkarika
3. Manas
4. Sunderbans

Which of the above are tiger reserves?

(a) Both 1 and 2 (b) 1, 3 and 4
(c) 2, 3 and 4 (d) All of these

55. Which of the following is/are Ramsar Site(s)?
(a) Bhitankarika mangroves
(b) Chilika Lake
(c) Deepor Beel
(d) All of these

56. Which among the following is not an elephant reserve?
(a) Mayurbhanj (b) Deomoli
(c) Panna (d) Rayala

57. Match the following lists:

List I (National Park/Sanctuary)	**List II (State)**
A. Kangar Ghats	1. Chhattisgarh
B. Nagarhole	2. Haryana
C. Kugti	3. Himachal Pradesh
D. Sultanpur	4. Karnataka

Code:

	A	B	C	D
(a)	3	2	1	4
(b)	1	4	3	2
(c)	3	4	1	2
(d)	1	2	3	4

58. Consider the following pairs:
1. Bhiterkarika Odisha – Salt Water Crocodile
2. Desert National Park, Rajasthan – Great Indian Bustard
3. Eravikulam, Kerala – Hoolak Gibbon

Which of these pairs given above is/are correctly matched?

(a) Only 1 (b) 1 and 2 Only
(c) Only 2 (d) 1, 2 and 3

59. Which of the following groups of animals belongs to the category of endangered species?
(a) Great Indian Bustard, Musk Deer, Red Panda and Asiatic Wild Ass.
(b) Kashmir Stag, Cheetal, Blue Bull, and Great Indian Bustard
(c) Snow leopard, Swamp Deer, Rhesus Monkey and Saras (crane)
(d) Lion tailed macaque, Blue Bull, Hanuman Langur and Cheetal

60. Consider the following pairs:
1. Indravati Tiger Reserve– Chhatisgarh
2. Kundarmukh Tiger Reserve – Karnataka
3. Ratapani Tiger Reserve– Maharashtra
4. Satkosia Tiger Reserve – Odisha

Which of these pairs given above is/are correctly matched?

(a) Only 1 (b) Only 2
(c) 2 and 3 (d) 1, 2 and 4

61. Which of the following Tiger Reserves is located in Rajasthan?
(a) Suhelwa (b) Sariska
(c) Bandipur (d) Valmiki

62. In which year Biodiversity Act was introduced in India?
(a) 1986 (b) 1990
(c) 1992 (d) 2002

63. Consider the following statements:
1. Nagarjunsagar–Srisailam is the largest tiger reserve in terms of area.
2. Panda Tiger Reserve in Maharashtra in minimum in area.
3. As per website of NTCA, Sunderbans in West Bengal has the largest number of tigers.

Which of the above statements is/are incorrect?
(a) Only 1 (b) 1 and 2
(c) All of these (d) None of them

64. Gambhir river is related to which of the following National Parks:
(a) Keoladeo (b) Corbett
(c) Bhiarkanika (d) Dachigam

65. In which biosphere reserve is Indian Wild Ass found?
(a) Simipal (b) Rann of Kutch
(c) Manas (d) Panna

66. In which biosphere reserve is Red Panda found?
(a) Dehang-Diband (b) Nilgiri
(c) Nokreak (d) Nanda Devi

67. Which among the following was the first national park established in India?
(a) Bandhavgarh National Park
(b) Corbett National Park
(c) Dudhwa National Park
(d) Kaziranga National Park

68. Which sanctuary is famous for 'Kashmiri Stag'?
(a) Simlipal sanctuary
(b) Dachigam Sanctuary
(c) Chandraprabha Sanctuary
(d) Bandipur Sanctuary

69. Which among the following is a correct pair?
(a) Goutam Buddha Sanctuary – Bihar
(b) Jaidapara Sanctuary – West Bengal
(c) Achanakmar Sanctuary– Karnataka
(d) Chandraprabha Sanctuary – Uttar Pradesh

70. Match the following lists and select the correct answer code given below:

	List I	List II
(a)	Gulf of Mannar	1. Assam
(b)	Pachmarhi	2. Sikkim
(c)	Dibru-Siakhonwa	3. Madhya Pradesh
(d)	Kanchenjunga	4. Tamil Nadu

Code:

	A	B	C	D
(a)	1	3	4	2
(b)	4	3	1	2
(c)	3	2	4	1
(d)	2	1	3	4

71. Which of the following pair is not correctly matched?

	Elephant Reserve	State
(a)	Lemru	Chhittisgsrh
(b)	Deomoli	Arunachal Pradesh
(c)	Nilambur	Kerala
(d)	Anaimudi	Tamil Nadu

72. Match the following Lists:

	List I (Tiger Reserve)	List II (State)
A.	Nandapha	1. Goa
B.	Mhadei	2. Rajasthan
C.	Kundermukh	3. Arunachal Pradesh
D.	Ranthambore	4. Karnataka

Code:

	A	B	C	D
(a)	2	3	1	4
(b)	1	2	4	3
(c)	3	1	4	2
(d)	4	2	3	1

73. Which of the following tiger reserve is not located in Madhya Pradesh?
(a) Suhelwa (b) Bandhavgarh
(c) Bori-Satpura (d) Ratapari

74. Which of the following is not a Bird Sanctuary in India?
(a) Ranganthitto (b) Panhmarhi
(c) Vedanthangal (d) Ghana

75. Which of the following is not a Ramsar Wetland Site?
(a) Kolleru lake (b) Wular lake
(c) Rudrasagar lake (d) Pushkar Lake

76. Which of the Sanctuary in India is famous for Rhinoceros?
(a) Sunderbans (b) Jaldapara
(c) Panchmarhi (d) Dandeli

77. Which one of the following state in India is not having Tiger Reserves?
(a) Himachal Pradesh (b) Rajasthan
(c) Mizoram (d) Goa

78. Which one of the following soils has the largest area in India?
(a) Lateritic soil (b) Alluvial soil
(c) Red soil (d) Regur soil

79. Laterite soils are not found in:
   (a) Aravalli
   (b) Brahmaputra valley
   (c) Odisha and West Bengal
   (d) Jharkhand and Western Ghats

80. Which one of the following states is not having black soil in India?
   (a) Madhya Pradesh (b) Maharashtra
   (c) Punjab (d) Telangana

81. In India, red soils cover an area of:
   (a) 25.2% (b) 27.6%
   (c) 28.5% (d) 26.8%

82. In India peaty and marshy soils are found in:
   (a) Rann of Kachchh (b) Tarai region
   (c) Himachal Pradesh (d) Jammu & Kashmir

83. Consider the following statements and select the correct answer using the code given below:
   1. Boggy soils are found in the Ganga Khadar and Assam valley
   2. Older alluviums are found in Punjab
   3. Serozen (Saline) soils are found in Rajasthan.
   4. Regur soils are found in Maharashtra and Gujarat.

   Code:
   (a) 1, 2 and 3 are correct (b) 1, 2 and 4 are correct
   (c) 1, 3 and 4 are correct (d) 1, 2, 3 and 4 are correct

84. With reference to soil conservation consider the following practices:
   1. Crop rotation 2. Sand fences
   3. Terracing 4. Wind breaks

   Which of the above are considered appropriate methods for soil conservation in India?
   (a) 1, 2 and 3 (b) 2 and 4
   (c) 1, 3, and 4 (d) All of these

85. When you travel in certain parts of India, you will notice red soil, what is the main reason for this color?
   (a) Abundance of magnesium
   (b) Accumulated humus
   (c) Presence of ferric oxides
   (d) Abundance of phosphates

Answer Key – SET 3

Que. No	Answer	Que. No	Answer	Que. No	Answer
1.	b	31.	b	61.	b
2.	a	32.	b	62.	d
3.	a	33.	a	63.	d
4.	d	34.	c	64.	a
5.	a	35.	d	65.	b
6.	b	36.	b	66.	c
7.	c	37.	a	67.	b
8.	a	38.	c	68.	b
9.	b	39.	d	69.	a
10.	c	40.	b	70.	b
11.	d	41.	a	71.	d
12.	a	42.	a	72.	c
13.	b	43.	c	73.	a
14.	d	44.	b	74.	b
15.	a	45.	a	75.	d
16.	b	46.	b	76.	b
17.	b	47.	c	77.	a
18.	a	48.	d	78.	b
19.	b	49.	c	79.	b
20.	c	50.	b	80.	c
21.	b	51.	d	81.	d
22.	d	52.	c	82.	a
23.	d	53.	b	83.	d
24.	a	54.	b	84.	c
25.	b	55.	d	85.	c
26.	c	56.	c		
27.	b	57.	b		
28.	d	58.	b		
29.	a	59.	a		
30.	b	60.	d		

SET – 4

INDIA: Social Geography and Population

1. Which one of the following is not a characteristic of Indian tribe?
(a) Kinship as an instrument of social bounds
(b) Segmentary character
(c) All are engaged as labourers
(d) Absence of strong, complex, formal organizations

2. What is the percentage of schedule tribe population in India in 2011?
(a) 9.8% (b) 8.6%
(c) 10.5% (d) 7.6%

3. The decadal growth rate percentage of schedule tribe population in India during 2001–2011 was:
(a) 23.7% (b) 24.5%
(c) 31.2% (d) 26.2%

4. The highest decadal growth rate of schedule tribe population was recorded in 1971–81. What was that percentage?
(a) 31.2% (b) 26.2%
(c) 24.5% (d) 35.8%

5. Which one of the following pairs of states and tribes is not correctly matched?

	States	Tribes
(a)	Assam	Miri
(b)	Nagaland	Konyab
(c)	Arunachal Pradesh	Apatini
(d)	Madhya Pradesh	Lambada

6. Which one of the following is correctly matched?

	Tribes	States
(a)	Tharu	Uttarakhand
(b)	Bhotia	Uttar Pradesh
(c)	Munda	Bihar
(d)	Kol	Rajasthan

7. Which one of the following states of India has no tribal population?
(a) Andhra Pradesh (b) Punjab
(c) Uttar Pradesh (d) West Bengal

8. Match List I with List II and select the correct answer using the code given below:

List I (Tribe)	List II (State)
A. Angami	1. Kerala
B. Khasi	2. Meghalaya
C. Mopala	3. Nagaland
D. Toda	4. Tamil Nadu

Code:

	A	B	C	D
(a)	4	3	2	1
(b)	3	2	4	1
(c)	3	2	1	4
(d)	4	3	1	2

9. Which one of the following is not correctly matched?
(a) Lanbudi – Tamil Nadu
(b) Munda – Jharkhand
(c) Bhil – Rajasthan
(d) Jarawa – Andaman

10. Which one of the following is not correctly matched?
(a) Tharu – Uttar Pradesh
(b) Garasiya – Rajasthan
(c) Gaddi – Himachal Pradesh
(d) Bakkarwal – Rajasthan

11. Match List I with List II and select the correct answer using the code given below:

List I (Tribe)	List II (Area)
A. Adivasis	Kerala
B. Moplaha	Nilgiri Hills
C. Todas	Manipur
D. Angamis	Madhya Pradesh

Code:

	A	B	C	D
(a)	4	2	1	3
(b)	4	1	2	3
(c)	4	1	3	2
(d)	4	3	2	1

12. Which state is inhabited by the Khasi tribe?
(a) Nagaland (b) Tripura
(c) Meghalaya (d) Arunachal Pradesh

13. Which tribe of India has no contact with outsiders?
(a) Zarawa (b) Birhor
(c) Oraom (d) Munda

14. Which are the largest two tribes of India?
(a) Naga and Thaurs (b) Gonds and Bhil
(c) Mizo and Khasi (d) Bhil and Munda

15. Which one of the following is not a tribal region of India according to B.K. Burmon?
(a) North-East India
(b) Sub-Himalayas
(c) Ganga-Yamuna Doab
(d) Central and East India

16. Consider the following statement and select the correct answer using code given below:
1. Tribe in India always prefers to live in areas unfavorable to agriculture.
2. The tribal people have now shifted to sub-urban areas of big cities.
3. The Western India has the highest concentration of tribal population.

Code:
(a) Only 1 is correct
(b) 1 and 2 are correct
(c) 2 and 3 are correct
(d) 1, 2 and 3 are correct

17. In which states of India tribal population is more than 80 per cent?
(a) Mizoram (b) Nagaland
(c) Meghalaya (d) All of these

18. Which one of the following is not a hunting tribe of India?
(a) Chandi of Andhra Pradesh
(b) Kolam of Maharashtra
(c) Juang of Odisha
(d) Kurumba of Kerala

19. In which state are Lusai, Maat and Paki tribes living?
(a) Mizoram (b) Meghalaya
(c) Odisha (d) Assam

20. Which one of the following is not a tribe of Madhya Pradesh?
(a) Baiga (b) Bhumia
(c) Gaddi (d) Damor

21. Which of the following is not correctly matched?

	Tribe	State
(a)	Mishmi	Arunachal Pradesh
(b)	Mikir	Meghalaya
(c)	Gaddi	Jammu & Kashmir
(d)	Agariya	Madhya Pradesh

22. Which one of the following is not correctly matched?

	State	Tribe
(a)	Rajasthan	Rawari
(b)	Karnataka	Mudugur
(c)	Kerala	Kudia
(d)	Odisha	Dhanka

23. Which one of the following is not a tribe of West Bengal?
(a) Sainsi (b) Baiga
(c) Lepcha (d) Mahali

24. Which one of the following is not a tribe of north-east states of India?
(a) Kuki (b) Khond
(c) Lashai (d) Sema

25. Which one of the following is not a tribe of Odisha?
(a) Bhumia (b) Kora
(c) Kolam (d) Gondia

26. Consider the following statements and select the correct answer using code given below:
1. A caste is an endogamous group.
2. Caste is a major reference point of social interaction in India.
3. Caste is an indication of economic status.

Code:
(a) Only 1 is correct
(b) 1 and 2 are correct
(c) 2 and 3 are correct
(d) 1, 2 and 3 are correct

27. Which is an example of sectarian caste?
(a) Rajbhar (b) Paraiyan
(c) Nair (d) Lingayats

28. Which one of the following is not a caste based on occupation?
(a) Potters (b) Barbers
(c) Nair (d) Goldsmiths

29. In India, which religion is having the highest percentage of followers?
(a) Hinduism (b) Islam
(c) Jainism (d) Buddhists

30. What was the percentage of Christian population in total population of India in 2001?
(a) 3.2% (b) 2.3%
(c) 1.9% (d) 1.6%

31. Which is the correct sequence in descending order of the followers of the following religion in India?
(a) Sikhs, Jains, Buddhist and Christians
(b) Buddhist, Sikhs, Jains, and Christians
(c) Christians, Sikhs, Buddhists and Jains
(d) Jains, Buddhists, Christians and Sikhs

32. Among these Indian languages, which one is spoken maximum in the world after Hindi?
(a) Telugu (b) Tamil
(c) Bengali (d) Malayalam

33. Which one of the following languages belongs to the Austric group?
(a) Marathi (b) Ladakhi
(c) Khasi (d) Tamil

34. Which one of the following statements is not correct?
(a) The largest Buddhist Monastery in India is Assam.
(b) The Language of Konyak is spoken in Nagaland.
(c) The percentage of Jains population in India's total population is less than 0.5 per cent.
(d) Himachal Pradesh is having the highest percentage of Hindu population.

35. One of the areas of concentration of Buddhist population in India is:
(a) Bihar (b) Uttar Pradesh
(c) Uttarakhand (d) Arunachal Pradesh

36. Which state has the highest proportion of Christian population?
(a) Kerala (b) Nagaland
(c) Andhra Pradesh (d) Goa

37. Who are the pastoralists of Tamil Nadu?
(a) Sathal (b) Gonds
(c) Todas (d) Bhils

38. Which of the following states is having the lowest Muslim population?
(a) Jharkhand (b) Mizoram
(c) Meghalaya (d) None of the above

39. The main concentration of Buddhists is in:
(a) Maharashtra (b) Jammu & Kashmir
(c) Tripura (d) Sikkim

40. How many languages have been recorded in Linguistic Survey of India conducted by Garrison?
(a) 160 (b) 165
(c) 170 (d) 179

41. The number of languages recorded in India in 1961 Census was:
(a) 182 (b) 187
(c) 194 (d) 198

42. Which is the dominant language group in India?
(a) Indo-Aryan (b) Dravidian
(c) Tibeto-Burman (d) Austroasiatic

43. Which one of the following is not a language family?
(a) Dravidian (b) Sino-Tibetan
(c) Dardic (d) Indo-European

44. Which is not a speech area of Indo-Aryan group of language?
(a) Maharashtra (b) Bihar
(c) Uttar Pradesh (d) Rajasthan

45. Which is not a correct match?

	State	Official Language
(a)	Andhra Pradesh	Telgu
(b)	Karnataka	Kannada
(c)	Gujarat	Gujarati
(d)	Jammu & Kashmir	Hindi/Punjabi

46. Arrange the following languages in descending order according to their number of speakers:
(a) Bengali, Gujarati, Telugu and Punjabi
(b) Bengali, Telugu, Gujarati and Punjabi
(c) Telugu, Bengali, Punjabi and Gujarati
(d) Gujarati, Bengali, Telugu and Punjabi

47. Which of the following languages is spoken by the least number of people in India?
(a) Nepali (b) Manipuri
(c) Sanskrit (d) Santali

48. In which area of India is Santali language spoken?
(a) Uttarakhand
(b) Karnataka
(c) Malwa Plateau
(d) Chotanagpur Plateau

49. Which one of the following is not a classical language in Ind'a?
(a) Urdu (b) Sanskrit
(c) Telugu (d) Malayalam

50. How many languages are included as classical languages of India till 2014?
(a) 4 (b) 5
(c) 6 (d) 7

51. Which one of the following is a regional language in India?
(a) Magalia (b) Sindhi
(c) Bagri (d) Tulu

52. India has 2.4 per cent of world's area, what is its share in world's population?
(a) 17.64% (b) 16.54%
(c) 15.38% (d) 16.92%

53. When was the first population Census carried out in India?
(a) 1851 (b) 1862
(c) 1872 (d) 1891

54. The decadal growth rate of India's population during 2001–11 was:
(a) 21.54% (b) 17.64%
(c) 21.64% (d) 18.52%

55. The highest decadal growth rate (24.80%) in India's Population has been recorded in:
(a) 1991–2000 (b) 1981–1991
(c) 1971–1981 (d) 1961–1971

56. In India's population which period is considered as a period of rapid growth?
(a) 1981–2011 (b) 1951–1981
(c) 1921–1951 (d) 1901–1921

57. Out of the following states the lowest decadal growth of population between 2001 and 2011 was recorded in:
(a) Punjab
(b) Arunachal Pradesh
(c) Nagaland
(d) Kerala

58. What is the percentage of the total population of India that lives in Uttar Pradesh?
(a) 16.5% (b) 15.5%
(c) 14.5% (d) 13.5%

59. The descending order of top three states in India according to their total population in 2011 is:
(a) Uttar Pradesh, West Bengal, Madhya Pradesh
(b) Uttar Pradesh, Maharashtra, Bihar
(c) Uttar Pradesh, Tamil Nadu, Madhya Pradesh
(d) Uttar Pradesh, Madhya Pradesh, West Bengal

60. Which one of the following states had recorded highest population in 2011?
(a) Maharashtra (b) Andhra Pradesh
(c) West Bengal (d) Bihar

61. Which of the following States in India is having the lowest total population in 2011?
(a) Goa (b) Arunachal Pradesh
(c) Mizoram (d) Sikkim

62. Which of the Union Territory is having the second largest population in UTs in 2011?
(a) Chandigarh
(b) Pondicherry
(c) Daman & Diu
(d) Andaman & Nicobar Islands

63. Consider the following statements and select the correct answer using the code given below:
1. The decadal growth rate of population between 2001–2011 has registered the sharpest decline since independence.
2. Dadra and Nagar Haveli has recorded the highest growth rate among UTs in 2011.
3. India's Population was 1.21 billion in 2011.

4. Between 2001–11 the male population grew more than the female population.

Code:

(a) 1, 2 and 3 are correct
(b) 2, 3 and 4 are correct
(c) 1, 2 and 4 are correct
(d) 1, 2, 3 and 4 are correct

64. Which of the following state(s) were having density of more than 1,000 persons in 2011?
(a) Kerala and Uttar Pradesh
(b) West Bengal and Bihar
(c) Uttar Pradesh and Punjab
(d) Kerala and Bihar

65. The lowest density of population in 2011 has been recorded in:
(a) Sikkim (b) Manipur
(c) Arunachal Pradesh (d) Mizoram

66. Which of the following UTs was having the highest density of population in 2011?
(a) Pondicherry (b) Daman & Diu
(c) Lakshadweep (d) Chandigarh

67. Which one of the following states was having the lowest density of population in 2011?
(a) Rajasthan (b) Uttarakhand
(c) Chhattisgarh (d) Himachal Pradesh

68. Consider the following statements-
1. There is a constant increase in population density of India since 1901.
2. There is a constant decline in sex-ratio since 1901.

Which of the above statements is/are correct?
(a) Only 1 (b) Only 2
(c) Both 1 and 2 (d) Neither 1 nor 2

69. In which Census year was the lowest sex-ratio recorded?
(a) 1991 (b) 1981
(c) 1971 (d) 1961

70. The state having the lowest sex-ratio in 2011 was:
(a) Punjab (b) Haryana
(c) Jammu & Kashmir (d) Sikkim

71. Among the states and UTs of India, which has recorded the lowest sex-ratio in 2011?
(a) Haryana (b) Chandigarh
(c) Daman and Diu (d) Sikkim

72. Which of the following states are having lowest child sex-ratio in the country?
(a) Punjab, Himachal Pradesh, Uttarakhand
(b) Haryana, Jharkhand, Gujarat
(c) Jammu & Kashmir, Rajasthan, Odisha
(d) Haryana, Punjab, Jammu & Kashmir

73. Among the Union Territories of India, which is having the lowest child sex-ratio?
(a) Chandigarh
(b) Delhi
(c) Lakshadweep
(d) Daman and Diu

74. Consider the following statements and select the correct answer using code given below:
1. At Independence, literacy rate of India was approximately one-fifth only.
2. In 2011, approximately three-fourth population of India was literate.

Code:

(a) Only 1 (b) Only 2
(c) Both 1 and 2 (d) Neither 1 nor 2

75. The highest sex-rationt in children (0–6 years) in 2011 is recorded in:
(a) Mizoram
(b) Arunachal Pradesh
(c) Meghalaya
(d) Manipur

76. Which one of the following states has the highest literacy rate in the country in 2011 after Kerala?
(a) Tripura (b) Mizoram
(c) Himachal Pradesh (d) Meghalaya

77. Consider the following statements and select the correct answer using the code given below:
1. Bihar, Arunachal Pradesh and Rajasthan are having the lowest literacy rate in the country in 2011.
2. The literacy rate in urban areas is 84.9 per cent, while in rural area it is 68.9 per cent.

Code:

(a) Only 1 is correct
(b) Only 2 is correct
(c) Both 1 and 2 are correct
(d) Neither 1 nor 2 is correct

78. According to 2011 Census, which city is having the highest population in India?
   (a) Mumbai (b) Delhi
   (c) Kolkata (d) Chennai

79. Consider the following statements:
   1. The proportion of rural population has declined from 72.19 per cent in 2001 to 68.84 per cent in 2011.
   2. In 2011 the sex-ratio in rural areas is lower than in urban areas.
   3. The growth of rural population in India during 2001–11 has been 12.18 per cent.

   Select the correct answer using code given below:
   (a) 1 and 2 are correct
   (b) 2 and 3 are correct
   (c) 1 and 3 are correct
   (d) 1, 2 and 3 are correct

80. In which one of the following total metropolitan areas, the lowest total population in 2011 was recorded?
   (a) Pune (b) Jaipur
   (c) Patna (d) Kochi

Answer Key – SET 4

Que. No.	Answer	Que. No.	Answer	Que. No.	Answer
1	c	28	c	55	d
2	b	29	a	56	b
3	a	30	b	57	c
4	d	31	c	58	a
5	d	32	c	59	b
6	a	33	c	60	c
7	b	34	a	61	d
8	c	35	d	62	b
9	a	36	b	63	a
10	d	37	c	64	b
11	b	38	b	65	c
12	c	39	a	66	d
13	a	40	d	67	d
14	b	41	b	68	d
15	c	42	a	69	a
16	a	43	c	70	b
17	d	44	a	71	c
18	b	45	d	72	d
19	a	46	b	73	b
20	c	47	c	74	c
21	b	48	d	75	a
22	d	49	a	76	b
23	a	50	c	77	c
24	b	51	b	78	a
25	c	52	a	79	c
26	b	53	c	80	c
27	d	54	b		

SET – 5

INDIA: Agriculture, Livestock, Fisheries and Irrigation

1. Consider the following statements and select the correct answer using code given below:
   1. In Indian agriculture there is predominance of food crops.
   2. The Indian agriculture is labor intensive.
   3. More than 50 per cent agricultural products are exported from India to neighboring countries.

   Code:
   (a) Only 1 is correct
   (b) Only 2 is correct
   (c) Only 2 and 3 are correct
   (d) 1 and 2 are correct

2. Which one of the following is not a characteristic of Indian agriculture?
   (a) Variety of crops and mixed cropping
   (b) Low productivity
   (c) Based on foreign capital
   (d) labour intensive agriculture

3. Which one of the following agricultural types is eco-friendly?
   (a) Commercial Agriculture
   (b) Organic farming
   (c) Subsistence farming
   (d) Shifting cultivation

4. The lower Ganetic plain is characterized by humid climate with high temperature throughout the year. Which one among the following pair of crops is most suitable for this region?
   (a) Paddy and Cotton
   (b) Wheat and Jute
   (c) Paddy and Jute
   (d) Wheat and Cotton

5. Which one of the following state is not having shifting cultivation?
   (a) Uttarakhand (b) Mizoram
   (c) Tripura (d) Odisha

6. Which is not a correct match?

	Name of the Cultivation	State
(a)	Pondu	Andhra Pradesh
(b)	Onam	Kerala
(c)	Masha	Maharashtra
(d)	Jhoom	Assam

7. In which region do Zaid crops grow?
   (a) Arid regions (b) Irrigated areas
   (c) Mountains areas (d) Delta regions

8. Arrange the following states of India in respect of net cultivated area and select the correct answer using code given below:
   1. Uttar Pradesh 2. Haryana
   3. Punjab 4. West Bengal

   Code:
   (a) Uttar Pradesh, Haryana, Punjab, West Bengal
   (b) Punjab, Uttar Pradesh, West Bengal, Haryana
   (c) Punjab, Haryana, West Bengal, Uttar Pradesh
   (d) Haryana, Punjab, Uttar Pradesh, West Bengal

9. Which region is called 'granary of South India'?
   (a) Thanjavur (b) Tiruchirapalli
   (c) Kerala (d) Coimbatore

10. Operation Flood is associated with:
    (a) Green Revolution (b) Blue Revolution
    (c) Brown Revolution (d) White Revolution

11. Which state is called the 'agricultural epitome' of India?
    (a) Madhya Pradesh (b) Tamil Nadu
    (c) Uttar Pradesh (d) West Bengal

12. The correct descending order of the top rice producing states in India is:
    (a) Uttar Pradesh, Bihar, Odisha
    (b) West Bengal, Punjab, Uttar Pradesh
    (c) West Bengal, Tamil Nadu, Bihar
    (d) Uttar Pradesh, Punjab, Haryana

13. *'Aman'*, *'Aus'* and *'Bora'* are the types of which crop?
    (a) Rice (b) Wheat
    (c) Gram (d) Mustard

14. Which one of following states has the highest areas under wheat cultivation?
(a) Punjab (b) Haryana
(c) Uttar Pradesh (d) Madhya Pradesh

15. Which state of India has the highest yields of wheat per unit area?
(a) Bihar (b) Haryana
(c) Uttar Pradesh (d) Punjab

16. Which one of the following is not a maize producing state of India?
(a) Rajasthan (b) Maharashtra
(c) Madhya Pradesh (d) Karnataka

17. Which one of the following states produce more than 50 per cent of the total Jowar production of India?
(a) Maharashtra (b) Karnataka
(c) Madhya Pradesh (d) Andhra Pradesh

18. The top two Bajra producing states in India are:
(a) Rajasthan and Gujarat
(b) Maharashtra and Gujarat
(c) Gujarat and Madhya Pradesh
(d) Gujarat and Uttar Pradesh

19. The top two Barley producing states in India are:
(a) Uttar Pradesh, Gujarat
(b) Bihar and Odisha
(c) Uttar Pradesh and Rajasthan
(d) Madhya Pradesh and Maharashtra

20. The top three pulses producing states of India in descending order are:
(a) Madhya Pradesh, Maharashtra, Haryana
(b) Uttar Pradesh, Madhya Pradesh, Punjab,
(c) Uttar Pradesh, Madhya Pradesh, Rajasthan
(d) Madhya Pradesh, Uttar Pradesh, Maharashtra

21. The highest groundnut producing state of India is:
(a) Gujarat (b) Andhra Pradesh
(c) Tamil Nadu (d) Karnataka

22. Which one of the following is the highest rapeseed and mustard producing state in India?
(a) Rajasthan (b) Madhya Pradesh
(c) Maharashtra (d) Punjab

23. The highest soybean producing state in India is:
(a) Gujarat (b) Maharashtra
(c) Madhya Pradesh (d) Rajasthan

24. The top three sugarcane producing states in India in descending order are:
(a) Maharashtra, Karnataka, Bihar
(b) Uttar Pradesh, Maharashtra, Karnataka
(c) Karnataka, Andhra Pradesh, Uttar Pradesh
(d) Maharashtra, Karnataka, Uttar Pradesh

25. The leading cotton producing state in India is:
(a) Andhra Pradesh (b) Maharashtra
(c) Haryana (d) Gujarat

26. Which state of India produces more than seventy per cent Jute in India?
(a) Assam (b) Bihar
(c) West Bengal (d) Odisha

27. Which one of the following is a correct match?

	Crops	Highest Producing State in India
(a)	Sugarcane	Maharashtra
(b)	Cotton	Gujarat
(c)	Soyabean	Madhya Pradesh
(d)	Rapeseed & Mustard	Rajasthan

28. In which state of India is cashew nut produced?
(a) Tamil Nadu (b) Karnataka
(c) Kerala (d) Telengana

29. The only state which produces saffron in India is:
(a) Uttarakhand (b) Jammu & Kashmir
(c) Himachal Pradesh (d) Sikkim

30. Which one of the following is not a tea growing state in India?
(a) Odisha (b) Assam
(c) Karnataka (d) Tripura

31. Which is not a correct match

	Crop	State
(a)	Rubber	Kerala
(b)	Tobacco	Andhra Pradesh
(c)	Ragi	Goa
(d)	Coffee	Karnataka

32. Which one of the following is not a main component of 'Green Revolution?'
    (a) Use of High Yielding varieties (HYV) seed
    (b) Use of Chemical Fertilizers
    (c) Mechanization of agriculture
    (d) Literacy Programme for farmers

33. The Farming of the species of algae is known as:
    (a) Apiculture (b) Algaculture
    (c) Horticulture (d) Sericulture

34. Which is not a correct match?

	Culture Type	Purpose/Product
(a)	Sericulture	Silk Farming
(b)	Apiculture	Beekeeping
(c)	Aquaculture	Mushroom Farming
(d)	Horticulture	Fruit Farming

35. Following are the characteristics of an area in India:
    1. Hot and Humid Climate
    2. Annual rainfall 200 cm
    3. Hill slopes upto an altitude of 1,000 m
    4. Annual range of temperature 15°c to 30°c
    Which one among the following crops is most likely to be found in the area described above?
    (a) Mustard (b) Cotton
    (c) Pepper (d) Tea

36. Which of the following is the chief characteristic of mixed farming?
    (a) Cultivation of both cash crops and food crops
    (b) Cultivation of two or more crops in the same field
    (c) Rearing of animals and cultivation of crops together
    (d) None of the above

37. Consider the following crops of India:
    1. Groundnut
    2. Sesamum
    3. Pearl Millet
    Which of the above is/are predominantly rain fed crop/crops?
    (a) 1 and 2 (b) 2 and 3
    (c) Only 3 (d) All of these

38. Which one of the following states has the highest yield of sugarcane per hectare?
    (a) Maharashtra (b) Uttar Pradesh
    (c) Tamil Nadu (d) Karnataka

39. Which one of the following crops has the highest percentage of irrigated area?
    (a) Wheat (b) Sugarcane
    (c) Rice (d) Oilseeds

40. Which of the following state has recorded the highest per hectare yield?
    (a) Uttar Pradesh (b) Haryana
    (c) Punjab (d) Madhya Pradesh

41. Which one of the following states is the leading producer of coconut?
    (a) Tamil Nadu (b) Karnataka
    (c) Goa (d) Kerala

42. Arrange the tea production of following states in a descending order using the code given below:
    (a) Assam (b) Karnataka
    (c) Kerala (d) West Bengal

Code:

	A	B	C	D
(a)	1	2	3	4
(b)	1	4	2	3
(c)	1	4	3	2
(d)	1	3	4	2

43. The chief cause of low yields of crops in India is:
    1. Small size of holdings
    2. Traditional methods of farming
    3. Illiteracy among farmers
    4. Low level of farm mechanization
    Select the correct answer using code given below:
    (a) 1, 2 and 3 (b) 1, 2 and 4
    (c) 2, 3 and 4 (d) 1, 3 and 4

44. The high quality Virginia tobacco is produced in:
    (a) Andhra Pradesh
    (b) Tamil Nadu
    (c) Maharashtra
    (d) Karnataka

45. **Assertion (A)** The eastern coast of India produces more rice than the western coast.
    Reason (R) The eastern coast receives more rainfall than the western coast
    (a) Both A and R are true and R is the correct explanation of A
    (b) Both A and R are true but R is not the correct explanation of A

(c) A is true but R is false
(d) A is false but R is true

46. Though coffee and tea both are cultivated on hill slopes, there is some difference between them regarding their cultivation. In this context consider the following statements:
1. Coffee plant requires a hot and humid climate of tropical areas, whereas tea can be cultivated in both tropical and subtropical areas.
2. Coffee is propagated by seeds but tea is propagated by stem cutting only.
(a) Only 1 (b) Only 2
(c) Both 1 and 2 (d) Neither 1 nor 2

47. If rain falls only for two months, the crops best suited to the resulting conditions will be:
(a) Sugarcane (b) Tea
(c) Rice (d) Pulses

48. The water-logging condition during initial stage is suitable for which of the following crops?
(a) Mustard (b) Rice
(c) Tea (d) Coffee

49. Which of the following are not grown in the *kharif* season?
(a) Barley and Mustard (b) Bajra and Rice
(c) Maize and Jowar (d) Jowar and Rice

50. In dry farming the most important crop produced is:
(a) Wheat (b) Cotton
(c) Paddy (d) Bajara

51. The total number of livestock in India according to 2007 livestock Census was:
(a) 559.69 Million (b) 529.70 Million
(c) 527.06 Million (d) 611.80 Million

52. Which one of the following states has the largest number of cattle in India?
(a) Haryana (b) Madhya Pradesh
(c) Punjab (d) Uttar Pradesh

53. Arrange the following states in descending order according to the total number of cattle:
1. Maharashtra 2. Rajasthan
3. Uttar Pradesh 4. Bihar
Select the correct answer using the code given below:

Code:

(a)	1	2	3	4
(b)	2	3	1	4
(c)	3	1	4	2
(d)	3	4	2	1

54. Which is not a correct match?

	Cattle Breed	**State**
(a)	Alambadi	Tamil Nadu
(b)	Kankrej	Gujarat
(c)	Mewati	Rajasthan
(d)	Dangi	Bihar

55. The cattle breed(s) of Rajasthan is/are:
(a) Rathi (b) Tharparskar
(c) Gir (d) All of these

56. In which state is Khillari and Dangi cattle breed found?
(a) Maharashtra
(b) Karnataka
(c) Andhra Pradesh
(d) Bihar

57. Which one of the following is not a breed of buffalo?
(a) Mehsana (b) Ongole
(c) Surti (d) Murrah

58. Which is not a correct match?

	Buffalo Breed	**State**
(a)	Zalwadi	Karnataka
(b)	Jaffarabadi	Gujarat
(c)	Nagpuri	Maharashtra
(d)	Murrah	Punjab

59. In 2007 the total number of sheep in India was:
(a) 140.54 Million (b) 105.34 Million
(c) 71.56 Million (d) 11.13 Million

60. Where is Central Sheep Breeding Farm located in India?
(a) Bhatinda (b) Ranchi
(c) Nagpur (d) Hissar

61. Which is not a correct match?

	Goat Breed	**State**
(a)	Pashmina	Jammu & Kashmir
(b)	Beetal	Punjab
(c)	Angora	Karnataka
(d)	Jamunapuri	Uttar Pradesh

62. India's position in fish production in the world is:
(a) Third (b) Fourth
(c) Fifth (d) Sixth

63. The leading state in production of marine fish is:
(a) Gujarat (b) Kerala
(c) Tamil Nadu (d) Andhra Pradesh

64. The leading inland fish producing state in India is:
(a) Kerala (b) Andhra Pradesh
(c) West Bengal (d) Assam

65. In which year 'blue revolution' was started for the development of fisheries?
(a) 1958 (b) 1960
(c) 1965 (d) 1970

66. Which district of Andhra Pradesh is known as 'Shrimp Capital of India'?
(a) Vijaywada (b) Nellore
(c) Guntur (d) Cuddapah

67. **Assertion (A)** Tropical seas in India are rich in fish species but there are only few fishing grounds.
Reason (R) Fishing in India is not commercialy exploitable due to lack of quantity a particular fish species.
Code:
(a) Both A and R are true and R is the correct explanation of A
(b) Both A and R are true but R is not a correct explanation of A
(c) A is true, but R is false
(d) A is false, but R is true

68. The main canal irrigated areas of India are located in:
(a) Northern Plains of India
(b) Western Plains of India
(c) Coastal region of India
(d) Southern part of the peninsular India

69. In which state of India is the Western Yamuna canal located?
(a) Uttar Pradesh (b) Haryana
(c) Madhya Pradesh (d) Rajasthan

70. Which is not a correct match?

	Canal	State
(a)	Sir Hind canal	Punjab
(b)	Kosi canal	Bihar
(c)	Mettur canal	Karnataka
(d)	Pravana canal	Maharashtra

71. In which state is Mayurakshi project located?
(a) Andhra Pradesh (b) Odisha
(c) Karnataka (d) West Bengal

72. Which one of the following is not correctly matched?
(a) Panchet Dam – Damodar River
(b) Rihand Dam – Son River
(c) Gobind B. Pant Sagar – Rihand River
(d) Hirakund Dam – Mahanadi

73. Mahi Dam is located is Rajasthan but most of its water is utilized in other state? In which state is its water used for irrigation?
(a) Madhya Pradesh (b) Maharashtra
(c) Gujarat (d) All of these states

74. Which one of the following is not correctly matched?
(a) Indira Gandhi Nahar Project – Beas River
(b) Pochampal Project – Godavari River
(c) Tehri Dam – Bhagirathi River
(d) Mayurakshi Dam – Subernrekha River

75. Which means of irrigation accounts for the largest proportion of the irrigated area of India?
(a) Wells and Tubewells
(b) Canals
(c) Tanks
(d) Other means

76. Visveswaraya canal is located in the state of:
(a) Odisha
(b) Karnataka
(c) Andhra Pradesh
(d) Bihar

77. Which is not a correct match?

	Multipurpose Project	**State**
(a)	Malpralbha Project	Karnataka
(b)	Kanjan Project	Gujarat
(c)	Ukai Project	Maharashtra
(d)	Thein Project	Punjab

78. In which state is Tista valley Project located?
(a) Sikkim (b) Jharkhand
(c) Odisha (d) West Bengal

79. Consider the following statements:
1. Canals provide water for irrigation as and when required.
2. Canal water is a free source of irrigation.
3. Apart from irrigation, canal water is also used for drinking and other purposes.
4. All the canal irrigation areas have problem of water-logging.

Select the correct answer using the code given below:
Code:
(a) 1, 2 and 3 (b) 2, 3 and 4
(c) 1 and 3 (d) 3 and 4

80. Match List I with List II

List I (Rivers)	**List II (Dames)**
A. Cauvari	1. Alamatti
B. Krishna	2. Mettur
C. Narmada	3. Gandhi Sagar
D. Chambal	4. Sardar Sarovar

Code:

	A	B	C	D
(a)	1	4	2	3
(b)	2	1	4	3
(c)	2	1	3	4
(d)	1	3	4	2

81. The percentage of canal irrigation area to total area is highest in:
(a) Rajasthan (b) Haryana
(c) Bihar (d) Andhra Pradesh

82. Gandhi Sagar Dam is a part of which of the following?
(a) Chambal Project
(b) Damodar Valley Project
(c) Bhakra Nagal Project
(d) Damodar Valley

83. The greater part of peninsular India is not suitable for well irrigation because of:
1. Rocky structure
2. Lack of and/or deeper underground water
3. Brackish ground water

Which of the above statements is/are correct?
(a) 1, 2 and 3 (b) 1 and 2
(c) 2 and 3 (d) only 2

84. In which year was the first tubewell constructed in India?
(a) 1911 (b) 1921
(c) 1931 (d) 1941

85. Tanks are the most important means of irrigation in the state of:
(a) Tamil Nadu (b) Maharashtra
(c) Andhra Pradesh (d) Karnataka

Answer Key – SET 5

Que. No.	Answer	Que. No.	Answer	Que. No.	Answer
1	d	30	a	59	c
2	c	31	c	60	d
3	b	32	d	61	c
4	c	33	b	62	a
5	a	34	c	63	b
6	c	35	d	64	c
7	b	36	c	65	d
8	c	37	d	66	b
9	b	38	a	67	c
10	d	39	b	68	a
11	c	40	c	69	b
12	b	41	d	70	c
13	a	42	b	71	d
14	c	43	b	72	b
15	d	44	a	73	c
16	b	45	c	74	d
17	a	46	c	75	a
18	b	47	d	76	b
19	c	48	b	77	c
20	d	49	a	78	d
21	b	50	d	79	d
22	a	51	b	80	b
23	c	52	b	81	d
24	b	53	c	82	a
25	d	54	d	83	b
26	c	55	d	84	c
27	a	56	a	85	c
28	c	57	b		
29	b	58	a		

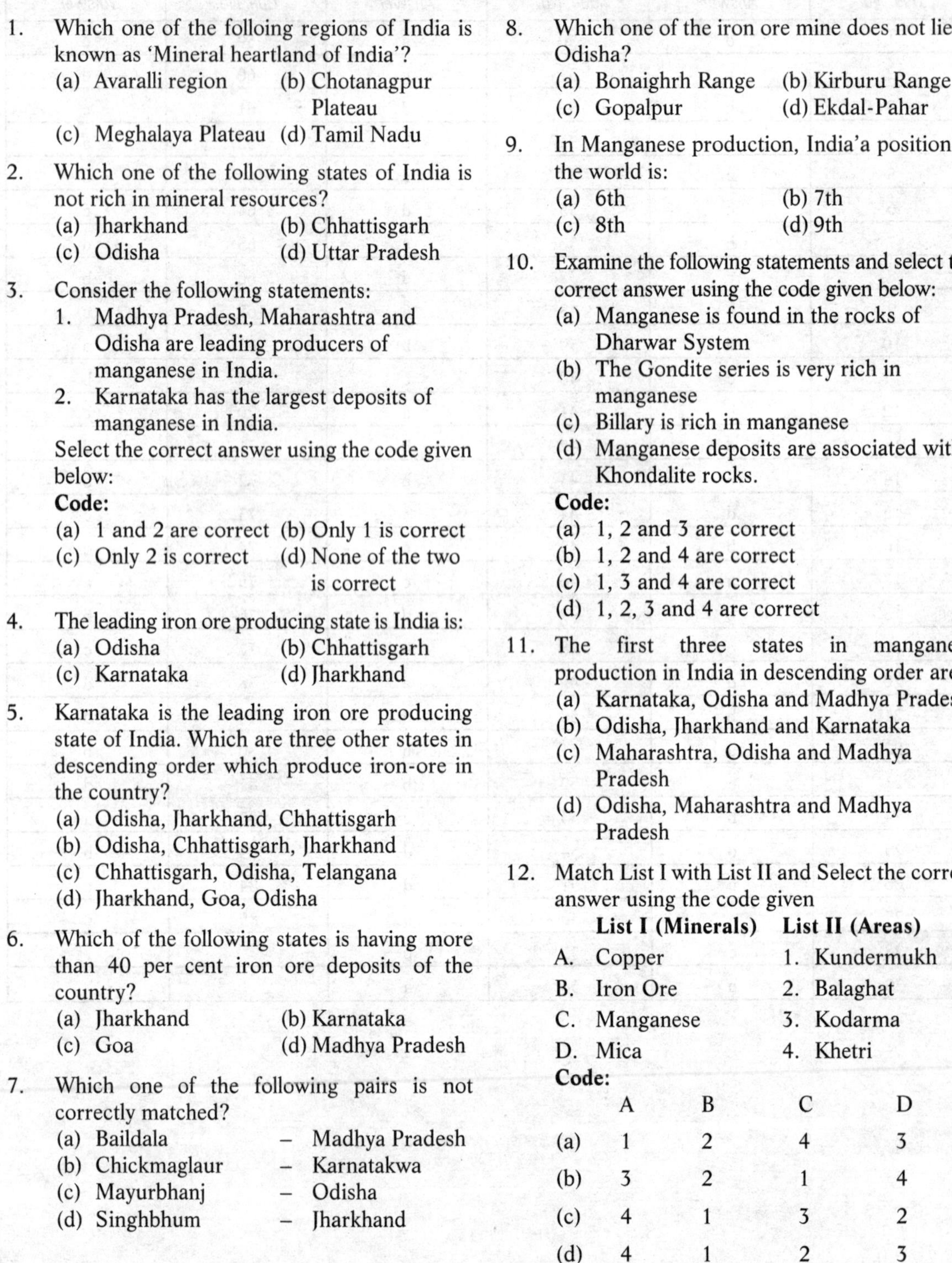

SET – 6

INDIA: Mineral and Power Resources

1. Which one of the folloing regions of India is known as 'Mineral heartland of India'?
   (a) Avaralli region (b) Chotanagpur Plateau
   (c) Meghalaya Plateau (d) Tamil Nadu

2. Which one of the following states of India is not rich in mineral resources?
   (a) Jharkhand (b) Chhattisgarh
   (c) Odisha (d) Uttar Pradesh

3. Consider the following statements:
   1. Madhya Pradesh, Maharashtra and Odisha are leading producers of manganese in India.
   2. Karnataka has the largest deposits of manganese in India.

   Select the correct answer using the code given below:

   Code:
   (a) 1 and 2 are correct (b) Only 1 is correct
   (c) Only 2 is correct (d) None of the two is correct

4. The leading iron ore producing state is India is:
   (a) Odisha (b) Chhattisgarh
   (c) Karnataka (d) Jharkhand

5. Karnataka is the leading iron ore producing state of India. Which are three other states in descending order which produce iron-ore in the country?
   (a) Odisha, Jharkhand, Chhattisgarh
   (b) Odisha, Chhattisgarh, Jharkhand
   (c) Chhattisgarh, Odisha, Telangana
   (d) Jharkhand, Goa, Odisha

6. Which of the following states is having more than 40 per cent iron ore deposits of the country?
   (a) Jharkhand (b) Karnataka
   (c) Goa (d) Madhya Pradesh

7. Which one of the following pairs is not correctly matched?
   (a) Baildala – Madhya Pradesh
   (b) Chickmaglaur – Karnatakwa
   (c) Mayurbhanj – Odisha
   (d) Singhbhum – Jharkhand

8. Which one of the iron ore mine does not lie in Odisha?
   (a) Bonaighrh Range (b) Kirburu Range
   (c) Gopalpur (d) Ekdal-Pahar

9. In Manganese production, India'a position in the world is:
   (a) 6th (b) 7th
   (c) 8th (d) 9th

10. Examine the following statements and select the correct answer using the code given below:
    (a) Manganese is found in the rocks of Dharwar System
    (b) The Gondite series is very rich in manganese
    (c) Billary is rich in manganese
    (d) Manganese deposits are associated with Khondalite rocks.

    Code:
    (a) 1, 2 and 3 are correct
    (b) 1, 2 and 4 are correct
    (c) 1, 3 and 4 are correct
    (d) 1, 2, 3 and 4 are correct

11. The first three states in manganese production in India in descending order are:
    (a) Karnataka, Odisha and Madhya Pradesh
    (b) Odisha, Jharkhand and Karnataka
    (c) Maharashtra, Odisha and Madhya Pradesh
    (d) Odisha, Maharashtra and Madhya Pradesh

12. Match List I with List II and Select the correct answer using the code given

List I (Minerals)	**List II (Areas)**
A. Copper	1. Kundermukh
B. Iron Ore	2. Balaghat
C. Manganese	3. Kodarma
D. Mica	4. Khetri

    Code:

	A	B	C	D
(a)	1	2	4	3
(b)	3	2	1	4
(c)	4	1	3	2
(d)	4	1	2	3

13. Examine the following statements:
    1. Bailadela lies in Chhattisgarh.
    2. Japan is the leading importer of Indian Iron ore.
    3. Badampahar lies in Odisha.
    4. Haryana state has no deposits of iron ore.

    Code:
    (a) 1, 2 and 3 are correct
    (b) 1, 2 and 4 are correct
    (c) 1, 3 and 4 are correct
    (d) 1, 2, 3 and 4 are correct

14. Singhbhum and Keonjhar are the leading producer centres of:
    (a) Mica (b) Iron-ore
    (c) Coal (d) Manganese

15. Which one of the following states has the richest deposits of Mica?
    (a) Andhra Pradesh (b) Jharkhand
    (c) Rajasthan (d) Bihar

16. Consider the following pairs and select the correct answer using code given below

Minerals	90% of the Indian Natural Resources
1. Copper	Jharkhand
2. Nickel	Odisha
3. Tungsten	Kerala

    Code:
    (a) 1 and 2 (b) only 2
    (c) 1 and 3 (d) All of these

17. Consider the following statements:
    1. Balaghat is known for its diamond mines.
    2. Majhgawan is known for its manganese deposits.

    Which of the above statements given above is/are correct?
    (a) Only 1 (b) Only 2
    (c) Both 1 and 2 (d) Neither 1 nor 2

18. In which of the following districts has reserves of diamond bearing kimberlie been discovered?
    (a) Warangal (b) Hoshangabad
    (c) Raipur (d) Sambalpur

19. Which of the following minerals are found in a natural way in the state of Chhattisgarh?
    1. Bauxite 2. Dolomite
    3. Iron-ore 4. Tin

    Select the correct answer using the code given below:
    (a) 1, 2 and 3 (b) Only 3
    (c) Only 2 (d) Only 1

20. Consider the following pairs:
    1. Barabil-Koira Valley – Haematite deposits
    2. Salem district – Magnetite deposits
    3. Amar-kantak area – Bauxite deposits

    Which of the following pair(s) given above is/are correctly matched?
    (a) Only 1 (b) 1 and 2
    (c) only 3 (d) all of these

21. In India 90 per cent copper is produced in two states. Which are these two states?
    (a) Jharkhand and Chhattisgarh
    (b) Madhya Pradesh and Rajasthan
    (c) Rajasthan and Bihar
    (d) Maharashtra and Karnataka

22. Which one of the following states produce about 50 per cent Bauxite of India?
    (a) Maharashtra (b) Gujarat
    (c) Odisha (d) Chhattisgarh

23. Which one of the following mining areas/centres is not correctly matched?

Mining Region	State
(a) Naomudu	Jharkhand
(b) Hosepet	Andhra Pradesh
(c) Chikmaglur	Karnataka
(d) Bailadila	Chhattisgarh

24. For which mineral is Dandakarnya known?
    (a) Chromite (b) Mica
    (c) Iron ore (d) Bauxite

25. Match List I with List II and select the correct answer using the code given below:

List I (Minerals)	List II (Region/Area)
A. Gold, Iron Ore, Chromite	1. Chotanagpur
B. Manganese, Bauxite, Mica	2. Midland Belt
C. Coal, Mica, Manganese	3. Southern Belt
D. Copper, Lead and Zinc	4. Western Belt

Code:

	A	B	C	D
(a)	3	2	1	4
(b)	3	2	4	1
(c)	2	3	4	1
(d)	2	3	1	4

26. Which state of India has a monopoly (about 98 per cent of total production) in Zinc production?
(a) Rajasthan (b) Madhya Pradesh
(c) Jammu & Kashmir (d) Tamil Nadu

27. Which one of the following is not a correct match?

	Mineral	Mining Area
(a)	Tungsten	Degana
(b)	Lead	Rajpura-Dariba
(c)	Bauxite	Koraput
(d)	Copper	Bellary

28. In which state are Kolar and Hutti mines located?
(a) Odisha (b) Madhya Pradesh
(c) Karnataka (d) Telangana

29. After Andhra Pradesh, which are the second and third mica producing states of India?
(a) Bihar and Madhya Pradesh
(b) Rajasthan and Jharkhand
(c) Rajasthan and Bihar
(d) Karnataka and Jharkhand

30. The two leading dolomite producing states of India in descending order are:
(a) Jharkhand and Rajasthan
(b) Andhra Pradesh
(c) Odisha and Karnataka
(d) Odisha and Chhattisgarh

31. The state producing 99 per cent gypsum of the country is:
(a) Rajasthan (b) Tamil Nadu
(c) Gujarat (d) Uttar Pradesh

32. Which is not a correct match?

	Mineral	Leading Producing State
(a)	Asbestos	Rajasthan
(b)	Chromite	Karnataka
(c)	Zinc	Rajasthan
(d)	Bauxite	Odisha

33. The largest producer of Kyanite in India is:
(a) Karnataka (b) Maharashtra
(c) Jharkhand (d) Odisha

34. Which is the largest Apatite producing state in India?
(a) Andhra Pradesh (b) West Bengal
(c) Rajasthan (d) Madhya Pradesh

35. Where is rock salt found in India?
(a) Jammu area of Jammu & Kashmir
(b) Alwar district of Rajasthan
(c) Abohar district of Punjab
(d) Mandi District of Himachal Pradesh

36. Consider the following statements:
1. The monazite sand of Kerala contains Uranium.
2. India does not have any deposits of Thorium.

Which of the statements given above is/are correct?
(a) Only 1 (b) Only 2
(c) Both 1 and 2 (d) None of the above

37. Which of the following substances are found in the beach sands of many parts of Kerala?
1. Ilmenite 2. Zircon
3. Silimenite 4. Tungsten

Select the correct answer using the code given below:
(a) 1 and 2 (b) 1, 2 and 3
(c) 3 and 4 (d) All of these

38. Aluminum usually occurs in the form of-
(a) Pyrite (b) Ilmenite
(c) Bauxite (d) Pure Metal

39. Which state in India is the major producer of antimony?
(a) Odisha (b) Tamil Nadu
(c) Bihar (d) Punjab

40. Copper–gold–iron–coal are connected with:
(a) Kolar–Kundermukh–Khetri–Jharia
(b) Khetri–Kolar–Kundermukh–Jharia
(c) Kundermukh–Kolar–Khetri–Jharia
(d) Kolar–Khetri–Jharia–Kundermukh

41. Which type of coal is mostly found in India?
(a) Bituminous (b) Anthracite
(c) Lignite (d) Peat

42. Which state of India is having the highest coal reserves?
(a) West Bengal (b) Odisha
(c) Chhattisgarh (d) Jharkhand

43. Where are Jharia, Bokaro and Giridih coal fields located?
(a) Bihar (b) Madhya Pradesh
(c) Jharkhand (d) Odisha

44. Which one of the following is not a coal field of Chhattisgarh?
(a) Surguja (b) Sambalpur
(c) Korba (d) Raigarh

45. Arrange the following states in descending order according to their coal reserves:
1. Jharkhand 2. West Bengal
3. Odisha 4. Chhattisgarh

Code:

(a)	1	3	4	2
(b)	3	1	2	4
(c)	1	4	3	2
(d)	4	2	1	3

46. Which one of the following coal deposits does not belong to the Tertiary Era?
(a) Tamil Nadu (b) Arunachal Pradesh
(c) Jammu & Kashmir (d) Kutch-Gujarat

47. About 80 per cent of the coal reverses of India lie in the:
(a) Mahanadi valley (b) Godavari valley
(c) Damodar valley (d) Son valley

48. Which one of the following ores is abundant in the region Balghat–Bhandara–Nagpur?
(a) Manganese (b) Bauxite
(c) Iron (d) Mica

49. In which year for the first time oil in India was discorded in Uppar Assam?
(a) 1855 (b) 1860
(c) 1875 (d) 1880

50. Match List I with List II and select the correct answer using the codes given below:

List I (Minerals)	List II (Major Producing States)
A. Coal	1. Karnataka, Madhya Pradesh
B. Petroleum	2. Bihar, Andhra Pradesh
C. Manganese	3. Bihar, West Bengal
D. Mica	4. Maharashtra, Karnataka
	5. Assam, Gujarat

Code:

	A	B	C	D
(a)	3	5	1	2
(b)	2	1	4	3
(c)	3	1	4	2
(d)	2	5	1	3

51. Which one of the following is not an oilfield of north-east India?
(a) Naharkatiaya (b) Moran-Hugrijan
(c) Kalol (d) Rudrasagar

52. Which one of the following is the largest discovered oil field in East Coast region?
(a) Mahanadi Basin
(b) Krishna-Godavari Basin
(c) Cauvery Basin
(d) Sunderbans

53. Where are Mangla, Guda and Bayatu oil fields located?
(a) Coromandal Coast
(b) Off-shore areas of Kachchh
(c) Gulf of Khambhat
(d) Barmer–Sanchore Basin

54. Which is not a correct match?
(a) West Bengal – Haldia
(b) Bihar – Barauni
(c) Uttar Pradesh – Bina
(d) Haryana – Panipat

55. Which one of the following in not a centre of oil refinery?
(a) Koyali (b) Mangalore
(c) Rajkot (d) Mathura

56. Which one of the following is not correctly matched?
(a) Koyali – Gujarat
(b) Kochi – Karnataka
(c) Haldia – West Bengal
(d) Bongaigaon – Assam

57. In India, which of the following fuels contribute maximum in power generation?
(a) Coal (b) Gas
(c) Hydro (d) Nuclear

58. In thermal power generation, which state ranks first in India?
(a) Uttar Pradesh (b) Gujarat
(c) West Bengal (d) Maharastra

59. Which is not a centre of thermal power station?
(a) Vijaywada (b) Surat
(c) Korba (d) Suratgarh

60. Where was the first hydro-electric plant established in India?
(a) Darjeeling (b) Matura
(c) Koyana (d) Rihand

61. In which year was the first hydro-electric plant established in India?
(a) 1850 (b) 1898
(c) 1910 (d) 1908

62. On which river is Tehri Hydro-power station located?
(a) Mandakini (b) Alaknanda
(c) Banganga (d) Bhagirathi

63. Which one of the following is Satara well known for?
(a) Thermal Power Station
(b) Wind Energy Plant
(c) Hydro-electric Plant
(d) Nuclear Power Plant

64. Tilaiya, Konar, Maithon and Panchet dams are the part of
(a) Bhakra Nagal Project
(b) Chambal Project
(c) Damodar Valley Project
(d) Nagarjun Sagar Project

65. Jawahar Sagar and Ranapratap Sagar Hydro-electric Plants are located on river:
(a) Son (b) Chambal
(c) Damodar (d) Mahi

66. Which is not a correct match?
(a) Sardar Sarovar – Narmada River
(b) Salal Project – Chenab River
(c) Rampad Sagar – Kaveri River
(d) Ukai Project – Tapi River

67. Match List I with List II and select the correct answer using the code given below:

	List I (Atomic Power)	**List II (State)**
A.	Rawatbhata	1. Karnataka
B.	Kaiga	2. Gujarat
C.	Narora	3. Rajasthan
D.	Kakrapara	4. Gujarat

Code:

	A	B	C	D
(a)	3	1	4	2
(b)	3	2	1	4
(c)	1	3	4	2
(d)	2	4	3	1

68. Which one of the following states is using the highest quantity of solar energy in India?
(a) Punjab (b) Haryana
(c) Uttar Pradesh (d) Gujarat

69. In Installed wind power capacity, the first four states in India in descending order are:
(a) Gujarat, Rajasthan, Karnataka and Andhra Pradesh
(b) Tamil Nadu, Gujarat, Maharashtra, and Rajasthan
(c) Rajasthan, Gujarat, Tamil Nadu and Kerala
(d) Maharashtra, Rajasthan, Madhya Pradesh and Gujarat

70. In which state does the maximum potential of wind power lie?
(a) Gujarat (b) Andhra Pradesh
(c) Karnataka (d) Rajasthan

71. Where was the first Indian Ocean energy plant installed?
(a) Durgaduani Creek (b) Gulf of Cambay
(c) Panji (d) Port Blair

72. Which one of the following is not correctly matched?
(a) Kalpakkam – Tamil Nadu
(b) Kundankulam – Kerala
(c) Kakrapara – Odisha
(d) Kaiga – Karnataka

73. Which one of the following is the most ideal location for the generations of tidal energy?
(a) Gulf of Mannar (b) Gulf of Cambay
(c) Kanyakumari (d) Rann of Kachchh

74. Consider the following statements:
    1. About 90 per cent of coal production in India is of non-coking coal.
    2. The best quality of coal in India is found in the Raniganj mines.
    3. Neyveli (Tamil Nadu) has the largest deposits of lignite.
    4. Singrauli coal mines lie in Madhya Pradesh.

    Select the correct answer using code given below:

    Code:

    (a) 1 and 2 are correct
    (b) 2 and 3 are correct
    (c) 1, 2 and 3 are correct
    (d) 1, 3 and 4 are correct

75. Match the following lists and select the correct answer using codes given below:

	List I (Region)	**List II (Renewable Energy Type)**
A.	Puga	1. Tidal
B.	Sunderbans	2. Wind
C.	Mappandal	3. Nuclear
D.	Narora	4. Geothermal

Code:

	A	B	C	D
(a)	1	2	3	4
(b)	3	2	1	4
(c)	2	1	3	4
(d)	4	1	2	3

Answer Key – SET 6

Que. No.	Answer	Que. No.	Answer	Que. No.	Answer
1	b	26	a	51	c
2	d	27	d	52	b
3	a	28	c	53	d
4	c	29	b	54	c
5	a	30	d	55	c
6	b	31	a	56	b
7	a	32	b	57	a
8	a	33	c	58	d
9	c	34	b	59	b
10	a	35	d	60	a
11	d	36	a	61	b
12	d	37	b	62	d
13	c	38	c	63	c
14	b	39	d	64	c
15	a	40	b	65	b
16	b	41	a	66	c
17	d	42	d	67	a
18	c	43	c	68	d
19	a	44	b	69	b
20	d	45	a	70	c
21	b	46	d	71	a
22	c	47	c	72	c
23	b	48	a	73	b
24	c	49	b	74	d
25	b	50	a	75	d

SET – 7

INDIA: Industries, Transportation and International Trade

1. What is the percentage of textile sectors in total industrial production of the country?
(a) 12% (b) 13%
(c) 14% (d) 16%

2. In which year was the first cotton textile mill established?
(a) 1818 (b) 1820
(c) 1925 (d) 1851

3. Which one of the following states contributes in the highest production of cotton cloth in India?
(a) Gujarat (b) Punjab
(c) Tamil Nadu (d) Maharashtra

4. Which one of the following is not a centre of cotton textile in Maharashtra?
(a) Sangali (b) Latur
(c) Jalgoan (d) Kolhapur

5. The largest centre of cotton textile industry in Gujarat is:
(a) Surat (b) Bharuch
(c) Ahmedabad (d) Varodara

6. Which of the following does not match correctly?

Cotton Textile Centre	**State**
(a) Dewas	Madhya Pradesh
(b) Amravati	Gujarat
(c) Coimbatore	Tamil Nadu
(d) Bhiwani	Haryana

7. Which city is known as 'Manchester of South India'?
(a) Salem (b) Alleppey
(c) Guntur (d) Coimbatore

8. In Silk production, India's position in the world is:
(a) 2nd (b) 3rd
(c) 4th (d) 5th

9. Which one the following is not a major silk producing state in India?
(a) Karnataka (b) West Bengal
(c) Maharashtra (d) Jammu & Kashmir

10. The largest silk textile producing state in India is:
(a) Jammu & Kashmir (b) Karnataka
(c) West Bengal (d) Tamil Nadu

11. Where was the first woollen textile mill in India established in 1876?
(a) Thane (b) Ratanagiri
(c) Kanpur (d) Ludhiana

12. Consider the following statements:
1. The Indian textile industry is labour intensive.
2. The textile industry is export oriented.
3. India exports more textile goods than any other Asian country.

Select the correct answer using the code given below:
(a) 1 and 2 are correct
(b) 2 and 3 are correct
(c) 1 and 3 are correct
(d) 1, 2 and 3 are correct

13. The first jute factory in India was established at:
(a) Titagarh (b) Serampore
(c) Budge Budge (d) Agarpara

14. In which of the following regions there is high concentration of jute mills?
(a) Chotanagpur region
(b) Godaveri delta region
(c) Hooghly region
(d) Southern Odisha

15. In India first iron and steel unit was established in 1830 and the second one in 1966 but real beginning of modern iron and steel industry was made in:
(a) 1892 (b) 1907
(c) 1910 (d) 1923

16. Where was Tata Iron and steel Company (TISCO) set up:
(a) Sakchi (b) Burnpur
(c) Bhadravati (d) Durgapur

17. Match List I with List II and select the correct answer using the code given below:

	List I (Industry)	List II (Location)
A.	IISCO	1. Bokaro
B.	HSL	2. Bhadravati
C.	BSL	3. Burnpur
D.	VISL	4. Rourkela

Code:

	A	B	C	D
(a)	3	4	2	1
(b)	3	4	1	2
(c)	4	3	1	2
(d)	4	3	2	1

18. Which one of the following in not correctly matched?

Steel Plant	Technical Assistance From
(a) Bhilai	Russains
(b) Bokaro	Americans
(c) Raurekla	Germans
(d) Durgapur	British

19. Which of the following is not a centre of steel plant?
(a) Salem (b) Dolvi
(c) Veraval (d) Paradwip

20. Which one of the following steel plants is not located in West Bengal?
(a) Bhillai
(b) Durgapur
(c) Kulti
(d) Burnpur

21. Match List I with List II and select the correct answer using codes given below:

List I (Places)	List II (Industries)
(a) Jamnagar	1. Aluminium
(b) Hospet	2. Woolen Textile
(c) Korba	3. Festilizers
(d) Haldia	4. Cement
	5. Iron and Steel

Code:

	A	B	C	D
(a)	4	3	1	2
(b)	2	5	1	3
(c)	4	5	2	1
(d)	2	1	4	3

22. Which one of the following statements is not correct?
(a) Selem Steel Plant is a premier producer of stainless steel in India.
(b) Rourkela Steel Plant, the first integrated steel plant in the public sector of India was set up with the Soviet Union collaboration.
(c) Maharashtra Elertrosmelt Limited is a subsidiary of the Steel Authority of India.
(d) Vishakhapatnam Steel Plant is a unit of the Rashtriya Ispat Nigam Limited.

23. Where are the important plants of BHEL located?
(a) Haridwar, Tiruchirapalli, Srinagar
(b) Bhopal, Hyderabad, Pinjore
(c) Bhopal, Hyderabad, Tiruchirapalli
(d) Delhi, Mumbai, Kolkata

24. Where were aluminum plants located in pre-independence India?
(a) Alupuram and Jaykaynagar
(b) Renukot and Alupuram
(c) Korba and Renukot
(d) Renukot and Mettur

25. Which is not a correct match?

Aluminium Company	Location
(a) INDAL	Belgaum
(b) MINDALCO	Renukot
(c) NALCO	Koraput
(d) MALCO	Korba

26. The largest soda ash producing centre of India is located at:
(a) Porbandar (b) Mithapur
(c) Tuticoran (d) Naya Nagal

27. Which is not a correct match?
(a) Hindustan Organic Chemiclas Ltd – Rasayani
(b) Madras Aluminium Co. Ltd – Mettur
(c) Tata Chemical Ltd – Veraval
(d) Indian Drugs and Pharmaceuticals Ltd – Gurgoan

28. Match List I with List II and select the correct answer using code given below:

	List I (Industries)	List II (Centres)
A.	Glass Industry	1. Pune
B.	Automobiles	2. Firozabad
C.	Ship Building	3. Pinjore
D.	Engineering Goods	4. Marmagoan

Code:

	A	B	C	D
(a)	2	1	4	3
(b)	2	1	3	4
(c)	1	2	4	3
(d)	1	2	3	4

29. Which one of the following is not a centre of chemical industry?
(a) Pimpri (b) Pune
(c) Bareilly (d) Rishikesh

30. Where was the first Super Phosphate factory established in India?
(a) Sindri (b) Ranipat
(c) Panipat (d) Thal

31. Which one of the following is not a public sector fertilizer unit?
(a) FCI – Sindhi
(b) NFL – Nagal
(c) HFC – Kochi
(d) RCF – Trombay

32. Which one of the following is not a centre of Indian Farmer's Fertilizer Co-operation Ltd. (IFFCO)?
(a) Kalol (b) Phulpur
(c) Kandla (d) Barauni

33. The first paper mill in India was established in 1832 at:
(a) Serampore (b) Ballygunj
(c) Titagarh (d) Lucknow

34. Which is not correctly matched?

State	Centres of Paper Mill
(a) West Bengal	Titagarh and Raniganj
(b) Odisha	Rajamundri and Sirpur
(c) Madhya Pradesh	Schadol and Nepanagar
(d) Bihar	Barauni and Samastiput

35. At which place is security paper mill located?
(a) Bangalore (b) Naini
(c) Hoshangabad (d) Jalgaon

36. The highest sugar producing state of India is:
(a) Uttar Pradesh (b) Maharashtra
(c) Tamil Nadu (d) Karnataka

37. India's position in cement production in the world is:
1. 2nd 2. 3rd
3. 4th 4. 5th

38. Which one of the following is not a cement production state in India?
(a) Telangana (b) Rajasthan
(c) Chhattisgarh (d) Assam

39. Which one of the following centre is not having units of Hindustan Machine Tools (HMT)?
(a) Srinagar (b) Kanpur
(c) Pinjori (d) Ajmer

40. Where is National Instrument Factory located?
(a) Ranchi (b) Hyderabad
(c) Jadavpur (d) Channai

41. Which one of the following is not known for pharmaceutical industries?
(a) Kanpur (b) Hyderabad
(c) Muzaffarpur (d) Pune

42. Pasco Steel Plant is located at:
(a) Paradwip (b) Dolvi
(c) Jalgoan (d) Ludhiana

43. The INDAL aluminum plant is located at:
(a) Mettur (b) Koraput
(c) Angul (d) Hirakund

44. Which belt has a concentration of heavy engineering industry in India?
(a) Kolkata–Asansol
(b) Jamshedpur–Asansol
(c) Kanpur–Lucknow
(d) Mumbai–Pune

45. The Indian Telephone Industries is located at:
(a) Mumbai (b) Bangalore
(c) Noida (d) Kolkata

46. Match the List I with List II and select the correct answer using codes given below:

List I	List II
(A) SAIL	1. Chemicals
(B) BALCO	2. Iron and Steel
(C) BPCL	3. Electronics
(D) BHEL	4. Aluminum

Code:

	A	B	C	D
(a)	2	4	1	3
(b)	2	1	4	3
(c)	4	2	1	3
(d)	2	1	3	4

47. Match the List I with List II and select the correct answer using codes given below:

List I (Industry)	List II (Producing Centre)
(A) Automobiles	1. Titagarh
(B) Diesel Locomotive	2. Travancore
(C) Paper	3. Channai
(D) Fertilizer	4. Varanasi

Code:

	A	B	C	D
(a)	2	1	4	3
(b)	1	2	3	4
(c)	3	4	1	2
(d)	3	4	2	1

48. Which one of the following is not a main cluster of automobile Industry in India?
(a) Chennai
(b) Mumbai-Pune
(c) National Capital Region
(d) Jamshedpur-Bhilai

49. Examine the following statements and select the correct answer using the code given below:
1. The first iron-smelting plant in India was established in 1853.
2. The first cotton textile industry in India was established in 1854.
3. The first jute industry in India was established in 1855.

Code:
(a) 1 and 2 are correct
(b) 1, 2 and 3 are correct
(c) 1 and 3 are correct
(d) 2 and 3 are correct

50. In India main bicycle production units are located at:
(a) Sonepat – Ludhina
(b) Hisar – Gurgaon
(c) Meerut – Kanpur
(d) Ahmedabad – Surat

51. Which one of the following is not a main centre of electronic industry?
(a) Palakkad (b) Coimbatore
(c) Mohali (d) Cochin

52. Which one of the following is not a major industrial region of India?
(a) Vishakhapatnam–Guntur
(b) Gurgaon–Delhi–Meerut
(c) Kanpur–Agra
(d) Mumbai–Pune

53. The Rail coaches are manufactured at:
(a) Kapurthala and Perambur
(b) Jamshedpur and Perambur
(c) Hederabad and Varanasi
(d) Varanasi and Perembur

54. India's position in total road network in the world is:
(a) 5th (b) 4th
(c) 3rd (d) 2nd

55. In India, the roads per square kilometer of land is:
(a) 0.66 km (b) 0.75 km
(c) 0.96 km (d) 1.25 km

56. The famous highway of medieval India, i.e. 'Grand Trunk Road' was built between:
(a) 1640–1645
(b) 1540–1545
(c) 1740–1745
(d) 1840–1845

57. Which of the following city was not on Grand Trunk Road?
(a) Patna (b) Agra
(c) Amritsar (d) Jammu

58. In which year was the National Highway Authority of India setup?
(a) 1969 (b) 1979
(c) 1989 (d) 1992

59. The total length of the National Highway in India in May 2014 was:
(a) 92,861 km (b) 1,63,896 km
(c) 85,280 km (d) 96,875 km

60. The new numbering scheme of National Highway of India was adopted on;
(a) 31 March 2010 (b) 28 April 2010
(c) 1 July 2011 (d) 15 August 2012

61. The new numbering scheme of National Highway network is based on:
(a) their total length
(b) their Traffic flow and importance
(c) the orientation and the geographic location
(d) on all the above factors

62. Which is the single longest national highway of India in the new number scheme?
(a) NH 48 (b) NH 44
(c) NH 02 (d) NH 52

63. The 'Quadrilateral Highways' connect:
(a) Delhi–Kolkata–Bangalore–Mumbai–Delhi
(b) Mumbai–Bangalore–Chennai–Mumbai
(c) Delhi–Mumbai–Chennai–Kolkata–Delhi
(d) Delhi–Ahmedabad–Chennai–Kolkata–Delhi

64. Which state of India is having the highest total length of state highways?
(a) Maharashtra (b) Uttar Pradesh
(c) Gujarat (d) Madhya Pradesh

65. In which year was Border Roads Organization (BRO) established in India?
(a) 1950 (b) 1955
(c) 1958 (d) 1960

66. Consider the following statements:
1. The road network of India carries 80 per cent of the freight traffic.
2. The longest National Highway runs from Srinagar to Kanyakumari.
3. The Jawahar Tunnel connects Jammu and Srinagar.
4. The highest roadway of the world is Manali-Leh Highway.

Select the correct answer using the code given below:
(a) 1 and 2 are correct
(b) 1 and 3 are correct
(c) 1, 2 and 4 are correct
(d) 2, 3 and 4 are correct

67. When was the railway system started in India?
(a) 1st May 1854 (b) 16th April 1853
(c) 12th May 1856 (d) 14th June 1858

68. In how many zones has railway network been divided in India?
(a) 12 (b) 15
(c) 17 (d) 19

69. The head quarter of West-Central Zone is:
(a) Bhopal (b) Ratlam
(c) Jabalpur (d) Kota

70. Secunderabad is the head quarter of:
(a) South Central Zone
(b) Southern Railway
(c) South Western Railway
(d) South East Central Railway

71. Which sites of the Indian railways are the UNESCO world heritage sites?
(a) Kalka–Shimala Railway
(b) Nilgili Mountain Railway
(c) Derjeeling Himalayan Railway
(d) All of these

72. Deccan Odyssey Train covers the tourist centres of:
(a) Kerala and Karnataka
(b) Maharashtra and Goa
(c) Odisha and Andhra Pradesh
(d) Tamil Nadu and Kerala

73. The length of National Waterway No. 1 (Allahabad to Haldia) is:
(a) 1,480 km (b) 1,540 km
(c) 1,620 km (d) 1,660 km

74. In which year was The Central Water Transport Corporation constituted?
(a) 1967 (b) 1972
(c) 1975 (d) 1987

75. Which one of the following is not a regular route of coastal shipping?
a. Kandla – Mumbai
b. Kochi – Nagapattinam
c. Nellore – New Tuticorin
d. Kolkata – Chennai

76. Which one of the following is not an eastern coast port of India?
(a) Ennore (b) Paradwip
(c) Vishakhapatnam (d) Cochin

77. Which one of the following is the largest container port of India?
(a) Vishakhapatnam
(b) Jawaharlal Nehru Port
(c) Kochi
(d) Chennai

78. Which one of the following is not correct?
(a) Air transport was born in India on 18 February 1911.
(b) The first air flight in India was between Allahabad and Naini.
(c) The scheduled Air Transport was started in India in 1912.
(d) Air Cargo Service means air transportation of cargo and mail.

79. The India's First Metro Project was commissioned at:
(a) Kolkata (b) Delhi
(c) Mumbai (d) Chennai

80. Which one of the following statements is not true of Konkan Railway?
(a) The total length is about 760 km.
(b) It runs through the states of Karnataka, Goa, Maharashtra and Kerala.
(c) It is the only rail route that cuts across the Western Ghats.
(d) The Konkan Railway construction Co., which came into being, raised money through public Issues.

81. **Assertion (A)** The foreign trade of India is controlled by WTO.
Reason (R) India is a founder member of WTO. Select your answer using the codes given below:
(a) Both A and R are true and R is the correct explanation of A
(b) Both A and R are true but R is not the correct explanation of A
(c) A is true R is false
(d) A is false R is true

82. Examine the following statements and select the correct answer using the code given below:
1. India's trade balance is always negative.
2. Both export and import of India is increasing year after year.
3. India's trade policy is based on free trade with neighboring countries.
(a) 1 and 2 are correct
(b) 2 and 3 are correct
(c) 1, 2 and 3 are correct
(d) Only 1 is correct

83. The main exports of India are:
(a) Tea, coffee, Newsprint, Fertilizer
(b) Tea, Iron ore, Cotton Fabrics, Gems and Jewellery
(c) Petroleum products, Engineering goods, Tea, Electronics
(d) Wheat, Handicraft, Electronics, Medicines, Fuel

84. To which of the following regions/countries India's export is the highest?
(a) European Union (b) USA
(c) China (d) African Countries

85. With which country India's import share is the highest?
(a) Saudi Arabia (b) Iraq
(c) China (d) Indonesia

86. Consider the following statements:
1. India's share in international trade is less than one per cent.
2. Foreign trade through selected ports.
3. Most of the overseas trade is done in private sector.
4. Petroleum and other petro-products are having the highest share of import.

Select the correct answer using code given below:
(a) Only 1 and 2 are correct
(b) Only 2 and 4 are correct
(c) 1, 2 and 3 are correct
(d) 2, 3 and 4 are correct

Answer Key – SET 7

Que. No.	*Answer*	*Que. No.*	*Answer*	*Que. No.*	*Answer*
1	c	30	b	59	a
2	a	31	c	60	b
3	d	32	d	61	c
4	b	33	a	62	b
5	c	34	b	63	c
6	b	35	c	64	a
7	d	36	b	65	d
8	a	37	a	66	d
9	c	38	d	67	b
10	b	39	b	68	c
11	c	40	c	69	c
12	a	41	a	70	a
13	b	42	a	71	d
14	c	43	d	72	b
15	b	44	b	73	c
16	a	45	b	74	a
17	b	46	a	75	c
18	b	47	c	76	d
19	c	48	d	77	b
20	a	49	b	78	c
21	a	50	a	79	a
22	b	51	d	80	b
23	c	52	c	81	d
24	a	53	a	82	a
25	d	54	d	83	b
26	b	55	a	84	a
27	c	56	b	85	c
28	a	57	d	86	c
29	c	58	c		

SET – 8

INDIA: Map Based Questions

1. The shaded area shown on the map indicates which type of geological formations?

(a) Dharwar (b) Archaean
(c) Vindhyan (d) Mesozoic

2. The shaded area on the map indicates which type of geological formation?

(a) Archaean (b) Vindhyan
(c) Quaternary (d) Cretaceous

3. The given map of India shows four mountain ranges marked 1, 2, 3 and 4 respectively. Select the correct answer:

(a) Vindhyana, Maikal, Satpura, Aravalli
(b) Maikal, Aravalli, Satpura, Vindhyan
(c) Satpura, Vindhyan, Maial, Aravalli
(d) Maikal, Vindhyan, Satpura, Aravalli

4. The given map of India shows four mountain ranges marked 1, 2, 3 and 4. Which of the following is No. 1 marked range?

(a) Naga (b) Jaintia
(c) Khasi (d) Nilgiri

5. On the map of India four rivers of Penisular India are shown and marked 1, 2, 3 and 4. Which one of the following is not marked correctly?

(a) 1. Krishna (b) 2. Narmada
(c) 3. Godavari (d) 4. Kavari

6. The shaded areas on the map of India shows areas of which of the following type annual average rainfall?
(a) Rainfall over 300 cm
(b) Rainfall between 200 to 300 cm

(c) Rainfall between 100 to 200 cm
(d) Rainfall below 100 cm

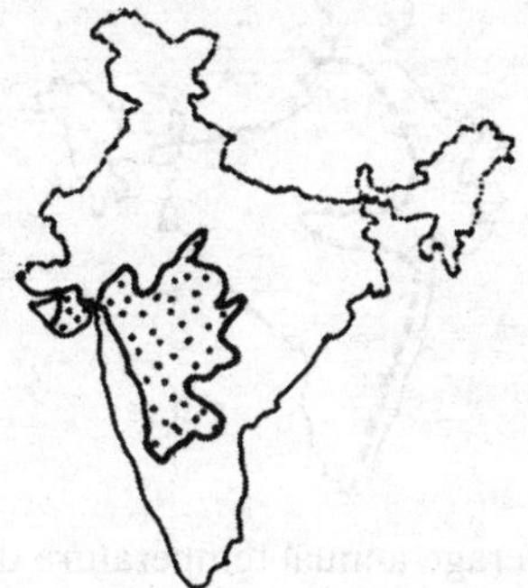

7. The shaded area on the map of India shows which type of soil?

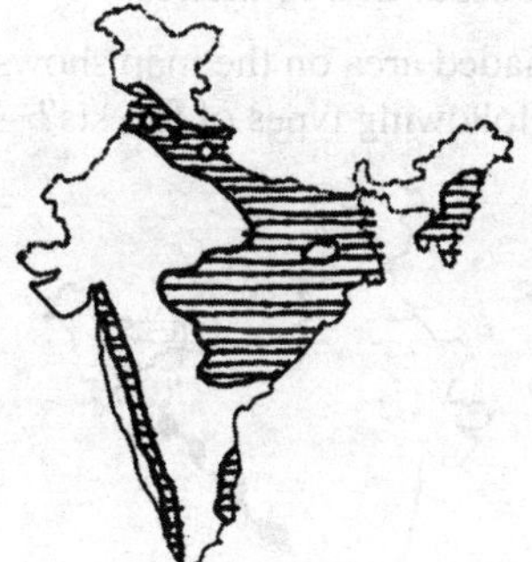

(a) Black soil (b) Red soil
(c) Laterite soil (d) Grey and brown soil

8. Which one of the following national park is not correctly shown on the map?
(a) 1. Dachigam (b) 2. Kaziranga

(c) 3. Panna (d) 4. Chandoli

9. Four Tiger Reserves are shown on the map as 1, 2, 3 and 4. Arrange them in correct sequence and select the correct answer as given below the map:
(a) Dudhwa, Bandipur, Ranthambor, Kahna
(b) Kanha, Ranthanbore, Dudhwa, Bandipur
(c) Bandipur, Dudhwa, Ranthambor, Kanha
(d) Kanha, Dudwa, Ranthambor, Bandipur

10. Which one of the following Biosphere Reserves is not shown correctly on the map?

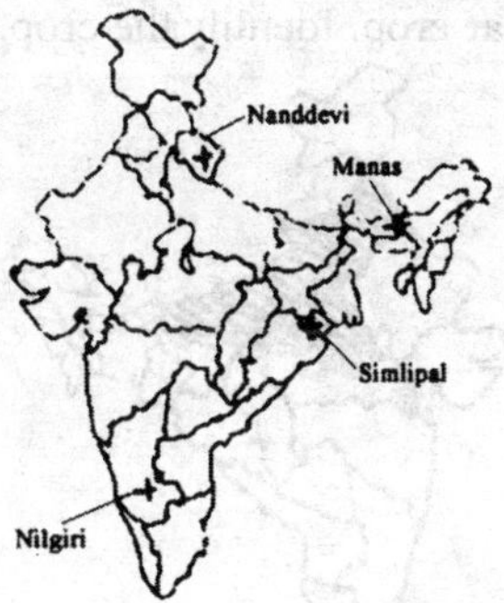

(a) Nandadevi (b) Nilgiri
(c) Simlipal (d) Manas

11. Select the correct sequence of dams/projects numbering 1 to 4.

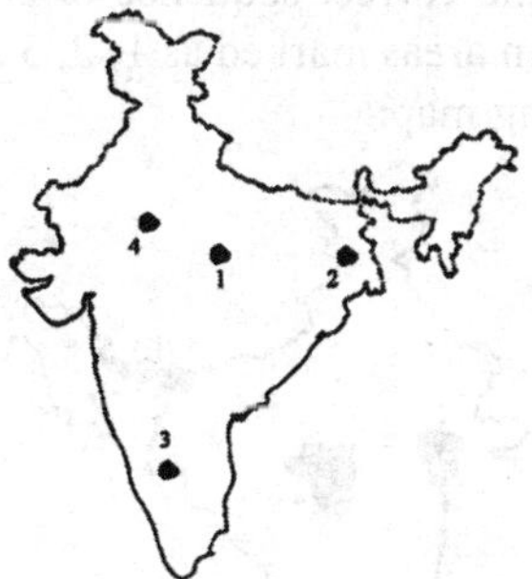

(a) Rana Pratap Sagar, Mayurakshi, Sivasamudram, Twa
(b) Twa, Sivasamudram, Mayurakshi, Rana Pratap Sagar
(c) Twa, Mayurakshi, Sivasamudram, Rana Pratap Sagar
(d) Rana Pratap Sagar, Twa, Mayurakshi, Sivasamudram

12. Which one of the following irrigation projects is not shown correctly on the map?

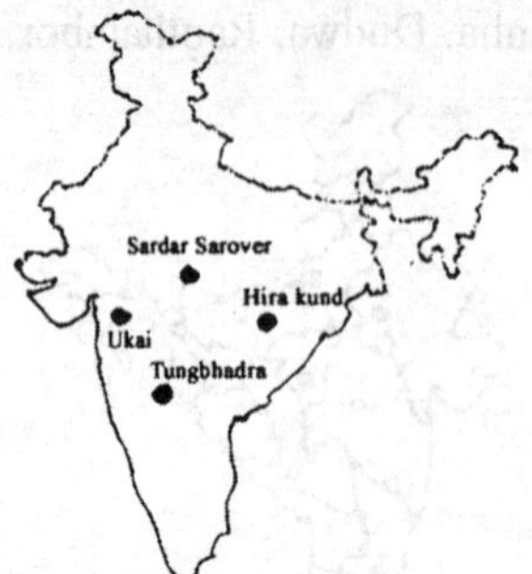

(a) Hiraund (b) Tungbhadra
(c) Ukai (d) Sardar Sarover

13. The map given below shows distribution of a particular crop. Identify the crop.

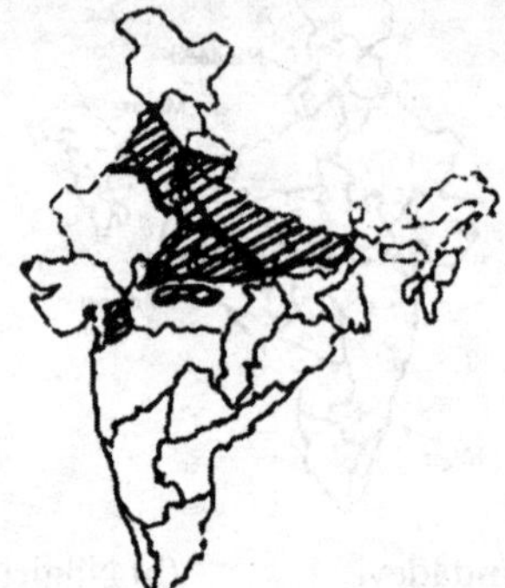

(a) Wheat (b) Cotton
(c) Soyabean (d) Groundnut

14. Select the correct sequence of crops that are grown in areas marked as 1, 2, 3 and 4 on the following map:

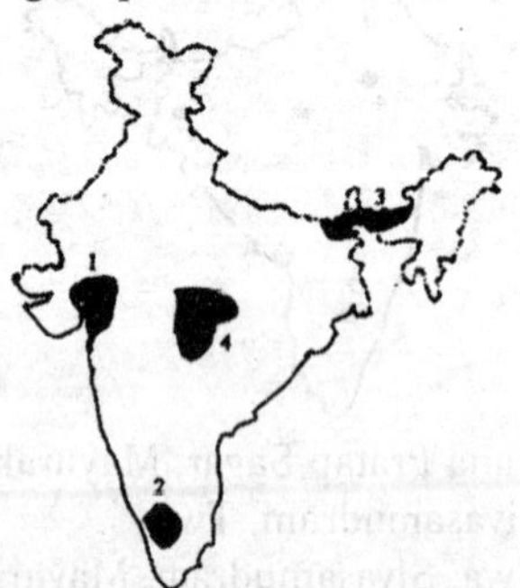

(a) Cotton, Tea, Coffee, Tobacco
(b) Tobacco, Tea, Cotton, Coffee
(c) Tabacco, Coffee, Tea, Cotton
(d) Cotton, Coffee, Tea, Tabacco

15. The dotted line marked on the map indicates:

(a) Average annual temperature divide
(b) Isohyet of 200 cm
(c) Indian water divide
(d) Contour of 300 metres

16. The shaded area on the map shows which one of the following types of forests?

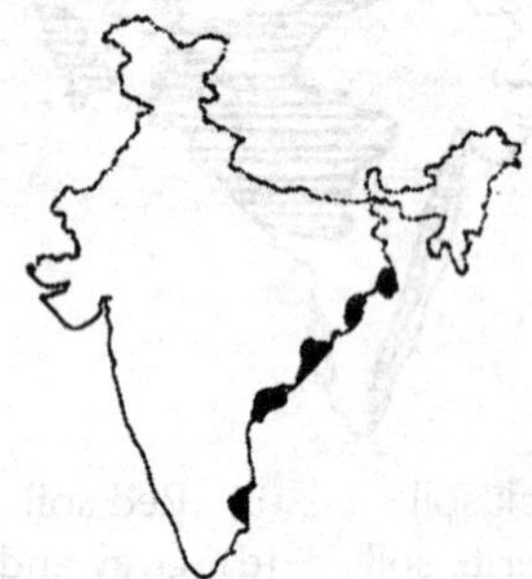

(a) Mixed deciduous
(b) Tropical Savannah
(c) Semi-arid scrubland
(d) Mangrove forests

17. Which type of climate is indicated in the shaded area on the map according to Koppen's classification?

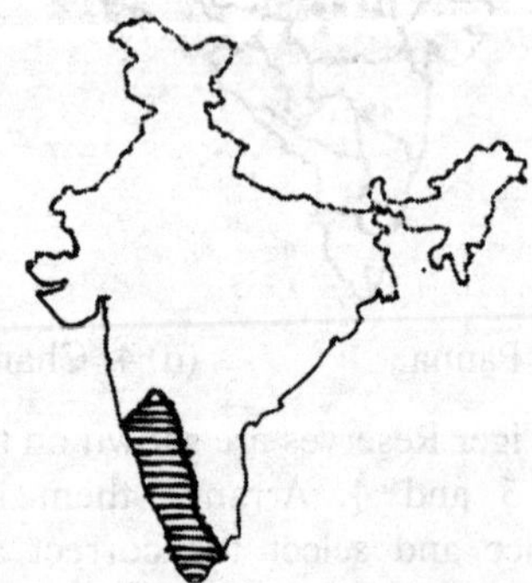

(a) As (b) Amw
(c) Aw (d) Cwg

18. On the given map 1, 2, 3 and 4 indicates the location of Oil Refineries. Select the correct location using code given below:

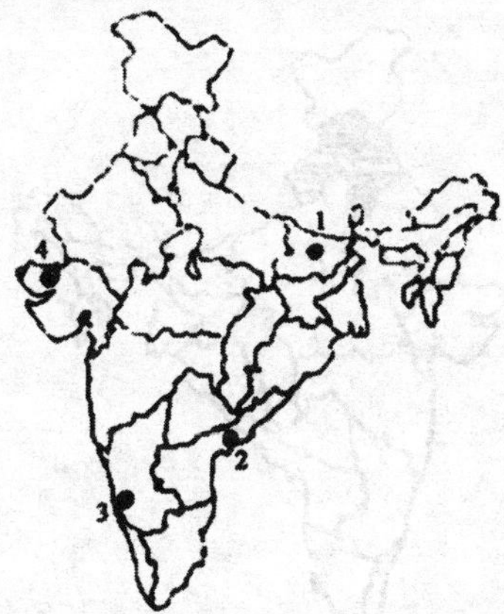

Code:
(a) Tatipaka, Manglore, Jamnagar, Barani
(b) Barauni, Manglore, Jamnagar, Tatipaka
(c) Barauni, Tatipaka, Manglore, Jamnagar
(d) Jamnagar, Manglore, Titipaka, Barauni

19. Which one of the following location of a thermal prower project is not correctly shown on the map?

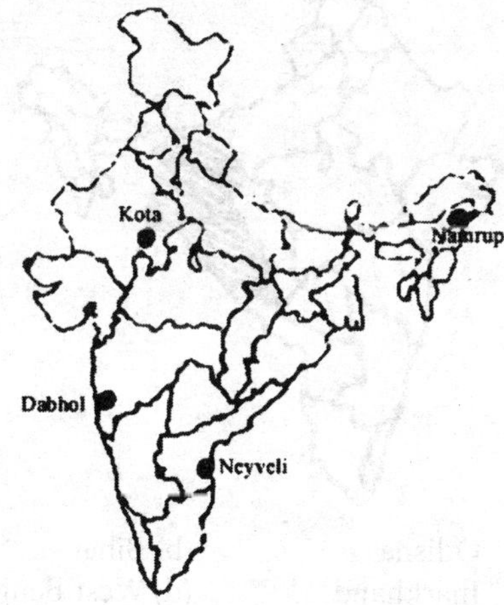

(a) Neyveli (b) Dabhol
(c) Kota (d) Namrup

20. The location of which nuclear plant is not correctly shown on the map?

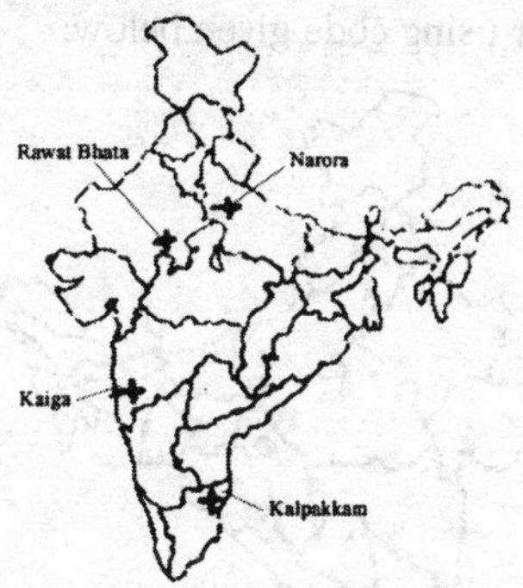

(a) Rawat Bhata (b) Kaiga
(c) Narora (d) Kalpakkam

21. Which minerals are found in areas marked A and B on the map?

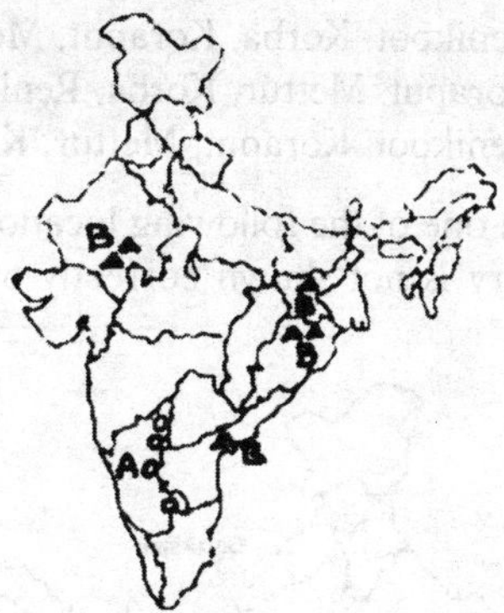

(a) Gold and Copper (b) Silver and Mica
(c) Gold and Lead (d) Lead and Bauxite

22. Which one of the following centres of iron and steel industry is not shown correctly on the map?

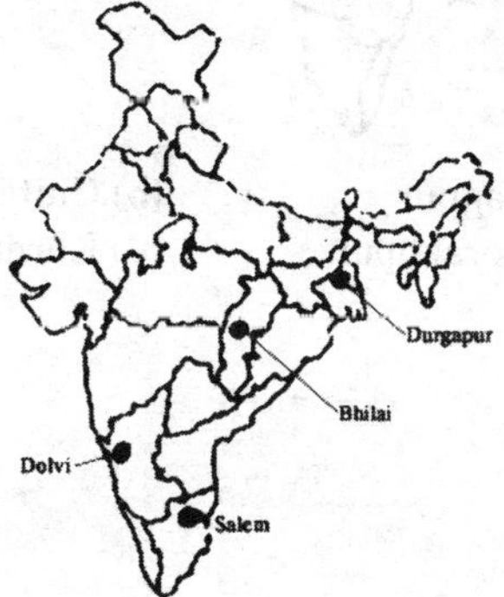

(a) Dolvi (b) Salem
(c) Bhilai (d) Durgapur

23. Identify the centres of Aluminium Industry shown as 1, 2, 3 and 4. Select the correct answer using code given below:

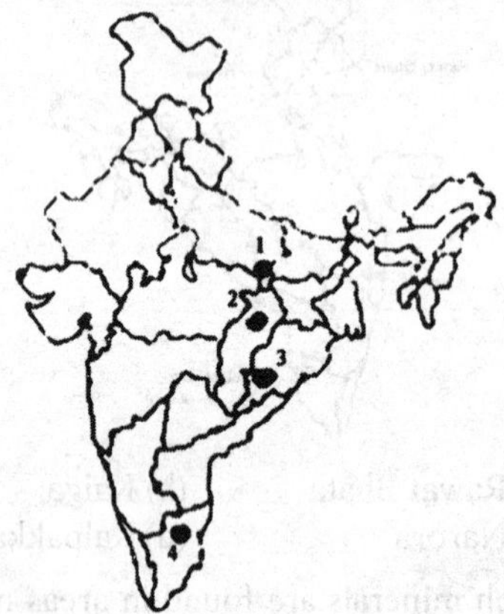

(a) Korba, Mettur, Renikoot, Koraput
(b) Renikoot, Korba, Koraput, Mettur
(c) Koraput, Mettur, Korba, Renikoot
(d) Renikoot, Koraput, Mettur, Korba

24. Which one of the following locations of paper industry is not shown correctly on the given map?

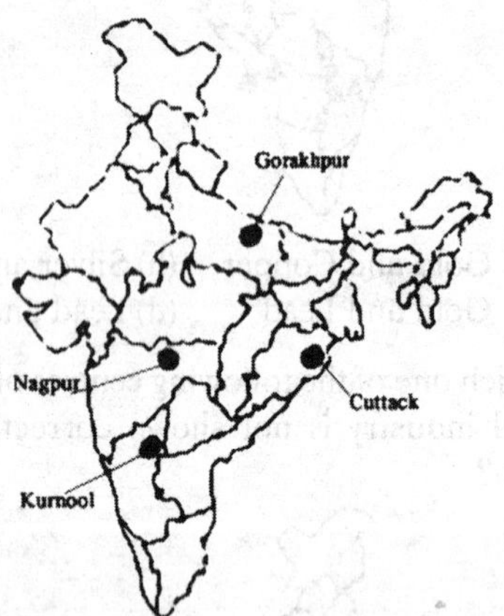

(a) Nagpur (b) Cuttack
(c) Gorakhpur (d) Kurnool

25. Which one among the following states shaded on the map is having the highest density of population in 2011?

(a) Punjab (b) Himachal Pradesh
(c) Haryana (d) Uttarakhand

26. Among the shaded states shown on the map, which state is having the lowest sex-ratio in 2011?

(a) Odisha (b) Bihar
(c) Jharkhand (d) West Bengal

27. Name the ports marked as 1, 2, 3 and 4 on the map in correct sequence according to the serial number given on the map:

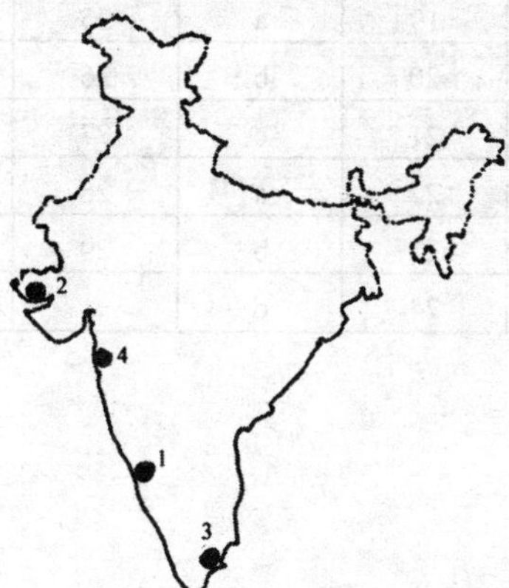

(a) New Mangalore, Kandla, Tuticorin, Nhova-Shiva
(b) Nohova-Shiva, Tutikorin, New Mangalore, Tuticorin
(c) Tuticorin, New Mangalore, Kandla, Nohoa-Shiva
(d) Nohova-Shiva, New Mangalore, Tutikorin, Kandla

28. The given map shows some air ports on air routes in India. Identify the air ports using the code given below:

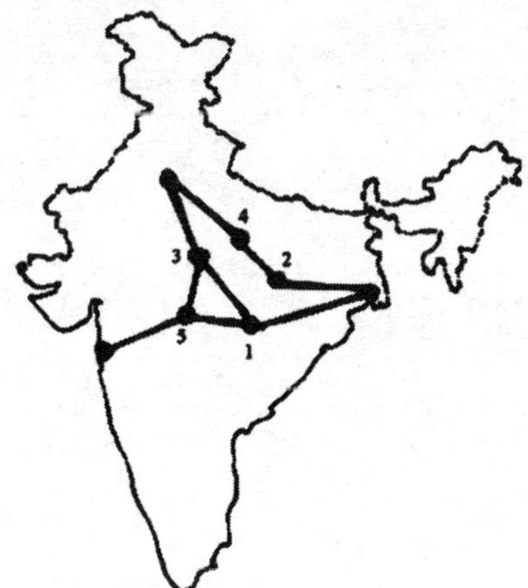

	Ranchi	Jabalpur	Varanasi	Raipur
(a)	1	5	3	2
(b)	1	2	3	4
(c)	2	3	4	1
(d)	2	1	4	3

29. On the given map Kozikhode fishing centre is shown by:

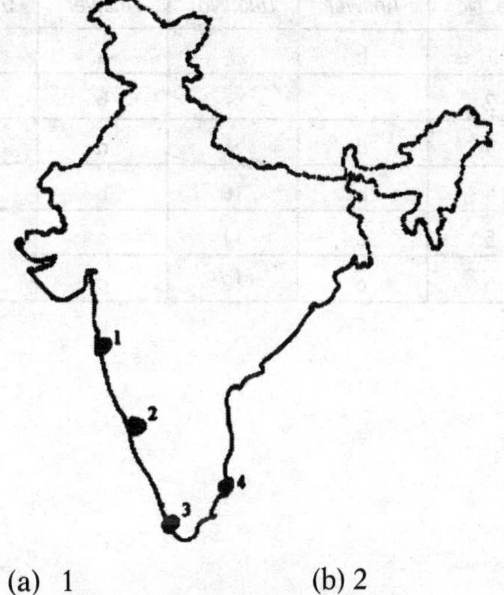

(a) 1 (b) 2
(c) 3 (d) 4

Answer Key – Map Based Questions

Que. No.	Answer	Que. No.	Answer	Que. No.	Answer	Que. No.	Answer	Que. No.	Answer
1	b	7	a	13	a	19	a	25	c
2	c	8	b	14	c	20	b	26	b
3	d	9	d	15	c	21	c	27	a
4	b	10	b	16	d	22	a	28	c
5	c	11	c	17	b	23	b	29	b
6	c	12	d	18	c	24	d		

Section II

WORLD

Physical, Social & Economic Geography

16

Planet Earth: An Introduction

The Earth is unique in the solar system in that it supports life. Its size, gravitational pull and distance from the Sun have created the optimum conditions for the evolution of life. Before we discuss the world's geographical pattern it is necessary to have a clear idea of what the earth is, as a whole.

The Solar System: Earth is a part of the solar system. The solar system is made up around it. The Sun is a star and the centre of the solar system. The Sun's great mass creates the gravitation that keeps the other objects revolving around it.

The planets are the largest objects that revolve around the Sun. There are eight planets in the solar system, each with its own orbit, or path, around the Sun. Planets vary in their distance from the Sun. **Mercury**, **Venus**, **Earth**, and **Mars** are nearest to the Sun, while **Jupiter**, **Saturn**, **Uranus**, and **Neptune** are farthest away. The planets vary in size also; Jupiter is the largest planet, while Mercury is the smallest. Earth ranks fifth in size among the planets. All the planets except Mercury and Venus have moons. Earth has one moon, while Saturn has perhaps as many as 26 moons.

Figure 16.1 *Solar System*

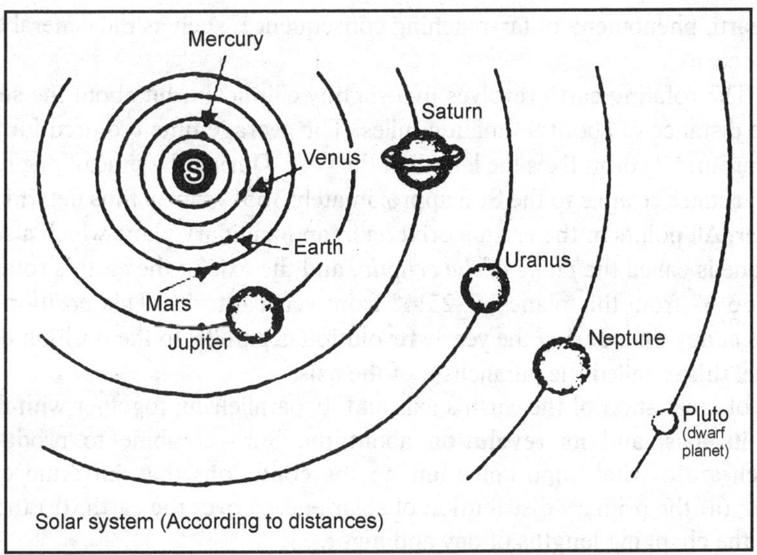

Solar system (According to distances)

Pluto – discovered in 1930 and until recently considered as ninth planet of the solar system. In a highly controversial decision, the International Astronomical Union ultimately decided in 2006 to call Pluto a *dwarf planet* reducing the list of 'real planets' in our solar system to eight.

There are thousands of smaller objects that revolve around the Sun. **Asteroids** are planet-like objects. These small, irregularly shaped objects are found mainly between the orbits of Mars and Jupiter. **Comets** are bodies of dust and frozen gases. They look like bright balls with long tails. **Meteoroids** are pieces of rocks and iron. Many meteoroids falls into the earth's atmosphere, which is the air that surrounds the earth, but most are burned up by friction before they reach the surface.

Earth's Measurements

The earth is approximately spherical in shape, with an equatorial circumference of 40,074.06 kilometres (24,901.55 miles) and a polar circumference of 40,006.90 kilometres (24,859.82 miles); the diameter of Earth at the equator is 12,755.97 kilometres (7,926.41 miles).

Earth is about 150 million kilometres from the Sun. Earth is the third planet from the Sun. Only Mercury and Venus are closer.

Rotation and Revolution of the Earth

The earth moves in two ways, first, it rotates on its axis; one rotation defines a day. Second, it orbits around the sun; one revolution defines a year. Rotation imposes a diurnal cycle on much of the life on earth, and Revolution, in conjunction with the axis of his earth tilted at 23.5° produces seasonal climatic changes and variation in length of daylight.

Rotation: The earth rotates about an axis which, owing to the polar flattening, is its shortest diameter. The ends of the axis of rotation are the earth poles. The time required for the earth to rotate once with respect to the Sun is about 24 hr, or 1 day, although is varies a bit from one part of a year to another. During that 24 hour period most places on the sphere turn alternately toward and away from the sun, thus experiencing a period of energy and light receipt and a period of darkness, and are twice passed through the boundary between light and dark, the circle of illumination, once at dawn and again at twilight. The earth rotates toward the east, and this fact has broad significances. Not only does it determine the direction in which the Sun, moon and stars appear to move across the sky, but is related to other earth phenomena of far-reaching consequence, such as the general circulation of the atmosphere and oceans.

Revolution: The rotating earth revolves in a slightly elliptical orbit about the sun, from which it keeps an average distance of about 93 million miles. The average time required for the earth to pass once completely around its orbit fixes the length of the year. During the time of one revolution around the Sun the earth rotates relative to the Sun approximately 365¼ times, thus determining the number of days in the year. All points in the earth's orbit lie in an imaginary plane which also passes through the Sun. This plane is called the *plane of the ecliptic*, and the axis of the earth's rotation has an inclination of about 66½° from this plane (or 23½° from vertical to it). This position is constant, and therefore the axis at any time during the yearly revolution is parallel to the position that it occupied at any previous time, this is called the parallelism of the axis.

The degree of inclination of the earth's axis and its parallelism, together with the earth's shape, its **rotation** on its axis, and its **revolution** about the Sun, combine to produce several earth phenomena which are of vital importance among the conditions that surround earth inhabitants, Some of these are (a) the primary distribution of solar energy over the earth, (b) the changing of the seasons, and (c) the changing lengths of day and night.

Figure 16.2 *The Revolution of the Earth Around the Sun*

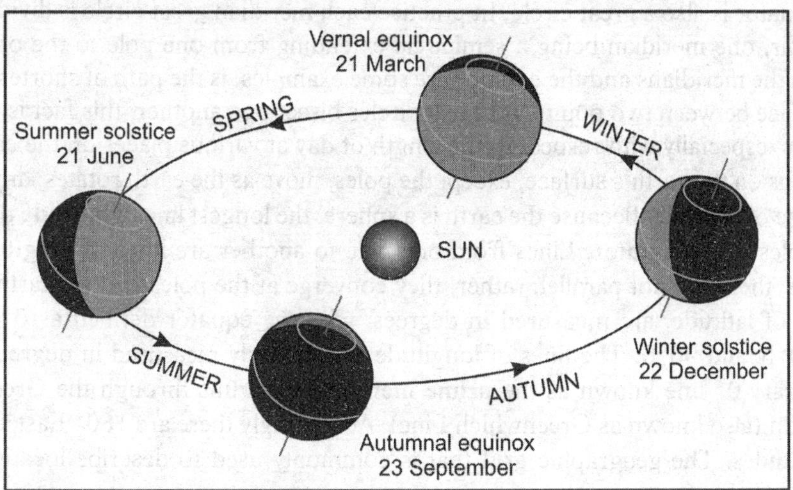

The four seasonal positions of the earth relative to the Sun introduces the terms **'equinox'** and **'solstice'**. The summer solstice occurs on 21st or 22nd June in Northern Hemisphere and on 21st or 22nd December in Southern Hemisphere. This in the time when the hemisphere reaches the maximum inclination towards the Sun directly overhead, the 23.5° North latitude is the **Tropic of Cancer;** while 23.5° South is the **Tropic of Capricorn.** The **winter solstice** occurs on 21st or 22nd December in Northern Hemisphere and on 21st or 22 June in Southern Hemisphere. This is the time when the hemisphere reaches the maximum inclination away from the sun. The two **equinoxes, Vernal** and **autumnal** (spring and fall), are the halfway points between the solstices, and on both occasions the axial inclination is at exactly 90° with respect to the Sun. The earth's revolution, or trip around the Sun, takes one year, or 365¼ days.

Location on the Earth

The Earth Grid

On a sphere there is neither beginning nor ending, no natural point or lines of reference from which to begin to measure the relative positions on the earth, therefore, a coordinate system called earth grid, has been evolved by which directions and locations are determined. The coordinate system on the earth is similar to the familiar rectangular coordinate system on cross-section paper, except for modifications necessary to fit the spherical earth. A line around the earth lying everywhere midway between the poles is called the **equator.** Any number of other lines parallel to the equator may be placed on the earth, these lines, together with the equator, are called **parallels,** and the earth directions east and west are determined by their orientation. Since each of these parallels is parallel with the equator it follows that every point on a given parallel will be the same distance to the north or to the south of the equator. The concept of the distance north or south on the earth coordinates system is called **latitude** and is expressed by identifying the appropriate parallel. Distance east and west, called longitude, is expressed by the second set of lines at right angles to parallels. In the normal flat coordinate system these lines are also parallel with one another, but on a sphere they must converge and meet at the pole. They are, however, equally spaced on any parallel. These lines, called **meridians,** determine the directions north and south; they are all the same length and each bisects the equator and all other parallels. They are the largest circles one can draw on a sphere and are called **great circles.** The equator is also a great circle. In practice each meridian great circle is divided at the poles and forms a pair, one meridian being a semicircle extending from one pole to the other. The great circle, of which the meridians and the equator are some examples, is the path of shortest distance over the earth's surface between two points. All great circles bisect one another, this fact is significant in a number of ways, especially with respect to the length of day at various places on the earth.

All locations on the earth's surface, except the poles, move as the earth rotates; any one circuit of the earth is a line of **latitude.** Because the earth is a sphere, the longest line of **latitude** that lies midway between the poles is the **equator**. Lines from one pole to another are lines of **longitude** and unlike lines of latitude, these are not parallel, rather, they converge at the poles and are farther apart at the equator. Lines of latitude, are measured in degrees, with the equator defined as 0° and the poles therefore as 90°N and 90°S. The lines of longitude are similarly measured in degrees, in this case, from and arbitrary 0° line known as the **prime meridian** that runs through the Greenwich Observatory in London (also known as Greenwhich Line). Accordingly there are 180° East and 180° West, total 360 longitudes. The geographic grid that is commonly used to describe location is based on latitude and longitude.

Land, Water and Air

About 29 per cent of the earth's surface is land. Land makes up a part of the earth called the **lithosphere**. Continents and ocean basins are the largest bodies of land. Another part of earth is the **hydrosphere**, i.e. water bodies. About 71 per cent of the earth surface is covered by water.

The **atmosphere** is that part of the earth which extends about 1609 kms above the earth's surface. Atmosphere consists of several gases in which 78 per cent is nitrogen and 21 per cent is oxygen. The remaining one per cent of the atmosphere consists of argon and small amount of other gases.

All the spheres are integrated and supports the life on the earth. The part of the earth where life is found is called **biosphere**.

Figure 16.3 *Inside the Earth*

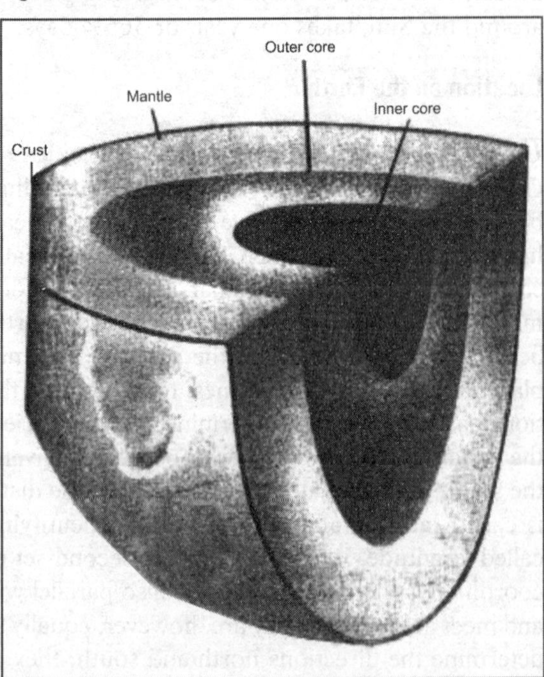

Inside the Earth

The interior of the earth is composed of three layers – the **core**, the **mantle** and the **crust**.

The central part of the earth below a depth of 2900 km is known as **core**. One part of the core is the inner core, which makes the centre of the earth. The inner core lies about 6400 km below the surface of the earth and it consists of a dense nickel-iron alloy (NiFe) with a temperature at about 2700°C. The outer part of the core begins about 2900 km below the surface of the earth and is made up of melted iron and nickel.

Next to the outer core is a thick layer of dense, hot rock called **mantle**. The rock in the mantle is made up of silicon, oxygen, aluminum, iron and magnesium. The depth of mantle varies between 35 km to 2900 km and density ranges between 2.9 and 3.3.

Next to the mantle is the earth's **crust**. This is relatively thin layer, extending perhaps 3 to 35 km below surface. The crust consists of huge plate-like sections of rock that rest on partially melted layer in the upper mantle. The crust includes the continents and ocean basins.

Continental Drift

The continental drift means movement of continents relative to each other across the earth's surface. Basically this is related with the origin of the continents and ocean basins. This subject has been widely discussed and several hypothesis/theories have been proposed. But a comprehensive and systematic theory of 'Continental Drift' was presented by a German meteorologist, Alfred Wegener in 1914.

Wegener suggested that all the Earth's land had once been joined into a single supercontinent known as '**Pangaea**' and a ocean surrounding to it was named as '**Panthalassa**'. Figure 16.4 illustrates the positions of the continents since outward movements began roughly 225 million years ago.

Figure 16.4 *Stages of the Movement of Continents*

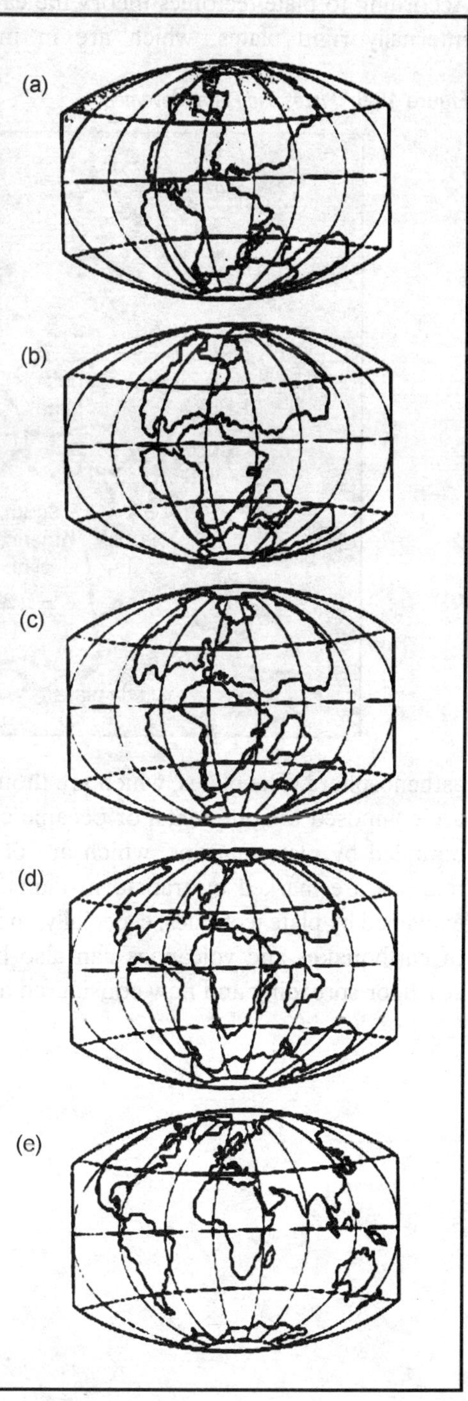

Stages of the Movement of Continents

(a) Beginning 222 million years ago, there is one landmass, a super continent called 'Pangaea'.
(b) 180 million years ago, Pangaea is breaking up, Laurasia is moving north, and the southern land mass Gondwana land, is also breaking away.
(c) 135 million years ago, the process continues, with the beginnings of the north and south Atlantic oceans.
(d) 65 million years ago Madagascar has separated from Africa and India is moving towards Asia, but Australia remains linked to Antarctica.
(e) Shows the present distribution of land and sea, i.e. continents and ocean basins.

Wegner cited various evidences in support of his theory. The evidence included the 'jig-saw fit' of opposing continental coast lines, similarity of geological structures on separate continental masses, the anomalous location of ancient deposits, indicating specific climatic conditions and distribution of fossil species through time. Although the theory was rejected first by most geologists due to lack of viable mechanism but it was considered and accepted by the scholars during 1950s and 1960s through evidence from Palaeomagnetism. In 1967 the theory of 'Plate Tectonics' was incorporated.

Plate Tectonics

An important theory, which proposes that the earth's crust consists of a number of mobile rigid plates. Over a long period of geological time, these plates move relative to each other, causing (i) ocean basins to change in size and form, (ii) the continental drift, and (iii) formation of major landforms. According to plate tectonics theory the earth's surface is divided into eight major and several minor internally rigid plates, which are in motion with respect to each other and the underlying

Figure 16.5 *Plates and Plate Movement*

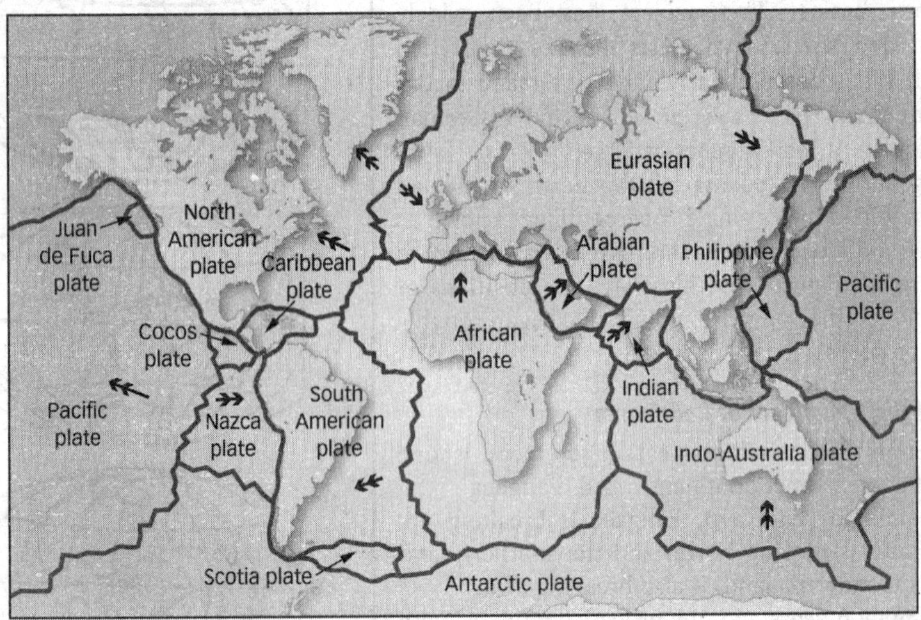

asthenosphere. The plates, which are thought to be moved by large-scale thermal conviction current, are composed of continental or oceanic crust or a combination of both, i.e. sial and sima. They are bounded by plate margins, which are of three types: constructive, destructive and neutral. These margins are marked by transform faults. It has been claimed that many earth's structures can be explained by plate tectonics, especially, mountain belts, island arcs, ocean deeps, while global patterns of earthquakes and volcanoes can also be explained. This theory also includes the hypotheses of sea-floor spreading and now considered as replacement of the old theories of continental drift.

17

Continents, Islands, Oceans, Seas and Lakes

CONTINENTS

The earth surface of the globe presents the two clearly marked, but irregular divisions of sea and land. While sea covers nearly 71 per cent, the land is about 29 per cent of the whole surface of the planet. Geographers divide most of the land surface of the earth into seven large landmasses called **continents.** Continents are one of the main continuous bodies of land on the earth's surface. The seven continents are Africa, Asia, Europe, North America, South America, Australia and Antarctica. North America and South America are the continents in Western hemisphere. Australia and Asia are in Eastern hemisphere. Most of the Europe and Africa are in the Eastern hemisphere, but a small part of their western coastal areas is in the Western hemisphere. Antarctica, which includes the South Pole, is located in both the Western and Eastern hemisphere. Asia is the largest continent, and Australia is the smallest.

A **subcontinent** is a large landmass forming part of a continent, e.g. the Indian subcontinent. The neighbouring **islands** (land surrounded by water) are also included as part of continents, thus, Britain is a part of Europe, Japan is a part of Asia, etc. Some of the continents, like Australia and Antarctica stand alone, while others are joined in some way. An **isthmus** is a narrow piece of land that connects to large land masses. The Isthmus of Panama joins North America and South America. Europe and Asia are actually parts of one huge land mass. Some people consider this land mass to be one continent and call it Eurasia. The continent wise area and population is given in Table 17.1.

Table 17.1 *Area and Population of the Continents*

Continents	*Land Areas in sq km*	*World Land Areas %*	*Total Population*	*Percentage of World Population*
Asia	44,009,000	29.4	3,957,700,000	60.6
Africa	30,251,000	20.2	915,700,000	14.0
Europe	10,445,000	7.0	727,700,000	11.2
North America	24,211,000	16.2	518,300,000	7.9
South America	17,832,000	11.9	375,700,000	5.8
Australia and Oceania	8,985,000	6.0	33,100,000	0.5
Antarctica	14,000,000	9.3	Nil	0.0

As per new population estimates (2011), the world population is 7.1 billion. The continent-wise population in 2011 is as follows:

Asia	–	4,140,336,501
Africa	–	994,527,534
Europe	–	738,523,843
N. America	–	528,720,588
S. America	–	385,742,554
Oceania	–	36,102,071

A brief review of the geographical characteristics of all the continents is as follows:

Asia

Asia is the largest continent having an area of 44,009,00 sq km which covers 29.4 per cent of the total land area of the world. The total population of Asia in 2011 was 4,140,336,501, thus accommodating about 61 per cent of the world population. Asia is situated east of Europe and there is no clear boundary between Asia and Europe. However, the Ural mountains are considered to be the dividing line between Europe and Asia. The boundary between Asia and Africa is the Suez Canal. Asia continent is considered as **'Cradle of Mankind'**, **'Cradle of Civilization'**, **'Orient Culture Area'**, **'Mysterious East'**, **'Far East'**, etc. In fact, Asia is a continent having geographical and cultural diversities as well as unity.

The five main geo-political and cultural divisions of Asia are as follows:

1. South-West Asia – which includes Turkey, Iran, Iraq, Israel, Jordan, Kuwait, Lebanon, Oman, Qatar, Palestine, Saudi Arabia, UAE, Yemen, Bahrain, Cyprus and Afghanistan.
2. South Asia – which includes India, Pakistan, Bangladesh, Nepal, Bhutan, Sri Lanka and Maldives.
3. South-East Asia – including countries of Myanmar (Burma), Cambodia (Kampuchea), Indonesia, Laos, Malaysia, Philippines, Singapore, Thailand, Vietnam and Brunei.
4. East Asia – which includes China, Japan, North Korea, South Korea and Taiwan.
5. Central Asia – including Sịberia, Russia, Kazakhstan, Kirglizia, Tajikistan, Turkmenistan, Uzbekistan and Mongolia.

Physiography

Four distinct geological blocks are found in Asia:

1. Ancient Block of North – comprising ancient hard, crystalline rocks. These are Russian Platform, Angarland, Chinese Massif and Sardinian Massif.
2. Ancient Blocks of South – Gondwana land includes Arabian and Indian Peninsula.
3. New Folded Mountain Ranges – of Himalayas, Karakoram, Anatolia, Suleman, Kirthar, etc.
4. Residual Parts – located in between the above mentioned regions.

Topography of Asia comprises of series of high mountain belts, plateaus, river basins, island arcs and alluvial lowlands. The highest mountain peak not only of Asia but of the world is the Mount Everest (8848 m), while the lowest point is Dead sea (–392 m).

The highest plateau is the **Tibet Plateau** also known as **'roof of the world'**. Pamir is the centre of the mountain system of Asia known as *'knot'* of mountains. **Himalayas** is the largest mountain system extending about 3500 km like a mountain wall. Mountain ranges like **Taurus**, **Zagros**, **Hindu-Kush**, **Tien-Shan**, **Altai**, **Da-Hinggan**, **Kunlun**, **Karakoram** and **Himalayas**, etc. all converge at Pamir. Numerous plateaus and basins are located within or along the margins of these mountain ranges.

Apart from Tibet, other plateaus of Asia are **Gobi, Loess, Anatolian, Arabian, Deccan Plateau, Patom** (North of Baykal Lake), **Yakut, Vitim** and **Aldan Plateaus** in Russia.

There are vast plain areas in Asia. The important plain areas are – **Indo-Gangetic Plain, North China Plain, Chang Jiang** (Yangtze kiang) **Huang Ho Plain, Irrawaddy Plain** (Myanmar), **Mekong Plain** (Cambodia), **Western Siberian Plain, Kirghiz Steppe**, in Iraq and Iran, etc.

The main deserts of Asia are-**Dasht-e-Kavir** (Iran), **Dasht-e-Lat** (Iran) **Gobi** (Mongolia), **Karakum** (Turkmenistan), **Kyzylkum** (Kazakhastan and Uzbekistan), **Ordos** (China), **Registan** (Afghanistan), **Taklamakan** (China), **Thar** (India and Pakistan).

Rivers: The main rivers of Asia are – **Chang Jiang** (Yangtze Kiang), **Hwang Ho** (Yellow), Amur, Lena, **Mekong, Yenisey, Ob Irtysh, Indus, Ganga, Brahmaputra, Salween, Euphrates, Amu Darya, Kolyma, Syr-Darya, Irrawaddy, Tarim, Tigris**.

Climate: In Asia variety of climatic conditions are found, ranging from equatorial to the polar climate. Three broad climatic regions of Asia are (i) **Monsoon climatic region** (ii) **Sub-tropical and arid climatic region**, and (iii) **Polar or cold climatic region**. In Asia the highest temperature recorded at Tirat-Tsvi (Israel) being 54°C, while the lowest temperature – 69°C was recorded at Verkhoyansk (Siberia).

Natural Vegetation – Equatorial regions of Asia are having tropical rain forests and vast areas of south and south-east Asia have tropical deciduous forests. The arid and sub-arid parts of the continent are having limited vegetation that too of short grasses and thorny shrubs. '**Steppe**' grasslands dominate in Mongolia, parts of central Asia and south-west Russia. In Russian part of Asia coniferous florists known as 'taiga' are found, while the Siberian region is having little or no vegetation.

Economy – The basis of Asian economy is the agriculture. Asian agriculture is a combination of traditional subsistence to modern mechanized form. Shifting cultivation is still prevalent in some areas, such as north-east India-**Jhuming**, Malaysia and Indonesia-**Ladang,** Phillippines-**Caingin**, Burma or Myanmar-**Taungya**, Indochina-**Rai**, Thailand-**Tamori** and Sri Lanka – **Chena**. Intensive subsistence agriculture is a characteristic feature of Monsoon Asia while plantation agriculture is done in many parts of south-east and Southern Asia. Asia is a leading producer of Rice, Wheat, Tea, Rubber, Coffee, Sugarcane, Cotton, Jute, Spices as well as of Tobacco, Date palm, Coconut, etc.

Asia is rich in mineral deposits of iron ore, chromium, manganese, mercury, graphite, magnesite, mica, pyrite, talc, etc. Asia's coal deposits are largest in the world. The countries of West-Asia are the leading oil producers of the world. In many parts of Asia vast hydroelectric power potential exists. China, Japan, India, south and north Korea are now developing as industrial nations.

Cultural Characteristics

Asia has a long cultural heritage of great diversity. Asia is the birth place of Judaism, Christianity, Islam, Confucianism, Hinduism, Buddhism, Jainism and Sikhism. Hinduism is the main religion of India and Nepal. Islam is the principal religion in South-West, South-East and parts of Central Asia. While Buddhism is principal religion in Myanmar, Thailand, Sri Lanka, Laos, and Cambodia, Buddhism also dominate in Tibet, Bhutan, Japan, etc. Asian countries are having too much diversity in language. In Asian countries Chinese is the language of one billion people of China. Hindi, Urdu, Arabic are also the main languages. In most of the Commonwealth countries of Asia, English is the corresponding language. All the Asian countries are having their own social structure which is distinct to each other. In fact, Asia is a highly diversified continent in terms of its culture and society.

Europe

Europe lies between 35° to 71°8′ north latitudes, and the Arctic circle passes through the northern regions of the continent, while its major land portion lies in the temperate zone; the northern region, a minor land portion, lies in the frigid zone. This continent also lies between 10° W to 65°E longitude. The important line of longitude 0°, the meridian of Greenwich passes, of course, through Greenwich which is a part of London. Its breadth from East to West is 5300 km while length from North to South is 4500 km and the total areas of the continent is 738,523,843 kms. The Arctic ocean is to the north of Europe while the Atlantic is to its west. To its south are the Mediterranean Sea, and the Caspian Sea and to the east lies Asia.

Physiography

The four physiographic regions of Europe are:
(i) The North-Western Highlands.
(ii) The Central Lowland or the Great European Plain.
(iii) The Central Plateau.
(iv) The Southern Mountain region of Alps.

The North-Western Highland: The region includes the mountains and plateaus of Norway and Sweden and is extended into Finland, Scotland and northern mountain regions of Ireland. The average height of this mountain region is about 2000 metres. It is a region of igneous rocks. At places, like the coast of Norway long narrow and deep arms of sea between high cliffs called fiords penetrate into the land.

The Great European Plain or the Central Low Land: The Great European Plain stretches the Atlantic coast in the west and the Ural mountain in the east. Northern France, Belgium, Netherlands, North Germany, Denmark and Poland are included in this region. It steadily increases its width eastwards. This great European plain is drained by a number of rivers like the Rhine, the Seine, the Elbe, the Vistula, the Don, and the Volga. All these rivers have generously spread silt in this plain so it has become a fertile plain of the continent.

The Central Plateau: A series of plateaus and the mountains lie close to the southern border of the Great plain and to the north of the Alps mountain range which includes Spain, the Central Plateau of France, the Rhine Highland and the Bohemian plateau. The average height of this plateau is between 700 to 1500 metres. River Rhine flows from the Rhine rift valley, which is a part of the Rhine lands.

The Southern Mountain Range of the Alps: This is the biggest and the highest mountain range of Europe. It is situated in the southern Europe; its average height is about 2500 metres and the highest peak of this mountain range is Mount Blanc, 5000 metres in height. Many other branches spread out in different directions from the Alps. To the west of the Alps is Pyrenees, to the South, Apennine, to the north-east, Carpathain and the Caucasus. The well-known volcanoes like Vesuvius, Etna and Stromboli are situated in the Apennine mountain range. Europe is a big peninsula.

Climate

Most of the Europe is having temperate climate. Europe can be divided into four climatic regions:
(i) North-West Europe – having maritime climate, it extends from the coast of Norway to northern Spain and inland towards the central Europe. In this region winters are mild, summers cool, and rainfall between 50 cm to 100 cm.
(ii) Continental climate in the East and North – includes western and central Europe. This climate is mainly found in Poland, Slovakia, Czech Republic, Hungary, Romania, Serbia and Bulgaria. In this region, winters become coldest and summers, hotter.
(iii) Mediterranean Climatic Region of Southern Europe – characterized by dry and warm summers and wet mild winter, winter rainfall is common with an average rainfall of 100 cm. This climate is

found in Southern Europe along the coast of Mediterranean sea and suitable for the production of citrus fruits.

(iv) Mountainous Climate – is found in Alps and the Caucasus region. Generally, temperatures range from 4 °C in January to 16 °C in July and annual rainfall averages from 50 cm on the leeward side of hills to more than 200 cm on the windward side.

The northernmost Europe is having very cold conditions.

Natural Vegetation

In northernmost Europe, due to cold condition and snow cover, vegetation is almost absent except some plants, like short grasses which grow during summers. To the south of the Northern Europe from west to east lies the belt of the coniferous forest region, where the shape of the trees is conical. It is a region of tall and straight trees like pine, spruce and fir. They are generally needle-shaped, thick and oily or glossy, which can store moisture for a long time. The wood of such trees is soft which is very useful for manufacturing paper. Further south lies a belt of the deciduous forest. The trees of this forest have broad leaves which are shaded in winter. These broad-leaved trees like oak, beach, poplar, ashe, walnut, and elms have very hardwood. In the extreme cold of winter the trees shed their leaves. That is why these forests are called the deciduous forests. On the coastal regions of the Mediterranean sea where it rains heavily, we find important trees of orange, sweet lemon, apple, figure walnut, apricot, and grapes olive. To the south-west of Europe due to scarce rainfall, only a short grass grows.

Economy

The economy of Europe is a combination of mineral resources, industries, agriculture and trade. Europe possesses large mineral deposits of iron ore, coal, bauxite, and phosphorous. Coal is found from Ruhr and Westphalia in Germany. Moreover UK, France, Poland, Russia are having the other coalfields of Europe. Important iron areas are situated in Krivirog, Magnitogorsk, Kurs in Russia, Loren in France, Geliwar, Kiruna, Gonyesburg in Sweden, Cliveland, Niceland in UK, Bilbaso in Spain, and Scotland. Bauxite mines are available in Russia, Yugoslavia, Italy, Greece and Hungary while mineral oil and gas are found from the Ural region and Rumania. Besides these minerals, zinc, lead, mercury, platinum, sulphur and manganese are available from different countries.

Europe is the foremost continent in the industrial development. Most of the countries of Europe are industrial countries. Various industries such as iron and steel, machine and machine tools, transport goods, colour, chemicals, medicines have developed on a large scale. Iron and steel industries have developed in Germany, Russia, France and UK. The motor manufacturing industry has developed in Germany. Shipbuilding industry has developed in UK. West Germany, Sweden, Netherlands, France, Italy, Norway and Russia planes Aero are manufactured in UK, France and Germany. Railway engines are manufactured in UK and Russia; Industries manufacturing chemical fertilizers and sulphuric acid are developed in Germany, Russia, France and Italy. Cotton textile industry has developed in Germany and France. Manchester in UK and Moscow in Russia, Sweden and Norway are the foremost paper manufacturing countries of the world. Silk cloth manufacturing industry has developed in Italy.

Three distinct agricultural regions identified in Europe are: (i) Mediterranean (ii) The Northern and Central Europe, and (iii) The Eastern Europe.

Mediterranean Region: Greece, Southern France and Italy are included in this region where it rains in winter but the summer is dry and long; climate is suitable for fruits like grapes, oranges, figs, peaches, berries, sweetlime, citrons, apricots and dry plum. Some farming is also successfully done and crops such as wheat, barley, maize, etc. are produced.

The Northern and the Central European Region: England, Northern France, Belgium, Holland, Denmark, and Western Germany are included in this region, from where crops like wheat, barley, oat, rye, sugar beat, etc. are taken. Flax, a fibre crop, is grown in France and Belgium. Intensive and rotational farming is a peculiar feature of the farming in this region.

The Eastern European Region: Instead of intensive and rotational farming here in this region they take only one crop throughout the year. They grow barley, rye and flax. By far the most important crop of Europe is wheat.

Europe had adopted methods of farming with the help of machines, chemical fertilizers, insecticides and scientific techniques. They are trying to procure more and more crops per hectare. Another aspect of economy of Europe is the fisheries. There are many favourable conditions for developing fishery in Europe. There are vast and wide areas like the "Dogger Banks" and "The Great Fisher Banks" having continental shelf where they get plenty of fish. The coast of Norway which is full of fiords is considered to be a "storehouse of fish". The Arctic ocean is very shallow and many "sandruns" are formed where different types of fishes like Card, Herring and Sardine are found in large proportion.

Other Features

Europe has a complex and diverse physical, economic, socio-cultural and political settings and also has great diversity. In brief, these are:

(i) In 2011, the total population of Europe was 738,523,843.
(ii) The geographical location of Europe in the world is unique between America and Asia and north of Africa, which makes it favourable in terms of transportation, trade and commerce.
(iii) European Economies are largely dependent on industries and tertiary activities.
(iv) European culture is known as western culture and have had its impact on developing countries.
(v) In Europe various dialects and languages are spoken. About twenty languages are important in Europe, but English French, Germany and Russian are more widespread.
(vi) Majority of the people of Europe are Christians. There is a sizeable number of jews and followers of Islam.
(vii) In spite of diversity in language, race, and religion, European culture is also an example of 'unity in diversity'.

North America

North America lies between 10° to 80° North latitude and between 60° to 160° west longitude. The Tropic of Cancer passes through the north, the waters of the Arctic Ocean wash the northern shore of America, the North Atlantic ocean touches its east coast; in the south, the Panama Canal separates it from south America, while Pacific ocean touches its western boundary. North America has about 16.2 per cent of the total land area of the world. In 2011, the total population was 528,720,588. It includes two of the world's largest countries, i.e. Canada and United States of America (USA).

Physiography

North America can be divided into four physiographic regions:

(1) The Western Mountain Region (2) The Central Plain (3) The Eastern Highland (4) The Coastal Plain.

1. *The Western Mountain Region:* The western mountain region stretches from north to south, which is divided into two parts: The mountain range along the Pacific coast and the Rocky mountain range. The pacific coastal mountain range extends from the high plateau of Alaska to

the cordilleras in the south. The highest mountain peak of north America is the Mount McKinley (6,194 metres) which is situated in the mountain range of Alaska. In the Western Highlands, the Rocky Mountains have the greatest height. They have been spread parallel to the coastal mountain range of the Pacific and covers the area from Alaska in the North to the Central America in the south, which includes the Alaska plateau. They are the widest in USA and get narrower while passing through Mexico. The Cascades and the Sierra Mardeni form the western edge of this range. There are many dead and active volcanoes in the Rocky mountain range; among them Popokestipetal is famous.

2. *The Central Plain:* The central plain lies between the Rockies mountain range and the Eastern Highland. The lowland begins from the North Pole, stretches from the Hudson bay in the north to the gulf of Mexico in the south while proceeding eastward. The height of the plain gradually decreases. The delta region of the river Mississippi is included in this region. This region is made up of the silt spread by the rivers Mississippi, Missouri, Mackenzie, Red and St. Lawrence, that is why, this plain is considered as the topmost fertile plain of this region.
3. *The Eastern Highland:* The Eastern Highland region of North America is made up of disconnected land portion. They are divided into two parts by the river St. Lawrence – the old disconnected plateau of Labrador in the North and the Appalachian in the south. The Labrador plateau is formed from the remains of Lawrentian mountain which acquires the land portion up to the Bay of Hudson. The average height of the plateaus is 300 metres. The Appalachian high plateau lies to the south of the St. Lawrence river valley region which is known as the high Appalachian mountain.
4. *The Coastal Plain:* The coastal plain slopes gradually from a high level in the east to the Hudson bay lowlands. Between the southern part of the Appalachian Mountain and the sea is an important coastal plain.

Rivers

The rivers of North America are mainly classified into (1) the rivers that meet the Arctic ocean (2) the rivers that meet the Pacific ocean, and (3) the rivers that meet the Atlantic ocean. The Mackenzie flows towards the north and meets the Arctic Ocean and the Nelson flows into the Hudson Bay. These rivers are not useful for navigation because they remain frozen for the most part of the year.

The Columbia and the Colorado rise from the Rocky Mountains and flow into the Pacific Ocean. Grand Canyon of America is formed by the River Colorado which is 300 kilometres long and 1.5 kms deep. This river is also not useful as waterway.

The Mississippi, the St. Lawrence, and the Rio-Grand Delanort flow into the Atlantic ocean. These rivers flow across the plains and their flow is steady, so they are navigable. The Mississippi rises from a small lake and on the way the rivers Missouri and Arkansas from the west and Ohio and Tennessy from the east join the Mississippi and its tributaries. These rivers and their tributaries flow across the central plains. Mississippi is the longest river of North America, which is useful for navigation for about 2000 kilometres from its mouth. The St. Lawrence river, made up of lakes, rises up, form a lake near the rising place of Mississippi, and passes through Superior, Michigan, Huron, Erie, and Ontario. It is known by different names near each of these lakes. Between lake Erie and Ontario it is known as the Niagara, which has the famous "Niagra falls". With the help of this waterfall hydro-electricity is generated on a large scale.

The Lakes

North America is called the **"region of lakes"**. There are number of lakes around the Hudson Bay. The lake Superior stands second among the lakes of the world. The other lakes of this region are

Michigan, Huron, Eire and Ontario. There are many other lakes like the Great Bear, the Great Slab, the Winnipeg and Athabaska, etc. over and above the lakes through which the St. Lawrence flows. The only large salt water lake of North America is the Great salt lake lying in the Western Cordilleras.

Climate

North America has extended from tropical to pole latitudes and its climate is different from place to place. A major land portion of the North America is included in the temperate zone but its climate is not temperate because many factors like the relief, distance from the sea, the ocean currents, and the effects of the winds have affected the climate of the continent. They remain warm due to the westerlies which blow from the warm Californian Current through the Pacific Ocean, so extreme climate is not experienced in this region. Through the upper northern region of North America the Arctic Circle passes. The area north of the Arctic Circle is a frigid zone which remains under snow throughout winter which causes extreme cold; so the summer is short and warm in this region. North America has great diversity in its temperature and rainfall distribution, the highest temperature 57°C was recorded in the Death Valley (California), while the lowest temperature was recorded –66°C at Northice (Greenland). Rainfall generally tends to decline towards the west as a result of the rain shadow effect the eastern half of North America is more humid. The distinct climatic regions of North America are:

1. The Tundra climatic region in northern coastal parts of Canada, northern islands and Greenland.
2. The Sub-Arctic climatic region in northern parts of Canada on both sides of the Hudson Bay and southern and central parts of Alaska.
3. The Humid continental Region confined in northern parts of USA and the southern parts of Canada on both the sides of the great lakes.
4. The Humid-Sub-Tropical Region found in greater southern parts of USA up to the Gulf of Mexico.
5. The Mediterranean climatic region – In central and southern California.
6. The Semi-Arid Steppe Region – Occupies area between the pacific and littoral of USA and the lowland margins of humid east, and extends north into Canada.
7. The Desert climatic Region is found along Mexico border.
8. The Mountain climatic region is in Rockies and Sierra Nevada.

Natural Vegetation

Hot wet evergreen forests occupy the wetter parts of the tropical Central America and the West Indies. Thick forests have developed in the region so it is called the region of evergreen forests. Mahogany, Logwood, Ebony, Rosewood trees are grown in plenty while the trees of coconut, betel nut and rubber trees are cultivated in this region. The deciduous forests are found in the west coast of America. The higher parts of the Appalachian Mountain are covered with Coniferous Forests as well as cool birch, oak, emu, chestnut, pine, fir, spruce, which grow in plenty. The desert region of Mexico and California is situated to the North of the evergreen forest region and to the south of the grassland region. The thorny bushes, cactus and short grass grow in this desert land region. The Californian valley region which is a region of the winter rainfall has the Mediterranean type of climate. The well-known fruit trees like grape, orange, peach, dry palm, apple, apricots, almonds, grow in this region. The temperate grassland regions are found throughout West Central North America which is known as the Prairies. The grass meadows have been cleared of and the vast farmlands have been prepared for developing agriculture. A belt of the coniferous is situated to the north of the Prairies in which cone-shaped trees are grown in plenty. The wood of the pine, fir, spruce and cedar tree is very soft. These forests are seen especially in Alaska and central Canada.

The northern region of North America is a snow covered region. Nothing grows here except moss and lichen. During summer, when the snow melts plants having flowers of varied colours and short grass are seen for a short time.

Economy

Economically, North America is considered as well developed; especially the economy of USA is highly developed. Agriculture, industries and mineral resources are the key components of the North American economy. Vast fields, fertile land, a favourable climate, sufficient rainfall, scientific methods and agricultural tools have helped a lot in the development of agriculture in this region in comparison with other countries. Where there is a scarcity of rainfall but sufficient facilities for the development of irrigation are available, they have developed agriculture with the help of machines. Besides, the farmers of this region are educated, hard working and economically prosperous. Moreover, being familiar with scientific methods, they use hybrid seeds, insecticides, and chemical fertilizers for the development of agriculture. USA is having well developed agricultural regions. These regions are famous for the production of particular crops. Such agricultural belts are cotton belt, wheat belt, maize belt, etc. Wheat, maize, cotton, tobacco and paddy are grown in North America. Moreover, fruit orchards are also developed in plenty. USA and Canada have vast prairies fields where wheat is grown on a large scale. They have been changed into "granaries of wheat". Canada stands fifth and USA second in the wheat-growing countries of the world. Maize originally is an American plant and it is grown in the USA, Canada and Mexico in plenty. More than fifty per cent product of maize is produced only in North America.

Cotton is also grown in plenty in the USA and Mexico. The USA stands second among the cotton producing countries of the world. Here long stapled and mercerized cotton is grown and exported to foreign countries. The West Indies islands, Cuba, and the Central America produces sugarcane in plenty. Cuba stands first among the countries of the world in manufacturing sugar. The USA stands first in producing tobacco and exports it.

North America is also rich in minerals. USA is a leading producer of iron ore and produces about four per cent of the world iron ore. Canada is also an iron ore producing country. Zinc, lead, copper, gold, silver, uranium, thorium, bauxite and mica are also found. Large coal deposits exist in USA and USA produces about 17 per cent of the world coal production. USA and Canada also produce petroleum. North America stands foremost in the industrial development and production among the countries of the world. All the important factors for industrial development such as varieties of mineral resources, agriculture, and forests products, animal rearing, well developed transport facilities, technically trained workers, mechanical instruments, capital, land, vast and extensive market are present in this continent.

Owing to the above mentioned facilities the north-east industrial region, the southern industrial region, the western industrial region and the south-east industrial region have flourished. The north-east industrial region is the most important industrial region amongst all. In this region, industries like iron-steel, petrochemical, chemical fertilizers and metal refineries as also industries such as machine and machine tools, motors, locomotives, aeroplanes, and steamships have developed. Moreover cotton textile, cigarette, and sugar industries which are depending on the agricultural products have also developed. Some industries like woollen and leather, which depend on animal rearing, have also developed. Besides all these industries, glass, camera, watch, plastic, electronic, soap and perfumery manufacturing industries have also flourished. The main industrial regions are New York, New England, Middle Atlantic States, Pittsburgh, Detroit, Michigan, Cincinnati, Indianapolis, Niagra-Ontario, the Pacific coast region and Southern region.

Population

The total population of North America in 2011 was 528,720,588, which is about 7.9 per cent of the world population. Barbados is having the highest density of 651 persons per sq km, while Greenland is having only 0.0 persons per sq. km. A major part of the population of North America lives in the northern coastal region because agriculture, industries and transport conditions are favourable for human inhabitation. On the west coast of North America the population is thin because of unfavourable conditions. Central America, Cuba, and the West Islands are also densely populated. The mountain region of North America, plateaus, forests, and the semi desert regions are thinly populated. The Tundras and the deserts have the least population.

The aboriginals of North America are known as the Red Indians who now-a-days live in the coniferous forests and the semi-desert regions of Western America. Their chief occupations are farming and trapping of the wild animals. European colonists like English, Dutch, German and French live on the east coast of North America in large numbers. The Spanish and the Portuguese mostly live in Mexico and Central America.

Lacs of Negros from Africa were brought to USA and Central America and made to work in cotton plantation and sugarcane fields. The Eskimos of Alaska, and North Canada live on hunting and fishing. The Chinese and other Asians have come here to work in mines. Some Indians went to West Indies and settled there.

People of different countries, who speak different languages have come and settled here. The Spanish and the Portuguese are spoken in Central America; English in USA, French and other European languages are spoken in Canada.

South America

South America lies between 12°N and 55°S latitudes and between 35° and 81° west longitude. A major land portion of this continent lies in the Southern hemisphere and Equator passes through the northern part of this continent. The Tropic of Capricorn passes roughly through the middle of the continent. It is broad in the north while towards the south it gets narrower. Two-third part of the continent is included in the torrid zone while the remaining is in the temperate zone. To the north of the continent is the northern Atlantic Ocean, to the east is the southern Atlantic while to the south is the Capehorn cape and to the west is the Pacific Ocean. Total area of South America is 17,832,000 sq km. It is the fourth largest continent from the point of area and covers 11.9 per cent of the total land area of the world.

Physiography

Physiographically, South America can be divided into four regions. These are Western Highlands, Central Plains, Eastern Highlands and Coastal Regions.

1. *The Western Highlands:* The Western Highlands consisting of the Andes mountain like the mountain range of North America which run along the west coast from north to south, has the length of about 7200 km. The Rocky mountain range of North America is divided while this mountain range is undivided and consists of several ranges with plateaus. The Bolivian high plateau is the broadest among them. Aconcagua (7000 metres) is the highest peak of the Andes Mountains which has about 130 volcanoes. Cotopaxi in Equador is an active volcano while Chimborazo is an inactive one.
2. *The Central Plain:* The Central Plain lies between the Eastern and the Western highlands. They occupy half the land portion of South America. Three important rivers and their tributaries have formed this plain. They are divided into three sub-divisions: (1) The northern basin of the

Orinoco River in which tall grass is grown. This basin is known as the Llanos. (2) The Central plain of the Amazon rivers. (3) The southern basin of the La plata. The basin is situated on the banks of the Parana and the Paraguay rivers and their tributaries. This basin is known as "The Pampas", a short grass plain. South of the Pampas lie the Bleak, barren steeps of "Patagonia", which is a semidesert region.

3. *The Eastern Highlands:* These highlands are made up of igneous rocks which contain plenty of minerals. The height of the Brazilian Highland varies from 300 to 1000 metres. The San Francisco has formed a vast valley in this region. This highland is like a wall of the coastline in Brazil. Tall grass meadows known as "**Campos**" are situated in some parts of this region.
4. *The Coastal Plain:* This is a narrow coastal strip 4800 kms long bordering the Pacific Ocean like the coastal plain of North America. It is considered to be the longest coastal plain among all the coastal plains of the world. Here there are fiords on the sea coast and small islands near the sea coast while mountain regions are situated near the coastline.

Rivers

The rivers of South America rise either from the Andes mountain or from the Eastern Highland and flows into Atlantic ocean. The Amazon, the Orinoco, the Laplata and the San Francisco are the important rivers of South America. Among them the Amazon is the largest. It is 6437 kms long. Due to its large volume of water the sea flows into the ocean and does not have any salinity for about 300 kilometres. Innumerable tributaries flow in this river among which Rio Negro and Madira are important. Due to the unhygienic climatic conditions and thick forests, people are not able to take any advantage of this river. The Orinoco is a navigable river for small boats and steamers. La Plata is the united stream of the Parana and the Paraguay rivers. There is irregularity in the course of the San Francisco River, so it is not navigable.

Climate

The climate of South America is controlled by its physical features, location, winds and ocean currents. The Equator passes through the northern part of South America, so a major part of the continent is situated in the southern hemisphere, and the season cycle of this continent is inverse compared to northern hemisphere, i.e. summers is in January while winters in July.

A greater part of South America lies within the Tropics; its climate is consequently hot. The thick forests of the Amazon get the convectional rainfall throughout the year. The climate remains unchanged throughout the year. On the whole the climate of this region is hot, moist and unhygienic. Both the coastal regions narrows towards the south of the Tropic of Capricorn, so the climate is temperate. Only Tierra-Del-Fuego, the southernmost part of South America, experiences extreme cold in winter.

The rainfall is heavy in the forests of the Amazon, more than 250 cms. Trade winds from the Atlantic Ocean flowing unhindered cause rainfall up to the interior parts of the continent. Here the average annual rainfall is between 100 to 200 cms; of course, towards the west it gradually decreases. The moist laden winds do not penetrate in the interior of the Andes. They are unable to proceed to the west coast of the continent. Due to this the Atacama Desert has come into existence. The Westerlies blowing from the Pacific Ocean in the southern regions of South America are obstructed by the Andes. As a result, it rains throughout the year on the southern coast, but Patagonia, lying on the east of the Andes, is semidesert. Similarly is rains scarcely in Argentina.

Central Chile of South America has rain in winter and is hot and dry in winter, so this region experiences the Mediterranean climate.

The region around the Amazon, which is known as the Amazon basin, falls within the belt of equatorial rainfall and has a high temperature throughout the year. It is covered with dense Equatorial Forests known as "**Selvas**" where the trees and plants of rubber, palm, mahogany, and ebonite are grown. The natural vegetation, on both the sides of the equatorial evergreen forests, is mainly coarse grass. The grass of this coarse grassland is known as the elephant grass which is 3 to 5 metre in height. The northern region of the grassland is known as the **Llanos** and the southern region as the **Campos**. The trees of hard, as well as soft wood grow in the lowland regions of Parana and Paraguay rivers to the north of Argentina. This region is known as "Gran-Chaco". Pine trees grow to the south of this region and in the southern Argentina where there are little rains, a short grass growing region has developed, which is known as the **Pampas**. In middle Chile where it rains in winter, we find numerous fruit orchards in which Chile regions of cold deserts are situated. In the desert region of Patagonia and Atacama, cactus, thorny bushes and short grasses are grown.

Economy

South American economy is based on agriculture, mineral resources, industries and trade. Although, industrial development is very limited, in some parts of the continent, large-scale animal husbandry is done on commercial basis. South America is the leading producer of coffee, cotton, wheat, sugarcane, soyabean and sunflower. Brazil is the leading producer of coffee in the world and produces about 25 per cent of world coffee production. In Brazil coffee plantation has developed mainly in south-eastern part. Sao Paulo is the main coffee producing province; other provinces are Rio-de-janeiro, geras Mains and Espiratosanto. This whole part is known as '**Coffee Region**'. In this region coffee plantations are known as 'Fazenda'. In some parts of Amazon, forests have been cleared and crops are suitable to hot and moist climate such as rubber, coco, cinchona, and spices. Argentina, situated in the temperate zone, has a cool and dry climate which is suitable for growing wheat. Argentina is considered as a foremost country in producing wheat. On the coastal regions they grow maize, paddy, sugarcane, and fruits. Argentina and Uruguay are the developed countries only from the point of view of agriculture.

In South America animal husbandry has developed on a large scale along with agriculture. "Alfalfa" is a kind of grass which is a good nutrient feed for the animals. In grassland vast "ranches" are set up and innumerable animals – sheep, alpaca, bikuna and lama – are reared on the slope of Andes. The chief aim for rearing animals is to get meat, leather, wool and milk, which are exported to UK, Italy and other European countries. Among the countries of the world Argentina stands first in exporting meat. South America is rich in minerals. Mineral oil, gold, manganese, copper and tin are the chief minerals. Brazil, Argentina, Chile, Peru, Venezuela and Guinea are the important regions from the point of view of mineral resources. The west, the east and the north of South America are the important regions where they get most of the minerals. The mountain regions and the high plateaus are considered treasures of minerals. Venezuela stands third among the mineral producing countries. Moreover Argentina, Columbia and Equador are also the countries where they get mineral oil. Brazil, Peru and Chile are having iron ore fields. There is a long strip of copper-mine from Sirodi of Peru, from Pasko to Bolivia and Chile. Bolivia stands second in the production of tin. Besides all these minerals, others like coal, gold, silver, bauxite, nitrate, platinum, zinc, nickel, tungsten, lead and chromite are also available in South America. South America exports minerals.

South America is not well developed industrially due to some geographical and economical factors. They are as under: (1) Thick forests, high mountain regions and plateaus and dry desert regions (2) Unhygienic and extreme climate of the region (3) Sparse population (4) Most of the mines are situated either in the mountain region or very high plateaus. (5) Lack of technical knowledge (6) Underdeveloped transport facilities (7) Most of the countries were under the heels of foreign rulers prior to their independence.

The above mentioned reasons are held responsible for the late and limited industrial development. During the second world war when the import of foreign goods was totally stopped, they had started local industries. As a result of this Brazil, Peru, Chile, Uruguay, Equador, Columbia, and Venezuela started their march towards industrialization. Most of the industries depend on forest products, agricultural products and mineral products.

The stress is laid on manufacturing agricultural machinery and industrial machines and machine tools. Sao paulo, Rio-de-Janeiro, Salvador, Soracaba are the important centres of textile production. Iron and steel industry has developed in Brazil because most of the iron mines are situated there. Oil refineries have been established in Brazil and Venezuela to refine mineral oil. Along with them many other petrochemical industries have also developed in these regions. Besides, the manufacture of cement, glass, sugar, paper, chemicals and medicine have also been started.

Population and Society

The total population of South America in 2011 was 3,85,742,554 which is 5.8 per cent of the total world population. South America's population remains concentrated along the continent's periphery. Most of the interior parts are still sparsely populated. The east, the west and the northern coastal regions of South America are thickly populated. Central Chile, agricultural and animal rearing regions of Argentina, coffee-producing regions of Brazil and industrial regions are populated regions. The main urban centres are Rio-de-Janeiro, Sao paulo, Buenos Aires, Montevideo, Santiago, Bogote and Caracas, etc.

The Amazon forests' region is sparsely populated due to thick forest, unhygienic climate and underdevelopment of agriculture and industries. The density of population in this region is only one person per square kilometre. The desert regions of Atakama and Patagonia are also sparsely populated. Different races like Negroes of Africa who were brought here as slaves are the aboriginals of this continent. Spaniards and Portugals live in this continent. These people have been living in this continent for ages and cross-race marriages have produced newly mixed generations.

Africa

Africa lies between 37° north and 34° south latitudes and between 17° west and 51° east longitudes. The Equator passes through the middle of the continent which divides it into two equal parts. Moreover, the Tropic of Cancer passes through the northern and the Tropic of Capricorn through the southern part of the continent. Thus a major part of the continent is situated in the tropical zone.

Africa is surrounded on all four sides by sea. To the north, the Mediterranean Sea separates it from the continent of Europe by the strait of Gibraltar. The distance between the two continents is only 14 kilometres. In the east the Red sea separates Africa from Asia. The water of the Indian Ocean wash the eastern coast of Africa while the Atlantic ocean touches its western coast. Africa is the second largest continent of the world having an area of 30,251,000 sq km, which is about 20.2 per cent of the total world area. In 2011 the population of Africa was 994,527,534.

Physical Features

Looking at the physical features of Africa we find that the whole continent is a huge plateau. The average height of the land portion of Africa is more than that of any other continent of the world except Asia. Only on the coastal land we do find narrow plains. In the northern part of Africa there is the Sahara desert while in the small part of south-west Africa is the Kalahari Desert.

Mountains: There is no long and high mountain range in Africa. To the north-west is the Atlas which is a fold mountain, the highest peak Toubkal is about 4165 metres high. It is made of soft rocks. Most of the other mountains of Africa are formed of hard rocks. The "Cameroon" mountain is to the west of the continent near the Gulf of Guinea, while the Drakensberg mountain range forms the eastern edge in the south, which runs north to south and parallel to the coastline, and acts as a barrier and, prevents easy access to the interior parts of Africa. The Mount Kilimanjaro (5895 metres) is the highest peak of Africa; Mount Kenya and Mount Rwenzori are volcanic mountains rising to great heights. Though these volcanic mountains are very near to the equator, they are covered with snow throughout the year.

The Plateau: The height of the plateau in Africa is not the same throughout the continent. If we draw a straight line from Port Said to the mouth of the Congo, we find that the plateau can be divided into two parts, the northern and the southern. The height of the southern plateau is more than that of the northern plateau. This plateau rises straight up from the coastal plains. The highest part of the plateau lies in the south of the continent and on the eastern coast. The plateau of Ethiopia is about 3000 metres high and the plateau of the East Africa is 1800 metres high.

The Great Rift Valley of Africa: This Rift Valley is situated in the continent. Due to the movement in the interior of the earth, cracks occur in the crust of the earth. It is about 5000 kilometres in length and is very narrow and deep. The deeper parts of the valley are filled with water and thus lakes have been formed. Right at the northern end of the valley is a Lake Rudolph. There is an entire range of lakes all across the valley, up to the other end in the South where there is Lake Nyasa. Just near the valley but on the plateaus is the round and shallow lake called Lake Victoria.

Rivers

In proportion to the size of Africa there are very few big rivers. Most of the rivers have their source in the central plateau. As they flow down the plateau there are many water falls and rapids in their course. Most of the rivers practically dry up in the hot season, with the result that the water of these rivers cannot be used either for irrigation or navigation. Nile, Congo, Niger and Zambezi are the important rivers of Africa. The Congo (Zair) river is considered to be the biggest in Africa, from the point of view of the quantity of water. Stanley and Livingstone falls are on this river. The Nile is the longest river of Africa. The prosperity of Egypt is due to the river Nile. The River Niger is also one of the important rivers of Africa. The River Zambezi falls from a height of about 122 metres into a deep narrow valley, forming the world famous Victoria Falls. Besides these rivers, there are other smaller rivers like Orange, Senegal, Gambia, Limpopo and Vaal. There are so many crocodiles in the river Limpopo that it is also known as the "Crocodile River".

Climate

Africa being a very big continent, there is lot of difference in the climate. The most important factors that affect the climate of Africa are latitudes, altitudes, ocean currents and winds. Africa does not have the same season at the same time. Northern Africa and Southern Africa have different seasons.

The Equator passes through the centre of Africa and the rays of the sun fall vertically. As a result, on either side of the Equator the temperature remains high throughout the year, and winter is not experienced at high temperature because of their altitudes. For instance, the peak of Kilimanjaro is covered with snow throughout the year even though it is exactly on the equator.

In the Sahara desert, temperature is extremely high in summer and extremely low in winter. In the interior of the desert it is between 50° and 54°C during the day in summer. On the whole the climate of North Africa and that of the desert region is more extreme.

Southern Africa is at a high level and narrow, so its climate is less hot and is pleasant. The cold Benguela current which flows on the western coast, reduces the temperature of the places on the coast, while on the eastern coast, the warm Mozambique current flows and increases the temperature of the places on the eastern coast.

The equatorial climatic regions, the Congo and the coastal regions of Guinea get the convectional type of rainfall. The highest rainfall in this area is in the mountain regions of Cameroon. Towards north or south of the equator, the rainfall decreases and the rainy season also becomes shorter. It rains about 100cms in the tropical high grassland regions in which Kenya, Uganda, Tanzania and Rhodesia are included. In the interior of the continent, the proportion of rainfall decreases.

In the southern parts of Africa, the south-east winds blow from Indian Ocean, and as they blow over the warm currents on the shore they become more moist. The clouds go inwards and cause heavy rainfall in the coastal regions of the Drakensberg. It rains very little on the other side of this region, which is a rain shadow region. The rainfall gradually decreases from the east to the west.

The least rainfall in Africa is in the Sahara Desert. In summer dry trade winds blow here from across the continent, so there is very little rainfall here even in summer. There are some parts in the desert where only a few showers of rain during a period of two to three years. But the Kalahari desert, in the south-eastern Africa, is not so dry as the Sahara Desert.

In the south-west and the northern coastal regions of Africa the Mediterranean type of climate is experienced.

Natural Vegetation

Due to high temperature and constant heavy rainfall in the equatorial regions of Africa we find ever green forests. The trunks of the trees in these forests have a big circumference and they are 60 to 90 metres in height. Trees mahogany, ebony, logwood, iron wood etc found in these forests have hard wood. The region on either side of the equator between 5° to 20° latitudes (north and south) grows a peculiar type of grass. Therefore, this region is known as the Tropical Grassland or the Savannah Grassland where the grass is 3 to 5 metres in height. Scattered trees grow in this region. The cactus, thorny bushes and small plants grow in the Sahara as well as the Kalahari Desert region due to scarce rainfall.

The Mediterranean climate is found in the region of North Africa where it rains in winter. In this region, cork, chestnut, oak, and other trees grow in large numbers. Flowers and fruits are the special vegetation of this region. Olive is a special fruit of this region.

Economy

Although, the economy of Africa is based on agriculture but it is still not fully developed. A major land portion of the continent is covered by thick forests, deserts, mountains and plateaus. Moreover, rainfall is limited and irrigation facilities are also limited. The farmers are poor so they cannot take any advantage of mechanized and scientific methods of agriculture. They, of course, take different crops in keeping with the climate, land and a peculiar irrigation system.

In the equatorial climatic region of Africa, cocoa, palm, coconut, sugarcane, coffee, bananas, etc. are grown especially in Zaire, Ghana, Ivory Coast, Nigeria and Guinea. Ghana stands third in the production of cocoa among the countries of the world.

The palm tree has been considered be a tree of wealth in the equatorial region. Oil is extracted from it. Zaire has the largest number of the palm trees. There are two types of coffee produced in Africa namely Robusta Coffee and Arabica coffee. Banana is the chief fruit of the equatorial climate region and plenty of bananas are exported from Zaire. Tanzania produces the largest quantity of clove in the world. Memba and Malagasi are also known for the clove.

The outskirts of the grasslands have been cleared at places. After summer people cultivate the land. Important crops of this region are paddy, cotton, groundnut, sugarcane, coffee and sisal. Maize, jowar, millet, oilseed, tobacco and cassava are produced too. These crops are produced in Sudan, Kenya, Uganda, Tanzania, Rhodesia, and Angola. Sudan produces cotton of the highest quality. Sisal is a crop which is similar to jute. If the agriculture is developed scientifically in this region it can also be a granary of the world.

Jowar, millet and pulses are produced in the desert regions where there is a scarce rainfall. Date is the important crop of the Sahara desert. The bank of the river Nile and its delta region are fertile regions of Egypt. Moreover the Aswan Dam has facilitated irrigation. Here paddy, cotton, tobacco, wheat, barley, jovar and millet are cultivated. Long stapled, mercerized cotton of the best quality is grown in Egypt.

Mediterranean climate is very suitable for various types of fruit orchards. Algeria, Tunisia, and Morocco are on the Mediterranean Sea while Natal in Southern Africa experiences the Mediterranean climate. Here they cultivate wheat, barley and tobacco.

Africa is a rich continent in mineral resources, but in many countries of the continent proper resource utilization of these resources has not been done due to unfavourable backwardness. The leading producers of important minerals in Africa are:

Antimony	—	South Africa
Bauxite	—	Guinea
Chromium	—	Guinea
Copper	—	Zaire
Diamond	—	Botswana and Zaire
Gold	—	South Africa
Manganese	—	South Africa
Petroleum	—	Nigeria
Phosphate	—	Morocco
Platinum	—	South Africa
Tungsten	—	Zaire
Vanadium	—	South Africa

Africa is not having much power resources, although, prospects of hydroelectricity development is too much in this continent. Coal is available in South Africa and Algeria, while mineral oil is available in Sahara desert, Egypt, Libya and Algeria.

Industries have not developed in Africa due to unfavourable geographical circumstances, and scarcity of iron, and other important minerals, capital, technology and transport facilities. When foreign imports were prohibited at the time of the First World War, South African countries had made some efforts to develop industries.

Certain industries depending upon the agricultural products and mineral products have developed here. Cotton textile industry has flourished in Egypt, Sudan, Kenya, Zaire and South Africa because Egypt produces the best quality of cotton. The important centres of cotton textile industry are Cairo, Alexandria, Khartoum and Durban. Sugar manufacturing industry has developed in the countries like Egypt, Uganda, South Africa and Malagasy. Minerals are refined in South Africa, Ghana, Zaire and Algeria. Oil refineries have been started in Azaryu of Algeria in Mombasa in Kenya, in Durban in South

Africa and Tema in Ghana. Chemical fertilizer factories have been started in Suez, Oran and Kesai while cement factories have been started in Sudan, Kenya, Uganda, Egypt and Zaire.

Population and Society

The total population of Africa in 2011 was 994,527,534, which is about 14 per cent of the world population. The population is not equally distributed throughout the continent. The following regions of Africa are sparsely populated: (1) the equatorial forests owing to their unhygienic climate, (2) the Sahara desert due to its extreme climate and (3) the mountain region and high plateaus. Human inhabitation has not developed due to unfavourable geographical conditions and underdeveloped economic activities. The Savannah grassland often faces drought conditions due to irregularity of rainfall. The tse-tse fly is a scourge to the most thickly populated part of Africa in the plains of the river Nile. This part has a rich fertile soil and sufficient irrigation facilities, for the development of agriculture. Therefore, it is thickly populated. Places having Mediterranean type of climate are also well populated due to moderate and pleasant climate. South Africa which is an industrially developed region is also a thickly populated region.

The coastal region of Guinea with its good agricultural facilities and sufficient mineral resources, and Kenya with its fertile land, moderate climate and sufficient rainfall have a moderate population. In these regions natives of Africa Pygmies, Bushman, Negroes, and Caucasians live. In a major part of Africa, Negroes live. Indians from Asia and the English people form European countries have come here in search of trade, occupation and minerals.

Most of the people of North Africa are the followers of Islam and they use the Arabic dialects while most of the people of the South Africa are the followers of Christianity, and they use English as a language of communication. The natives of Africa worship nature, and use their local dialects. They are economically and culturally backward. The standard of living of the people of Africa is low except that of the Europeans.

Australia

Australia is the smallest of the three continents in the southern hemisphere but the biggest island of the world, and for this reason it is known as "island continent". Australia is quite different from the other continents due to its strange animals, birds, plants and grass lands. As Australia is situated in the southern hemisphere its climate is quite reverse to the climate of the northern hemisphere. It is the oldest of all the continents but the world came to know about it very late. The European explorers William Johnson, Abdtasman, Captain James Cook and Captain Arthur Phillips have played an important role in discovering this continent.

Australia is known as the museum of strange vegetation and animals due to peculiarity of the plants and animals of this land.

Australia is situated on the confluence of the Pacific Ocean and the Indian Ocean in the southern hemisphere. It lies between 10° and 48° south latitudes and between 113° and 153° east longitudes. The Tropic of Capricorn passes through its centre. The total area of Australia and Oceania is 8,985,000 sq km and population in 2011 was 36,102,071.

Physical Features

Among all the continents of the world Australia is the only continent having the lowest lowland region and it is also the only the continent whose physical features can be easily distinguished. Australia may be divided into three physical divisions: (1) the Eastern Mountain Region (2) The Central Lowlands (3) the Western High Plateau.

1. *The Eastern Mountain Region:* In Australia, this mountain range lying roughly parallel to the east coast, stretches from Cape York in North Queensland to the southern coast land. It is called "The Great Dividing Range". Different parts of Australia have different names for this mountain range. They are known as Blue Mountains in New South Wales, the Australian Alps in Victoria and as Darling Downs in Queensland. The highest mountain peak in this mountain range is Mount Kosciusko (2228 metres). It drops steeply to the east but slopes gently to the west. In the interiors of the continent, there are highlands, steep hills, plains and river basins. Most of the rivers that flow in Australia rise from this mountain range. Murray and Darling are important among them.
2. *The Central Plain:* The Central Plain is situated between the eastern mountain region and the western high plateau. The northern region of this plain is made up of the small rivers viz. Flanders, Norman, Gilbert, and Mitchel. Most of these rivers dry up in summer; so this part is not fertile. The central part of these regions is known as the Lake Eyre basin which is a saucer-shaped basin. There are many salt water lakes in this part including Lake Eyre which is the biggest among them. The whole region is dry, salty and barren. The southernmost part of this region is a rich river plain in which the river Murray and its tributaries Darling and Laklan spread silt and make it a fertile plain.
3. *The Western High Plateau:* There are many high plateaus in the Western Australia. Their height is between 200 to 500 metres above sea-level. The most important of these are Hammersley and Musgrave Range. The peaks of these ranges – Mount Bruce and Mount Wodroffe are 1226 and 1440 metres high respectively. Besides these, other mountain ranges in this region are Macdonnell, Peterman and Zeil. There are three deserts in the Western High land plateau, the Great Victoria Desert, the Great Sandy Desert and Gibson Desert.

The Great Barrier Reef: The long line of coral reef in the north east of Australia parallel to the sea coast is 30 kilometres away. This reef known as the Great Barrier Reef in made up of the world's biggest coral islands. It is about 2000 kms in length and 15 to 150 kms in breadth. There is a sea 20 to 160 kms in breadth and calm between these coral islands and the eastern coast. One observes here a special type of fish, the aquatic creature, platypus, colourful corals and beautiful shells. This region is very rich in mineral resources.

Rivers

The rivers of Australia can be divided into two parts: (1) The rivers flowing in the Pacific Ocean (2) The rivers flowing into the Indian Ocean. The rivers which are flowing into the Pacific Ocean are raising from "The Great Dividing Range", the rivers Murray and Darling flowing throughout the year are useful for navigation and irrigation. The rivers such as Murumbidgee and Lachlan are their tributaries. The rivers Hunter, Gulbarn, Bili yando and other rivers flow towards the east and into the sea. Filanders, Norman, Gilbert and Mitchel are such rivers which flow to the east for flowing in the Gulf of Carpentaria.

The rivers, which flow across the western plateau are Swan, Murchison, Ashburton, Fitzroy and Gascoyne either flow towards the north or the west but flow in the Indian Ocean.

There are no rivers in the Central Australia but there are many big and small lakes in this region. Some of them are Lake Eyre, The Lake Barberton, the Lake Georgena, the Lake Frome, the Lake Torens, the Lake Gardener, the Lake Hores and the Lake Edward.

Climate

As Australia is situated in the southern hemisphere all the seasons of the continent are quite reverse to ours. The Tropic of Capricorn passes through the middle of the continent, and the rays of the sun fall straight on it, so summers are extremely hot. Moreover, there are vast desert regions in the central and western parts of the continent which cause high temperature during summer. Australia is known as the "Dry Continent" due to its extreme temperature. Some local factors also affect the climate of Australia. As this continent is a big island, all the coastal regions have temperate climate, and the climate of Central Australia is extreme because of the mountain range of the east and the high plateau of the west. The Northern Australia experiences the monsoon type of climate while the regions of south-west and north-west experience the Mediterranean type of climate. On the whole, the climate of Australia is varied.

The winds coming form the north-west through the sea are moist-laden monsoon winds. They give heavy rainfall to the northern coast. The south-east trade winds coming from the south give more rain to the eastern coast. These moisturous winds cannot penetrate the Great Dividing Range, so a rain shadow region is formed to the west of the range. Most of the central parts of Australia do not get any rainfall, therefore, they are dry. The regions around the lake Edyre receive the least rainfall in Australia. The south-west and south-east parts of Australia enjoy Mediterranean climate, so is has rain in winter. The Westerlies bring rain to these regions. Tasmania receives rainfall throughout the year. Therefore it is said that rainfall in Australia is uncertain and irregular.

Vegetation

The variety of climate in Australia is clearly reflected in its vegetation, growing in five different belts. In the north-east and the northern coastal parts of Australia where the rainfall is heavy, there grows mangrow swamps. Passing inland, we come to monsoon or evergreen forests, further in the interior the evergreen forests give way to grassland with scattered trees of Savanna and finally to short dry grassland. Cactus and thorn bushes are seen in the desert regions of the Central Australia. To the north and north-east of Australia is the monsoon or tropical evergreen forests, full of trees like palm, ash, cedar and beach. Towards the eastern coast and the south west, we find different kinds of eucalyptus trees. The eucalyptus tree is considered a national tree of Australia. The wood of the trees like Kari and Jarrah is hard and durable, which does not purify and is not eaten by termites. Pine trees grow in Queensland and Tasmania.

In the interior areas of Australia there are vast plains of tall or short grass. In the desert regions of the Central and Western Australia Brigola and Spinifex are grown respectively. They are the plants of the semi desert and desert regions respectively.

Wildlife

Australia stands out from the countries of the world in the matters of animal world. Kangaroo is a herbivorous animal amidst the wild carnivorous animals. Though Kangaroo is a quadruped animal it jumps along on its hind legs at a speed of 40kms per hour. The female has pouch to carry its young one. While running, its tail maintains balance, and when it rests it acts as its third leg. Platypus, Koala and dingo (a wild dog) are among the peculiar animals of this region. These animals are not found in any continent of the world. Emu (Australian Ostrich), and Jack (when it laughs) makes a noise akin to a donkey's bray). Lyrebird and Black Swan are some of the strange birds of Australia. Only for this Australia has been considered as a "Museum of living animals and birds".

Economy

Agriculture, animal husbandry and minerals are the components of the Australian economy. The plains where short grass grows due to less rainfall are utilized as agricultural land in Australia. They cultivate wheat on a large scale. Most of the work in the wheat field is done with the help of machines and scientific methods. Most of the wheat producing areas of Australia are in the south-west part of

West Australia, South Australia, New South Wales and Victoria state which are the regions in southern temperate zone. Plenty of wheat is exported from Australia. The cultivation of soyabean and sunflower seeds has been taken up. Inferior kind of grains like sorghum, oat, barley, and maize are grown in plenty to feed sheep, pigs and other animals. Queensland is foremost in growing cotton and sugarcane. Fruit orchards are found in places having the Mediterranean type climate.

Australia is considered as a continent of shepherds. Here agriculture and animal rearing activities go together. Short grass grows on the slopes of Darling where shepherds rear "Marino" and other types of sheep on a scientific basis. Marino sheep is world famous for its wool. Marino means a bale of wool. Thirty per cent of the total product of wool in the world is produced by Australia. The Hereford, Aberdeen Angus, Devon, Red-Pole and other high breed of cattle are reared for meat. Australia is next to Argentina in exporting meat. Animal rearing has been developed not only for meat and wool but also for milk, butter, paneer, cheese and leather. Queensland is famous for animal rearing. New South Wales and Victoria are famous especially for sheep rearing. It is said that Australia has grown rich and powerful "on the back of sheep".

Dairy industry has flourished in Australia because in Queensland, Jersey, Ayreshire and Ilwara breeds of cow are reared. Besides milk, meat and dairy products like butter, cheese and condensed milk, leather is exported to UK and USA. Grass meant for these milch cattle is stored in special places called "silo".

Mineral resources are limited in Australia. Coal and gold are important minerals of this continent. Coal mines are situated in New South Wales, and Queensland which supply 70 per cent of fuel to the industries of Australia. Australia stands fifth among the gold producing countries of the world. Important gold mines are situated in Kalgoorlie of the Western Australia. Moreover, other gold mines are found in Victoria, Queensland, New South Wales and in Tasmania. The biggest gold mine of the world is situated in Broken Hill in New South Wales. Besides gold, iron, bauxite and uranium, mineral oil and gas are also mined. A precious stone called "Opal" is also found in Australia.

In the 19th century, agriculture, cattle rearing and gold mining were the only important industries. During the 20th century agriculture and animal rearing developed on a large scale. As soon as they got coal fields they developed hydro-electric power houses and sea routes for exporting goods, and started industries manufacturing tractors, agricultural tools, machinery and means of transport. The important centres for iron and steel industries are Port Kembla, New Castle, Latham and Brisbane. Shipbuilding docks have developed in Adelaide, New Castle, Sydney, and Melbourne. Industries manufacturing cotton, woollen cloth, electric materials, colour and chemicals have developed specially in Sydney and Melbourne. Industries manufacturing agricultural implement and machinery, plastic goods are established in Bellarat, Cherbourg, and Coalmines of Victoria state. There are oil refineries in Kwinana, Kawana and Geelong. Other smaller industries like those manufacturing soap, paper, crockery, leather goods, blankets, carpets and saw mills have developed on a large scale. Preservation and fruit canning for export have fully flourished in this continent. The extensive sea coast of Australia has helped to develop fishing industry of Hobart Port and Port Lincoln.

Population

The total population of Australia and Oceania in 2011 was 36,102,071. Majority of the population lives in six big cities. The well populated areas of Australia are the south west, the south-east and the eastern coastal regions. The factors that are responsible for a thick population of the cities in these regions are the temperate climate, industrial development and easy means of transport. The central, western and northern regions of Australia are thinly populated due to the extreme climate, less rainfall, undeveloped industries and scarcity of means of transport.

There are about 50 thousand aborigines living in Australia. A majority of the population consists of Europeans the whites, and for this reason it is called a "Colony of the whites". Asians or other coloured people were not allowed to settle here. Even Europeans were not allowed to settle here till recently. With a view to increasing their population, people of the European descent are now allowed to settle as "New Australians".

Most of the Australians are Christians, while some are Jews, Muslims and Buddhists. Canberra is the Federal Capital of Australia, while Sydney, Melbourne, Brisbane, Adelaide, Perth, New Castle, Wollongong, Hobart, Geelong, and Townsville are the important cities of Australia.

Antarctica

Antarctica has spread in all the four directions of the South Pole which lies between 60° to 90° south latitudes. The total area of the continent is 14 million sq km, which is 9.3 per cent of he world area. Antarctica continent is surrounded by the southern ocean or the Antarctic Ocean, which is connected with the waters of the Atlantic, the Pacific and the Indian Ocean. Antarctica continent is nearer to South America than any other continent. It is quite near to Capehorn of America.

Antarctica was discovered in the 20th century. Ninety-eight per cent of the land portion is covered with a blanket of ice. The thickness of these snowy mountain ranges rises above the surface to a height about 6000 metres. We experience here the days and nights of six months duration. The ocean around this continent is known as the southern Polar Ocean or the Antarctic Ocean.

Before the beginning of the 20th century there was no information about this continent. Now Argentina, Australia, Chile, Belgium, France, Japan, New Zealand, Norway, South Africa, Russia, UK, USA and India have established research centres here for scientific research. That is why Antarctica is known as "laboratory of the world". Research stations of various countries have been established. The largest research station is McMurdo of USA. India has also established research station, e.g. Maitri, Dakshin Gangotri and Mount Indra for scientific studies.

Antarctic is mainly a plateau, covered with a blanket of ice varying from a few hundred to several thousand metres in thickness except where snowy mountain ranges rise above the surface to a height of up to 5185 metres. Cape North in Victoria Land is the highest mountain region of this continent. This snow-covered continent is like an even level plain. Physically this continent can be divided into two parts, the east and the west. There is a valley which separates these two regions. These two regions would not have been clearly separated had this valley not been covered with snow. The western part of this continent which is made up of different snowy islands is in the western hemisphere towards the Pacific Ocean. The range of its height from the sea level is from 1220 to 1830 feet. There are two seas namely Ross and Wedell. The eastern part of the continent is situated in the eastern hemisphere. The mountain range of this region gradually raises towards the polar region. The coastline of Antarctica gradually gets pressed and lowered by the heavy weight of the snow. The icebergal continental shelf near the consoling is a characteristic feature of this continent. Mount Erebus of Ross Island is an active volcano. Snow increases every year in Antarctica at the rate of 6.60 cubic metres.

There is possibility of having mineral resources in Antarctica. It is considered that this continent is rich in copper, gold, nickel, petroleum and platinum. Certain animals like penguin and white bear are found in this continent otherwise Antarctica is not having any population.

ISLANDS

Island is an 'area of land completely surrounded by water'. Oceanic Islands are built up from the ocean floor and are part of the basal structure, not attached to continents, as in the example of the Hawaiian group. Continental islands are part of the neighbouring continental geological structure. As exemplified by the British Isles there are several thousands of islands in the world. A list of the largest islands is given in Table 17.2.

Table 17.2 *Largest Islands of the World*

Islands/Status	*Location*	*Area in sq km*
1. Greenland (integral part of Denmark)	Atlantic Ocean	2,175,600
2. New Guinea (western section is called Irian Jaya and is part of Indonesia; eastern section forms most of Papua New Guinea, which gained its independence in 1975)	Pacific Ocean	8,08,000
3. Borneo (southern part is part of Indonesia; northern part is made up of Sultanate of Brunei and two states of Malaysia)	Pacific Ocean	7,43,300
4. Madagascar (Independent country)	Indian Ocean	5,87,041
5. Baffin	Arctic Ocean	4,76,067
6. Sumatra (part of Indonesia)	Indian Ocean	4,73,607
7. Great Britain (consisting of England, Scotland and Wales)	Atlantic Ocean	2,29,895
8. Honshu (one of the main islands of Japan)	Pacific Ocean	2,28,000
9. Ellesmere (part of Canada)	Arctic Ocean	2,12,688
10. Victoria (part of Northwest Territories, Canada)	Arctic Ocean	2,12,199
11. Sulawesi or Celebes (part of Indonesia)	Pacific Ocean	1,89,036
12. South Island (part of New Zealand)	Pacific Ocean	1,50,461
13. Java (part of Indonesia)	Pacific Ocean	1,26,392
14. North Island (part of New Zealand)	Pacific Ocean	1,14,688
15. Cuba (independent country)	Caribbean Sea	1,14,525
16. Newfoundland (part of Canada)	Atlantic Ocean	1,12,300
17. Luzon (one of the islands of Philippines)	Pacific Ocean	1,06,200
18. Iceland (independent country)	Atlantic Ocean	1,03,000
19. Mindanao (one of the main islands of the Philippines)	Pacific Ocean	95,800
20. Ireland (southern part and independent country; northern part of the United Kingdom)	Atlantic Ocean	84,426
21. Hokkaido (one of the main islands of Japan)	Pacific Ocean	77,899
22. Hispaniola (eastern part is the Dominican Republic; western part is Haiti)	Caribbean Sea	76,484
23. Sri Lanka (independent country)	Indian Ocean	65,610
24. Tasmania (Australian state)	Pacific Ocean	63,326

OCEANS

Oceans are the large water bodies of saline water that occupy around 71 per cent of the earth's surface. Oceans constitute the largest storage of water on the earth, although its water is not directly usable because of the saltiness. But oceans provide food to a large sections of the humanity besides proving a good and cheap means to transportation and also form a part of world ecosystem.

The oceans cover 361 million sq km of the Earth surface. The average depth of the ocean is about 3,070 metres (10,085 feet) and the volume of water is 1.37 billion cubic km. All the oceans are interconnected, although geographical subdivisions are made for convenience, thus the oceans of the world are – Pacific, Atlantic, Indian and Arctic. A fifth ocean has been identified as Antarctic or the Southern Ocean. The three major oceans the Pacific, the Atlantic and the Indian ocean cover almost 89 per cent of total oceanic area of the earth's surface. The boundaries between the three main oceans are rather artificial because all the oceans are in fact interconnected. Oceans occupy about 61 per cent area of northern and 81 per cent area of the southern hemisphere. Because of this fact southern hemisphere is also called part of ocean hemisphere. Brief characteristics of the oceans are as follows:

Pacific Ocean

The Pacific Ocean is the largest ocean and occupies about a one-third of the total area of the globe. It forms more or less a broad triangle in shape, with its apex in the north at the Bearing Strait. The greater part of the floor of the pacific comprises the deep sea having average depth of 4,820 metres.

There are about 2000 islands located in this ocean. In the east are the Aleutians, the island of British Columbia, and the Chilean island while in the west are the most extrusive island arcs of eastern Asia – Kurilles, the Japanese Archipelago, the Philippines, Indonesia and New Zealand. There are numerous high volcanic islands, and the low coral islands are also located.

The Pacific Ocean is characterized by deep elongated ocean trenches. The greatest depth is that of Mariana trench –11,033 metres, which is the deepest ocean point of the world. Other great deeps include Emden Deep of Philippines (10,475 m), Kurile Trench (10,498 m) and Ramapo Deep of Japan. Several marginal seas in the Pacific Ocean are located in western side in western pacific. The marginal seas are Bering Sea, Sea of Okhotsk, the sea of Japan Yellow sea, China sea, South China Sea, Celebes and Banda Seas, Gulf of Carpentaria, Arafura Sea and the Bass Trail. While in eastern side the Gulf of California is the only pivotally enclosed water.

Atlantic Ocean

Atlantic Ocean is the second largest ocean in the world, occupying about one-sixth of the total area of the world. Its shape is that of the English alphabet 'S'. The ocean is narrows towards the equator, while the South Atlantic Ocean opens broadly into the Antarctic Ocean, but the northern part of it is much more enclosed by Greenland and Iceland. Its mean depth is 3930 metres.

The unique feature of floor of Atlantic is the Mid-Atlantic Ridge. This was discovered by the Challenger in 1873. The Ridge is in 'S' Shaped with an average depth of about 3000m. In the north Atlantic the ridge widens to form the Telegraph plateau, which extends across the ocean from Ireland to Labrador. A few ocean trenches of the ocean are Puerto Rico (8385 m.) lies off the island arcs of west Indians Romanche Trench (7631 m) lies across the Mid-Atlantic Ridge and the south sandwich islands.

The marginal seas and bays of Atlantic Oceans are Mediterranean, Baltic, Caribbean Sea, Gulf of Mexico and Hudson bay. The number of islands is group of West Indies islands, Iceland, Faeroes, Falklands, South Orkneys, South Shetlands, Georgia, Sandwich, Azores, St. Helena, Cauaries and Cape Verde.

Indian Ocean

The Indian Ocean is the third largest ocean after the Pacific and Atlantic. It's is enclosed by land in the north and just extends beyond the Tropic of Cancer. Its shore consists of Africa, Arabia, the Deccan and the Western Australia. In the north-east it is bounded by the islands of Indonesia and Coast of Myanmar.

About 60 per cent of the ocean bottom of Indian Ocean forms the deep-sea plains having depth between 3,600 m and 5,500 m. Trenches are almost absent except Sunda Tranch (7,454 m) which lies south of and parallelled the Java Lesson Sunda Islands. The Indian Ocean has little marginal seas; these are Arabian Sea, Bay of Bengal, Red sea and the Persian Gulf. The Red Sea is separated from Indian Ocean by the submerged rock hill across the strait of Bab-el-Mandeb, over which the there is only 370 metres of water.

The Madagascar and Sri Lanka (Ceylon) are the big islands located in Indian Ocean. Other islands include Socotra, Zanzibar, Comoros, Andaman and Nicobar, Lakshadweep and Maldives, Kerguelen, Mauritius and Reunion islands.

Arctic Ocean

Arctic Ocean is almost circular in shape, spreading in an area of about 14 million sq km. It is surrounded by continents of Asia, Europe and North America. The Arctic Ocean is virtually a frozen ocean, located directly below the North Pole. The Ocean is 4087 m. deep and knowledge about this ocean is also limited. Two main ridges of the ocean are Lomonosov and Mendeleev. The marginal seas of the ocean are Beaufort Sea of Alaska, the East Siberian Sea, Kara Sea, and the Barent Sea.

Antarctic Ocean

The Antarctic Ocean is the fifth ocean; was demarcated by the International Hydrographic Organization only in year 2000. It encompasses all the waters surrounding Antarctica. In fact the northern boundary of this ocean is imperceptible and it is considered as the southward extrusion of the three major oceans. Parts of this ocean also remain frozen during the winter season.

SEAS

Sea is a large body of usually saline water which is smaller in size than ocean. The seas adjacent to oceans are known as marginal seas. The marginal seas of the various oceans are as follows:

Marginal Seas of Atlantic Ocean

Caribbean Sea, Celtic Sea, Labrador Sea, Marmara Sea, Norwegian Sea, Bothnian Sea, Barents Sea, Branford Sea, Amundsen Sea, Laptev Sea, White Sea, Pechora Sea, Kara Sea, East Siberian Sea, Greenland Sea, Prince Gustav Adolf Sea.

Marginal Seas of Pacific Ocean

Arafura Sea, Bering Sea, Celebes Sea, Coral Sea, Java Sea, Yellow Sea, Sea of Japan, Bohol Sea, Moluccas Sea, South China Sea, Gulf of Alaska, Gulf of Carpentaria.

Marginal Seas of Indian Ocean

Arabian Sea, Andaman Sea, Bay of Bengal, Timor Sea, Red Sea, Laccadive Sea, Gulf of Oman, Gulf of Aden.

Mediterranean Sea is a large sea covering an area of 2,499, 350 sq km and also having marginal seas of Sargasso Sea, Aegean Sea, Adriatic Sea, Sea of Sicily, Tyrrhenian Sea and Black Sea. The major seas of the world are given in following Table 17.3.

Table 17.3 *Major Seas of the World*

Major Seas	*Area sq km*
South China Sea	2,849,000
Mediterranean Sea	2,499, 350
Bering Sea	2, 274, 020
Caribbean Sea	1,942,500
Gulf of Mexico	1, 813, 000
Sea of Okhotsk	1,528, 100
Sea of Japan	1,295,000
Hudson Bay	11,230, 250
Andaman Sea	777,000
North Sea	574, 980
Rea Sea	440, 300
Caspian Sea	424, 242
Baltic Sea	422,170
Black Sea	411, 810

Oceanic Trenches

Oceanic trenches are the long and narrow linear, very deep depressions on the ocean floors. Some of the important ocean trenches are as follows:

Table 17.4 *Major Ocean Trenches*

Trench/Ocean	*Depth (Metres)*
1. Mariana Trench (Pacific Ocean)	11,022
2. Tongu Trench (Pacific Ocean)	10,882
3. Japan Trench (Pacific Ocean)	10,554
4. Kuril Trench (Pacific Ocean)	10,542
5. Mindano Trench (Pacific Ocean)	10,497
6. Kermode Trench (Pacific Ocean)	10,047
7. Puerto Rico Trench (Pacific Ocean)	9,220
8. Peru – Chile Trench (Pacific Ocean)	8,050
9. Aleutian Trench (Pacific ocean)	7,822
10. Cayman Trench (Atlantic Ocean)	7,680
11. Java Trench (Indian Ocean)	7,450

LAKES

Lakes are the water bodies surrounded by land; it may be a fresh water lake or a salt lake. Lakes are widespread distributed throughout the world. The major lakes of the world, their location and areas are as follows:

Table 17.5 *Major Lakes of the World*

Large Lakes	*Location*	*Area (sq km)*
Caspian Sea	Europe-Asia (Russia-Azerbaijan-Iran-Turkmenistan-Kazakhstan)	424,242
Superior	North America (United States-Canada)	82,414
Victoria	Africa (Tanzania-Uganda-Kenya)	67,340
Aral Sea	Asia (Kazakhstan-Uzbekistan)	63,805
Huron	North America (United States-Canada)	59,596
Michigan	North America (United States)	58,016
Tanganyika	Africa (Zaire-Tanzania)	32,893
Baikal	Asia (Russia)	31, 469
Great Bear	North America (Canada)	31,080
Great Slave	North America (Canada)	28,930
Malawi	Africa (Malawi-Tanzania-Mozambique)	28,490
Erie	North America (United States-Canada)	25,745
Winnipeg	North America (Canada)	24,341
Chad	Africa (Chad-Niger-Nigeria-Cameroon)	20,720
Ontario	North America (Canada-United States)	19,529
Ladoga	Europe (Russia)	18, 389
Balkhash	Asia (Kazakhstan)	17,301
Maracaibo	South America (Venezuela)	12,950
Onega	Europe (Russia)	9,842
Eyre	Australia	8,884
Titicaca	South America (Peru-Bolivia)	8,288
Nicaragua	Central America (Nicaragua)	8,029
Athabasca	North America (Canada)	7,920

Notes: 1. Does not include the Black Sea, often regarded as an arm of the Mediterranean Sea.
2. The Caspian and Aral Seas are considered lakes because they are inland bodies of water.

18

Major Landforms: Mountains, Plateaus and Plains

The landforms of the earth vary greatly from place to place. The three major forms of landforms are mountains/hills, plateaus and plains. These land forms can be described in terms of their shapes and elevations.

Mountains

Ordinarily a land block that stands conspicuously above the surrounding areas is called a mountain. According to Salisbury a mountain is a highland whose crest area is narrow; while Finch defined a mountain is that land block whose height is at least 900 metres above sea level and its slope makes an angle of 25° to 35° with the horizontal plane. The lower mountains are called hills. In fact mountain and hills are relative terms.

Terms Related to Mountains

Mountain System: A group of mountain ranges which are similar in their form, structure and extension, e.g. Appalachian and Rocky Mountain system.

Ridge: It is a high and narrow continuation of hills and rises abruptly from the adjoining regions, e.g. Blue Ridge mountain (USA)

Mountain Range: It is a series of ridges which are mutually related and form a compact unit and stand over an extensive area, e.g. Sierra Nevada in Eastern California and the Front Range of Colorado.

Mountain Chain: An elongated unit of the mountain ranges or chains whose size and age need not necessarily be similar.

Cordillera: A spanish word meaning highest unit of mountain. In North America, the mountain complex extending from eastern boundary of the Rocky mountains to the coast of the Pacific Ocean is called a *cordillera* of North America

Characteristics

The main characteristics of mountains are:

(i) The formation of mountain is due to compression or tension. Due to compression, the pressure on the rocks is so much that rocks are folded and metamorphosed. In the areas subjected to tension, the rocks are not subjected to tension.

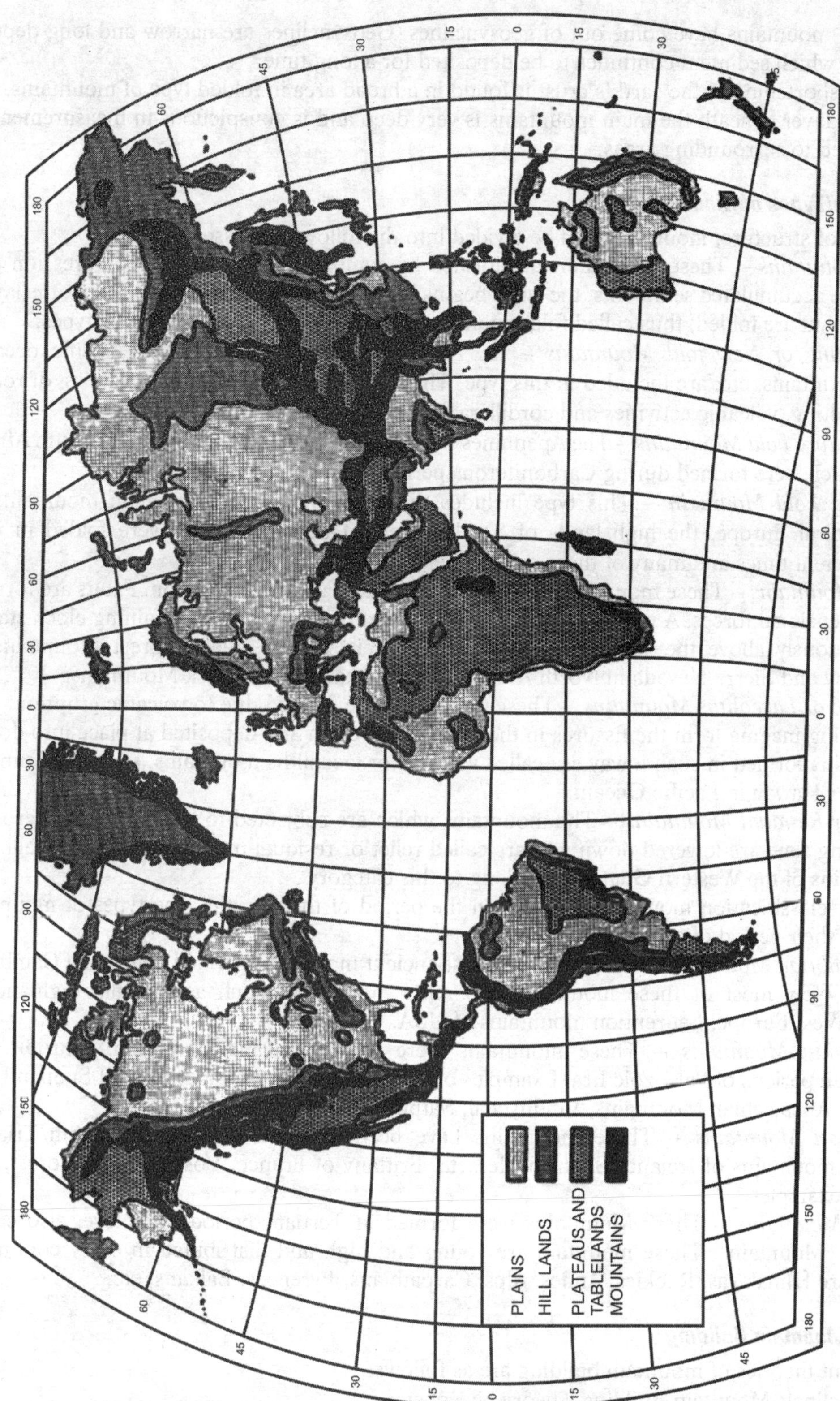

Figure 18.1 *The World Distribution of Major Landforms*

(ii) Most of mountains have come out of geosynclines. Geosynclines are narrow and long depressions in which sediment continues to be deposited for a long time.
(iii) Crustal shortening of the earth's crust is found in a broad area in folded type of mountains.
(iv) The sial layer beneath the main mountains is very deep and is conspicuous in measurement as compared to surrounding areas.

Classification/Types of Mountains

On the basis of structure, mountains can be divided into the following classes:

1. *Fold Mountains* – These Mountains are formed by compression. When the compression acts upon the accumulated sediments, the latter begins to rise. In the process of upliftment, the layers of sediment are folded, thus called fold mountains. The fold mountains are of two types.
   (a) *Young or New fold Mountains* – the Alps, Himalayas, the Circum – Pacific oceanic mountains, etc. are included in this type. These mountains have complex folding of rocks, faulting, volcanic activities and cordillera form.
   (b) *Mature Fold Mountains* – The Apennines the Appalachians and Cape range of South Africa, which were formed during Carboniferous period are included in this category.
   (c) *Old Fold Mountain* – This type includes the Caledonian and Hercynian mountains of central Europe, the high lands of Scotland, etc. These mountains were folded in very ancient times and many of these exist as relicts due to erosion.
2. *Block Mountains* – These mountains were formed on the crust due to tension. Faults are formed due to tensional forces. A part of the land block is depressed so that the remaining block stands conspicuously above the surroundings. The Vogues in France, Black Forest Mountains in Germany and Sierra Nevada in North America are the examples of Block Mountains.
3. *Volcanic or Laccoliths Mountains* – These mountains are formed due to volcanic eruption. The outrushing magma form the fissures in the crust of the earth and deposited at place into dome; mountains formed in such a way are called volcanic or laccoliths mountains, such as Mountain Kea over Hawaii in Pacific Ocean.
4. *Relict or Residual Mountains* – The mountains which are subjected to weathering and erosion for a long time are lowered down and are called relict or residual mountains. The Aravalli, the mountains of the Western Ghats, etc. belong to this category.

Another classification mountain is based on the period of their origin. The types of mountain according to their period of origin are:

1. *Pre-Cambrian Mountains* – These are the most ancient mountains formed before the Cambrian period. Now most of these mountains are relict mountains, such as Aravalli, highland of North-West Europe, Laurention mountains of USA.
2. *Caledonian Mountains* – These mountains were formed during Cambrian, Silurion and Devonian periods of Paleozoic Era. Examples of such mountains are mountains of Scotland and Ireland, Appalachian Mountains, Vindhychal, Satpura, Mahadev of India.
3. *Hercynian Mountains* – These mountains have been originated during Permainn/Triassic period; mountains of Ireland, Spanish Messeta, Brittany of France, Vosges, Black Forest, etc, are the examples.
4. *Alpine Mountains* – These Mountains were formed in Tertiary period, therefore, also called Tertiary Mountains. These mountain are young and high and distributed in every continent. These are Himalayas, Rockies, Andes, Alps, Carpathians, Pyrenees, Balkans, etc.

Theories of Mountain Building

The important theories of mountain building are as follows:

1. Geosynclinals Mountain Building Theory by Kober

2. Thermal Convention Current Theory by Holmes
3. Radioactive Theory by Joly
4. Continent Drift Theory by Wegener
5. Plate Tectonic Theory by Harry Hess, Mckenzie, Parker and Morgan

Geographical Distribution

Mountains are distributed in all the continents of the world. The following is the list of the longest mountain ranges of the world:

Longest Mountain Ranges of the World

1. Andes (7242 km), South America
2. Rocky Mountains (6035 km), North America
3. Himalayas/Karakoram/Hindu Kush (3862 km), Asia
4. Great Dividing Range (3621 km) Australia
5. Trans-Antarctic (3541 km), Antarctica
6. Brazilian East Coast Range (3058 km) Brazil, (S. America)
7. Sumatran/Javan Range (2897 km), Indonesia (Asia)
8. Tien shan (2253km.), Asia
9. Altai (2012 km), Asia
10. Central New Guinean Range, (2012 km)
11. Urals (2012 km) Russia, Europe

Some other ranges are Atlas (1930km.), Western Ghats (1610km.), Caucausus (1200km.), Alaska (1130 km) and Alps (1050 km). Other mountain ranges are distributed throughout the world.

The highest peaks of the world are located in the **Himalaya** and **Karakoram** ranges to the south and west of the Tibetan plateau. In fact all 7,000 m (23,000 ft) peaks in the world are located in the centre of Asia (East Asia, Central Asia and South Asia) in a rectangle edged by Noshaq (7,492 m or 24,580 ft) on the Afghanistan-Pakistan border in the west, Jengish Chokusu (Tuomu er feng) (7,439, 24.406 ft.) on the Kyrgyzstan-Xinjiang border to the North, Gongga Shan (Minya koneke) (7,556 m or 24,790 ft) in Sichuan to the East, and Kabru (7,412 m or 24,318 ft) on the Sikkim (India)-Nepal Border to the south.

A list of top 25 mountain peaks is given below:

Rank	Mountain	Height (m)	Height (ft)	Range
1	Mount Everest/ Sagarmatha/Chomolungam	8,848	29,029	Mahalangur Himalaya
2	K2/Qogir/Godwin Austen	8,611	28,251	Baltoro Karakoram
3	Kanchenjunga	8,586	28,169	Kanchenjunga Himalaya
4	Lhotse	8,516	27,940	Mahalangur Himalaya
5	Makalu	8,485	27,838	Mahalangur Himalaya
6	Cho Oyu	8,188	26,864	Mahalangur
7	Dhaulagiri	8,167	26,781	Dhaulagiri Himalaya
8	Manaslu	8,163	26,781	Manaslu Himalaya
9	Nanga Parbat	8,126	26,660	Nanga Parbat Himalaya
10	Annapurana I	8091	26,545	Annapurna Himalaya
11	Gasherbrum I/Hidden Peak/K5	8,080	26,509	Baltoro karakoram

Cont'd...

...Cont'd

12	Broad Peak/K3	8,051	26,414	Baltoro Karakoram
13	Gasherbrum II/K4	8,034	26,362	Baltoro Karakoram
14	Shishapangma	8,027	26,335	Jugal Himalaya
15	GyachungKang	7,952	26,089	Mahalangur Himalaya
	Gasherbrum III	7,946	26,070	Baltoro Karakoram
16	Annapurna II	7,937	26,040	Annapurna Himalaya
17	Gasherbrum IV	7,932	26,024	Baltoro Karakoram
18	Himalchuli	7,893	25,896	Manaslu Himalaya`
19	Distaghil Sar	7,884	25,866	Hispar Karakoram
20	Ngadi Chuli	7,871	25,823	Manaslu Himalaya
	Nuptse	7,864	25,801	Mahalangur Himalaya
21	Khunyang Chhish	7,823	25,666	Hispar Karakoram
22	Masherbrum/K1	7,821	25,659	Masherbrum Karakoram
23	Nanda Devi	7,816	25,643	Garhwal Himalaya
24	Chomo Lonzo	7,804	25,604	Mahalangur Himalaya
25	Batura Sar	7,795	25,574	Batura Karakoram

It becomes clear from the above list that all the top 25 mountain peaks of the world are located in Himalayan mountain range. Not only above mentioned peaks but more than hundred peaks above 7000 metres high are located in Himalaya ranges. In Andes the highest peak is **Mount Aconcagua** (6962 m), while in rocky Mountains, **Mt. Elbert** is the highest with the height of 4401 m. The highest mountain peak of Europe is **Mt. Elbrus** (5633 m.) while **Mt. Blanc** in of 4807 m. high. Appalachian Mountain range is located in eastern region of USA starting from Albama province to New England in a length of about 1440 km. In Alaska region, **Mt. McKinley** is the highest peak (6194m). In Australian Great Dividing Range the highest peak is **Mt. Kosciuszko** (2,228 m.) **Mt. Kilimanjaro** is the highest mountain peak of Africa having the height of 5895 m. followed by **Mt. Kenya** of 5199 m. height.

The Great Cordilleran Belts: Most of the mountains of the earth and many areas of hill country as well, are found in a few broad and long cordilleran belts. Each continent contains at least a part of one of these complex arrays of mountain system, though the apportionment among the continents is very unequal. A glance at Figure 18.2 will reveal that all of the cordilleran belts are in fact connected, forming a nearly complete ring around the Pacific Ocean, and extending westward across southern Eurasia.

The cordilleran belts vary greatly from section to section in breadth, height, and complexity. In the western United States, Northern Mexico, and Alaska the Northern American cordillera is conspicuously wide and "frayed out", with many plains and tablelands interspersed among the loosely linked ranges. In Canada and in central America, however, the belt narrows down markedly, and is less broken, though generally no higher. The Andean Cordillera of South America, by contrast is noteworthy of its narrowness, its continuity and height.

The Eurasian cordillera in its western half some what similar to that of western North America. Loosely arrayed system and ranges are separated by wide valleys and basins, some of those in Europe being at very low elevation (e.g. the Po valley of northern Italy and the Hungarian Plain). There are several gaps affording easy passage across the belt, notably the Rhone Valley of southern France and the straits of the Dardanelles and Bosporus that lead from the Mediterranean into the Black Sea. In the Pamir region, to the northwest of India, however, the ranges come together in tight and high "knot", from which they then fan out eastward to cover practically the entire eastern part of the continent and even the great island systems offshore. This is the broadest and most complex cordilleran section

Figure 18.2 *The Great Cordilleran Belts of the World*

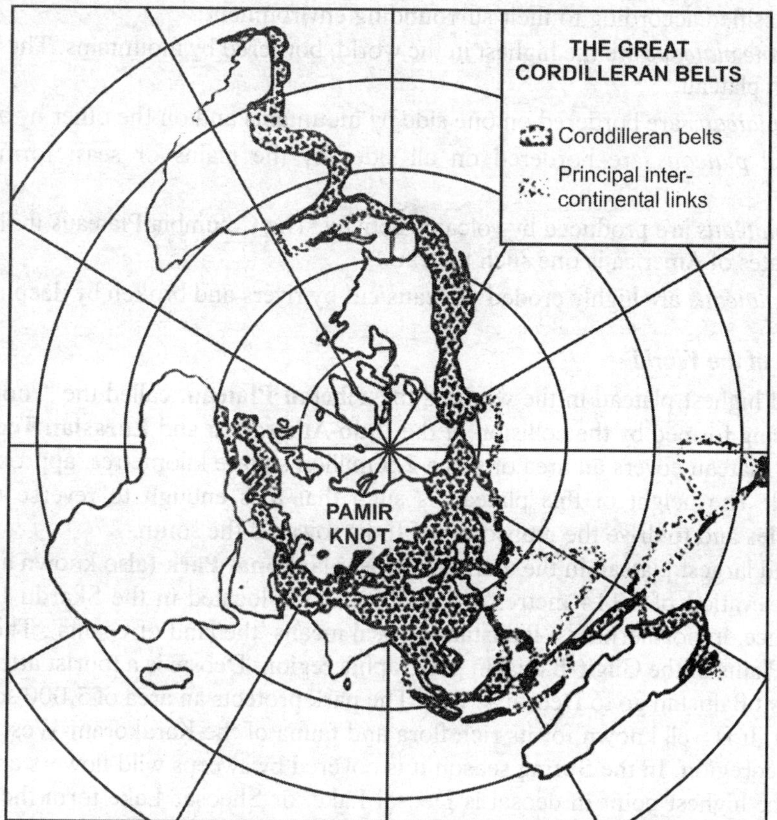

anywhere on the earth. From this tangle of ranges streamers extend northeastward across the Bering Strait and the Aleutian Island into Alaska to connect with the North American belt. To the southeast other lines follow through the East Indies into the mountain systems of Australia and New Zealand.

Africa, strictly speaking, does not have a cordillera of its own. A few strands of the Eurasian belt cross the Mediterranean to form the Atlas system in the extreme northwest. With that accepting, however, Africa's mountain is scattered and discontinuous.

Plateaus

Plateau is an extensive landform with a plain crest area and is sufficiently high at least on one side of its land or water surrounding. The term plateau also applies to a flat high land region – a tableland or elevated tract of comparatively flat or level land.

Plateaus can be formed by a number of processes, including upwelling of volcanic magma, extrusion of lava, and erosion by water and glaciers. Magma rises from the mantle causing the ground to swell upward, in this way large, flat areas of rock are uplifted. Plateaus can also be built up by lava spreading outward from cracks and weak areas in the crust. Plateaus can also be formed by the erosion processes of glaciers on mountain ranges, leaving them sitting between the mountain ranges. Water can also erode mountains and other landforms. The feedback between Computer modeling studies suggests that high plateaus may also be partially a result form the feedback between tectonic deformation and dry climatic conditions created at the lee side of growing orogens.

Classification

Plateaus are classified according to their surrounding environment:

- *Intermontane plateaus* are the highest in the world, bordered by mountains. The Tibetan Plateau is one such plateau.
- *Piedmont plateaus* are bordered on one side by mountains and on the other by a plain or sea.
- *Continental plateaus* are bordered on all sides by the plains or seas; forming away from mountains.
- *Volcanic plateaus* are produced by volcanic activity. The Columbia Plateaus in the northwestern United States of America is one such plateau.
- *Dissected plateaus* are highly eroded plateaus cut by rivers and broken by deep narrow valleys.

Major Plateaus of the World

The largest and highest plateau in the world is the **Tibetan Plateau,** called the "**roof of the world**", which is still being formed by the collision of the **Indo-Australian** and **Eurasian Tectonic plates**. In all, the Tibetan plateau covers an area of some 2.5 million square kilometres, approximately 5000 m above sea level. The height of this plateau is such that it is enough to reverse the **Hadley cell convection cycles** and to drive the monsoons of India towards the south.

The second largest plateau in the world is **Deosai National Park** (also known as Deosai Plains) at an average elevation of 4,114 metres (13,497 ft) and is located in the Skardu District of Gilgit Baltistan province, in northernmost Pakistan. Deosai means 'the land of Giants'. The park is located on the Deosai Plains of the Gilgit-Baltistan geographic region. Deosai is a tourist attraction and lot of tourists who visit Baltistan go to Deosai as well. The park protects an area of 3,000 square kilometres (1.2000 sq m.). It is well known for its rich flora and fauna of the Karakoram-West Tibetan Plateau alpine steppe ecoregion. In the Spring season it is covered by sweeps wild flowers and a wide variety of butterflies the highest point in deosai is Deosai Lake, or Sheosar Lake form the Shina language meaning "Blind lake" (Sheo – Blind, Sar-lake) is in the park. The lake, at an elevation of 4,142 metres (13,589ft.) is one of the highest lakes in the world. Its length is 2.3. kilometres (7,500 ft.), width 1.8 kilometre (5,900 ft) and average depth is 40 metre (130 ft). It is located near the Chilim valley on the Dosai Plains. The third largest plateau in the world is the **Antarctic Plateau,** which covers most of the central part of **Antarctica.** In the region of Antarctica, there are no mountains, rather 3000 metre of more of ice which very slowly spreads toward the coastline via enormous glaciers. This ice cap is so massive that echolocation sound measurements of the thickness of the ice have shown that large part of the "dry land" surface of Antarctica has been pressed below sea level. Thus, if the icecaps were somehow removed, large areas of Antarctic would be flooded by the oceans. On the other hand – more realistically – where the icecaps togradually melt away, the surface of the land beneath it would gradually rebound away form the centre of the Earth, and that land would ultimately rise above sea level. The biggest plateau in Europe is the **Hardangervidda** in **Norway**. In North America, the largest plateau is the **Colorado Plateau** covering an area of about 3,37,000 square kilometre (1,30,000 sq m) in Colorado, Utah, Arizona and New Mexico.

The Colorado Plateau in northern Arizona and southern Utah is bisected by the valley of the **Colorado River** and the **Grand Canyon.** How this came to be is that over 10 million years ago? A river was already there, though not necessarily on exactly the same course. Then, Subterranean Geological forces caused the land in that part of North America to gradually rise by about a centimetre per year for millions of years. An unusual balance occurred. The river that would become the Colorado River was able to erode into the crust of the earth at a nearly rate to the uplift of the plateau.

Now, millions of years later, the North Rim of the Grand Canyon is at an elevation of about 2450 metres (9800 ft) above **sea level** and the south Rim of the Grand Canyon is about 2150 metre (8200 ft) above sea level. At its deepest, the Colorado River is about 1830 metre (6000 ft) below the level of the North Rim.

Other Important plateaus are Deccan Plateau (India), Arabian Plateau (South west Asia), Brazilian Plateau (South America), Plateau of Mexico (Mexico), Alaska Plateau (North west, North America), Bolivia Plateau (South America), Columbia Plateau (USA), Patagonia (South America), etc.

Plains

Plains are the extensive areas of comparatively level to gently undulating land surface. Plains are found on all the continents except Antarctica and occupies more than one-third of the land surface. Plains are the birth places of ancient civilizations and are the best areas for agricultural development. The plains of Ganga, Sindh, Nile, Yangtze Kiang, Tigris-Euphrates, Great Plains of USA, etc. are famous for agricultural development. In fact, plains are the regions of economic development and also the regions of high concentration of population. Some plains are tree-covered and others are grassy.

Types of Plains

The three broad categories of plains are:
(i) Structural Plains
(ii) Erosional Plains
(iii) Depositional Plains

Structural Plains

Structural Plains were formed by the endogenetic forces. These are structurally depressed areas. The two types of the structural plains are (a) Plains of horizontal layers such as Russian platform, the central west plains of North America, etc. and (b) Plains of Emergence – the plains which have emerged out as a result of the uplift of sea coast, such as Eastern and Gulf coast of USA, the Western coast of India, etc.

Erosional Plains

Erosional Plains are the result of the erosional activities performed by running water, glaciers, winds or sea waves.

Plains of Fluvial Erosion or Riverine Plains – are formed by river erosion. It will form peneplains, which according to W.M. Davis is the final stage of the erosion. The Peneplains are having very gentle slopes and produced by the excessive erosion. Although a few mounds of hard rocks are also found in these plains, which are known as Monadnocks. There are few peneplains in the world; the Guinea Plain in the north-east of South America is a good example.

Glaciated Plains – These plains have been formed by movement of glaciers when ice sheet melted and area eroded by ice was exposed. The plains of North America and North-West Europe are of this type. In these plains rounded hills whose covering of till is removed are found.

Aeolian Plains or Wind Eroded Plains – In arid and semi-arid regions, these plains have been created by wind deflation, abrasion, etc. The plains produced by wind actions are Reg (Sahara), Serir and Hamada.

Plains at Continental Edges – These plains have evolved at sea coast by the action of sea waves and currents and were later uplifted. The flat plains situated at the coast of Norway fall into this category.

Karst Plains – These plains are found in limestone areas. The underground water removes the limestone layers by the process of solution. The coastal plains of the Adriatic Sea and the Karst plain of Florida are its examples.

Depositional Plains

Deposition is carried on by rivers, glaciers, winds, sea waves, etc and plains resulted by such deposition are known as depositional plains. Alluvial plains, Flood Plains, Deltaic Plains, Plains of Glacial Deposition, Desert Plains and Plains of Marine Deposition are included in this category. These plains are very important and fertile plains of the world, famous for their productive activities.

Major plains of the world have been created by rivers, in other words, river basins are also the major plain areas. The world major plains are Nile Plain, Amazon Basin, Yangtze Plain, Mississippi-Missouri Plain, Yenisey Plain, Huang-Ho Plain, Indo-Gangetic Plain, Parana, Amur, Ob, Lena, Mekong, Mackenzie, Niger, Zambezi, Volga, Maderia, Jurua, Purus, Yukon Plains, Murroy-Darling Plain, Congo, Po, Rhine, Seine, Volgo, Rio-Grande, Tenessee, Paraguay Plains, etc. Apart from these, there are several plains of smaller size located in various parts of the world.

19

Climate of the World

Climate is the most basic element of the natural environment. Climate, not merely an average, is the composite of weather, including extremes, frequencies, and the annual march of events in the atmosphere. The basic elements of climate are, (i) air temperature, (ii) atmospheric moisture including clouds, fog, humidity, and precipitation, and (iii) pressure and winds. The varying occurrence and proportion of these elements provide a variety of climate, ranging from the monotony of the tropics to the extreme diversity of the middle latitudes.

The climate of the world can be classified in a variety of ways and each of these has merit. There are a number of good classifications of climate. The broadest and most general classification of climates is the one devised by ancient Greeks, who divided each hemisphere into three broad belts or zones, e.g. tropical or torrid, polar or frigid and temperate zone. This three-fold classification of earth's climates does not take into consideration of other great climatic element, precipitation, since within the both low and middle latitudes there are very wet as well as very dry climates and still others which are intermediate in their amounts of rainfall.

On the earth's surface, there are broad climatic zones or regions and also subdivisions within these regions. There is a common characteristic of these climatic regions that they occur in corresponding latitudinal and continental locations as shown in a hypothetical representation of a continent (Figure 19.1), which shows the great climatic groups.

Figure 19.1 *The Great Climatic Groups*

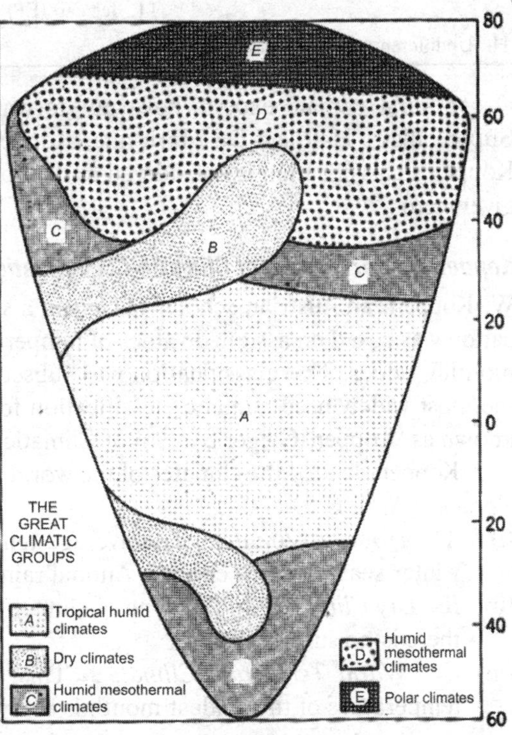

The broad and general groups and types of the climate of the world are as follows:

Types of Climate

Groups	Types
	I. Low Latitudes (The Tropics)
A. Topical humid climates	1. Tropical wet (Af, constantly wet) (Am, monsoon variety)
	2. Tropical wet-and-dry (Aw)
	3. Low-latitude dry climates
	a. Low-latitude desert (B, Wh, arid)
	b. Low-latitude steps (B.Sh, semiarid)
B. Dry climates	II. Middle Latitudes (Intermediate Zones)
	4. Middle-latitude dry climates
	a. Middle-latitude desert (BWk, arid)
	b. Middle-latitude steppe (BSk, semiarid)
C. Humid mesothermal climates	5. Dry-summer subtropical (Cs)
	6. Humid subtropical (Co)
	7. Marine (Cb, Cc)
	8. Humid continental climates
D. Humid microthermal climates	a. Humid continental, warm summer (Da)
	b. Humid Continental, cool Summer (Db)
	9. Subarctic (Dc, Dd)
	III. High Latitudes (Polar caps) or High Altitudes
E. Polar Climates	10. Tundra (ET)
	11. Ice cap (EF)
H. Undifferentiated highlands	

Several climate experts have put forward climatic classification of the world. Among them Supan, Blair, Hattner, De-Martonne, Koppen, Thornthwaite's classification are well known. But Koppen's classification is considered to be more popular and acceptable; therefore, its brief review is given here.

Koppen-Geiger System of Climatic Classification

W. Koppen, an Austrian scholar has given a systematic climatic classification in 1918. This classification was based on assigned values of temperature and precipitation computed in terms of annual or monthly values. This classification was subsequently revised by Koppen and his students to become the most widely used climatic classification for geographical purposes. This revised classification is known as 'Koppen-Geiger System of Climatic Classification'.

Koppen divided the climates of the world into major groups and represented them by the English alphabets A, B, C, D and E.

(i) *A – Tropical Climate* – The average temperature of every month is above 18°C. There is no winter season in this climate. Annual rainfall is heavy and exceeds the annual evaporation.

(ii) *B – Dry Climate* – In this type of climate the annual evaporation exceeds annual rainfall. Hence there is no surplus water.

(iii) *C – Warm Temperate Climate* – This climate is also known as Mesothermal. The average temperature of the coldest month is below 18°C but above – 3°C. In at least one month during the year the average temperature remains above 10°C. Hence both summer and winter seasons are found in this climate.

(iv) *D – Cold Temperature Climate* – This climate is known as snow climate and is also called Microthermal. The average temperature of the coldest month is below – 3°C and that of the hottest month is above 10°C. The isothermal of 10°C represents the limit of trees towards the poles.

(v) *E – Polar Climate* – The average temperature of the hottest month is below 10°C. In fact, there is no summer season in this climate.

It is to be noted that only B group is represented by rainfall to evaporation ratio. Other groups (A, C, D, E) have been separated on the basis of temperature.

Sub-groups – In addition to these groups, sub-groups have also been described and represented by English alphabets. Another alphabet following the name of the climate represents the sub-group.

Sub-groups of the Main Group B – The main B group is divided into two sub-groups (S and W). Hence the nomenclature of the sub-groups is BS and BW.

(i) *BS – (Steppe Climate)* – This semi-arid climate is found in lower latitudes with rainfall ranging from 38 cm (15") to 75 cm (30").

(ii) *BW – (Desert Climate)* – The average annual rainfall is below 25cm (10"). The boundaries of BS and BW are found out with the help of a formula.

Sub-groups of other Main Groups (A, C, D and E) – Other groups have also been sub-divided. Other alphabets are added to the nomenclatures of the main-group. Other alphabets are described below:

(i) *f – Moist* – Every month has sufficient rainfall. Hence there is no dry season in this sub-type. With its help Af, Cf and Df sub-groups are formed.

(ii) *w* – When the cold seasons dry, alphabet w is used. Aw, Cw and Dw sub-groups are formed.

(iii) *s* – This alphabet is used when a summer season exists. Cs sub-group is used commonly.

(iv) *m* – When rainforest climate is found in Monsoonal rainfall areas which have a small summer season, the alphabet m is used. It is used only in A group and Am sub-group is formed.

The climatic divisions as descrambled in Koppen-Geiger classification are as follows:

(1) *Af – Tropical Rainforest* – Rainfall all through the year, no dry season.

(2) *Am – Tropical Rainforest* – In the monsoon climate, rainforests grow in spite of a small dry season.

(3) *Aw – Tropical Savannah* – Winter season is dry in tropical areas.

(4) *BS* – Steppe.

(5) *BW* – Desert Climate.

(6) *Cw – Warm Temperate Rainy Climate* – The winter season is dry. It is called Humid Mesothermal.

(7) *Cf – Warm Temperate Rainy Climate* – Every season has rainfall. It is also called Humid Mesothermal.

(8) *Cs – Warm Temperate Rainy Climate* – It has a dry summer. It is also called Humid Mesothermal.

(9) *Df – Cool Temperate Wet Forest* – Rainfall takes place all through the year.

(10) *Dw – Cold Temperate Forest* – It has a dry winter.

(11) *ET* – Tundra Climate.

(12) *EF* – Climate of Ice Caps – It is a always snow bound.

Finer Sub-groups – Due to the variations of temperature, rain, etc. the sub-groups as mentioned above are further sub-divided. There alphabets are given below:

(a) For only C and D Groups

a = The temperature of the hottest month is above 22°C (Summer is very hot).

b = The temperature of the hottest month is below 22°C (Summer is ordinary hot).

c = The temperature of less than four months is above 10°C (Summer is cool).

(b) For only D Group

d =The temperature of the coldest month is below – 36°C (Extremely Cold Winter).

(c) For only B Group

h =When average annual temperature is above 18°C (Dry hot).

K = When the average annual temperature is below 18°C (Dry cool).

Climate Classification of Thornthwaite

C.W., Thornthwaite presented two classifications of climate. First classification was published in 1931 and in 1933 and the other in 1948. Thornthwaite had made some more modifications in 1955 in his method of classification.

Though both the classifications appear similar, yet, Thornthwaite considers both of them basically different. For example Thornthwaite in his 1933 classification considered plants to be an instrument of climatology with which climatic elements can be measured. In his second classification (1948) Thornthwaite considered plants to be those physical mechanisms through which the moisture of the soil is transferred to the atmosphere. In other words a plant is considered to be the machinery of evaporation as the cloud is the machinery of precipitation. Details of this classification are not given here but its characteristics are included in general types.

World Climate Types

General types of the world climate regions and their characteristics are summarised as follows:

World Climate Types and their Characteristics

	Climatic Sub-Divisions		*Koppen's Symbol*	*Characteristics Related to Air Masses*	*Other Normal Climatic Characteristics*
1.	**Equatorial Rainforest Climate** (10°N-10°S but in Asia 10°N-20°N).	(i)	Af – Tropical Rain-forest.	Equatorial trough is situated in this region. Airmasses converge into this trough.	Average temperature 27°c, cumulonimbus clouds, annual range of temperature low, Selvas (Rainforest) 30–25 metre high trees, laterite soil, Abundant water in infertile river valleys.
	Main Countries Equator, West Columbia, Northern Peru, Brazil (Amazon), Sieria Leone, Liberia Congo, Malabar (India), Southern Burma, Malaysia, Indonesia, Philippines, Guinea.	(ii)	Am – Tropical Rain-forest (Monsoon).	Main Airmasses- mT and mE. These airmasses are hot and humid. Rainfall is caused by convectional storms.	
2.	**Trade Wind Littoral Climate** (15°–35° N and S). Main countries – Honduras, Nicaragua, Costa Rica, Brazil, Madagascar, Indo-China, the eastern coast of Indo-China.		Af and Am	Easterlies blow on the western coasts of oceans from the Tropical High pressure cells. The winds have mT airmasses.	High temperature all through the year, 27°C, 200 cm rainfall, some-what dry weather in February, March and April, Annual temperature range being comparatively high. This is the last boundary of the rainforests.

Cont'd...

...Cont'd

3.	**Hot Desert and Steppe Climate** (15°-30° N and S). Main countries – West Pakistan, Iran, Arab, Northern Africa, Northern Mexico, Desert of Australia, and SE of USA.		Subsidence due to high pressure areas over tropics of Cancer and Capricorn, areas dry domain of cT airmass.	Rainfall less than 25cm in BWh deserts, but rainfall between 38 and 75cm is BSh steppe climate, high evaporation in this climate. Evaporation may be 20 times the annual rainfall. Heavy rainfall possible occasionally due to sudden entry of mT airmass causing floods but has no permanent effect upon climate. Great variations in rainfall is the main characteristics of the climate. Temperature range is very high to 32°C but the daily temperature is higher (20°C). Ajizia Tripoli has the highest temperature 58°C recorded so far. Practically no vegetation, plants with some thick skinned leaves and stems are rarely seen, soil has no humus but is saltish. Steppe climate is better. Grass cover is common. Rain is comparatively higher but temperature lower. Soil is comparatively more fertile.
4.	**Western Coastal Desert Climate** (15°–30° N and S). Main countries – California, Western coast of South America, (Middle), S.W. of South America, Middle Australia, some parts of Africa, etc.	(i) BWK-Desert climate (cold). (ii) BWH – Desert climate (hot).	The maritime airmasses from high pressure cells are stable and dry.	Western coasts (15°–30°) are dry, rainfall less than 25 cm, temperature about 21°C, temperature is 3°C lower than that in interior deserts. Fog is common on the coast. Vegetation and soil similar to deserts described above.
5.	**Tropical Savannah Climate** (5° to 20°N and S in Africa and America and 10° to 30° in Asia) Main countries – Southern part of Mexico, Chad, Nigar, Mali, Ethiopia. Tanganyika, India, Northern Australia, Venezuela, Brazil, etc.	(i) Aw – Tropical humid and savannah climate. (ii) Cw a – Temperature humid climate.	Sometimes dry cT and sometimes humid mT or mE airmasses dominate the area.	A part of the area is dry and another part humid. Rainfall is 100cm and temperature is 21°C. Long grasses and plants which can tolerate aridity are found, yellow and red infertile soil but the river born soil is fertile.
6.	**Humid Semi-tropical Climate** (20°–35° N and south). Main countries – S.E. U.S.A. China, Japan, E. Australia, South Brazil etc.	Group II (This climate is tropical and is Cf a-Temperate humid climate.	Middle Latitude Climate controlled by polar airmasses mT airmass is born in the higher pressure cells of oceans.	Enough rain, higher rainfall in Summer, average temperature of 22°C in the hottest month, temperature range normal, polar airmass causes cold in winter, snowfall, less snowfall in Asia. Deciduous and coniferous trees, red-yellow soil containing iron and aluminum.

Cont'd...

...*Cont'd*

7.	**Maritime West Coast Climate** (40°–60° N and S) Alaska Coast upto Vancouver in N. America, southern coast of Chile, the whole of western coastal Europe, etc.	(i) Cf b – Temperate humid climate (Hot summer). (ii) Cf c – Temperate humid climate (cool and short summers)	mP airmass in Cyclones.	No dry season, the average temperature of the hottest month is 22°C and the average temperature of 4 month is 10° but is Cf c of such months are less than 4 in number. The climate is humid in spite of low rainfall because evaporation is low due to the low temperature. Winter cold is ordinary but summers are cold. Broad leaves forests in Europe and coniferous forest in America. Acidic soil on mountain slopes and humus soil in plains.
8.	**Mediterranean Climate** (30°C–45°N and S). Main countries – California, central Chile, Mediterranean counties, S-W. Australia, etc.	(i) Cs a – Temperate humid climate (dry and hot summers). (ii) Cs b – Temperate humid climate (Dry, less hot summers)	Domain of mP in winters and that of mT in summer.	Intense cyclonic activities due to mP in winters, dry weather due to mT in summers, Average temperature in summers in 16°C, and in winter 10°C, Average rainfall is 50 cm. Vegetation can withstand cry summers. Stunted trees and bushes are commonly seen. Oak, Cork, olive, etc. are important trees, Chestnut gray soils, red or yellow soils found.
9.	**Middle Latitude Desert and Steppe Climate** (25°–50° N and S). The middle parts of N. America and Asia.	(i) BWK – Desert climate (Dry). (ii) BSK – Desert Climate (Cooler). (iii) BSK - Steppe Climate (Cold) (iv) BSK – Steppe climate (Very cold).	Domain of cT airmass in summers and that of cP in winters.	Annual temperature-range high, very hot summers, very cold winters, Annual range of temperature is 18°C, the annual rainfall is 25 to 65 cm. Grass grows. Fertile soil is found. Soil erosion is high.
10.	**Humid Continental Climate** (35° – 60°N) Main countries – Central and eastern USA, Poland, Russia, Central Asia, Eastern China, Japan, etc.	(i) Df a – Humid Mesothermal cold, snow-bound forest climate, humid all through the year, hot summers. (ii) Df b – Similar to above, but hot summers (iii) Dw a – similar to above, Dry cold winter, Hot summers. (iv) Dw b – Similar to above, but ordinary hot summers.	mT and cP airmasses dominate.	The temperature usually falls below 0°C in winters but summer temperature rises to 27°C. Enough rainfall throughout the year. Rainfall averages 60 to 120cm. Coniferous and broad leaved trees, Podzol soil (full of humus), etc.

Cont'd...

...Cont'd

11. **Sub Arctic Climate** (55°–70°C) Main countries – Alaska, Canada, Norway, Sweden, Finland, N. Russia along with the Siberia, etc.	(i) Df e – Humid, low temperature climate, cold snow covered, humid all through the year, cold summers. (ii) Df d – Similar to above, winters very cold. (iii) Dw c – Cold, snow covered. Dry cold season, cold summers. (iv) Dw d – Similar to above, cold season very cold.	cP airmasses regains all through the year but mP airmass brought by cyclones also enters this region.	The temperature of winters ranges from – 50° to – 40°C but the summer temperature rises to +12°C. Summers of very short duration. Annual rainfall 25 cm. Coniferous trees (Mainly spruce and fir).
12. **Tundra Climate** (To the north of 55°N and to the south of 50°S)	ET – Polar Tundra climate.		
13. **Ice-cap Climate** Greenland and the Atlantic	EF – Polar climate with permanent snow-cover.		

20

Natural Vegetation

One of the most striking features of earth's land surface is the plant cover which varies greatly in kind and in density from region to region. This vegetation cover has geographical significance in a variety of ways, ranking from aesthetic contribution to a significant source of the several forest products. In fact, natural vegetation is an expression of the composite physical environment, and is the integration of all factors, past as well as present, it likewise serves as an indicator of the potentialities of an environment for human use. There is an especially close correlation between zones of climate and zones of vegetation.

Plant life is divided into three broad vegetative groups. Forests dominate humid areas, ranging from the broadleaf evergreens of the rainy tropics to the deciduous and coniferous forests of the middle and higher latitudes. Grasslands of the Savannah, Prairies and Steppe occupy the zones of lesser and seasonal precipitation completing the pattern is the sparse scrub and low vegetation of the dry and cold deserts. Although vegetative groups are similar in like environments, species may differ in the groups from continents to continents. (Figure 20.1)

Types and World Distribution

The distribution of natural vegetation over the earth reflects present climatic conditions more tham any other single element. Plant geographers have long recognized this fundamental principle which is evidenced by the general parallelism which exists between temperature and precipitation characteristics on one hand and vegetation characteristics on the other. Due to similar physical conditions plants are distributed in various continents. The broad types of natural vegetation of the world based on plant association are as follows:

1. Forest Associations

(A) Tropical forests
  (i) Tropical rainforest
  (ii) Lighter tropical forest (including semi deciduous, deciduous, scrub, and thorn forest)
(B) Middle latitude forests
  (i) Mediterranean, woodland and shrub
  (ii) Broadleaf forest
    (a) Deciduous
    (b) Evergreen

Figure: 20.1 *The World Distribution of Natural Vegetation*

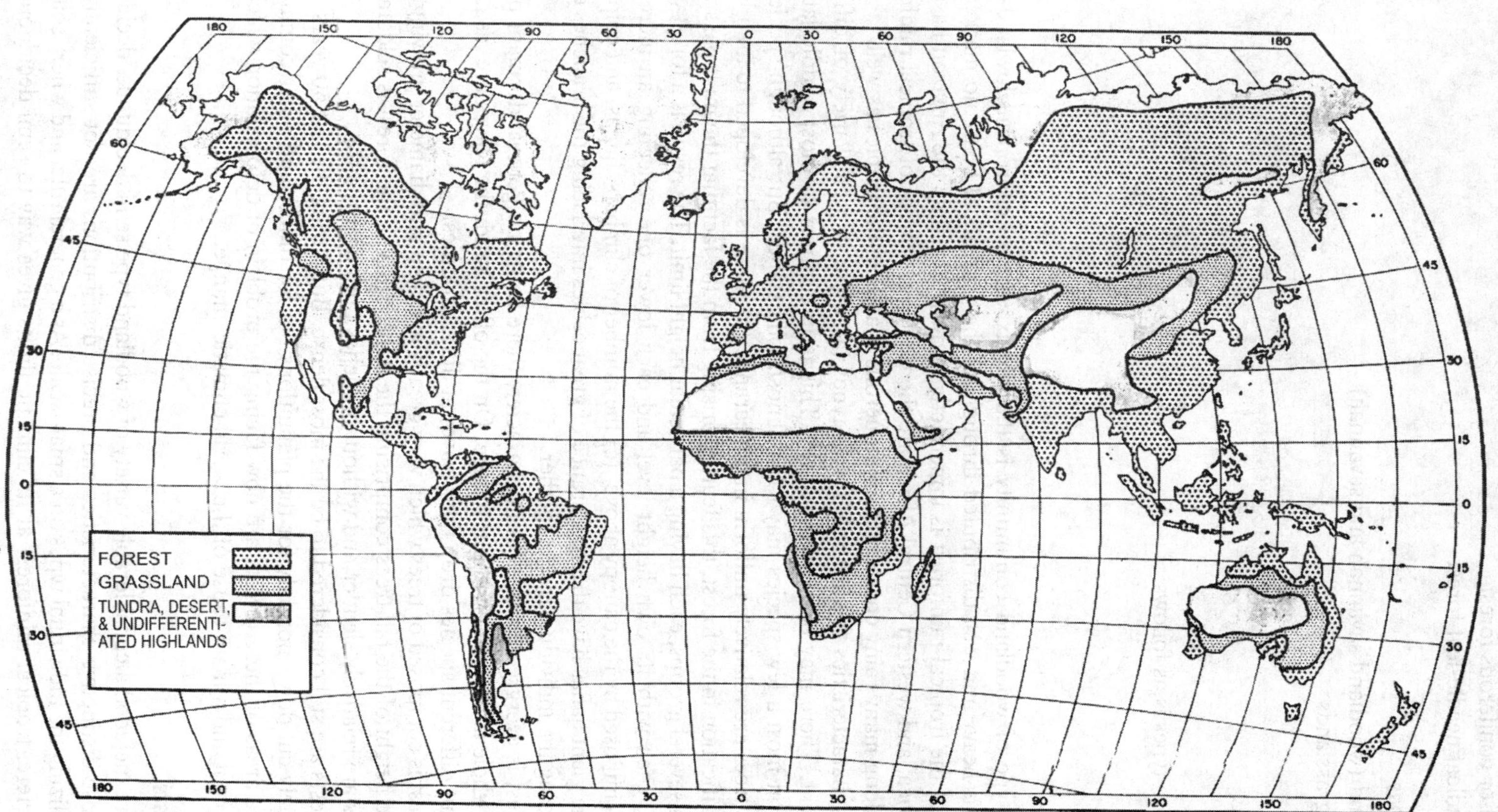

(iii) Needle-leaf or coniferous forest
(iv) Mixed broadleaf needle-leaf forest

2. Grassland Associations

(i) Tropical grassland (woodland savannah and savannah)
(ii) Middle latitude grasslands

3. Desert Shrub

4. Tundra

A brief analysis of these types is as follows:

Forest Associations

Tropical Rainforests

This most luxuriant type of woodland community is the climax vegetation tropical lowlands and slopes where rainfall is heavy and well distributed throughout the year, there being no marked dry season. Distribution of the tropical rainforest is imperfectly known. Certainly the Amazon basin, in northern South America, and western Central Africa are the two largest areas of tropical rainforest, although it is found along many rainy coasts and on numerous islands in the tropics as well.

Three principal characteristics feature this forest type: (a) there is great variety of different species of trees present. There may be as many as 40. This is in contrast to most middle-latitude forest, where one, or at most a few species may form almost solid stands. But although species are numerous, rainforest trees are relatively uniform in appearance (b) there is developed to an unusual degree, a vertical stratification in the forest, this feature arising from the fact that the multiple species arrange themselves in several groups, each having a particular height limit. The result is a forest with a number of tree tiers, each with its own height level and each lower one reflecting an increasing tolerance for shade as imposed by the canopy above. (c) the number of climbers, lianas, and epiphytes is unusually large. The giant lianas have the appearance of great cables interlacing the branches of the forest crown and binding the individual trees together.

Tropical rainforest is evergreen broadleaf in character, there being no general dormant period when the forest as a whole is bare and without foliage. On the other hand, the leaves are, as a rule, renewed each year, but different species drop their leaves at different times.

Tropical rainforest is composed of trees which vary greatly in height and diameter growing close together. The average height of the tall trees comprising the upper story of the forest is seldom over 150 to 180 ft. Trunks are relatively slender and without branches until near the top.

Tropical rainforests are still considered to be the most dense. But during the last 50 years, these forests have been rudely cut down not only for the plantations but for timber and other commercial purposes. Most of the areas of these forests are now facing the problem of deforestation which is a threat to the environment and a main cause of the world climatic change.

Lighter Tropical Forests

Under this heading are included a considerable variety of woodland type semi deciduous, deciduous, scrub, and thorn whose distinctive characteristics and precise distributions are not sufficiently well known to permit localizing the individual types on a small-scale. As a general rule, and where soils and drainage do not interfere, tropical rainforest at its climatic limits gives way to semi deciduous and deciduous forest and with declining rainfall this in turn passes over to savannah woodland and finally to thorn woodland and desert shrub.

Characteristically, these woodland of the tropics, other than the rainforest, are composed of smaller trees more widely spaced and with a denser undergrowth of shrubs or grass. More of the trees also are deciduous in character, although not all species are leafless during the drier dormant season. It is during the season of drought, nevertheless, that contrast with the rainforest is most marked. Scrub and thorn forest varies in density from an open park-like growth of low stunted trees and thorny plants to dense thicket of the same.

Middle-Latitude Forests

Mediterranean Woodlands and Shrub Forests

Mediterranean woodland and shrub is a relatively rare type, for seldom are broadleaf evergreen trees and at the same time adapted to regions with long, hot periods of summer drought. In those parts of the humid tropics where pronounced dry seasons are characteristic, trees protect themselves by shedding their leaves thereby becoming xerophytic during the dry season, although they are hygrophytic during the periods of rains. In Mediterranean woodlands, on the other hand, adjustment is made in other ways, by protective devices against rapid transpiration permitting the trees to retain their foliage, and consequently their evergreen characteristics, during the period of aridity. But although evergreen in character, there is more of a seasonal rhythm in vegetative and reproductive processes than is true of the tropical rainforest. This unique Mediterranean woodland is found in subtropical regions with mild, rainy winters and long, dry, hot summers. The climate as well as the vegetation is unusual, for the times of maximum temperature and maximum rainfall do not coincide. The largest representative area is the Mediterranean sea borderlands, with smaller areas is California, central Chile, southern Australia, and the Cape Town region of Africa.

Mediterranean woodland is predominantly a mixed forest of low, or even stunted, trees and wood shrubs. Tall trees are rare. Where climatic and soil conditions are most favourable the virgin forest is composed of low, widely spaced trees with massive trunks and gnarled branches. Between the trees the ground is completely or partially covered by a pale, dusty, bush vegetation.

Even more common than the woodland composed of low trees and shrubs described above is a vegetation mantle consisting principally of shrubs and bushes in which there may be some, stunted trees, this bush thicket is known as **chaparral** in California and **maqui** in lands bordering the Mediterranean basin. In places the woody shrubs form a thick and relatively tall cover; in others it is short and sparse. Chaparral in some regions may represent the original vegetation cover. In other sections it is the underwood remaining after the low trees of the original woodland have been destroyed. The chief economic importance of chaparral usually lies in its watershed protection.

Broadleaf Forests

Within the more humid parts of the middle latitudes are found two great forest groups (a) the broadleaf trees, and (b) the needle-leaf conifers. Over large areas they exist as mixed conifer-broadleaf forest. As a general rule, but with important exceptions, the coniferous forest occupies the colder continental locations and these are usually on the poleward side of the broadleaves. In regions of porous, and therefore water-deficient, sandy soils, such as the Atlantic and Gulf coastal plain of the Untied States, or on steep mountain slopes where soils are thin or rocky and temperatures lower, conifers may supplant broadleaves even in the lower middle latitudes. The latter condition is illustrated in the case of the southern Appalachians, which carry a long tongue of coniferous and mixed forest southward into the broadleaf belt. Outside the United States, other relatively large areas of temperate broadleaf or mixed forests are found in Japan, Korea, southeastern China, central Russia, Rumania, southwestern Siberia, western Europe, southern Chile, south-eastern Australia, and New Zealand.

Deciduous Broadleaves

By far, the greater part of the temperate broadleaf forest is deciduous in character, the trees dropping their leaves during the wither season. Except in the dormant season this forest is rather uniformly bright green in colour, and its profile is regular. The amount of underwood varies with density of tree stand, being much greater where an appreciable amount of light reaches the ground. Trunks of the deciduous trees are xerophytic in character, having a relatively thick bark. On the other hand, the leaves are thin and delicate, requiring no protective devices, since they remain on the tree only during the warmer part of the year. As a result of seasonal leaf fall, for the year as a whole, considerable sunlight reaches the soil under deciduous forests.

Evergreen Broadleaves

Only along the humid subtropical margins of the middle latitudes there are important evergreen broadleaf forests, but these are not nearly so extensive as the deciduous variety. Their principal regions are in southern Japan, southern and southeastern China, New Zealand, and southeastern Australia. In many respects these subtropical forests are akin to those of the wet tropics. The number of species composing the forest is likewise considerable. Oaks of various kinds are among the commonest trees. Eucalyptus and acacia are important elements of the southern hemisphere forests.

Needle-leaf or Coniferous Forests

Coniferous trees are predominantly evergreen, the addition and fall of the needles being a continuous process and not confined to any particular periods or season. In some species the leaves may remain on the trees for 5 years or more. Unlike broadleaves, however, the needles of conifers are xerophytic in character so that shedding is not necessary to protect against a cold or drought season. The two categories of coniferous forests are subarctic conifers and conifers of lower middle latitudes.

Subarctic Conifers

Conifers reach their maximum development as far as areal extent is concerned; in the severe subarctic regions of North America and Eurasia, where they form wide and continuous east-west forest belts stretching from coast to coast, to the subarctic coniferous forests have been given the name **taiga**. On their northern frontiers they make contact with the treeless tundra, a region thoroughly hostile to trees. The Eurasian taiga forms the single largest continuous forest area on the earth. Conifers (larch, spruce, fir, pine) predominate, although broadleaf trees (alder, willow, aspen, birch, mountain ash) are scattered throughout, individually as well as in thickets or clusters. The latter are characteristic of low swampy areas and of regions bearing a second growth. Species are few in number.

Conifers in Lower Middle Latitudes or Mixed Broadleaf

South of the great belts of subarctic conifers are other areas of needle trees which, although less extensive, are nevertheless more valuable forest regions. This comes about as a result of their being composed of larger trees and superior timber species and at the same time being more easily accessible.

In western North America broken belts of conifers extend southward from the taiga following the rainier highland chains (Pacific Coast Mountains and Rocky Mountains) to beyond the Mexican border. The forests of the American Pacific coast states, western Canada, and Alaska constitute the most extensive area of fine coniferous forest anywhere in the world. Large trees, dense stand, good-quality timber, all contribute to this high rank, most valuable of its trees is the douglas fir, which reaches a diameter of 6 ft. or more and a height exceeding 250 ft.

East of the Rockies conifers extend southward from the taiga into southeastern Canada and the northern portions of the northeastern tier of American states Minnesota, Wisconsin, Michigan, the Adirondacks in New York, and much of Maine. The most valuable timber trees from this eastern forest have been removed, leaving behind extensive areas of waste and cut over land of little value. South of the taiga in Eurasia, valuable coniferous forests occupy the slopes of the Alps, Carpathians, and other highland regions as well as certain sandy areas of coastal plains.

Separated from this northern belt of conifers, in the United Stated, by extensive broadleaf forests is the southern pine forest which occupies the Atlantic and Gulf coastal plain.

Grassland Associations

Two distinct types of grassland are: (i) Tropical grasslands, and (ii) Middle latitude grasslands

Tropical Grassland

Wooded Savannah and Savannah of both the wooded and the open variety are common. In places the trees are few and widely scattered while in others they are so numerous so as to form thickets. Patches or clumps of trees intermingling with patches of grass seem the most common. Tall trees are not absent, to be sure, but low and dwarf varieties are more common. As a usual thing they have twisted and gnarled trunks, a thick corky bark, and leathery leaves.

Savannah grasses are variable in height. In Africa there are high savannahs where the grasses range from 5 to 15 ft. height, but these do not seem to exist in Latin America. Much more common are the tall bunch grass (1 to 2 ft. high) and the short bunch grass (under 1 ft. high). Both exist in the form of slender tufts with much bare earth between individual tufts so that the grass may occupy not more than 60 per cent of the soil surface. As a consequence a sod is absent. The leaves of savannah grasses are stiff and leathery so that only fresh young shoots are palatable to grazing animals. Among the natives it is a common practice to burn off the grasses in the dry season in order to make room for new growth at the beginning of the rains. Throughout the tropical grassland dense galleria forest usually occupy the river floodplains.

Middle latitude Grassland

The grasslands of middle latitudes appear to be the climax vegetation of sub humid and semi arid regions. In this respect they stand in contrast to the savannahs of the general absence of trees in the grasslands of middle latitudes except along their contact zones with forest and in the vicinity of streams.

In the grasslands of interior North America, east of the rocky mountains, three large subdivisions are recognized: (a) the short grass occupying much of the semi arid great plains, (b) the mixed prairie occupying a somewhat more humid belt to the east of the short grass, and (c) the tall grass prairie, or true prairie, typical of the better watered lands between the mixed prairie and the eastern forest.

Steppe is also a short-grass, **Prairie** concentrated largely in the mid-latitude zone. In fact, the name steppe is traditionally given to the grasslands of Russia. The only noteworthy area under this grassland in southern hemisphere is the **Veld** region of South Africa.

Desert Shrub

Hot deserts are widespread in all the continents except Antarctica, characterized by high temperature and low rainfall. Because of this climatic characteristic, the arid and semi-arid regions of the world are having scanty vegetation.

Desert shrub characteristic of arid region, consists of woody xerophytic plants usually only a few feet in height. They are broadleaf in character and may be either evergreen or deciduous in their habits. Growth may be in the form of single isolated plants widely spaced or they may exist in groups or patches.

Most of the arid regions of both low and middle latitudes have vegetation of low bunch grass, widely spaced with bushes, or in places fleshy water-storing plants such as cacti, but much more commonly it is perennial xerophytic shrub.

Another class of desert plants, in contrast to the perennial shrubs, depends entirely upon the erratic rainfall, germinating with a rain, ripening seeds when the moisture is gone, and dying forthwith. Such annuals do not appear xerophytic, their adaption to drought being through a very rapid development and a short duration of life. Their stems and leaves are delicate, roots are thin and relatively shallow, and flowers are of considerable size. In contrast to the scanty, low, pale-green vegetation of the desert proper is the verdant colour of luxuriant vegetation around oases where water is abundant. Almost knife-edge boundaries frequently separate the two. In wet or sub irrigated land containing a high percentage of alkali, green, fleshy-leaved, salt-tolerant plants.

Tundra Vegetation

It consists largely of herbaceous or nonwoody plants other than grass, although the latter may not be absent. Sedges, mosses, and lichens predominate. This lowly plant cover is typical of those high-latitude portions of the continents where summers are too short and cool for even the hardy conifers to thrive. For much of the year the ground remains frozen so that while there may be adequate water in the soil, it exists in the solid form which is inaccessible to plants. On the southern margins of tundra, where, it merges into the taiga or coniferous forest, the vegetation cover is more complete and stunted and creeping forms of trees and bushes are conspicuous. Grasslands exist on the marine margins of the soil, which retards water absorption, as well as in the long winter period of physiological drought when the soil moisture is locked up in solid from, most tundra plants appear xerophytic, having stiff, hard leathery leaves, with thick cuticle, as a result of the short period between frosts, the vegetative period in the tundra is reduced to two months or less. For this reason plants are compelled to hurry through their vegetative cycle, and even then many of them are frozen while still in flower or fruit stage.

21

Soils

Tiny rock particles mixed with organic mater, water, air and living organisms form the thin veneer of soil that covers the earth's surface. Soil is an essential resource of the earth; it is the medium of plant growth, and all men are directly or indirectly dependent upon plants for food, clothing, and many other necessities of living. Soil is the upper weathered and biologically-molded portion of the crust – a complex of minerals and organic substances, varies greatly in quality as well as quantity from place to place. Depth, texture, structure, fertility elements, drainage and topography are the principal variables of soil; the regional differences in soils are based upon variations in bedrock of parent material, topography, climate, and vegetation. The parent material is the primary source of the chemical elements in a soil and a governing factor in the fertility of a given soil. The vegetation, temperature and moisture influence the rate at which soil is developed. The topography affects soil composition and retention.

Dominant Great Soil Groups and Soil Categories

The following outline (also see Figure 21.1) hows the broad descriptive categories and those of the major zonal great soil groups included in each:

Broad Descriptive Soil Categories (1 to 9) Based upon Zonal Soils and Dominant Great Soil Groups

Great Soil Groups	*Descriptive Soil Categories*
Podzol	1. Podzol Soils
Grey – Wooded	
Grey Podzolic Brown Podzolic Grey-brown Podzolic	2. Podzolic Soils
Red-yellow Podzolic	3. Podzol – Latosolic Soils
Reddish latosols and a variety of imperfectly known latosolic, latosolic podzolic, latosolic-chernozemic, and other soil groups	4. Latosolic Tropical Forest and Grassland Soils
Prairie Chernozem Degraded Chernozem Reddish Prairie Noncalcic Brown Rendzina	5. Chernozemic Soils

Cont'd...

...Cont'd

Brown Reddish-brown Chestnut Reddish chestnut Terra rossa	6. Chernozemic – Desertic Soils
Desert Red desert Sierozem	7. Desertic Soils
Alluvial	8. Alluvial Soils
Tundra	9. Tundra Soils
	10. Soils of Areas with High Local Relief with Complex Soil Associations.

Podzols

Podzols are the typical mature soils of regions having humid subarctic climate. Although occurring in the humid tropics and subtropics podzolization is much more favoured by long winters and short summers. Podzolic soils are different from flatosolic soils in structure, profile, and colour.

The mature podzol is developed under natural vegetation mainly of coniferous forest. Conifers do not require abundant basic soil elements for their growth, and since they are mainly shallow-rooted trees, they draw only small supplies of the soil bases to the surface, even where such are available in the parent material beneath. Thus, there is little chemical reaction to oppose the tendency towards soil acidity. The effects of the long cold winter, moderate summer temperatures, and a forest litter of resinous pine needles are to retard bacterial action and to permit the formation of a brown layer of raw humus or half-decomposed organic remains, which represent the accumulation of many years.

The main regions of podzol soils are in the higher middle latitudes. However, they are not strictly confined to those latitudes. Partially podzolized soils are abundant in the humid lower middle latitudes and to some extent are found in the red-soil regions, or even in the humid tropics where unusual conditions have permitted the accumulation of coverings of acid organic matter.

Podzolic Soils

These are a group of soils which are usually found in those regions of the world that have broadleaf deciduous forests and humid microthermal and climatic conditions. There is a surface accumulation of organic material which forms a dark layer 3 to 10 centimetres deep. It is not so abundant, so poorly decomposed, or so acidic as that associated with the podzols. Moreover, the organic material derived from broadleaf forest contains more lime, potash, and other basic elements than that from coniferous forest, and these are easily mixed by earthworms with the mineral components of the soil. Mature podzolic soils of United States and northwestern Europe and several other regions are intensively cultivated.

Podzolic-Latoslic Soils

These soils which have some characteristics of both. Latozation and podzolization occur in association with other in many areas of the tropical and subtropical regions. Their occurrence in areas of considerable agricultural importance such as the cotton belt of the United States and southern China renders them significant geographically. They have been subjected to the process of latozation, but either the process has not been so complete as in the case of a true latosol, or it has taken place under slightly different conditions of rainfall and drainage.

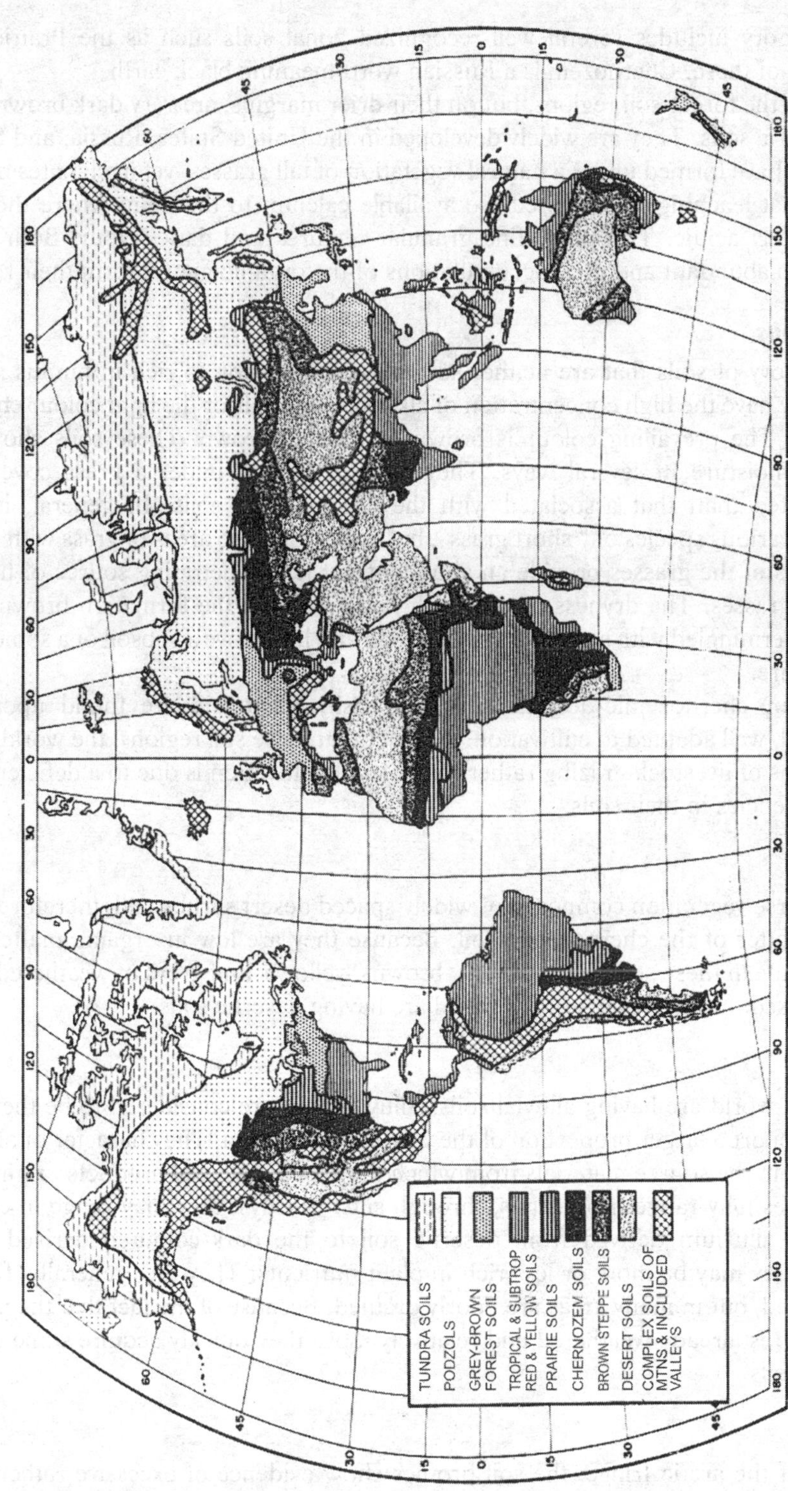

Figure 21.1 *The Great Soil Groups of the World*

Chernozemic Soils

The chernozenmic category includes several well-recognized zonal soils such as the Prairie, the chernozem, and variants of these. Chernozem is a Russian word meaning black earth.

Adjacent to some of the forest-soil regions, but on their drier margins, are very dark brown soils which are known as Prairie soils. They are widely developed in the United States, Russia, and South America. They appear to have formed under a natural vegetation of tall grasses, yet in climates having sufficient moisture so that leaching has lowered the available calcium to the point where they are neutral or even moderately acidic. They have fine granular textures and dark colour. Both these qualities are derived from abundant and deep accumulations of the organic matter form their roots.

Chernozemic-Desertic Soils

These constitute a category of soils that are neither so dark coloured nor so rich in humus as the chernozemic, nor do they have the high concentration of alkaline materials or the light colour characteristic of desertic soils. The prevailing colour is brown or reddish brown. These soils show the influence of decreased moisture in several ways. They have developed under a grass cover less luxuriant and deep rooted than that associated with the chernozemic soils. In general, it is a continuous sod cover of various species of " short grass", but it includes also areas of grass with intermingled shrub. The roots of the grasses provide an abundant but less penetrating source of humus than those of the taller grasses. The dryness of the earth has promoted the formation brown than black humus, which is intermingled with a powdery surface soil and lies above a subsoil of a somewhat coarse and lumpy structure.

In general, however, chernozemic-desertic soils are easily tilled and are found upon the undulating to rolling land, well adapted to cultivation. The fact that these soil regions, the world over, are predominantly regions of livestock grazing rather than of soil cultivation is due to a deficiency of rainfall rather than deficiencies in their soils.

Desertic Soils

These develop under sparse vegetation composed of widely spaced desert shrubs and, therefore, lack the abundant organic matter of the chernozemic soil. Because they are low in organic matter, the lighter colours predominate in these soils, and the reds, browns, yellows, and greys of weathered rock minerals are widely exposed. All the deserts of the world are having desertic soils.

Alluvial Soils

All the river plains of the world are having alluvial soils. Alluvial soils are considered to be the most fertile soils; therefore support a larger proportion of the world's population. The character of alluvial soils is largely derived from the source materials from which the specific alluvial products originated. Consequently, the textures may range from sands, through silts, to clays; they may range in colour from the light-coloured alluvium derived from desertic soil to the dark-coloured derived from chernozemic soils; and they may be more or less rich in plant nutrients. They are, generally, free of stones and easily cultivated, but in many areas are poorly drained. Because of the fact that the parent material, when deposited, is already weathered and relatively loose they quickly acquire some of the characteristics of zonal soils.

Tundra Soils

In the treeless regions of the arctic fringe, the soil profiles show evidence of excessive rather than deficient moisture. This is due to the low rate of surface evaporation and the presence of permanently frozen subsoil. The better drained sites of the arctic-fringe regions have soils somewhat like podzols. But the usual horizons include a brown peaty surface layer which is underlain by greyish horizons, one of them characteristically plastic or even fluid. A large part of the tundra is poorly drained and the prevailing soil conditions are those of bog and hummocky marshland. The soils are in several respects similar to the glacial marsh and bog soils found in middle latitudes, many of which are drained and cultivated.

USDA Classification of Soils

The classification of soils given by the US Department of Agriculture is one of the scientific classification of world soils. The following table indicates the global soil orders:

Global Soil Orders

Soil Order	*Percentage of Total World Soils*	*Characteristics*
Inceptisol	16%	Young soils found in glacial, alluvial, or volcanic deposits; productive for agriculture.
Aridisol	19%	Soils of dry areas having little organic matter and shallow depth.
Mollisol	9%	Mature soils often found under midlatitude grasslands; most agriculturally productive.
Spodosol	5%	Infertile, acidic, sand soils that occur mostly in cold-winter forested areas.
Alfisol	15%	Soils of wide geographic distribution, containing concentrations of aluminum (Al) and iron (Fe) compounds.
Ultisol	9%	Red soils due to concentrations of iron, acidic with poor fertility.
Entisol	13%	Soils with limited profile development due to erosion. Youth, or human disturbance.
Vertisol	2%	Clay-rich soils of tropical and subtropical areas.
Oxisol	9%	Soil of the wet tropics, cultivation results in severe nutrient depletion.
Histosol	1%	Very organic soils; found extensively in higher latitude tundra and bog environments.

22

Volcanoes and Earthquakes

VOLCANOES

Volcano is a conical hill built up by the ejection of igneous material from a vent. It is an opening in the crust out of which magma, ash and gases erupt. The shape of the volcano depends much upon on the type of lava, its chemical composition (**acid lava** rich in silica and relatively viscous, forms steep-sides dome like volcanoes, **basic lava** containing little quartz, is more mobile, giving rise to more gently sloping volcanic cone) and subsequently erosional history of the volcano. While **volcanism** is a process by which magma and gases are transferred form earth's interior to the surface.

A form of the movement of the magma is volcanic eruptions and the volcanic structures, or **volcanoes**, associated with them. During a volcanic eruption molten lava (temperature 900–1200°C) gases, water vapours, and also ash and other solid products emerge to the surface. An erruption may be quiet with the lava gradually flowing out of a vent and spreading over the volcano slopes, or it may be explosive, i.e. accompanied by mighty explosions and the blowing out of enormous quantities of gases and lava which spread in the atmosphere and are deposited right on the slopes of the volcanic mountain or in its vicinity. Naturally, eruptions do not strictly conform to type and mixed-type eruptions occur as well. General structure of volcanoes is shown in Figure 22.1

Figure 22.1 *Structure of Volcanoes*

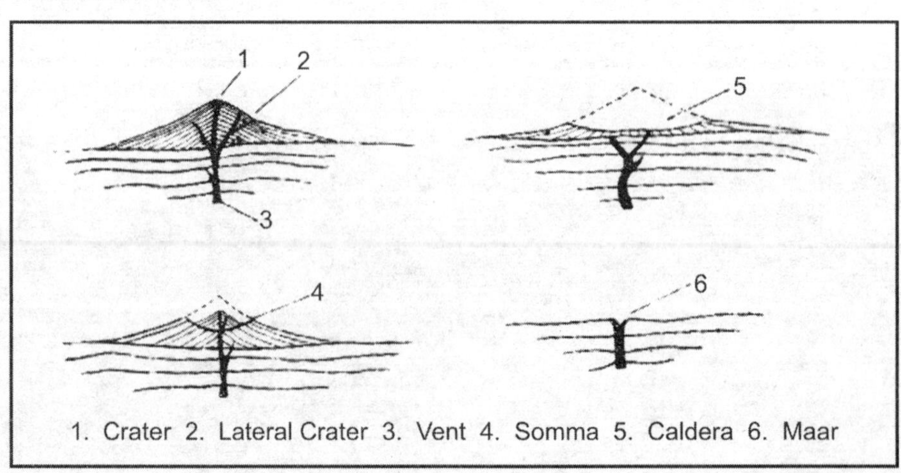

Every volcano has a **vent**, i.e. a duct piercing the bedrocks through which the eruption product reaches the surface. The mountain is crowned by a bow-like depression, called the **crater**, which is the terminal part of the chimney. In some volcanoes lateral conduits branch out from the main vent offering passages for the eruptions of lava, ashes and gases. Where they reach the surface, cones repeating the shape of volcanoes, though of smaller dimensions, appear; these so-called **parasitic** or **lateral** volcanoes are situated on the slopes of bigger volcanoes. Numerous parasitic cones are found on the slopes of Mt. Etna on Sicily and on the Klyuchevskaya volcano on Kamchatka.

With time the upper parts of volcanic cones may be destroyed and enormous (up to 25–30 km wide) circular depressions with steep inner walls and a flat floor are formed in their place. Such depressions are called calderas. Some volcanoes have so called somma (volcano-within-volcano) semi destroyed old creater at the summit of Vesuvius. The slopes of volcanoes which are composed of concealed lava and loose eruption material are often furrowed by gullies, steeply radiating from the creates in all directions called **barrancos**.

Classification of Volcanoes

Volcanoes are classified on various basis, these are;

(A) Classification based upon activity

(1) *Active Volcanoes* – the volcanoes which eject lava, gas, etc, are called active volcanoes. Some of them emit lava after every 15 or 20 minutes. Stromboli is most well known among them. Other volcanoes emit lava after years. Active volcanoes number about 300 in the world.

(2) *Dormant Volcanoes* – these are the volcanoes which have stopped ejecting lava but they can become active any time. Vesuvius is the such a dormant volcano. It becomes dangerously active after hundreds of years. In 1579, forest had grown in its crater and in the surrounding areas. The beautiful cities, **Pompeii**, **Herculaneum** and **Stabiae** were situated nearby. When the Vesuvius exploded, the three cities along with their inhabitants, houses and other property were buried under the 7 to 15 metre deep lava. It again exploded in 1631 and caused a great devastation.

(3) *Extinct Volcanoes* – these are the volcanoes which are quiet now. Though the volcanoes had exploded in the past yet there is no hope of any explosion now. It is difficult to differentiate between the extinct and dormant types of volcanoes. The Vesuvius was considered to be an extinct volcano. There are thousands of extinct volcanoes which are undergoing erosion by running water and wind and therefore look really extinct.

Classification Based Upon Exploded and Ejected Material

The explosion of a volcano has different styles. The ejected material is also not the same in all the volcanic explosions. The following is a classification based upon the factors mentioned above:

(1) *Hawaiian Type* – The main ejected material from these types of volcanoes is lava but it does not come out at the time of explosion. Lava continues coming later from main crater or secondary craters. When wind blows the lava spreads in the form of red threads. These threads are thought to be the hair of goddess Pelee. Mona Loa and Kilaura are respectively 4,168 and 1,232 metres high and are separated from each other by a distance of 12 km. The difference between their height indicates that they receive lava from different vents. The fountain of lava that gushes out of them rises to 244 metres.

(2) *Visuvian Type* – These types of volcanoes explode after a long time but the explosion is very devastating. The lava that comes out of it has a large quantity of gas in it. The erupted material

balances itself in air in the form of a cauliflower. Lava, ash, etc, are rained down form it. The Vesuvius is situated at a distance of 11.2 km south-east of the city of Naples. It is 1,220 metres high. The crater of the Vesuvius is situated inside Mt. Somma. This volcano destroyed three beautiful cities in 1579 and its lava dust reached Constantinople, 1,207 km away. Stromboli is a similar type of volcano which emits out volcanic bombs, ash, etc. The magma of this volcano is more viscous.

(3) *Pelean Type* – the magma from this type of volcano is very dense. This dense magma forms a plug and seals its crater. When the upsurge of magma takes place, the great pressure exerted by the upcoming magma breaks the seal with a devastating explosion. Some magma also comes out of the cracks situated in the weaker cracks around the volcano. Its explosion completely destroyed St. Pierre city in 1902. This burning lava flowed out with a speed of 3 km per minute. The molten lava, poisonous gases, etc, produced a colour which is known as fiery cloud (Nuee Ardent). Mt. Pelee of West Indies exploded in 1902. Mt. Katmai (Alaska) also exploded in the same way but it was fiercer than that of Mt. Pelee; the ash emitted by it spread over the whole of the world. Katmai valley is known as the valley of ten thousand smokes. Another volcano of Pelean type is named Hibok-Hibok and is situated in Philippines. It exploded in 1951.

(4) *Mud Volcano* – These are not the type of volcanoes as have been described earlier. In areas where petroleum or gas is found, sand and clay accompany the gas emitting out in those areas. This sand and clay is deposited at the vent and form a cone. If water also accompanies the gas, sand and clay form mud. This mud is deposited on the mound and gets hardened. It also stops the emission of gases. When the pressure of the gas increases, the top of the cone is exploded away. It appears as if a small volcanic explosion has taken place. These mounds of mud are called mud-volcanoes. An example of such a volcano is Bog-Boga and is situated in Baku area near Caspian Sea. Many of such volcanoes are situated far from real volcanoes. It does not mean that mud volcanoes have nothing to do with real volcanoes. At many places steam comes out of mud areas on account of volcanic action. For example, many mud volcanoes are found at the mouth of the Alamo River in California. They have been formed due to steam.

Causes of Volcanism

Several causes are responsible for the volcanic activities. It is also associated with sea-floor spreading, plate-tectonics and mountain building processes. The main causes are:

(1) *Water* – The water on the surface of the earth seeps into the interior through many cracks and fissures. When the water gets into contact with hot lava, it is dissolved in the latter. When this lava crystallizes, the dissolved water is freed. The freed water is at a high temperature and pressure. When this water (in the form of steam) escapes out of the many holes, explosion is caused. Later on, the lava also comes out of the holes, this principle is applicable to those volcanoes which contain a lot of steam in their gaseous emission.

(2) *The Effect of Fold* – The lava in the interior of the earth is subjected to a great pressure which does not allow the lava to melt. When orogenetic action affects the earth's surface, the sedimentary rocks rise in folds. The pressure on lava is reduced at places which are zones of weakness. At these places the lava is melted due to reduction of pressure over it and rises up. It explodes as it forces itself out of fissures. This is why volcanoes are found close to fold mountain ranges.

(3) *Radioactivity* – A study of magma brings out that radioactive particles that are present in it. Some scholars think that the heat generated by these radioactive particles melts the magma and forces it out of the fissures. This hypothesis has many objections to meet. If the temperature

continues to rise in orogenetic belts, it should take millions of years for the lava to melt. Hence, if the orogenesis of mountains has any thing to do with volcanic explosion, the latter should have other reasons. How is it that the amount of heat which should be conducted out does not come if the soil layer is considered to be in melting condition?

(4) *Plate Tectonic* – Now most of the geologists agreed that there is a close relationship between the movements of plates of the earth and volcanic eruptions. Most of the active volcanoes are found along mid-oceanic ridges, which represent the splitting zones of divergent boundries in the zones of convergent (destructive) plates are found the explosive type of volcanoes.

World Distribution

At present there are known more than five hundred volcanoes of varying degree of activity. Many a times, greater is the number of extinct volcanoes which preserved their conical form and distributed over the surface of our planet most unequally but there are some belts or zones, where their occurrence is more. The main volcanic belts of the world are:

(1) *Circum-Pacific Belt* – About two-third of the world's active volcanoes of the world and thousands of inactive volcanoes are found in this belt. The volcanic belt surrounds the Pacific Ocean and also known as **'circle of fire or fire ring of the Pacific'.** It extends across the Kamchatka Peninsula, Kurilb Islands, the Islands of Japan, Philippines, New guinea. New Zealand and the Solomon Islands.

(2) *Mid-Continental or Alpine Himalayan Belt* – Volcanoes are not large in number in this belt. The belt from Madeira and the Canary Islands passes through the Vesuvius, Lipari Islands, the Etna and the Aegian volcanoes of the Mediterranean Sea. The Belt continues further through Caucasus (Edlburz), Armenia (Ararat), Iran (Damvand) and reaches Bilochistan. After crossing the Himalayas the belt crosses Yunnan, Burma, Andamans and probably ends in Indonesian Islands.

(3) *African Rift Valley* – This starts from the Cape of Guinea to Red Sea passing through the middle of Africa. The most active volcano in this belt is Cameroon, Mt. Kilimanjaro; and Kenyan volcanic mountains are beyond this Rift Valley but Algon is within this belt.

Figure 22.2 *Geographic Distribution of Volcanoes*

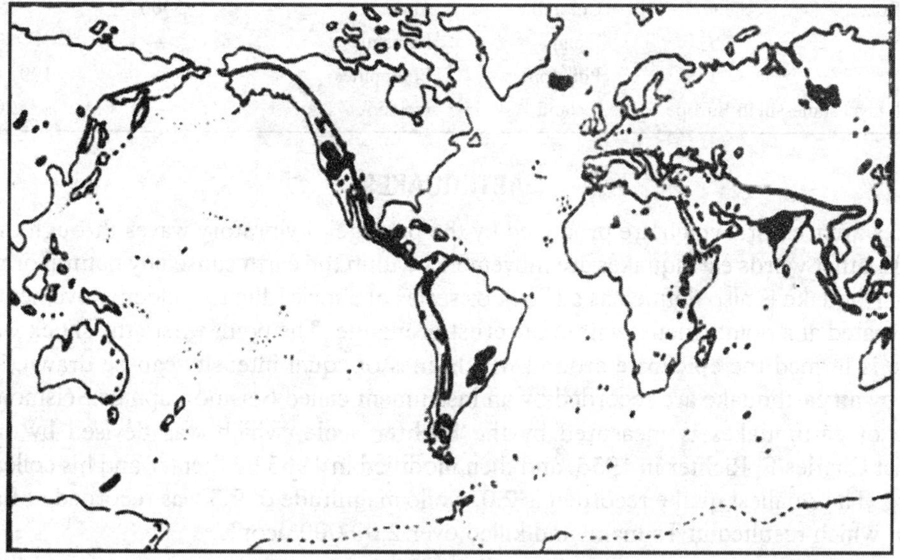

(4) *Other Volcanic Areas* – Some volcanoes are situated in mid-Atlantic. Cape verde is famous among them. Azores, St. Helina, and Ascencion are volcanic Islands but they are mostly inactive. The place where the ridge starting from Britain to Greenland meets the above mentioned belt had very active volcanoes. For example, Iceland has more than 20 very active volcanoes.

The volcanoes of the Indian Ocean are Comoro, Mauritius and Reunion which are situated close to Madagascar. There are many active volcanoes close to Antarctica which are situated at the boundary of Ross Sea. Arebus and Mount Terror are important volcanoes in it. The Hawaii, Galapagos, etc. are situated to the north and Juan Fernandez to the south of equator. Ester Islands in the east and Samao in the west are other important volcanic areas.

Main Active Volcanoes of the World

Name	*Country*	*Location*	*Height (m)*	*Eruption Year*
Ana	Indonesia	Karakotoa	155	1929
Guallatiri	Chile	Andes	606	1960
Mt. Cameroon	Cameroon	Monarch	278	1959
Fuego	Guatemala	Sierra Madre	3836	1962
Surtsey	Iceland	S.E. Iceland	173	1963
Agung	Indonesia	Bali	3142	1964
Tupangatiti	Chile	Andes	5640	1964
Lascar	Chile	Andes	5641	1968
Klya chevskaya	Russia	Sredinny Khsebet	4850	1974
Cotopax	Equador	Andes	5897	1975
Frebus	Antarctica	Ross Iceland	3795	1975
Sangay	Equador	Andes	5230	1976
Nyiragongo	Zaire	Virunga	3470	1977
Mt. Etna	Italy	Sicily	3308	1979
Mauna Loa	USA	Hawaii	4170	1978
Purace	Colombia	Andes	4590	1977
Ojos del Saldo	Argentina Chile	Andes	6885	1981
Navado del Ruiz	Colombia	Andes	5400	1985
Mt. Unzen	Japan	Honshu	—	1991
Mt. Mayon	Philippines	Philippines	—	1991 & 1993
Mt. Eyjafjorll tavel stanessill in Europe	Iceland	Iceland	—	2010

EARTHQUAKES

Earthquakes are tremors which are produced by the passage of vibratory waves through the rocks of the earth. In other words earthquakes are movements within the earth caused by natural or man-made stresses. Earthquake is also defined as a shock or series of shocks due to sudden movement of crystal rocks, generated at a point (focus) within the crust or mantle. The point where the shock waves reach the surface is termed the **epicentre** around which lines of equal intensity can be drawn. The waves generated by an earthquake are recorded by an instrument called Seismograph or Seismometer. The magnitude of earthquakes is measured by the **Richter scale**, which was devised by an eminent seismologist Charles F. Richter in 1935, and then modified in 1965 by Richter and his colleague Beno Guttenberg. The smallest quake recorded is 2.0, while magnitude of 9.3 was recorded at Banda Aceh (Sumatra), which resulted in Tsunami and killed over 2,00,000 people.

Causes of Earthquakes

The main causes of the origin of earthquakes are:

(i) Volcanic Activity
(ii) Faulting
(iii) Plate Tectonics
(iv) Anthropogenic Forces

Volcanic Activity

The volcanic eruption is one of the main causes of earthquakes. The crust of the earth which has fissures for the emission of lava and hot gases is subject to great pressure. The great pressure caused by emissions creates an earthquake wave in the surrounding areas. Such earthquakes occur either simultaneously with eruption or more commonly in the period proceeding on eruption. Many earthquake areas of the world are located in volcanic belts, which support the relation between earthquake and volcanic activity.

Faulting

A fault is a fracture in a rock and earthquakes often occur when movement of the rock takes place along a line of fracture called fault. When a new fault is formed or there is a movement along an old fault earthquake always results. The San Andreas Fault of California is a typical example which led to earthquakes in 1906. The 1975 earthquake in Guatemala resulted from movement along the Motagua Fault between the American and the Cariblam plates. The massive earthquake of 11 March 2011 in Japan was the result of subduction of the pacific plate underneath Japan (N. Honshu) at Japan Trench and dips to the west beneath the Eurasian plate. Several other examples also support faulting as a main cause of earthquakes.

Plate Tectonics

The earth's crust consists of seven major and several minor plates, which are not static but constantly moving relative to one another. Most of the earthquakes occur along the boundaries of major and minor plates. The major zones of earthquakes coincide almost with the boundaries between divergent plates.

Anthropogenic Causes

Apart from above mentioned causes, there are certain human activities which accelerate the occurrence of earthquakes in various parts of the world. The extraction of minerals, deep underground mining, blasting of rocks by dynamites for construction of roads, dams, nuclear explosions, etc. lead to the occurrence of earthquakes of various intensities and magnitudes. The Koyna earthquake of 1967 in Satara district of Maharashtra was mainly due to the Koyna Reservoir constructed in 1962. The other such examples are Mangla Dam (Pakistan), Kariba Dam (Zambia), Monuic Dam (Canada), Kurobe Dam (Japan), Nourek Dam (Russia), etc.

World Distribution

Earthquakes do not occur with a uniform frequency over the globe. In some places they are frequent and attain great intensity, in others they are rare and feeble. Strong and frequent earthquakes occur in such countries as Italy, Greece, Turkey, Iran, Pakistan, Northern India, Indochina, Indonesia, China, Japan, New Zealand, Philippines, the western mountain belt of North and South America, and in Central America.

In general, high seismic activity distinguishes in the following belts:

1. The Circum-Pacific Belt
2. The Mid-Atlantic Belt
3. The Mid-Continental Belt
4. Other Areas

Figure 22.3 *Major Foci of Earthquakes in the World*

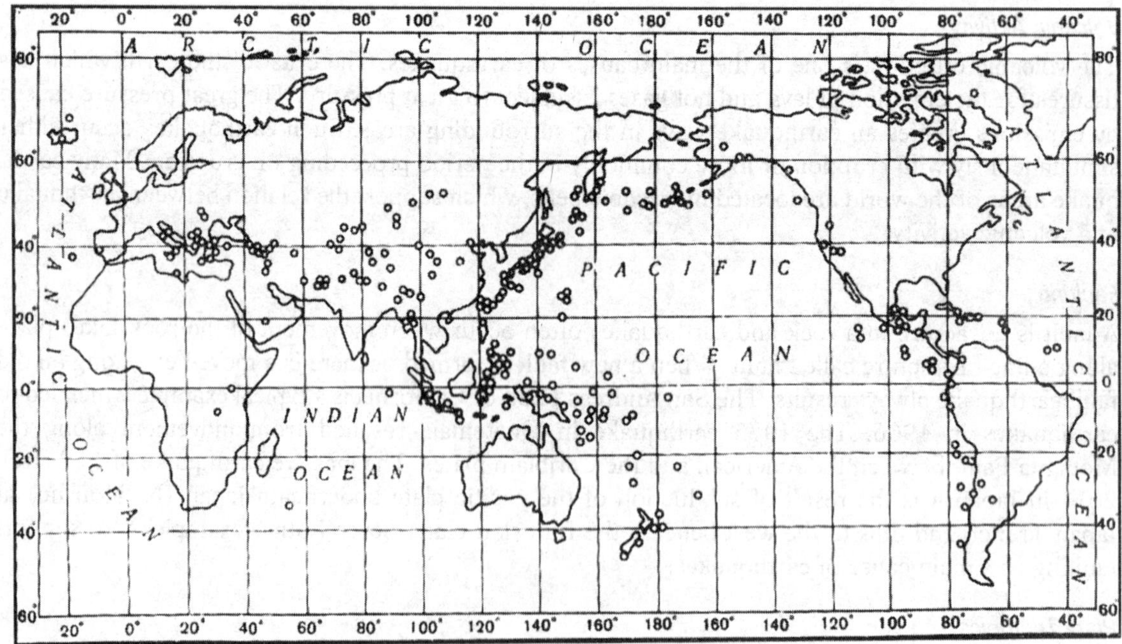

1. *The Circum-Pacific Belt:* This belt extends in the west of Alaska to Kurile, Japan, Marina and the Philippine trenches beyond which it is divided into two branches, one going towards Indonesia and other towards Keramac-Tonga trench to the north west of New Zealand. On eastern side of the Pacific, it follows the west coast of North America and continues southward along Peru, chief trench on the west coast of South America. It has been found that more than 75 per cent used to occur in this belt.
2. *The Mid-Atlantic Belt:* This earthquake belt extends along the mid-oceanic ridges and several islands near the ridges of the Atlantic Ocean. Earthquakes of moderate and mild intensity are recorded in this belt. The Rift Valley of East Africa and the Red Sea are considered as an extension of this belt.
3. *The Mid-Continental Belt:* This belt extends along Alpine Mountain system of Europe, North America, through Asia Minor, Caucasia, Iran, Afghanistan, Pakistan to the Himalayan mountain system including Tibet, Tienshan, Altai, the mountains of China, Myanmar and eastern Siberia.
4. *Other Areas:* The other areas of the occurrence of earthquakes are mostly associated with inter-plate seismicity. The zone of earthquakes in East Africa, Western United States and the Peninsular India are included in this category. In India three zones are considered as earthquake prone areas; these are the Himalayan region, Northern plains and the coastal areas of the southern Peninsula.

Important Earthquakes of the World
(List in Decanting Order as per Reported Deaths)

Date UTC	Location	Deaths	Magnitude
January 12, 2010	Haiti region	3,16,000	7.0
July 27, 1976	Tangshan, China	2,42,769	7.5
December 26, 2004	Sumatra	2,27,898	9.1
December 16, 1920	Haiyuan, Ningxia (Ning-hsia), China	2,00,000	7.8
September 1, 1923	Kanto (Kwanto), Japan	1,42,800	7.9
October 5, 1948	Ashgabat (Ashkhabad), Turkmenistan (Turkmeniya, USSR)	1,10,000	7.3
May 12, 2008	Eastern Sichuan, China	87,587	7.9
October 8, 2005	Pakistan	86,000	7.6
December 28, 1908	Messina, Italy	72,000	7.2
May 31, 1970	Chimbote, Peru9	70,000	7.9
June 20, 1990	Western Iran	50,000	7.4
May 22, 1927	Gulang, Gansu (Kansu), China	40,900	7.6
December 26, 1939	Erzincan, Turkey	32,700	7.8
January 13, 1915	Avezzano, Italy	32,610	7.0
December 26, 2003	Southeastern Iran	31,000	6.6
May 30, 1935	Quetta, Pakistan (Baluchistan, India)	30,000	7.6
January 25, 1939	Chillan, Chile	28,000	7.8
December 7, 1988	Spitak, Armenia	25,000	6.8
February 4, 1976	Guatemala	23,000	7.5
March 11, 2011	Japan	20,896	9.0
January 26, 2001	Gujarat, India	20,085	7.6
May 10, 1974	China	20,000	6.8
April 4, 1905	Kangra, India	19,000	7.5
August 17, 1999	Turkey	17,118	7.6
February 29, 1960	Agadir, Morocco	15,000	5.7
September 16, 1978	Iran	15,000	7.8
September 1, 1962	Bu'in Zahra, Qazvin, Iran	12,225	7.1
October 21, 1907	Qaratog (Karatag), Tajikistan (Turkestan, Russia)	12,000	8.0
July 10, 1949	Khait, Tajikistan (Tadzhikistan, USSR)	12,000	7.5
August 31, 1968	Dasht-e Bayaz, Iran	12,000	7.3
January 15, 1934	Bihar, India-Nepal	10,700	8.1
August 10, 1931	Near Fuyun (Koktokay), Xinjiang (Sinkiang), China	10,000	8.0
January 4, 1970	Tonghai, Yunnan Province, China	10,000	7.5
September 29, 1993	Latur-Killari, India	9,748	6.2
September 19, 1985	Mexico, Michoacan	9500	8.0
August 25, 1933	North of Maowen, Sichuan (Szechwan), China	9,300	7.5
April 25, 2015	Nepal	8,693	7.8
January 15, 1944	San Juan, Argentina	8,000	7.4
August 16, 1976	Mindanao, Philippines	8,000	7.9
January 23, 1909	Silakhor, Iran (Persia)	6,000	7.3
March 16, 1925	Near Dali (Talifu, Ta-li), Yunnan, China	5,800	7.0
May 26, 2006	Indonesia	5,749	6.3

Cont'd...

...Cont'd

Date UTC	Location	Deaths	Magnitude
January 16, 1995	Kobe, Japan	5,502	6.9
December 28, 1974	Pakistan	5,300	6.2
April 10, 1972	southern Iran	5,054	7.1
August 5, 1949	Ambato, Ecuador	5,050	6.8
December 23, 1972	Nicaragua, Managua	5,000	6.2
November 24, 1976	Turkey-Iran border region	5,000	7.3
October 10, 1980	El Asnam, Algeria (formerly Orleansville)	5,000	7.7
December 16, 1902	Andijon (Andizhan), Uzbekistan (Turkestan, Russia)	4,700	6.4
October 3, 1914	Burdur, Turkey (Ottoman Empire)	4,000	7.0
November 26, 1943	Ladik, Turkey	4,000	7.6
November 27, 1945	Makran Coast, Pakistan (Baluchistan, India)	4,000	8.0
May 30, 1998	Afghanistan-Tajikistan Border Region	4,000	6.6
August 17, 1906	Valparaiso, Chile	3,882	8.2
May 1, 1929	Koppeh Dagh, Iran (Persia)	3,800	7.2
June 28, 1948	Fukui, Japan	3,769	7.3
April 28, 1903	Malazgirt, Turkey (Ottoman Empire)	3,500	7.0
March 24, 1923	Near Luhuo, Sichuan (Szechwan), China	3,500	7.3
April 20, 1935	Miao-li, Taiwan (Formosa)	3,270	7.1
March 7, 1927	Tango, Japan	3,020	7.6
April 18, 1906	San Francisco, California	3,000	7.8
March 2, 1933	Sanriku, Japan	3,000	8.4
July 25, 1969	Yangjiang, Guangdong, China	3,000	5.9
June 11, 1981	southern Iran	3,000	6.9
August 9, 1912	Murefte, Turkey (Ottoman Empire)	2,800	7.4
April 27, 1931	Zangezur Mountains, Armenia-Azerbaijan border (Armeniya-Azerbaydzhan, USSR)	2,800	5.7
December 13, 1982	Yemen	2,800	6.0
February 1, 1944	Gerede, Turkey	2,790	7.4
July 16, 1935	Hsin-chu (Shinchiku), Taiwan (Formosa)	2,740	6.5
November 23, 1980	southern Italy	2735	6.5
August 19, 1966	Varto, Turkey	2,529	6.8
May 6, 1930	Salmas, Iran (Persia)	2,500	7.2
March 31, 1931	Managua, Nicaragua	2,500	6.0
December 12, 1992	Flores Region, Indonesia	2,500	7.5
September 20, 1999	Taiwan	2,400	7.6
February 4, 1998	Hindu Kush region, Afghanistan	2,323	5.9
September 6, 1975	Turkey	2,300	6.7
May 21, 2003	Northern Algeria	2,266	6.8
May 25, 1923	Torbat-e Heydariyeh, Iran (Persia)	2,200	5.7
April 13, 2010	Southern Qinghai, China	2,200	6.9
July 17, 1998	Papua New Guinea	2,183	7.0
April 19, 1902	Quezaltenango and San Marcos, Guatemala	2,000	7.5
February 4, 1975	Haicheng, China	2.000	7.0

Cont'd...

...Cont'd

Date UTC	Location	Deaths	Magnitude
October 19, 1991	Northern India	2,000	7.0
May 27, 1995	Sakhalin Island	1,989	7.5
January 12, 1945	Mikawa, Japan	1,961	7.1
July 30, 1917	North of Daguan, Yunnan, China	1,800	7.5
May 22, 1960	Temuco-Valdivia, Chile	1,655	9.5
July 16, 1990	Luzon, Philippine Islands	1,621	7.7
May 10, 1997	Northern Iran	1,567	7.3
August 15, 1950	Near Zhamo (Rima), Xizang (Tibet), China"Assam-Tibet" Earthquake	1,526	8.6
January 20, 1917	Bali, Indonesia	1,500	
March 4, 1977	Romania	1,500	7.2
July 28, 1981	southern Iran	1,500	7.3
July 23, 1930	Irpinia, Italy	1,404	6.5
November 10, 1946	Ancash, Peru	1,400	7.3
December 20, 1946	Nankaido, Japan	1,362	8.1
October 30, 1983	Turkey	1,342	6.9
March 28, 2005	Northern Sumatra, Indonesia	1,313	8.6
May 31, 1946	Ustukran, Turkey	1,300	5.9
March 16, 1906	Chia-i, Taiwan	1,250	6.8
September 9, 1954	Chlef (Orleansville, El Asnam), Algeria	1,250	6.8
July 2, 1957	Near Sang Chai, Mazandaran, Iran	1,200	7.1
September 10, 1943	Tottori, Japan	1,190	7.4
January 25, 1999	Colombia	1,185	6.1
December 13, 1957	Sahneh, Iran	1,130	7.1
September 30, 2009	Southern Sumatra, Indonesia	1,117	7.5
December 20, 1942	Erbaa, Turkey	1,100	7.3
July 26, 1963	Skopje, Former Yugoslav Rep. of Macedonia (Makedonija, Yugoslavia)	1,100	6.0
March 28, 1970	Gediz, Turkey	1,086	6.9
March 18, 1953	Yenice-Gonen, Turkey	1070	7.3
May 28, 1903	Gole, Turkey (Ottoman Empire)	1,000	5.8
January 31, 1906	Off coast of Esmeraldas, Ecuador	1,000	8.8
January 14, 1907	Kingston, Jamaica	1,000	6.5
February 13, 1918	Nan'ao, Guangdong (Kwangtung), China	1,000	7.4
November 10, 1940	Vrancea, Romania (Rumania)	1,000	7.3
August 2, 1951	Cosiguina, Nicaragua	1,000	5.8
March 7, 1966	East of Longyao, Hebei (Hopeh), China	1,000	7.0
March 22, 1966	Southeast of Ningjin, Hebei (Hopeh), China	1,000	6.9
May 22, 1971	Turkey	1,000	6.9
May 6, 1976	Northeastern Italy	1,000	6.5
October 10, 1986	El Salvador	1,000	5.5
March 6, 1987	Colombia-Ecuador	1,000	7.0
August 20, 1988	Nepal-India border region	1,000	6.8
March 25, 2002	Hindu Kush Region, Afghanistan	1,000	6.1

Source: USGS website.

23

Social Geography of the World

A Note on Culture, Society and Geography

In popular perception the distinction between 'Culture' and 'Society' is not very clear. To some extent these two terms although different but denote similarity. The emerging disciplines of anthropology and sociology are focused on 'Culture' and 'Society' respectively, but their definitions of these concepts do not differ in any significant way. According to Tylor (1871) culture is, 'that complex whole which includes knowledge, belief, art, morals, law, customs and any other capabilities and habits acquired by man as a member of society'. While society or social system on the other hand, would refer to the 'specifically relational system of interaction among individuals and collectivities' (Krolbes and Parsons, 1958). In fact, these two terms are distinct but similar in several ways. In Geography both these terms are used in more or less similar ways. North American geographers who adhere to Sauer's landscape school have employed the term 'Culture' rather than 'Society' in their analysis of human landscape; British and European geographers have done the reverse using 'Society' rather 'Culture'. In fact, there is no clear and absolute distinction between these two terms. The social geography is the analysis of the social patterns and processes arising from the distribution of, and access to, scarce resources and – an examination of social causes of and suggested solutions to, social and environmental problems (Eyles, 1974).

In this chapter, the three aspects of social geography will be discussed in brief, these are Language, Religion and Tribes.

LANGUAGE

Language is defined as a systematic means of communication between people, in other words, it is a source of social contact among people and communities. Most societies transmit their language in vocal and written form. Language, probably the single most important human achievement, is of interest to human geographers because (i) language is a cultural variable, (ii) it is useful in delimiting groups and hence regions, and, (iii) it helps in interactions with environment, both physical and human. Spatial variations in language are caused in part by variations in physical and human environments.

For many groups, language is the primary basis of identity – hence the close links between language and nationalism, the desire to preserve minority languages, and even the various efforts that have been made to create a universal language. A common language facilitates communications; different languages create barriers and frictions between groups, further dividing our divided world.

World Distribution of Language Families

Like many cultural variables, language began as a single entity (or at most a few different ones) and diversified into many. Of the roughly 7,000 distinct languages (not counting minor dialects) that existed four hundred years ago, approximately 1,000 have disappeared, leaving about 6,000 languages today. Languages of the world show clearly some language families or groups of closely related languages that suggest a common origin and their gradual spreading with the passage of time. As human groups moved, their language changed. Table 23.1 provides a numerical summary of the number of speakers in each of the 19 languages families, while Table 23.2 identifies the eight languages spoken by the most people.

Table 23.1 *Language Families of the World*

Language Family	*Number of Languages*	*Number of Speakers (000)*
Indo-European	144	2,300,000
Sino-Tibetan	258	1,200,000
Altaic	63	250,000
Niger-Kordofanian	1,064	1,81,000
Austronesian	959	1,80,000
Afro-Asiatic	241	1,75,000
Dravidian	28	1,45,000
Daic	57	50,000
Uralic-Yukaghir	24	22,000
Amerind	583	18,000
Nilo-Saharan	138	11,000
Austric	4	7,000
Caucasian	38	5,000
Indo-Pacific	731	2,735
Na-Dene	34	202
Khoisan	31	120
Lnuit-Aleut	9	85
Australian	170	30
Chukchi-Kamchatkan	5	23

Source: Adapted and updated from M.C. Ruhlen, a guide to the world's languages: volume 1, classification (Stanford: Stanford University Press, 1987)

Table 23.2 *Numbers of Speakers, Major Languages in the World*

Language	*Number of Speakers (million)*	*Language*	*Number of Speakers (million)*
Mandarin	1052	Russian	277
English	508	Arabic	246
Hindi	487	Bengali	211
Spanish	417	Portuguese	191

Source: The world almanac and book of facts, 1999

The largest language family is the Indo-European, followed by the Sino-Tibetan. Mandarin has more speakers than any other language, with English ranking second and Hindi third. The other languages which are significant in terms of number of speakers are Spanish, Russian, Arabic, Bengali and Portuguese.

The Indo-European family is spatially dispersed today, while the Sino-Tibetan is essentially limited to one is the difference which reflects the fact that many Indo-European speaking countries colonized other parts of the world, whereas Sino-Tibetan speaking countries did not. Thus the numerical importance of Mandarin is attributable simply to the large population of China. Similarly, the numerical importance of Hindi reflects the large population of India. A brief continent-wise review of language pattern is as follows:

Europe

Except for Germany and Austria, and the miniature states, no two countries in Europe have the same language structure. In some instances, a language group extends beyond the political boundary line, but in such cases the country into which it penetrates has more than one official language or the penetration is of only minor significance.

Within the main groups, the languages are often so similar that written or sometimes even spoken words of one language can be understood by persons having other mother tongues within the same group. The pattern is fairly simple: Germanic languages in the northwestern and central parts of Europe; Romanic languages in the west, southwest, and south-central sections; and Slavic languages in the east.

The only complication is that Romanian language of the Latin stock (a remnant of the Roman Empire) and Hungarian (which was brought in from the East) separate a north Slavic wing, which reaches to the Order River and to Bohemia, from a southern wing, which reaches to the coast of the Adriatic. The language of the Magyars is related to that of Finland and Estonia, both brought by Asiatic invaders in historic times. Gaelic, a Celtic tongue, is still spoken in parts of Wales, Western Scotland, Ireland, and Western Brittany; it has been pushed westward by the Germanic and Latin group. The existence of other, smaller linguistic areas, such as the Greek, the Albanian, the Basque which is located *a cheval* of the Pyrenees, the Lapps, and the Baltic group combining Lithuanian and Latvian. A special case, too small to be shown, is Malta where the influence of the Phoenician dialect still exists.

Switzerland and Belgium are examples of European linguistic complexity. In both these countries two or more unrelated languages are used. Switzerland has four official languages – German, French, Italian and Rhaeto-Roman. In Belgium, language boundary separates the Germanic Flemish from the Latinized Walloons.

American Countries

Simplicity of the linguistic pattern has been an important factor in the international contact between the American countries. From Hudson Bay to Cape Horn, three chief languages are spoken nationally. English is spoken in the north, Spanish and Portuguese in the south; the United States – Mexico boundary line marks the sharp lingual divide. These three are the languages of the original invaders. They were adopted by the native Americans who survived the colonial struggles. In addition each of the three languages, within its own sphere, was in general strong enough to dominate the languages of all immigrants who came later. In the United States, for example, the offspring of foreign parents have generally lost command of their parents' original tongue; in many cases they do not understand it at all.

There are a few exceptions to these generalizations. In the Province of Quebece, French has maintained itself as a remnant of early colonial enterprise and migration. Similarly, French, English, or Dutch is spoken on various West Indian Islands and in the Guianas, while group of Germans in Brazil and Chile and Italians in Argentina have kept the use of their mother tongue; in the latter case, Italian has, to some extent, influenced Argentine Spanish sufficiently to make it a mixture of the two languages. These are, however, minor exceptions to the overwhelming linguistic uniformity on the two American continents.

The present Anglo-Latin linguistic divide shows every evidence of maturity. Efforts have failed to make Puerto Rico, as an American colony, bilingual through the use of English in the schools, although English is widely understood.

Asia

No attempt has been made to show the language pattern of Asia because of its great complexity. For example, in India and Pakistan alone there are 222 different languages, not to mention the numerous dialects. In Indonesia, where a similar linguistic confusion exists, Malay is recognized as the *lingua-franca,* that is, a common or commercial tongue used among peoples of different speech. In the Philippines, where 64 natives' languages and dialects are spoken, Tagalog is now the official national language although English and even Spanish are used to a extent deal. An interesting situation exists in China. Chinese of various parts of the country can understand each other when they write, but not when they speak. The reason is that the written language uses symbols for each word, entirely independent from the spoken tongue.

Africa

The various colonies of Africa have a veneer of bilingualism, using to some extent the language of the governing countries. This does not affect the bulk of the native people, however, who continue to speak their own various languages.

In South Africa there is a mixture of Boers, who speak Afrikaans, which is related to Dutch, and of British, who speak English. Except in Natal, where the English element prevails, the population is a mixture of the two ethnic groups, with a Boer majority. Both languages, Afrikaans and English, are recognized by the state, and all state officials are required to speak both. The English find it rather difficult to accept this situation. After winning the Boer War they expected an all-English colony, and instead found themselves in the minority. But circumstances have forced them to cooperate, and out of this cooperation will come the real South Africans, who will speak neither Afrikaans nor British alone, but will use both languages with equal facility.

Language in Landscape and Landscape in Language

Language plays a key role in landscape, in the form of place names or **toponyms**. The place names are given to understand and give meaning to landscape. A landscape without meaning would be like a group of people without names. Place names are a significant feature of human-made landscapes often visible in the form of road signs and integral components of maps. Analysis of place names can provide information about both spatial and the social origins of settlers and also valuable for understanding the cultural history of an area.

Not only is language in landscape, but landscape is in language. First, because physical barriers tend to limit movement, there is a general relationship between language distribution and physical regions. Second, the vocabulary of any language necessarily reflects the physical environment in which its speakers live. As languages move to new environments, it also becomes necessary to add new words. Finally, the human landscape is in language to the extent that language reflects class and gender within a given language.

Religion

Religion, or at least some type of belief system, pervades almost all aspects of daily living from eating and working habits to marital relationships, child rearing, views regarding fertility and mortality, and attitudes concerning utilization and management of natural resources. Thus, religions provide most of the standards of basic human conduct. Religion can serve as a catalyst to encourage change in a society, or conversely it can be conservative and promote traditional customs and morals. Religion has affected the environment and vice versa.

Major Religions and their World Distribution

The four major religions of the world are Christianity, Islam, Hinduism and Buddhism. Table 23.3 shows a detailed continent-wise number of followers of the major world religions:

Table 23.3 *Major World Religions: Number of Adherents (Thousands), 2001*

Religion	*Africa*	*Asia*	*Europe*	*Latin America*	*North America*	*Oceania*	*Total*	*%*
Christianity	3,68,244	3,17,759	5,59,359	4,86,591	2,61,752	25,343	2,019,052	33.0
Islam	3,23,556	8,45,341	31,724	1,702	4,518	307	1,207,148	19.7
Hinduism	2,384	8,13,396	1,425	775	1,350	359	8,19,689	13.4
No religion	5,170	6,11,876	1,05,742	16,214	28,994	3,349	7,71,345	12.6
Chinese folk religions	33	3,85,758	258	197	857	64	3,87,167	6.3
Buddhism	139	3,56,533	1,570	660	2,777	307	3,61,985	5.9
Ethnic religions	97,762	1,29,005	1,258	1,288	446	267	2,30,026	3.8
Atheism	432	1,22,408	22,555	2,787	1,700	369	1,50,252	2.5
New religions	29	1,01,065	160	633	847	67	1,02,801	1.7
Sikhism	54	22,689	241	0	535	19	23,538	0.2
Judaism	215	4,476	2,506	1,145	6,045	98	14,484	0.2
Spiritism	3	2	134	12.169	152	7	12,466	0.2
Baha'i faith	1,779	3,538	132	893	799	113	7,254	0.1
Confucianism	0	6,277	11	0	0	24	6,313	0.1

It becomes clear from the table that Christianity is the leading religion numerically, with 33 per cent of the world population followed by Muslims (Islam) at 20 per cent Hindus at 13 per cent and Buddhist at 6 per cent. A high 13 per cent are classed as non-religions.

Figure 23.1 shows the distribution of the four great religions-Christianity, Islam, Hinduism and Buddhism in a general and much simplified way. It does not show the distribution of minorities.

The map however, pictures clearly the general religious pattern: Christianity in Europe, the Americans, South Africa, Australia, and the Philippines; the Mohammedan core extending from the Atlantic coast of northwest Africa into central Asia and towards Indonesia and the east coast of Africa; Hinduism in India and Buddhism in eastern and southeastern Asia.

There are sub-divisions within the main religious divisions. For example, the Roman Catholics, the Protestants, and the Orthodox, are all parts of the Christian faith; the Sunnite-Shiite schism split the Islam world; and the many forms of Buddhism subdivide religious sects in eastern Asia. Some of these subdivisions are highly centralized, such as the Roman Catholic under the Pope and Mohammedans under the Caliphs during the middle Ages. Others are broken up into many smaller groups, as is the case among the Protestants with their innumerable divisions.

Figure 23.1 *Distribution of Four Great Religions in the World*

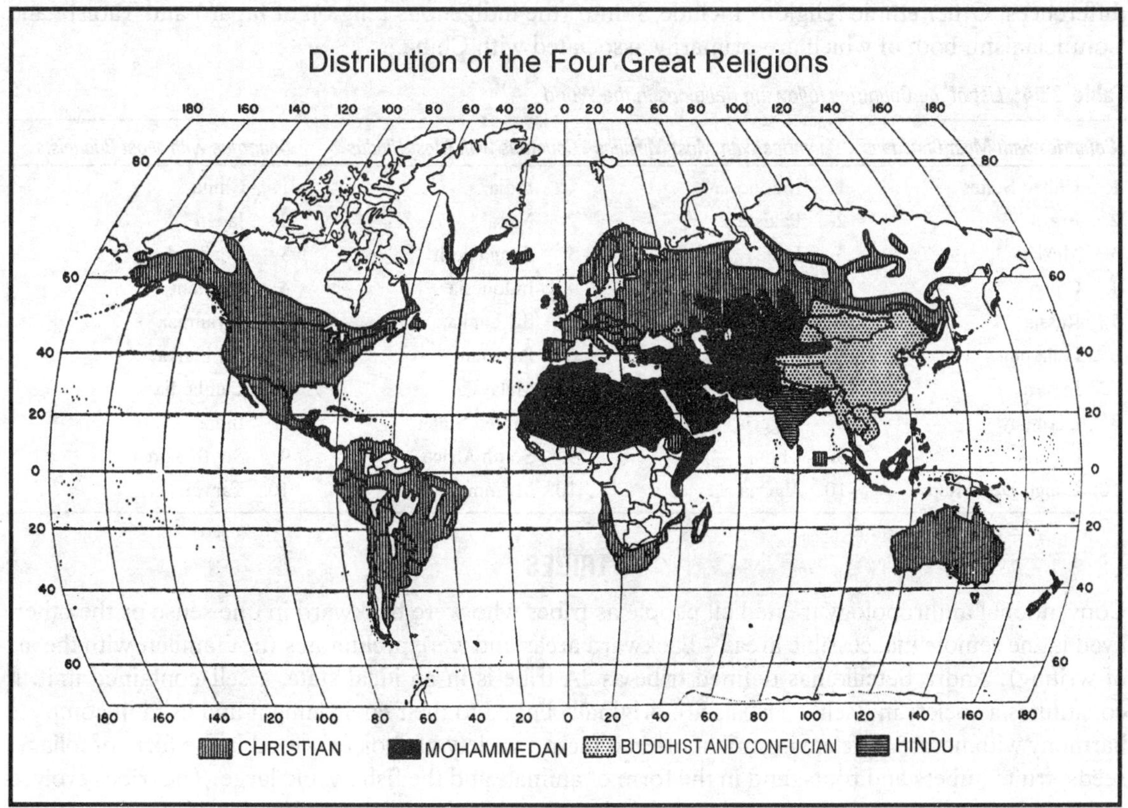

Ethnic Religions

The principal ethnic religions emerged earlier than their universalizing counterparts. They are associated with a particular group of people and do not actively seek to convert others. Of the several hundred religions in the category, the largest is Hinduism, which evolved in the Indo-Gangetic hearth-a lowland area of north India that drains into the Indus and Ganges rivers-about 2000 BC. Hinduism was the first major religion to evolve in this area, from which it spreads east down the Ganges and then south through India, eventually to dominate the entire region. As it diffused, it absorbed and blended with other religions beliefs. From India it was carried overseas, but it has not retained significant numbers of adherents outside India. Today, Hinduism in an Indian religion associated with a specific place and specific cultural group. It is polytheistic (worshipping more than one God) and the close ties to the rigid social stratification of the caste system. Like many other religions, Hinduism has spawned numerous offshoots; Jainism, for example, rejects Hindu rituals but shares many basic tenets of Hinduism, including the belief in reincarnation and *ahimsa* (the ethical doctrine that humans ought to avoid hurting any living creature). A more recent off shoots is Sikhism that arose about 500 years ago in the Punjab region in India.

A second ethnic religion is Judaism. The first monotheistic religion (worshipping one single God), Judaism originated about 2000 BCE in the Near East. Following the Romans' destruction of Jerusalem in 70 CE the Jews were driven out of their homeland and eventually dispersed throughout Europe, the entire body of Jews living outside Israel, in Europe and elsewhere, is known as the Diaspora. Only in 1948 was the long-term goal of a Jewish homeland achieved with the creation of the state of Israel. Judaism contains significant internal divisions reflecting theological and ideological differences. Other ethnic religions include Shinto (the indigenous religion of Japan) and Taoism and Confucianism, both of which are primarily associated with China.

Table 23.4 *List of the Countries following Religions in the World*

Countries with Most Christians	*Countries with Most Muslims*	*Countries with Most Hindus*	*Countries with Most Buddhists*
1. United States	1. Indonesia	1. India	1. China
2. Brazil	2. Pakistan	2. Nepal	2. Japan
3. Mexico	3. India	3. Bangladesh	3. Thailand
4. China	4. Bangladesh	4. Indonesia	4. Vietnam
5. Russia	5. Turkey	5. Sri Lanka	5. Myanmar
6. Philippines	6. Egypt	6. Pakistan	6. Sri Lanka
7. India	7. Iran	7. Malaysia	7. Cambodia
8. Germany	8. Nigeria	8. United States	8. India
9. Nigeria	9. China	9. South Africa	9. South Korea
10. Congo Dem. Rep.	10. Algeria	10. Myanmar	10. Taiwan

TRIBES

Conventional anthropology treated all people as tribes who were backward in one sense or the other, lived in the remote inaccessible areas – backward areas and were preliterates (not familiar with the art of writing). Andre Beteille has defined tribe as "A tribe is in an ideal state, a self-contained unit, it constitutes a society in itself". Tribals are originally linked to their environment and lived in complete harmony with nature. The environment offered them a variety of choices – food in the form of foliage, seeds, fruits, tubers and roots, and in the form of animals and the fish by and large. The tribes evolved a simple technology to deal with the natural situations. Their social organization is simple, without any stratified or hierarchical social order. These are the indigenous people and their estimated population is more than 250 million. In general a tribe can be defined as an aggregate of people linked together by means of social ties, a common language and the recognition of a common cultural heritage.

Tribal societies are distributed in all parts of the world and still have their own identity in socio-economic structure.

Table 23.5 *Indigenous People of the World*

Continent/Country	*Indigenous People*
1. Africa	Afar, Babinga, Bantu, Bongo, Bobo, Bushman, Dinka, Efe, Fulani, Hottentot, Kikuyus, Lese, Mabuti, Masai, Nuer, Oromo, San, Shilluk, Tigmyon, Tswa, Tuareg, Twa, Zulu
2. Asia	Ainu, Asmat, Badawins, Chakma, Chin, Dani, Dayak, Hamang, Haren, Iban, Ifugao, Kachin, Kalinga, Karen, Kayan, Kedang, Kurds, Lisu, Marmo, Pakhtoons, Sakai, Semang, Shah, Vedda
3. Australia	Arunta, Bidjandjara, Dieri, Gurindji, Kadaku, Kamilaroi, Murngin, Tiwi, Wailbri
4. Europe/Russia	Chuckchi, Eskimo, Koryak, Lapps, Nemat, Somoyed, Tungus, Saami, Yakuts, Yukaghir
5. North America a. Canada	Bela, Blackfoot, Coola, Deni, Cree, Haida, Innu, Inuit, Meti, Tlinit
b. USA/Mexico	Apache, Araphao, Chol, Comanachee, Crow, Dine, Hopi, Keres, Lecando, Maya, Navajo, Nez, Pawnee, Perce, Uti, Yucatec
6. South America	Achi, Akrori, Arawak, Aymara, Boro, Kareen, Kayapo, Lakona, Mapuchi, Maskoy, Secoya, Toba.

Detailed List of the Tribes of the World

Africa

Central Africa

Central Africa generally includes the lands mainly of the Congo River basin, south of the Sahara and west of the East African Rift.

- Pygmy Peoples: Central and Western Africa

East Africa

East Africa, including the African Great Lakes region and the Indian Ocean Islands.

- Acholi: Uganda
- Alur: Uganda
- Ambo: Zambia
- Ankole: Uganda
- Antalot: Comoros
- Aushi: Zambia
- Awee: Kenya
- Babongo: Gabon
- Baganda: Uganda
- Bahima/Ankole: Uganda, Rwanda
- Bagisu: Uganda
- Bagwere: Uganda
- Bakiga: Uganda
- Bakonjo: Uganda
- Basoga: Uganda
- Batoro: Uganda
- Bemba: Zambia
- Betsileo: Madagascar
- Bisa: Zambia
- Bunyoro: Uganda
- Cafre: Comoros
- Chagga: Tanzania
- Chewa: Malawi
- Chilkunda: Zambia
- Chokwe: Mozambique, Zambia
- Chopi: Mozambique
- Chisinga: Zambia
- Gova: Zambia
- Hadzabe: Tanzania
- Haya: Tanzania
- Hehe: Tanzania
- Hutu: Burundi, Rwanda
- Ila: Zambia
- Kunda: Zambia
- Kwandi: Zambia
- Kwandu: Zambia
- Kwangwa: Zambia
- Lala: Zambia
- Lamba: Zambia
- Lango: Zambia
- Lenje: Zambia
- Leva: Zambia
- Lima: Zambia
- Liyuwa: Zambia
- Lomwe: Malawi
- Lozi: Zambia
- Luano: Zambia
- Lucazi: Zambia
- Lugbara: Uganda
- Luhya: Kenya
- Lumbu: Zambia
- Lunda: Zambia
- Lundwe: Zambia
- Lungu: Zambia
- Lou: Kneya, Tanzania
- Luvale: Zambia
- Luunda: Zambia
- Maasai: Kenya and Tanzania
- Makoa: Comoros
- Makoma: Zambia
- Makonde: Mozambique
- Makua: Mozambique
- Mambwe: Zambia
- Manyika: Mozambique
- Mashasha: Zambia
- Mashi: Zambia

Cont'd...

...Cont'd

- Inamwanga: Zambia
- Iteso: Uganda
- Iwa: Zambia
- Jopadhola: Uganda
- Kabende: Zambia
- Kalenjin: Kenya
- Kamba: Kenya
- Kosa: Zambia
- Mulonga: Zambia
- Munyoyaya: Kenya
- Ndau: Mozambique
- Ndembu: Zambia
- Ng'umbo: Zambia
- Ngonde: Malawi
- Ngoni: Malawi
- Nguni: Mozambique
- Nkoya: Zambia
- Nsenga: Zambia
- Nyakyusa: Tanzania
- Nvamwezi: Tanzania
- Nvanja: Malawi, Zambia
- Nyengo: Zambia
- Nyiha: Zambia
- Ogiek: Kenya
- Oimatsaha: Comoros
- Pare: Tanzania
- Rundi: Uganda
- Rwanda: Uganda
- Sakalava: Uganda
- Seba: Zambia
- Sena: Malawi, Mozambique
- Senga: Zambia
- Sengwer: Kenya
- Shangana: Mozambique
- Shanjo: Zambia
- Shila: Zambia
- Shona: Mozambique
- Simaa: Zambia
- Soli: Zambia
- Subiya: Zambia
- Sukuma: Tanzania
- Swaka ; Zambia
- Mbowe: Zambia
- Mbukushu: Zambia
- Mbumi: Zambia
- Mbunda: Zambia
- Mbwela: Zambia
- Merina: Madagascar
- Meru: Kenya
- Mukulu: Zambia
- Unga: Zambia
- Wandya: Zambia
- Watha: Kenya
- Yao: Malawi, Mozambique
- Yiaku/Yaaku: Kenya
- Yombe: Zambia

Horn of Africa

The **Horn of Africa** includes Djibouti, Eritrea, Ethiopia and **Somalia**.

- Afar: Djibouti, Eritrea, Ethiopia
- Agaw: Ethiopia
- Akisho: Ethiopia
- Amhara: Ethiopia
- Anuak: Ethiopia
- Beja: Eritrea
- Berta: Ethiopia
- Bilen: Eritrea
- Borana: Ethiopia
- Daasanach: Ethiopia
- Dorze: Ethiopia
- Gumuz: Ethiopia
- Gurage: Ethiopia
- Hamar: Ethiopia
- Hedareb: Eritrea
- Jeberti: Ethiopia
- Kichepo: Ethiopia
- Kunama: Eritrea
- Me'en: Ethiopia
- Mursi: Ethiopia
- Nara: Eritrea
- Nuer: Ethiopia
- Nyangatom: Ethiopia
- Oromo: Ethiopia
- Qemant: Ethiopia
- Rashaida: Eritrea

Cont'd...

...Cont'd

- Swahili: Kenya, Tanzania, Mozambique, Uganda, Somalia, Comoros, Mayotte, Mauritius, Seychelles, Djiboti, Libya, South Africa, Zambia, Malawi, Madagascar, Eritrea, Ethiopia, Sudan, Rwanda, Burundi, Democratic Republic of Congo
- Tabwa: Zambia
- Tambo: Zambia
- Toka: Zambia
- Totela: Zambia
- Tumbuka: Malawi, Zambia
- Tonga: Malawi, Mozambique, Zambia
- Tutsi: Burundi, Rwanda
- Twa peoples (Pygmy): Angloa, Botswana, Burundi, Congo, Rwanda, Namibia, Zambia, Uganda
- Rer Bare: Ethiopia
- Saho: Eritrea
- Shanqella: Ethiopia
- Sidama: Ethiopia
- Silt'e: Ethiopia
- Somalis: Somalia, Djibouti, Ethiopia, Kenya
- Suri: Ethiopia
- Tigre: Eritrea
- Tigray-Tigrinya people (Tigrinya): Eritrea
- Tigray-Tigrinya people (Tigray): Ethiopia
- Tirma: Ethiopia
- Welayta: Ethiopia
- Zay: Ethiopia

North Africa

North Africa generally includes African countries with borders on the Mediterranean and northern Red Sea and Atlantic Ocean, bounded largely by the Sahara Desert to the south. Main article: Ethnic groups of North Africa are:

- Berber (or Amazigh): Morocco, Tunisia, Algeria, Libya, Egypt
- Dinkar: South Sudan
- Haratin: Morocco, Mauritania
- Nuba: Sudan
- Nubians: Egypt, Sudan
- Nuer: South Sudan
- Shilluk: South Sudan
- Tuareg: Southern Sahara and Sahel regions of Algeria, Libya, Morocco

South Africa

Southern Africa generally includes lands from the Cape of Good Hope northwards to the borders of Democratic Republic of the Congo and Tanzania, and islands such as Madagascar.

- Bushmen: Kalahari Desert, Botswana/Namibia
- Khoikhoi: South Africa
- Namaqua: South Africa

West Africa

West Africa generally includes the region bounded by the Sahara Desert to the north and the Gulf of Guinea to the south.

- Baka: Cameroon, Congo (Brazzaville), Gabon and Central African Republic
- Balengue: Equatorial Guinea
- Benga: Equatorial Guinea
- Bubi People: Bioko Island, Equatorial Guinea

- Bujeba: Equatorial Guinea
- Combe or Ndowe: Equatorial Guinea
- Duala people: Cameroon
- Beti-Pahuin#Fang: Equatorial Guinea
- Ogoni people: Nigeria
- Serer people: Senegal, the Gambia, Mauritania, Western Sahara,
- Tuareg: the Sahel
- Toubou: Southern Sahar

Americas

The Americas is the continent (or super-continent) comprising North and South America, and associated islands.

The Caribbean

The Caribbean, or West Indies, generally includes the island chains of the Caribbean.

- Garib: Lesser Antilles
- Galibi
- Taino: Amerindians who inhabited the Caribbean island of Hispanola and Puerto Rico, of Arawakan descent.
- Neo-Taino nations: Some scholars distinguish between the *Taino* and Neo-Taino groups. Neo-Taino groups were also Amerindians of the Antilles islands, but had distinctive languages and cultural practices that differed from the High Taino. These groups include:
  - — Ciboney: a term preferred in Cuban historical texts for the neo-Taino-Siboney nations of the islands of Cuba.
  - — Ciguayo: Eastern Hispaniola
  - — Lucaya: Based in Cuba and the Bahamas.
  - — Macorix: Hispaniola.
  - — Guanahatabey: Far Western Cuba, at the Guanahacabibes Peninsula.
  - — Everi: Often called Carib.

Central America

- Achi' (Mayan people of Guatemala)
- Bokota (Panama)
- Bribri (Costa Rica)
- Garifuna (Belize and Honduras)
- Chorotega (Honduras, Nicaragua, and Costa Rica)
- Ch'orti' (Mayan people of El Salvador)
- Chuj (Mayan people of Guatemala)
- Embera-Wounaan (Panama)
- Guaymi (Panama)
- Itza (Mayan people of Guatemala)
- Ixil (Mayan people of Guatemala)
- Jakaltek (Mayan people of Guatemala – also called Popti)

- Tolupan (Honduras)
- Kaqchiked (Mayan people of Guatemala)
- K'iche' (Mayan people of Guatemala)
- Kuna (Panama)
- Lenca (Honduras and El Salvador)
- Maleku (Costa Rica)
- Mam (Mayan people of Guatemala)
- Miskito (Honduras and Nicaragua)
- Mopan Maya (Mayan people of Guatemala and Belize)
- Naso (Panama)
- Ngobe-Bugle (Panama)
- Pech (Honduras)
- Pogomchi' (Mayan people of Guatemala)
- Pogomam (Mayan people of Guatemala)
- Q'anjob'al (Mayan people of Guatemala)
- Q'eqchi' (Mayan people of Guatemala)
- Rama (Nicaragua)
- Sumo (Nicaragua)
- Tojolabal (Mayan people of Guatemala)
- Tz'utujil (Mayan people of Guatemala)
- Xinca (Guatemala)

North America

- Aboriginal people in Canada
- Alaska Natives
- Indigenous people of California,
- Great Basin
- Great Plain
- North eastern
- Woodlands
- Northwest Plateau
- Northwest Coast
- Southeastern Woodlands
- Florida
- Subarctic, etc.
- North American Arctic: Aleut, Kalaallit/Inuit, Inupiat, Metis, Yup'ik
- Alaska Native: All of Alaska

Mexico

- Amuzao (Oaxaca, Mexico)
- Chocho (Oaxaca, Mexico)
- Cocopa (Baja California, Mexico)

- Guarijio (Sonora, Mexico)
- Kikapu (Coahuila, Mexico)
- Paipai (Baja California, Mexico)
- Tepehuan (Chihuahua and Durango, Mexico)
- Chontal de Oaxaca (Tequistlatecan people of Oaxaca, Mexico)
- Chatino (Oaxaca, Mexico)
- Chichimeca Jonaz (San Luis Potosi, Mexico)
- Chinantee (Oaxaca, Mexico)
- Ch'ol (Mayan language of Chiapas Mexico)
- Chontal Maya (Mayan language of Tabasco, Mexico)
- Cochimi (Baja California, Mexico)
- Cora (Jalisco and Nayarit Mexico)
- Cuicatec (Oaxaca, Mexico)
- Huastec (San Luis Potosi, Mexico)
- Huave (Oaxaca, Mexico)
- Huichol (Jalisco and Nayarit Mexico)
- Ixcatec (Oaxaca, Mexico)
- Kiliwa (Baja California, Mexico)
- Kumeyaay (Baja California, Mexico)
- Lacandon (Chiapas, Mexico)
- Matlatzinea (Mexico State, Mexico)
- Mayo (Sonora, Mexico)
- Mazahua (Mexico state, Mexico)
- Mazatec (Puebla and Oaxaca, Mexico)
- Mexicanero (Durango, Mexico)
- Mixe (Oaxaca, Mexico)
- Mixtec (Oaxaca, Mexico)
- Nahua (Mexico)
- Pame (San Luis Potosi, Mexico)
- Pima Bajo (Chihuahua, Mexico)
- Popoloca (Oaxaca, and Puebla, Mexico)
- P'urhepecha (Michoacan, Mexico)
- Seri (Sonora, Mexico)
- Tarahumara (Chihuahua, Mexico)
- Tlapanec (Me' phaa) (Guerrero, Mexico)
- Tatonac (Veracruz and Puebla, Mexico)
- Trique (Oaxaca, Mexico)
- Tzeltal (Chiapas, Mexico)
- Tzotzil (Chiapas, Mexico)
- Yaqui (Sonora, Mexico)
- Yucatec Maya (Yucatan, Quintana Roo and Campeche Mexico)
- Zapotec (Oaxaca, Mexico)

South America

- Ache: Paraguay
- Ashaninka: Peru
- Awa-Guaja: eastern Amazonian rainforest, Brazil
- Awa-Kwaiker: Northern Ecuador
- Aymara: Peru/Bolivia/Chile
- Ayoreo: the Chaco, Paraguay/Bilivia
- Bora: Colombia/Peru
- Bororo: Mato Grosso, Brazil
- Charrua: Uruguay, Brazil and Argentina
- Chayahuita: Loreto, Peru
- Cocama-Cocamilla: Loreto, Peru
- Embera: Colombia/Panama
- Enxet: Paraguay
- Jivaroan: Loreto and San Martin, Peru
- Guarani: Paraguay, Uruguay, Brazil, Bolivia and Argentina
- Karaja: Brazil
- Korubu: Brazil
- Kuna: Panama
- Mukuxi: Brazil/Guyana
- Mapuche: Chile/Argentina
- Matis: Brazil/Peru
- Matses: Brazil/Peru
- Nukak: Colombia
- Nasa: Colombia
- Secoya: Loreto, Peru/Ecuador
- Tapirape: Brazil
- Ticuna: Brazil/Peru/Colombia
- Tukano: Colombia
- Tupi: Paraguay, Brazil, Bolivia, Peru and Argentina
- Urarina: Chambira Basin, Loreto Peru
- U'wa: Colombia
- Yora: Amazon rainforest, southeast Peru
- Wichi: the Chaco, Argentina/Bolivia
- Warao: of Venezuela's Orinoco River delta region
- Wayuu: Venezuela/Colombia
- Yanomami: Venezuela/Brazil

Asia

Western Asia

Western Asia includes the region of Dead Sea Transform, the Arabian Peninsula, the Levant, Asia Minor, the Caucasus region and extending to the southern Caspian coast, Kopet Dag mountains and the eastern Dasht-e Lut desert.

- Assyrians – Aramaic-speaking people mostly found in northern Iraq. They are also found in Syria, Turkey, Iran and Armenia.
- Marsh Dwellers (Ma'dan) – Arabic-speaking group in the Tigris-Euphrates marshlands of southern Iraq/Iranian border
- Samaritans – An ethno-religious group of the Levant closely related genetically and culturally to the Jewish diaspora and is understood to have branched off from the latter around the time of the Assyrian exile. Religiously, the Samaritans are adherents of Samaritanism, an Abrahamic religion closely related to Judaism. Their sole norm of religious observance is the Pentateuch. (See also: Jews)

Central Asia

Central Asia generally includes the landlocked region east of the Caspian Sea, south of the Russian Taiga, to the Himalayas, and extending eastwards to Mongolia and the western Chinese provinces and autonomous regions.

- Oirats: West Mongolia
- Sartuul: Zavkhan, Mongolia
- Kyrgyz: Kyrgyz Republic
- Khoid: Mongolia
- Naimans: Inner Mongolia
- Mangud: Historically Mongolia
- Uyghur: Xinjiang, PRC
- Tajik: Tajikistan
- Yaghnobi: Tajikistan

East Asia

East Asia generally includes the People's Republic of China, the Korean Peninsula, and the associated Pacific islands, principally Japan and Taiwan.

- Ainu: Hokkaido, Japan and (until the end of World War II) on Sakhalin Island, Russia
- Ruykyuans: Ryukyu Kingdom, now Japan
- Taiwanese aborigines: the island of Taiwan
  - — Amis
  - — Atayal
  - — Bunun
  - — Kayalan
  - — Paiwan
  - — Puyuma
  - — Rukai
  - — Saisiyat
  - — Sakizaya
  - — Seediq
  - — Tao
  - — Thao
  - — Tsou
  - — Truku
- "Indigenous inhabitants": New Territories, Hong Kong
- Salar people: China
- Yamato: Japan

North Asia

North Asia generally includes the Russian Far Fast and the northern and eastern parts of Siberia.

- Northern indigenous peoples of Russia: over 40 distinct peoples, each with their own language and culture in Siberia and Russia.
- Sakha:
- Tuvans:
- Altayans: Titular nations of Altai Republic
- Buryats:
- Khakas:
- Tungus:
- Sami:

South Asia/Indian Subcontinent

South Asia generally includes the Indian subcontinental region, adjacent areas, and related islands of the Indian Ocean.

- Adivasi: Collective term for many indigenous peoples in India (See also List of Scheduled Tribes in India)
  - — Andamanese: indigenous peoples of the Andaman Islands, India which include:
  - — Great Andamanes: formerly at least 10 distinct groups living throughout Great Andaman, now confined to a single community on Strait Island, Andaman Is.
  - — Jarawa: South Andaman and Middle Andaman
  - — Onge: Little Andaman
  - — Jangil (Rutland Jarawa): now extinct, formerly of Rutland Island, Andamans
  - — Sentinelese: North Sentinel Island, Andaman Is.
  - — Indigenous people of Sikkim: India
  - — Bhutia
  - — Lepcha
  - — Kisan Tribals: indigenous peoples of the Orissa Sundergarh, India:
  - — Nicobari: Nicobar Islands, India
  - — Shompen: Nicobar Islands, India
- Vedda people: Sri Lanka
- Anniyala-Aetto: Sri Lanka.
- Ladakhi: India
- Kodava people: Kodagu, South India
- Toda: South India
- Kuruba: South India
- Badaga: South India
- Kota: South India
- Irula: South India
- Mech Kachari: North-East India
- Bodo: North-East India
- Naga: North-East India
- Gond: Central India (Gondwana Land, India)

- Kalasha of Chitral: Ancient pre-Muslim ethnic minority in Chitral District, Northern Pakistan
- Khasi-Jaintia: North-East India
- Tripuri: North-East India
- Raute: Nepal, North-East India
- Tharu: Nepal, North-East India
- Giraavaru people: Maldives
- Chakma people: Bangladesh, Arunachal Pradesh-India
- Marma people: Bangladesh
- Karbi people: North-East India

Southeast Asia

Mainland Southeast Asia

- Idu mishmi: of Northeast India, Arunachal Pradesh
- Akha: of Thailand, Laos, Burma and Chinese minority
- Degar: or Vietnam
- Hmong: subgroups of Thailand, Burma, Laos, Vietnam and a Chinese minority
- Karen: an alliance of hill tribes of Burma and Thailand
- Lahu: one of the hill tribes of Thailand, Burma, Laos and Yunnan.
- Karbi: of Karbi Anglong, Assam, Northeast India
- Lisu: one of the hill tribes of Burma, Thailand, Arunachal Pradesh, India & Yunnan and Sichuan, China
- Negrito: includes the Semang of the Malay peninsula, the Aeta of Luzon, the Ati of Panay, the Mani of Thailand, and the Andamanese.
- Orang Asli: in Peninsular Malaysia, They are divided into three main tribal groups – Semang (Negrito), Senoi, and Proto-Malay
- Tai: (Shan, Laotian, Isan, Lanna and formerly Siamese) are the predominant ethnicity of modern Thailand and Laso with a minor regional diaspora.

Maritime Southeast Asia

- Pribumi (Native Indonesians): of Indonesia
- Bajau: Borneo and the Sulu Archipelago (Malaysia, Indonesia, Philippines)
- Davak of Borneo
- Igorot: Cordillera mountains in Luzon in the Philippines
- Lumad: Mindanao in the Philippines
- Mangyan: Mindoro in the Philippines
- Tribes of Palawan: Palawan, Philippines
- Penan: Sarawak, Malaysia
- Batak: Palawan, Philippines

Europe

Europe generally refers to the mass of the Eurasian peninsula westwards of the Ural Mountains, the islands of the Mediterranean and North Atlantic Ocean.

- Basaques: Northern Spain and Southern France
- Crimean Karaites: Crimean Peninsula in Southern Ukraine

- Crimean Tatars: Crimean Peninsula in Southern Ukraine
- Izhorains: Northwest of Russia
- Komi: Komi Republic in Northeast of European Russia
- Mordvins: of the western Ural Mountains in Russia
- Nenets: Northeastern part of European Russia
- Sami: Northern and central Norway, Sweden, Finland and Kola Peninsula in the Northwest of Russia.
- Veps: Republic of Karelia, Northwest of Russia.

Circumpolar North

The Circumpolar North generally includes the lands surrounding the Article Circle.

- Evenks, China, Mongolia, Russia
- Inuit: Greenland, Northern Canada (Nunavut and Northwest Territories), Alska
  - Kalaalit, Greenland
- Korayaks, Russian Far East
- Chukchi, Siberia, Russia
- Sami: Northern Norway, Sweden, Finland, and Kola peninsula in Russia
- Yupik: Alaska and the Russian Far East
  - Alutiq, Alaska
  - Central Alaskan Yup'ik, Alaska
  - Cupik, Alaska
  - Siberian Yupik, Siberia, Russia
- Inupiat: Alska's Arctic and North Slope boroughs and the Bering Straits
- Northern Samoyedic peoples
  - Nenets, Russia
  - Enets, Western Siberia, Russia
  - Nganasan, Taymyr Peninsula, Siberia, Russia
- Ugric peoples, Yugra, Siberia, Russia
  - Khanty, Yugra, Siberia, Russia
  - Mansi, Yugra, Siberia, Russia
- Yukaghirs, East Siberia, Russia

Oceania

Oceania includes most islands of the Pacific Ocean, New Guinea and the continent of Australia.

Australia

Australia includes the continental landmass, and associated islands.

- Indigenous Australians: Australia
  - Australian Aborigines
  - Djabugay people Caine, Australia
  - Torres Strait Islanders: Australia
  - Palawa Tasmania, Australia
  - Tiwi Tiwi Islands, Northern Territories AU

Melanesia

Melanesia generally includes New Guinea and other (far-) western Pacific islands from the Arafura Sea out to Fiji.

- Fijian: Fiji
- Papuans: more than 250 distinct tribes or clans, each with their own language and culture. The main island of New Guinea and surrounding islands (territory forming independent state of Papua New Guinea and the Indonesian provinces of West Papua and Papua considered "Indigenous" these people are a subject to many debates.
  - — Wopkaimin: western PNG, Star Mountains.
  - — Hewa Southern Highlands, PNG
  - — Kaluli Great Papuan Plateau, Png
  - — Dhani Papua, Indonesia
- Ni-Vanuatu: Vanautu
- Malaitan people: Malaita, Solomon Islands
- Otong Javans

Micronesia

Micronesia generally includes the various small islands chains of the western and central Pacific.

- Chamorros: Northern Marianas
- Chuukese: Island of Chuuk – Federated States of Micronesia
- Kiribatese: Kiribati islands and elsewhere – Republic of Kiribati
- Kosraean: Island of Kosrae – Federated States of Micronesia
- Marshallese: Marshall Islands – Republic of the Marshall Islands
- Nauruan: Island of Nauru – Republic of Nauru
- Pohnpeian: Island of Pohnpei – Federated States of Micronesia
- Trukic and Ponapeic islanders – Federated States of Micronesia
- Yapese: Island of Yap – Federated States of Micronesia

Polynesia

Polynesia generally includes New Zealand and the islands of the central and southern Pacific Ocean.

- Kanka Maoli: Hawaii
- Maohi: Tahiti
- Maori: New Zealand Aotearoa
- Moriori: Chatham Islands
- Cook Islanders: Cook Islands
- Samoan: Samoa, American Samoa
- Tongan: Tonga
- Rapanui: Easter Island

24

Population

In geographical science, whose central theme is the variety which characterizes the earth's surface, population is the pivotal element, for geography, to a large extent, is man-centred. Population is the point of reference with which other elements are observed and from which they derive significance and meaning. It is the population which furnishes the focus. Population is human resources of great importance, because it determines the economic and social development of a country. Not only the quantity but quality of population is necessary for the growth of a region. All the aspects of population are discussed in a science known as 'demography' but the geographer's goal in any or all analysis of population is an understanding of the regional variations, in other words, it deals with the spatial or distributional aspects. This involves not only regional variations in numbers of people, including density and distribution patterns, but also the movements and migrations of population and its areal variation in quality as expressed through such attributes as health, sex-composition, literacy, religion, occupational status, stage of economic development, etc.

Growth of Population

In the initial phase of human history and pre-history, the human population grew at a snail speed because of hazardous climatic conditions and other unfavourable conditions for the growth of population. The abrupt change in human activity, also known as 'agricultural revolution' which assured supply of food and settled life provided conditions for the growth of population, consequently, the human population started increasing at a steady pace. The second time when the population increased faster was in 1779 when man harnessed fossil fuels and developed steam engine. This development is known as 'industrial revolution'. Both these developments in the human history had profound effects on the growth and spatial distribution of population. The population of the world is increasing continuously as shown in Table 24.1

According to United States Census Bureau (USCB) the total world population is 7,129 billion. The world population has experienced continuous growth since the end of the 'Great Famine' and the 'Black Death' in 1350, when it stood around 370 million. The highest rates of growth – global population increased above 1.8 per cent per year – were seen briefly during the 1950s and for a longer period during the 1960s and 1970s. The growth rate peaked at 2.2 per cent in 1963, then declined below 1.1 per cent by 2012. Total annual births were highest in the late 1980s at about 130 million and are now expected to remain constant at their 2011 level of 134 million, while deaths number 56 million per year and are expected to increase to 80 million per year by 2040.

Table 24.1 *Growth of World Population*

Year	*Population (Millions)*	*Year*	*Population (Millions)*
10,000 BC	5	1975	4000 (4 Billion)
1 AD	200	1987	5000 (5 Billion)
1000 AD	300	1999	6000 (6 Billion)
1750	800	2012	7000 (7 Billion)
1850	1000 (1 Billion)	2025	8000 (8 Billion)*
1930	2000 (2 Billion)	2050	About 9000 (9 Billion)*
1962	3000 (3 Billion)		

* Projected Figures

Source: United Nations Population Reference Bureau, 2010

Current UN projections show a continued increase in population in near future (but a steady decline in the population growth rate) with the global population expected to reach between 8.3 and 10.9 billion by 2050.

Distribution and Density of Population

The characteristic feature of the distribution pattern of world population is the concentration in limited areas. At all levels, i.e. continent, country and regional level, distribution of population is controlled by the following factors:

1. Climate
2. Physiography
3. Natural vegetation
4. Availability of arable land and water
5. Age of civilization
6. Accessibility of places
7. Resource availability
8. National policy

Table 24.2 indicates the continent-wise population and density:

Table 24.2 *Population and Density in Continents*

Continent	*Total Population (2013 Estimates)*	*Density per sq km*
Asia	4,298,723,000	96.4
Africa	1,110,635,000	36.7
Europe	742,452,000	72.9
North America	565,265,000	22.9
South America	406,740,000	22.8
Oceania	36,304,000	4.5

Asia is having the largest population of 42.98 millions, which is about 61 per cent of the total world population, followed by Africa with a population of 1,110 million, about 12 per cent of the world population. Oceania is having only 36 million population, which is lowest among the continents. The population of Europe is 742 million, North America 565 million and South America 406 million.

Not only the continents but among countries of the world, great variations are visible in total number and distribution of population. The following table shows population of the selected countries of the world.

Table 24.3 *Population of the Selected Countries (2011)*

Country	*Population (millions)*	*Country*	*Population (millions)*
China	1,360	Japan	128
India	1,210	Nigeria	137
USA	304	Mexico	109
Indonesia	229	Germany	82.7
Brazil	193	Iran	71.2
Pakistan	165	Egypt	76.6
Russian Federation	141	France	60.9
Bangladesh	147	UK	60.0

There are wide regional variations in the density of population. The average density of the world population is 44 persons per sq km. In Asia the density of population is 96.4 persons per sq km, followed by Europe 72.9 and Africa 36.7 persons per sq km respectively. The lowest density of population is 4.5 persons per sq km in Oceania. In fact, half of the world population is clustered on just 5 per cent of the land and about 33 per cent of the total land area is virtually uninhabited.

The four densely populated areas of the world having more than 100 persons per sq km are:

1. *East Asia* (China, Japan, South Korea, Taiwan)
2. *South Asia* (India, Pakistan, Bangladesh, Sri Lanka, Maldives and Nepal)
3. *North-West Europe* (United Kingdom, France, Germany, Netherlands, Poland, Belgium, Luxembourg, Ireland, Denmark, Spain, Portugal and Italy)
4. *Eastern North America* (North-East United States and South-East Canada)

The sparsely populated regions of the world are those which are not having suitable geographical conditions; such regions are: – (i) Deserts and arid regions, (ii) ice caps and cold regions, (iii) mountainous regions, and (iv) tropical rainforests regions. All the deserts of the world are having less than 5 persons density. Similarly Arctic region and Antarctica are the areas of severe cold. The continent of Antarctica is almost exclusively uninhabited. In Tundra regions of Arctic also only Eskimos are living and their number is very limited. The higher mountainous regions of Himalayas, Rockeys, Andese, Alps, Urals as well as other mountainous regions are having no population or limited population in lower levels. The Equatorial regions of Amazon and Congo Basin as well as islands of South-East Asia are the areas of tropical rainforest. The climate of these regions is hot and too humid not suitable for human habitation.

The Demographic Transition

Generally, nations go through a four-stage change process over time, known as the 'demographic transition' as depicted in Figure 24.1.

Stage I (no growth); Both birth and death rates are high, resulting in little or no population increase. There are no clear examples of such nations in present times.

Stage II (rapid growth): The birth rate is still high, but the death rate is falling due to advancement in public health. Many developing countries are in this category; prominent are India, Bolivia and Saudi Arabia.

Stage III (slow growth): The death rate is relatively low and the birth rate is declining such as in China and Indonesia.

Stage IV (stagnation): Birth and death rates are both low, giving a very low rate of increase. Most of the European countries are in this category.

Migration of Population

Migration is ordinarily defined as the permanent or semi-permanent change of residence of an individual or group of people over a significant distance. Migration is a geographical phenomenon that seems to be a human necessity in every age. Since man has a tendency to leave the areas in which life is difficult, he migrates to the areas where life may be easy and better. Migration for both permanent residence and employment is major ramification of population growth, development, and global interdependency. The movement of individuals results in a transfer of ideas, social customs, and financial resources between countries and regions.

Figure 24.1 *Demographic Transition*

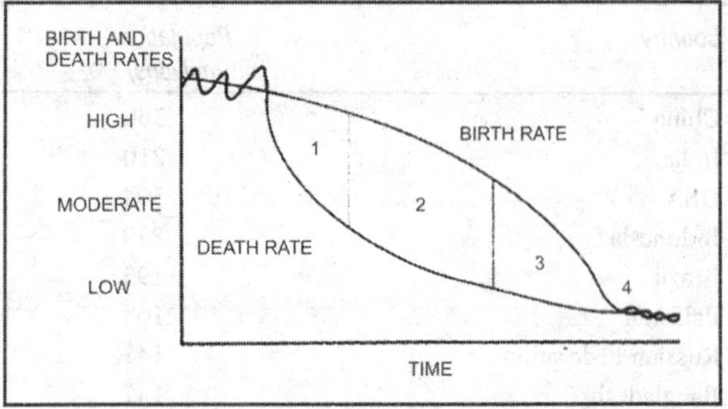

Migration together with 'fertility' and 'mortality' is a fundamental element determining population growth and population structure in an area. Migration is classified as international, inter-regional, inter-urban or intra-urban. On the basis of time's criteria, migration may be temporary or permanent, while on the basis of distance it may be long or short. On the basis of number, migration may be individual or in groups or mass. The migration can be politically sponsored, forced or voluntary; while on the basis of social organization, it may be that of family, clan or individual. Another basis of migration classification is the cause or objective, accordingly it may be economic, social, political or religious. There is also rural-urban migration.

Causes of Migration

Several causes are responsible for the migration; some of them are natural; while others are economic and political. These may range from natural calamities, climate change, epidemics, overpopulation, opportunities for jobs, religious and political causes, etc. These causes can be classified as:

(i) Physical or Natural causes
(ii) Economic and Religious causes
(iii) Social and Religious causes
(iv) Demographic causes
(v) Technology and diffusion of information
(vi) Political causes including war and political unrest

International Migration

On the basis of volume (number), distance and means of transport use, international migration may be classified into: intercontinental, and transoceanic. The international migration is quite older, and limited in volume and time. It is caused because of a number of special conditions at the place of origin and place of destination. International migration may be short migration of small groups of people or large steps travelling long distance. Contrary to this, the transoceanic migration has been observed largely in modern times, affecting the chief centres of population in Europe, America, India, China, Japan, East Asia and South-West Asia.

Reliable data and information on international migration are not available. In ancient times, there had been large-scale exodus of people to Canaan where Jews immigrated. The Mongols migrated to South China and Thailand around 2000 B.C. and the people from Central Asia arrived in waves in the subcontinent of India. In India, millions of people from peninsular India migrated to the tea plantations in Assam and Bengal. In the beginning of the 20th century about 30 million Chinese invaded Manchuria and settled there. Similarly, from China outmigrated and settled in the different countries. From South-West Asia, people migrated towards Sudan and the Muslim Fulani moved westwards to the oases of Chad, Algeria, Nigera, Niger, Mali, Central African Republic, Mauritania, Congo basin and Cameroon. Along the southern coast of the Mediterranean Sea, the Arabs, penetrated towards Libya, Algeria, Morocco, Spain, Portugal and France. In America, both before and since the arrival of the Europeans, there had been much migration among the Americans. Until recently, the Amerindian (Red Indian) tribes often migrated in the United States. Canada, Mexico, Central and South America.

There is much internal migration still in India, Indonesia, and America. In West Indies and Brazil, the seasonal labourers go to the centres of banana production. From Mexico there is a constant stream of workers to the United States. The Negro workers migrate from southern states to the industrial centres in the north-east and the Great Lakes regions. Similarly, in India, agricultural labourers from Bihar and eastern Uttar Pradesh seasonally migrate to Punjab, Haryana, Jammu & Kashmir and western Uttar Pradesh.

Emigration from the British Isles as a whole represents the greatest swarm of mankind that has ever crossed the sea. About 17 million persons left the island between 1815 and 1926. This steady flood of emigrants has not diminished the population of Britain where an excess of births over deaths has constantly kept up the total and even increased it.

The first half of the 20th century may be said as a period of great political turmoil in which the two world wars were fought. The First World War resulted into forced labour. The forced migration because of political crisis and wars resulted into great human tragedies. The significant characteristic of the forced labour is that the normal selectivity in migration by age, sex, skill and education is lacking and communities as a whole are uprooted. Over 20 million Russians migrated within and outside Russia after the Great Revolution of 1917, partition of the subcontinent of India resulted into the migration of about over 12 million people from one part of the subcontinent to another.

At present, the number of international migrations has declined because of strict immigration policies. The migration is controlled and regulated by the policies of the governments and the international regulation. There are very few countries in the world which allow or encourage the immigration of foreigners on a permanent basis. The number of those countries which still attract a sizeable immigrants is small, which include USA, Canada, Australia, New Zealand, Greece, Germany, Saudi Arabia, UAE, Central Africa and Israel.

During the last three decades, South-West Asia has emerged as an important region which is attracting technical, skilled and unskilled labourers. Among these countries, Bahrain, Iran, Iraq, Jordan, Kuwait, Libya, Oman, Qatar, Saudi Arabia, UAE and Yemen are important. There are refugees and illegal migrants also who cross the international boundaries. The emergence of Bangladesh and Bosnia as well as the Gulf War of 1991 resulted into enormous movement of refuges across the international borders.

25

Types of Agriculture

Agriculture is the primary form of economic activity, including within its denotation a wide range, from simple to highly complex, of systems of utilizing the soils for the production of foodstuffs or raw materials for human use. Agriculture not only includes cultivation but also stock ranching, dairy, forestry, lumbering and host of other economic activities. Etymologically, the word 'agriculture' is derived from two words; 'ager' means field or soil, and 'culture' means care of or the tilling of. But agriculture is now more than tilling of the field and now it has become a whole science and practice of farming. It implies a conscious and determined effort on the part of man to utilize the soil for his benefit. Agriculture is an example of man's modification of his environment. Agriculture is still the chief occupation of man, in which more than 50 per cent of the labour force is engaged.

Types of Agriculture

The classification of the agricultural types of the world is very difficult because of the complexity of farming activities due to varied geographical and economic conditions. Economic geographers have classified the types of agriculture depending on their expansion, availability of water, cropping pattern, volume of production, season variations, regional concentrations, social system, ownership of land, etc. On the basis of these factors following types of world agriculture have been mentioned:

1. *On the basis of the expansion of land*
   (a) Intensive agriculture
   (b) Extensive agriculture
2. *On the basis of availability of water*
   (a) Humid farming
   (b) Irrigated farming
   (c) Dry farming
3. *On the basis of cropping pattern*
   (a) Monoculture: single crop agriculture
   (b) Duoculture: double crop agriculture
   (c) Oligoculture: multiple crop agriculture
4. *On the basis of seasonal variations*
   (a) Winter crop agriculture
   (b) Summer crop agriculture
   (c) Autumn crop agriculture
   or as in India – (a) kharif crops, (b) rabi crops

5. *On the basis of volume of production and their nature*
   (a) Simple subsistence agriculture
   (b) Intensive subsistence agriculture
   (c) Commercial grain farming
   (d) Plantation agriculture
6. *On the basis of social system*
   (a) Capitalistic
   (b) Socialistic
   (c) Feudalistic
7. *On the basis of regional characteristics*
   (a) Monsoon type of agriculture
   (b) Mediterranean agriculture
   (c) Mixed farming of North-West Europe
   (d) Tropical and sub-tropical plantation agriculture
8. *On the basis of ownership of land*
   (a) Collective state farming
   (b) Cooperative farming
   (c) Individual farming

In above mentioned agricultural types, a great deal of overlapping is there. Therefore, it creates complexities in classification. To avoid all such complications, it is better to group the world's agricultural types according to the characteristic features both of organization and crops, which give each type a uniqueness. The main agricultural types of the world and their brief description are as follows:

Subsistence Farming

There are two types of subsistence farming:
(a) Primitive or Simple Subsistence Farming, and
(b) Intensive Subsistence Farming.

Primitive or Simple Subsistence Farming

Primitive farming is the oldest form of agriculture and still prevalent in some areas of the world. From primitive gathering, some people have taken a step 'upward' on the economic ladder by learning the art of domesticating plants and their economy has moved into primitive cultivation. This type of farming is done on self-sufficient basis and farmers grow food only for themselves and their families. Some small surpluses may be either exchanged by barter or sold for cash. The resultant economy is thus static with little chance for improvement, but there is a high degree of rural independence because farmers are not tied to landlords or to trading centres.

Location

This form of agriculture is widely practiced by many tribes of the tropics, especially in Africa, in tropical South and Central America, and in South-East Asia. It is better known as shifting cultivation (Figure 25.1). Shifting cultivation is practiced in the tropics by many different peoples and thus has many different names, e.g. *milpa* in Central America and parts of Africa, *conuco* in Venezuela, *roca* in Brazil, *masole* in Zaire, *ladang* in Malaysia, *humah* in Indonesia, *caingin* in the Phillippines, *taungya* in Burma, *tamarai* in Thailand, *bewar or pada* in India and *chena* in Sri Lanka.

The primitive subsistence agriculture or shifting cultivation is characterized by the following features:

Figure 25.1 *Areas of Primitive Subsistence Agriculture*

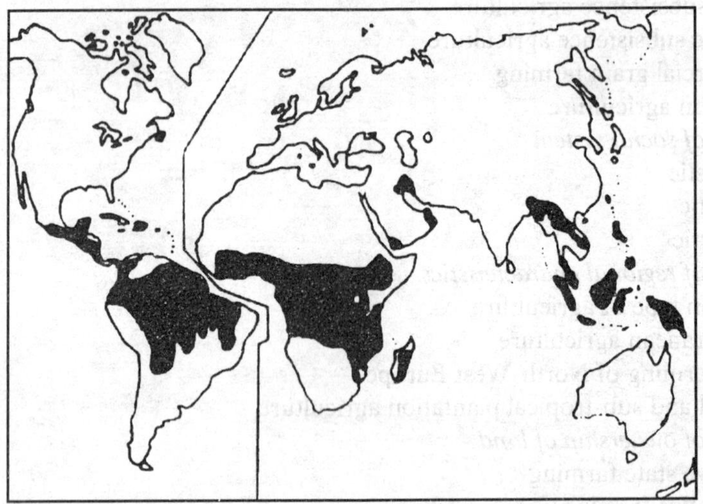

(i) Sites for the ladang are usually selected in the virgin forest by the experienced elders. Hill slopes are preferred because of better drainage. Many ladangs are located in the remote interiors, far from the main population centres.
(ii) The forests are usually cleared by fire and the ashes add to the fertility of the soil. Trees that are not burnt are hacked out by the men or left to decay naturally. Shifting cultivation is thus also called 'slash-and-burn agriculture'.
(iii) The cultivated patches are usually very small, about 0.5–1 hectare (1–3 acres) scattered in their distribution and separated from one another by dense forests or bush.
(iv) Cultivation is done with very primitive tools such as sticks and hoes, without the aid of machines or even drought animals. Much manual labour is needed in land clearance to produce food for a few people.
(v) Few crops are raised in the ladangs. The main crops are starchy foods, e.g. tapioca, cassava or manioc, yams, maize or corn, millet, upland rice, beans and bananas.
(vi) Short periods of crop occupance alternate with long periods of fallowing.
(vii) This form of 'migratory agriculture' still supports many of the aboriginal tribes of the tropical rain forest, despite the efforts made by the local governments to resettle them.

Intensive Subsistence Farming

The term, 'intensive subsistence agriculture' is used to describe a type of agriculture characterized by high output per unit of land and relatively low output per worker. Although the nature of this agriculture has changed and in many areas now it is no more subsistence. But despite changes the term 'intensive subsistence' is still used today to describe those agriculture systems which are clearly more sophisticated than primitive agriculture. Sometimes it is also known as 'monsoon type of agriculture'.

Location

This form of agriculture is best developed in and practically confined to the monsoon lands of Asia. It is found in China, Japan, Korea, India, Pakistan, Sri Lanka, the greater part of continental South-East Asia and parts of insular South-East Asia (Java, Luzon, Visayan Islands, coastal Sumatra and Malaysia) (Figure 25.2).

Figure 25.2 *Areas of Intensive Subsistence Farming*

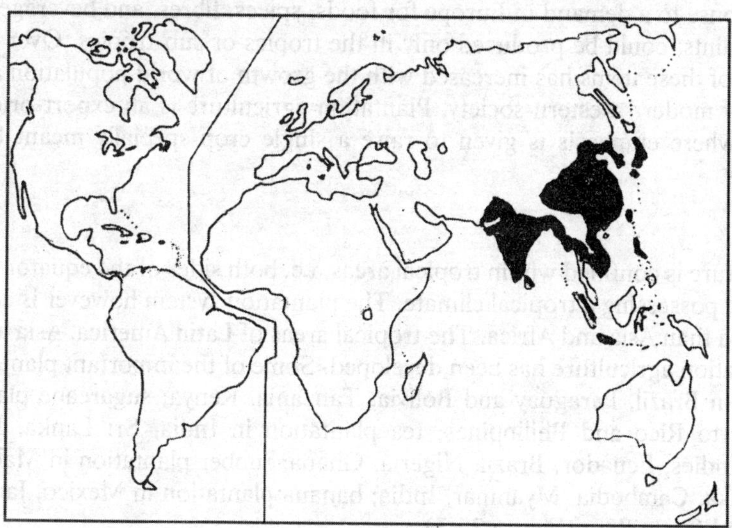

Characteristics

The main characteristics of the intensive subsistence agriculture are as follows:

(i) Very small holdings: Farms have been subdivided through many generations so they have become extremely small and often uneconomic to run. An average farm in Japan is approximately 0.6 hectare (about 1.5 acres) but in India and elsewhere in Asia, farms may be even smaller.

(ii) Farming is very intensive: In Monsoon Asia, the peasants are so 'land hungry' that every bit of tillable land is utilized for agriculture. Farming is so intensive that double or treble cropping is practiced, that is, several crops are grown on the same land during the course of a year. Where only one crop of paddy can be raised, the fields are normally used in the dry season to raise other food or cash crops such as sugar, tobacco or oil-seeds.

(iii) Much hand labour is entailed: Traditionally, much hand labour is required in wet paddy cultivation. Ploughing is done with the aid of buffaloes, the fields are raked by hand, the paddy is planted painstakingly in precise rows by the women, harvesting is done with sickles and threshing is done by hand. Farm implements are often still very simple. The basic tools are simple ploughs, the cangkul, a kind of spade, and hoes. Nowadays machinery has been developed which is capable of working in the flooded fields and separate machines can plough, plant and harvest the paddy.

(iv) Use of animal and plant manures: To ensure high yields and continued fertility, farmers make use of every available type of manure including farm wastes, rotten vegetables, clippings, fish wastes, guano, animal dung (especially those from the pig sties and poultry yards) and human excreta.

(v) Dominance of Paddy and other food crops: Paddy is the most dominating crop produced in intensive subsistence agriculture.

Plantation Agriculture

The tropical plantation is one of the world's oldest systems of commercial agriculture. Since 1500 AD, the products from over a dozen tropical crops have been in constant demand by people in the temperate regions. The specialized commercial cultivation of cash crops on estates or plantations is a very distinctive type of tropical agriculture and is found in many parts of Asia, Africa and tropical and sub-tropical America. Plantation agriculture is the product of colonialism. Plantations have been developed in response to a demand in Europe for foods, spices, fibres, and beverages, which because of climatic constraints, could be produced only in the tropics or sub-tropics. Over the centuries the demand for most of these items has increased with the growth of world population and also with the insatiable needs of modern western society. Plantation agriculture is an export-oriented specialized farming method where emphasis is given to raise a single crop specially meant for export to the overseas countries.

Location

Plantation agriculture is confined within tropical areas, i.e. both sides of the equator. Plantations exist on every continent possessing a tropical climate. The plantation system however is considerably older in tropical America than Asia and Africa. The tropical areas of Latin America, Asia and Africa are the areas where plantation agriculture has been developed. Some of the important plantation farming are coffee plantation in Brazil, Paraguay and Bolivia, Tanzania, Kenya; sugarcane plantation in Cuba, Brazil, Peru, Puerto Rico and Philippines; tea plantation in India, Sri Lanka, Indonesia; cocoa farming in West Indies, Ecuador, Brazil, Nigeria, Ghana; rubber plantation in Malaysia, Indonesia, Thailand, Sri Lanka, Cambodia, Myanmar, India; banana plantation in Mexico, Jamaica, Columbia, Brazil, Panama and Costa Rico (Figure 25.3).

Characteristics

Plantation agriculture is different than other forms of agriculture. The characteristic features of plantation agriculture are as follows:

(i) Plantation agriculture is highly sophisticated and scientific methods are used for large-scale production.
(ii) There is specialization of single crop in plantation agriculture, e.g. coffee in Brazil, tea in India, rubber in Malaysia, etc.
(iii) Estate Farming: Plantation crops are usually raised on large estates, of more than 40 hectares (100 acres) each, though the success of such crops has often encouraged other farmers to grow them so that small holdings exist side by side with the large estates.

Figure 25.3 *Major Plantation Areas of the World*

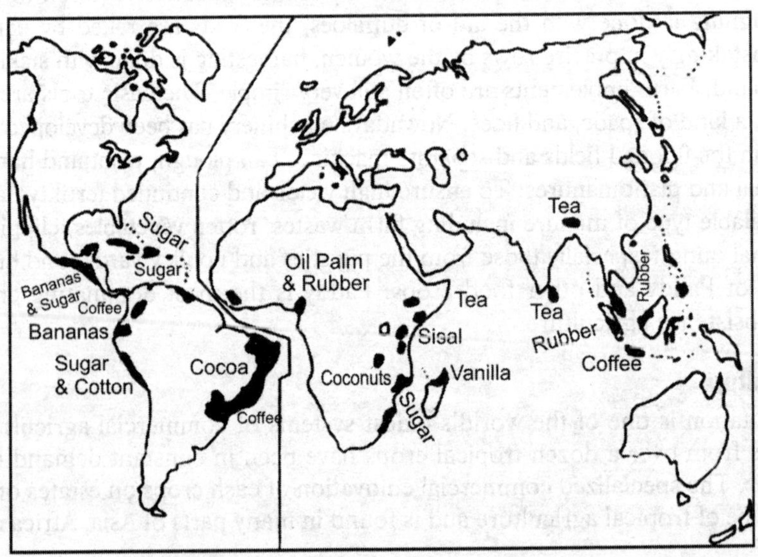

(iv) Foreign Ownership and Local Labour: Sugar plantations in Queensland and Australia, are unusual as they employ white labour.
(v) Farming in Estates in Scientifically Managed: Work in estates is executed with specialized skill, and wherever possible with the application of machinery and fertilizers. It aims at high yields, high quality production and a large output, most of which is exported.
(vi) Heavy Capital Outlay: To initiate and to maintain a tropical plantation, large sums of money are required. It is a very expensive undertaking especially under tropical conditions where maintenance costs are also high.
(vii) Plantation agriculture is an export-oriented agriculture. Nearly the entire product of plantation agriculture is generally exported to the international market. Therefore, it requires a better communication network, packaging and processing facilities.
(viii) Unlike other forms of agriculture, plantations are well-planned not only in terms of field structure but also having facilities like residence, transport, hospitals, retail markets, etc. for workers and associated people and their families.
(ix) Most of the crops grown in plantation agriculture have a life cycle of more than two years. Natural rubber, coconuts, oil palm, tea, cocoa, and coffee are all tree crops and take years to mature, but afterwards they are productive for long periods.

Mediterranean Agriculture

The term 'Mediterranean agriculture' applies to the agriculture done in those regions which are having Mediterranean type of climate. Mediterranean agriculture is unique because it is a mixture of diverse biocultural activities (both animal husbandry and crop farming) that have developed in five major world regions. This type of agriculture is determined by climatic conditions, which exert such an influence that both traditional and commercial agriculture flourish with a dominance of the agriculture of citrus fruits along with horticulture and floriculture.

Location

There are five major regions in the world having Mediterranean type of agriculture. The largest of these nearly encircles the Mediterranean Sea and it is from there the type derives its name. The regions having Mediterranean type of agriculture in the world are as follows:
(i) The Mediterranean Basin: France, Spain, Italy, Algeria, Morocco, Turkey, Tunisia, Israel, Northern Nile valley, i.e. all the regions around Mediterranean Sea
(ii) California in USA
(iii) Central Chile
(iv) Southern part of South Africa
(v) Lower Murray – Darling basin of South Australia

Characteristics

(i) The natural setting that lends itself to Mediterranean agriculture is distinguished by erratic rainfall, mild temperatures, irregular topography, and nearness to large water bodies.
(ii) In this type, farming is intensive, highly specialized and varied in the kinds of crops raised. Subsistence agriculture occurs side by side with commercial farming.
(iii) The land use in these areas is dependent on such factors as the total annual amount of rainfall, length of summer drought, availability of melting snow, local soil conditions, and price fluctuations in local and world market.
(iv) The four main aspects of Mediterranean agriculture are: (a) orchard farming, (b) viticulture, (c) cereal and vegetable cultivation, and (d) limited animal husbandry.

Figure 25.4 *Areas of Mediterranean Agriculture in the World*

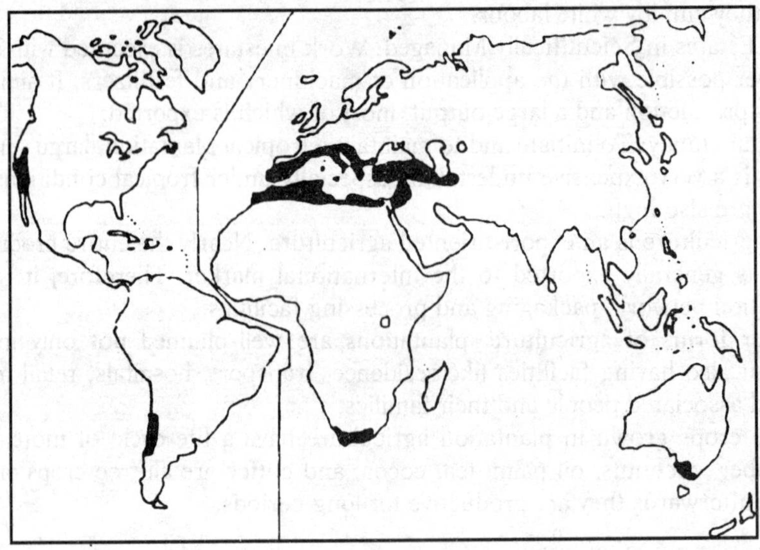

Mixed Farming

In mixed farming, a farmer combines the cultivation of crops and the domestication of animals and gets income from both. Mixed farming can, therefore, serve as a transition between the animal-raising economics and the crop-raising ones. This type of mixed farming provides greater security than the growing of a single crop as in the extensive prairie, wheat farms may suffer from market fluctuations and crop failures.

Location

The two most extensive regions of mixed farming are in Eurasia and the United States (Figure 25.5). This activity takes up more land than any other type of bio-culture in Europe and is prevalent eastward in an ever narrowing belt that stretches from the Atlantic to the Pacific, with only one interruption in Eastern Siberia. This belt is the thickest between Ukraine and central Finland.

Other regions of mixed farming are Mexico, South America and South Africa.

Characteristics

(i) The mixed farming is done for
   (a) Sustenance of animals,
   (b) For own consumption, and
   (c) For commercial sale.

(ii) In mixed farming about 90 per cent land is devoted to agriculture.

(iii) In mixed farming crop rotation is followed in order to maintain soil fertility.

(iv) The sequence of cultivation in mixed farming is cereal and vegetable production for own consumption, followed by hay, alfalfa, clover, etc. for livestock consumption, and finally, some amount of cereal production, wheat, maize, etc. for commercial sale.

(v) This farming is more mechanized. The use of heavy machines like tractors, harrowers, thrashers, etc. is very common.

(vi) There is a large-scale use of both organic and inorganic fertilizers.

Figure 25.5 *Mixed Farming Regions of the World*

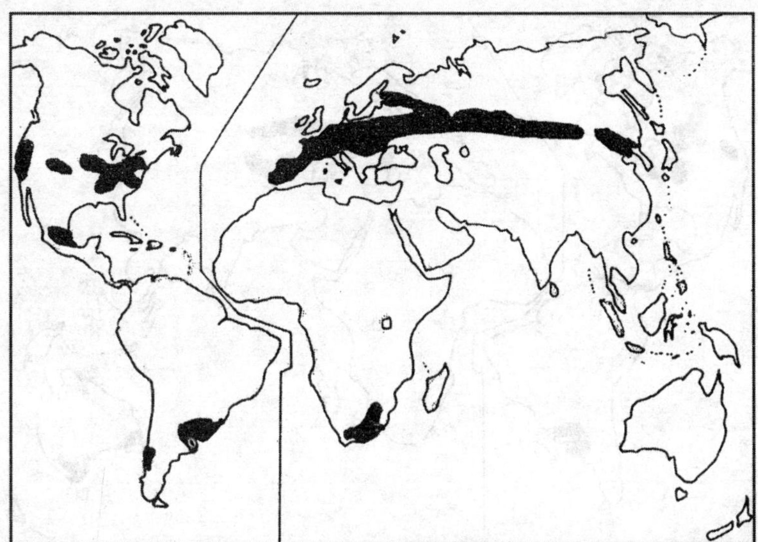

(vii) The degree of commercialism varies considerably. In west-central Europe, the northern United States, and Argentina, mixed farming is highly commercialized, while in other areas the commercialization is limited.

Commercial Grain Farming

Commercial grain farming is an extensive and mechanized form of agriculture. This is a development in the continental lands of the mid-latitudes, which were once roamed by nomadic herdsmen.

Location

The major world regions of commercial grain farming are shown in Figure 25.6. The largest one, in Eurasia, stretches from Kiew in southern Russia to Omsk in western Siberia in a width of about 1,000 km from Caucasus to Saratov on the Volga River.

In north America, there are several areas of commercial grain farming. The largest area runs from Alberta, through Saskatchewan and Manitoba to Dakotas. Another centre is in Kansas and spills over into neighbouring states. Smaller regions appear in eastern Washington and Oregon, eastern Illinois and northern Iowa.

In South America, Argentina has a large region of commercial grain farming. Australia has two areas, one in the south-west and another in the south-east. In fact, commercial grain farming is a mid-latitude activity and mostly done in between 30° to 55°N and S latitudes.

Characteristics

(i) Specialization in single crop.

(ii) Farms are very large: The wheat farms in mid-latitudes are very large, ranging from 240 to 16,000 hectares, though average size of the farm in the USA is about 400 hectares.

(iii) Highly mechanized: The commercial grain farming is highly mechanized. Cultivation from ploughing to harvesting is often entirely mechanized. The use of tractors, ploughs, drills and combine harvests which reap, thresh, winnow and sack the grain all in one operation is common. Other features of the commercial grain farming are:

Figure 25.6 *Areas of Commercial Grain Farming*

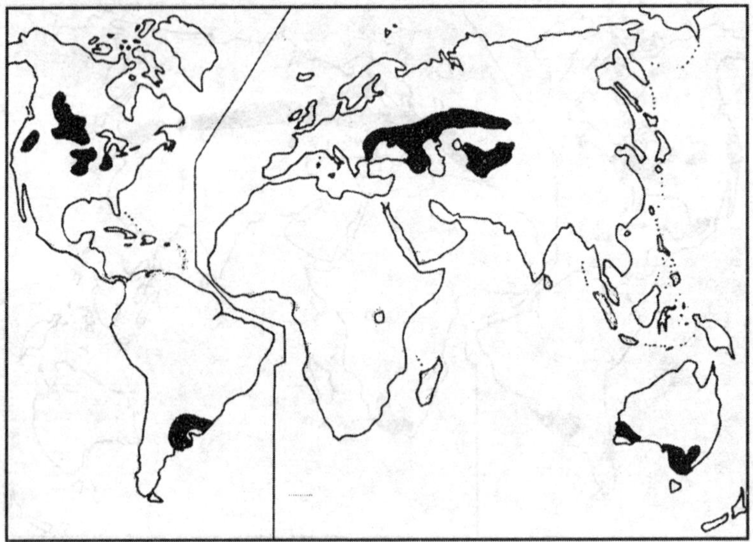

(a) Lack of irrigation
(b) Farm ownership
(c) Prone to climatic hazards
(d) Dependence on market fluctuations, etc.

Irrigation Farming

Irrigation farming is prevalent in areas where rainfall is seasonal and confined to a particular season. In all parts of the world there is a long dry season, when irrigation is practiced. It is, however, the most important in monsoon and the sub-tropical regions. Here rainfall is precarious, or is confined to one season only, and the temperatures are suitable for crop cultivation throughout the year. Also large populations exist here whose mainstay is mostly agriculture, which may produce nothing in drought years.

In fact, irrigation farming is not a special farming but it is simply farming through irrigation facilities. Although, construction of dams and canals is a costly affair but the same has been compensated by high production of cereals and other cash crops. This has been done in India and China. Irrigation enables cultivation throughout the year and the farming system is practiced intensively.

Dry Farming

The term 'dry farming' was first used in America to designate farming without irrigation in a section where irrigation was generally practiced. The use of the term has been extended and it has come to mean specifically the production of crops without irrigation in regions of deficient rainfall. The factor that distinguishes dry farming from humid farming is the limited water supply. Water being the vital thing for agriculture, its conservation and economic utilization becomes of primary importance in dry farming agricultural practice.

The essential part of dry farming is deep ploughing (20 to 30 cm), to enable greater and deeper sinking of water. This is done immediately after the harvest, so that any succeeding rain will soak in. The ploughed land also checks the drifting of snow that may fall in winter. The conservation of moisture is attained by the maintenance of a layer of dust or dust-mulch over the surface of the field.

This 'mulch' prevents water from getting evaporated. Leaving the land fallow in occasional years and not allowing any weeds to grow is also sometimes necessary. While the land is lying fallow, it is carefully and repeatedly tilled to prevent the growth of weeds which would use up soil moisture. The ratio of fallow years to crop years varies with the amount and distribution of rainfall. Sometimes a crop may be grown only once in three years.

Wheat, oats, barley, rye, sorghum and beans are other important crops grown through dry farming.

The most important dry farming area of the world is in the United States, where the largest area is located in the Great basin, Columbia river basin and the Snake river basin. Other regions where dry farming is practiced to some extent are in Australia, Canada, Western Asia and South Africa.

26

Agricultural Products

'Agro products' embraces a broad all-inclusive category of products related to agriculture, ranging from cereals, pulses, oilseeds, beverages, fibre crops, spices, and commercial crops such as sugarcane, rubber, sugar beet, fruits and vegetables, etc. In this chapter a brief review of agricultural products has been given.

CEREALS

Rice

Rice (*Oriza sativa*) has been cultivated since time immemorial. The origin of rice is not traceable, but it is believed that rice must have originated somewhere in China or India. Chinese literary works mentioned rice as long ago as 3000 BC. It has been the staple food of China for a long time and is consumed in many ways. Ancient Hindu scriptures also mentioned rice in connection with religious offerings as well as the peasants' daily food. From China and India, rice cultivation spread to neighbouring Asian countries. Now rice is a special crop of the monsoon region, where alone it finds almost ideal conditions for growth.

Geographical Conditions

Temperature: Rice is a tropical crop and grown where the average temperature during the growing season is between 20°C and 27°C. Abundant sunshine is essential during its four months of growth. The minimum temperature should not go below 15°C as germination cannot take place below that temperature.

Rainfall: Paddy requires more water than any other crop. As a result, paddy cultivation is done only in those areas where minimum rainfall is 115cm although the regions having average annual rainfall between 175–300 cm are the most suitable. Paddy also needs flooded condition with the depth of water varying over 25mm at the time of transplanting to as much as 150mm for 10 weeks of the growing period.

Soils: Paddy is grown in wide range of soil, from the podzolic alluvium of China to the impermeable heavy clay of central Thailand. Fertile riverine alluvial soil is best for rice cultivation. Clayey loam soil in monsoon land is considered to be the best for rice cultivation as water retention capacity of this soil is very high. Rice is also grown in saline areas of deltic region. Rice cultivation needs high fertilizer application.

Surface: Unlike other crops, paddy needs a levelled surface to enable the fields to be flooded at least during the growing period. Its ideal habitat is therefore in the great alluvial deltas and river basins of the world; the Ganges, Siking, Yangtzekiang, Irrawaddy, Menam Chao Phraya and Mekong, where there is practically no gradient.

Labour: Paddy cultivation is extremely labour-intensive, therefore, requires more labour in comparison to other cereal crops. Labour is necessary for preparing the field, weeding, sowing, transplanting, manuring, harvesting, threshing, winnowing and milling. For rice cultivation large number of cheap labour is required.

Although, there are several varieties of rice, but on the basis of their location two varieties have been identified. These are: (i) upland rice, which is grown on upland areas; and (ii) lowland rice, which grows in low lying and marshy areas. Another classification is based on origin and geographical concentration, and accordingly, there are two types:

(i) *Japonica:* Found in temperate climate and requires more fertilizer. Mostly found in Japan, Korea, Philippines, etc.
(ii) *Indian:* Grown in warmer regions like India, Bangladesh, Indonesia, Thailand, etc.

World Distribution and Production

The tropical region of Asia or Monsoon Asia is the largest rice-producing area. The countries of this region together produce 90 per cent of the global output of rice. Major rice-producing countries are China, India, Indonesia, Thailand, Bangladesh, Myanmar, Vietnam, Laos, Cambodia, Korea, Japan and Philippines. Besides these South-East Asian countries, rice is also produced in Brazil, USA, France, Venezuela, Columbia, Egypt, Spain, etc. The world distribution of rice producing areas has been depicted in Figure 26.1.

The total production of rice in the world during 2007–08 was 433.82 million metric tonnes which was increased in 2008–09 as 443.82 million metric tonnes. China is the largest producer of rice in the world with annual production of more than 130 million metric tonnes, followed by India with over 96 million metric tonnes. Bangladesh, Vietnam, Thailand, Myanmar and Philippines are the countries which produce more than 10 million metric tonnes of rice annually.

Figure 26.1 *Rice-Producing Areas of the World*

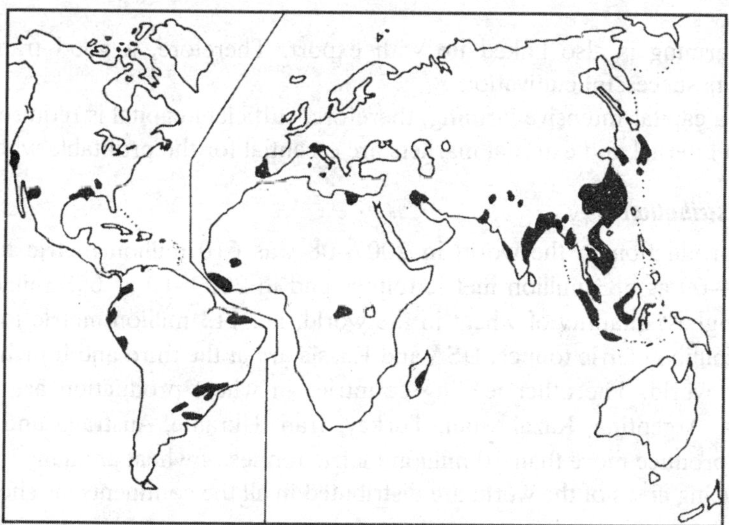

The other rice producing countries are Brazil, Japan, USA, Pakistan, Korea Republic, Egypt, Cambodia, Nigeria, Madagascar, Iran, Peru, Laos, Korea Demoratic, Malaysia, Italy and Spain.

Wheat

Wheat is the most important food crop. Its precise origin is lost in antiquity, and traces of its ancient cultivation have been found in many parts of the world. It is probably originated in Asia Minor and the Middle East, gradually spreading across the Mediterranean to the European countries. It is essentially a crop of temperate lands, but it is also grown in many other climatic regions, such as the warm temperate, western margin Mediterranean regions, the sub-tropical eastern margin regions, the tropical monsoon areas, the savannah lands, etc.

Geographical Conditions

Temperature: The temperature required for wheat during growing season is around 15.5°C. The weather should be warm and moist during the early stages of growth and sunny and dry in the later stages. The average temperature of the hottest month should not exceed 20°C. A frost-free period of 100 days is usually required but some fast-ripening varieties may mature only in 90 days.

Rainfall: The amount of rainfall required for wheat cultivation varies between 30cm and 100cm. The major wheat lands of the temperate regions have an annual rainfall of 38 cm to 80 cm.

Soils: The soil suitable for wheat is either light clay or heavy loam. The world's best wheat comes from the chernozem soils in the 'Black Earth' region of the Ukrainian Steppes, the dark brown soil of North America and also the grey brown podzolic soils of the deciduous forest region.

Topography: Wheat is grown in plain as well as in rolling topography, which provides adequate drainage and at the same time facilities of the use of machinery. The world's largest commercial wheat lands are the undulating temperate grasslands of Steppes, Prairies, Pampas and the Australian Downs.

Economic Factors

The economic factors that favour wheat cultivation are:

(i) *Technology:* In wheat cultivation maximum use of technology is possible. The commercial wheat farming is technology-intensive cultivation which requires tractors, harvesters, threshers, elevators, etc.

(ii) *Transport:* Farming is also linked up with export. Therefore, a good transport network is essential for its successful cultivation.

(iii) *Capital:* It is a capital-intensive farming, therefore, sufficient capital is required.

(iv) *Market:* Both internal and external markets are essential for the profitable wheat cultivation.

Production and Distribution

The total wheat production in the world in 2007–08 was 610 million metric tonnes, which was increased in 2008–09 as 683 million metric tonnes and in 2009–10 as 658 million metric tonnes. China produces highest quantity of wheat in the world, i.e. 113 million metric tonnes, followed by India with 77.50 million metric tonnes. USA and Russia are at the third and fourth position in wheat production in the world. The other leading countries in wheat production are France, Pakistan, Germany, Canada, Argentina, Kazakhstan, Turkey, Iran, Ukraine, Australia and Britain. All these countries used to produce more than 10 million metric tonnes of wheat annually.

Wheat-producing areas of the world are distributed in all the continents, as shown in Figure 26.2

Figure 26.2 *Wheat-Producing Areas of the World*

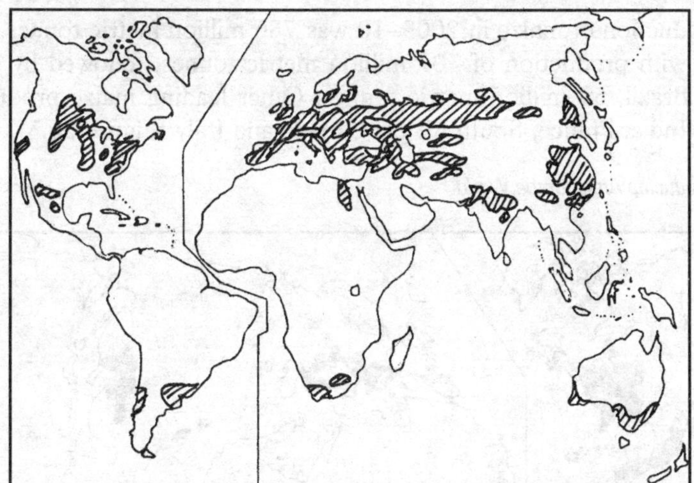

Apart from the above mentioned important countries, wheat is also produced in many other countries of the world. The other important wheat-producing countries are Morocco, Bulgaria, New Zealand, South Africa, Uruguay, Central Chile and Southern Brazil.

Maize

Maize is the third important cereal crop of the world after rice and wheat. Maize is often known as Indian corn or simply corn and is a widely cultivated cereal. It originated in America and has long been cultivated by the American Indians. It was grown by the people of ancient civilization of the Incas and Aztecs and is still the staple food of many Latin American people. Afterwards it was introduced in Europe, Africa and Asia and is now grown in wide range of climatic conditions in many countries.

Geographical Conditions

Maize is grown under divergent physical conditions. Maize is one of the most widely cultivated crops and grown in both tropical and warm temperate latitudes. The general requirement for maize cultivation is as follows:

Temperature: Maize is grown in temperatures between 18°C and 27°C during the day and around 14°C during the night. But the most important factor is the 140 frost-free days. The crop is very susceptible to frost, therefore, its cultivation in temperate latitudes is limited.

Rainfall: Maize is grown mostly in regions having annual rainfall between 60 cm to 110cm. But it is also grown in areas having rainfall of about 40 cm.

Soils: Maize grows in a wide range of soils, ranging from temperate podzols to the leached red soils of the tropics. But, the best suitable soil for maize is deep, rich soils of the sub-tropics, where there is abundant nitrogen.

Topography: The plain regions are most suitable for maize cultivation, because this helps in use of machines. Although, maize is also cultivated on undulating lands as well as on lower slopes of the hills. Maize is also grown as a major crop of shifting cultivation.

Economic Conditions: Unlike other crops maize can be cultivated with small capital. Use of machines have reduced the labour requirements. Most of the maize grown is utilized within the country, although its limited international trade is also there.

Production and Distribution

The total world production of maize in 2009–10 was 786 million metric tonnes. USA is the largest producer of maize with production of 307 million metric tonnes, followed by China, 162 million metric tonnes and Brazil, 54 million metric tonnes. Other leading maize producing countries are Mexico, Argentina, India, France, South Africa, Canada and Italy (Figure 26.3).

Figure 26.3 *Maize-Producing Areas of the World*

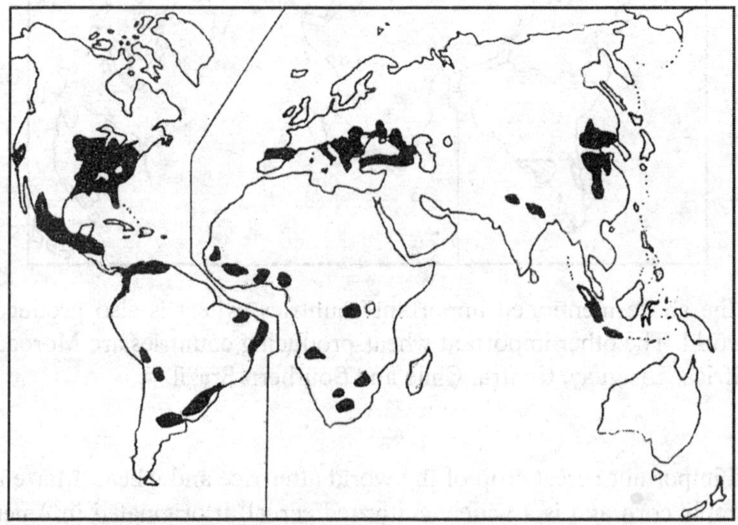

Apart from the above mentioned major maize-producing countries, the other maize producing countries are Ukraine, Philippines, Nigeria, Egypt, Vietnam, Serbia, Hungary, Russia, Romania, Thailand, Ethiopia, Turkey, Poland, Zimbabwe, etc.

Barley

Barley is believed to be one of the most ancient cereals. It was cultivated by the Chinese, the Egyptians and Mediterranean peoples in early times. Barley is now very extensively cultivated. Barley is unique among the food crops for its short growing season. It has been estimated that, barley matures in about 90 to 100 days and, thus, is very popular as a food crop in the regions where physical constraints, either in the form of low temperature or short supply of moisture, restricts the cultivation of other crops.

Geographical Conditions

The geographical conditions for barley cultivation are almost the same as for wheat, but barley is more tolerant of dry conditions and will ripen at lower temperatures. It is thus grown in a wide range of climatic, topographical and soil conditions, where wheat or other cereals fail to survive. The cultivation of barley is done in the following areas:

Sub-Arctic Regions: Barley can be grown in high latitudes, even beyond the Arctic Circle. It is grown in Finland up to 700 North latitudes where growing season is very short.

Semi-Arid Regions: Barley is grown in desert areas having semi-arid climatic conditions. Barley crop is having tolerance for arid conditions.

High Altitudes: Barley can be grown in upland regions also because it can ripen in cool temperatures.

Steep Hill Slopes: In mountainous areas of Japan, Norway and Switzerland, barley is grown on steep hill slopes.

Light, Loamy Soils: The most suitable soil for barley cultivation is the light, sandy or dry soil. But it is also grown on dry chalk and limestone regions in Europe.

Barley is used as human food, to feed animals and also used in malting. Barley is still used for bread-making, specially in those poor areas where wheat cannot be grown. Large quantities of barley are used nowadays to feed livestock. It is often cut green and is fed to cattle, pigs and poultry. The most important use of barley is for making beer and whisky.

Production and Distribution

In year 2007–08, the total world production of Barley was 133 million metric tonnes, which was increased to 154 million metric tonnes in 2008–09, while in 2009–10 the total Barley production was 160 million metric tonnes.

Russia is the largest producer of barley in the world and accounts for nearly 32 per cent of the world's barley output. Spain is the second largest barley producer in the world. Its average annual production is 11.50 million metric tonnes. Canada produces about 11 million metric tonnes, Germany, 11.90 million metric tonnes and France, 11.76 million metric tonnes in 2009–10. Turkey, Ukraine, Australia, Britain, USA, Poland and Denmark are also important producers of barley in the world.

The world distribution of barley is shown in Figure 26.4 below:

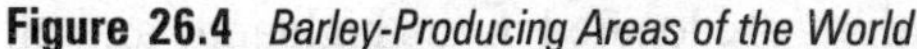

Figure 26.4 *Barley-Producing Areas of the World*

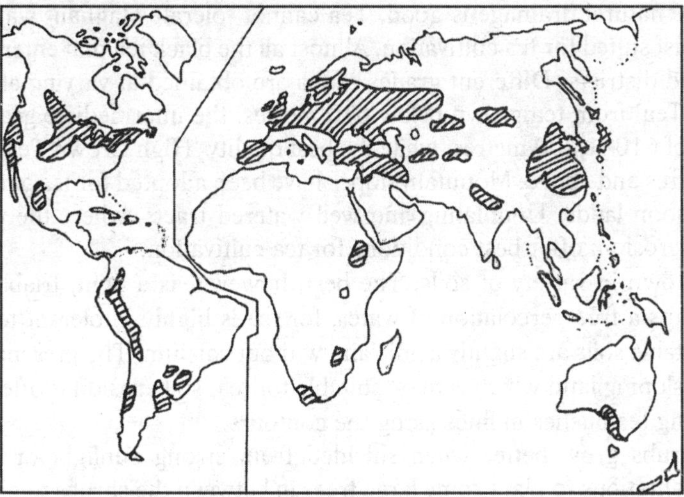

Barley is produced almost in all the continents. Apart from the countries mentioned above, the other barley-producing countries are Iran, China, Kazakhstan, Finland, Czech Republic, Belarus, Argentina, Sweden, Ethiopia, India, Italy, Iraq, Hungary, Mexico, Algeria, Austria, Morocco, etc. On the continental basis, Europe is the greatest producer accounting for about half of the world's barley.

COMMERCIAL AGRICULTURAL CROPS

The important commercial agricultural products are – tea, coffee, cocoa, rubber, cotton, jute, sugarcane, sugar beet and tobacco.

Tea

Tea is one of the most important beverage crops of the world. A much larger proportion of the world population drinks tea than coffee or cocoa. Tea is grown exclusively in the tropical and sub-tropical regions, but consumed in all parts of the world. Tea is made from the leaves of a tropical shrub, of the camellia family. Tea cultivation is believed to have originated in Yangtze valley of China as early as the 6th century AD. It has always been valued for its flavour and stimulant qualities and was even used for medical purposes. Tea was introduced in Britain and rest of Europe in mid-17th century. Because of its popularity as a common drink large-scale commercial cultivation in plantation has started in India, China, Bangladesh, Sri Lanka and the former British colonies of East Africa.

Conditions of Growth

Climate: Climatically, tea belongs to the monsoon lands where high temperatures, long growing season and heavy rainfall help in the growth of tea plants. A temperature of 21 °C during the growing season of not less than eight months is ideal. Warm summers and frequent rains promote rapid leaf reproduction and increase the number of annual picking. During the cool season, e.g. in China and Japan, the tea bushes lie dormant and there is no picking but where growth is not restricted by a distinct season, as in Sri Lanka, picking can be carried out all the year round. In India and Bangladesh, the highest yields are obtained from June to September when the weather is hot and rainy, but the best quality tea is derived from the earlier and later pickings when the climate is cooler and drier.

Relief: Almost all the commercially-managed tea plantations are located in the highlands and on hill slopes where the natural drainage is good. Tea cannot tolerate stagnant water and waterlogged lowlands are thus least suited for tea cultivation. Almost all the black tea that enters international trade comes from highland districts. Different grades of tea are obtained at varying altitudes, e.g. the low grade 'Low Crown Tea' from tea grown below 610 metres; the intermediate grade 'Medium Crown Tea' from altitudes of 610–1,200 metres, while the best quality 'High Crown Tea' comes from estates located at 1,220 metres and above. Mountain slopes have been adopted for tea gardens all over the tea regions of the monsoon lands. Undulating and well-watered tracts, where the water escapes freely without serious soil erosion, offer best conditions for tea cultivation.

Soils: Tea is grown in variety of soils. The best, however, is a light, friable loam with porous sub-soil which permits a free percolation of water, for tea is highly intolerant to stagnant water. In general the most suitable soils are slightly acidic and without calcium. The presence of iron in sub-soil is desirable. On the sloping land which is most suitable for tea, soil erosion is often a problem. This is combatted by planting tea bushes in lines along the contours.

Shade: Tea shrubs grow better when shielded from strong sunlight or violent winds. It is therefore usual in plantations to plant some large trees in between the shrubs to give shade not only to tea plants but also to the pickers.

Labour: Tea cultivation and processing are labour-intensive tasks. The work of field preparation, weeding, manuring, pruning and picking goes on all the year round on tea estates. Female labour is preferred for tea picking because it is a tedious job, requiring both skill and patience. Picking is done entirely by hand, and pickers are paid by piece rate, i.e. by the amount of tea picked during the day. Workers are also required in the tea processing plans where the leaves are dried, rolled, fermented, sifted, graded and packed.

Capital: Tea cultivation is a capital-intensive farming. Tea cultivation requires large finance for the development of plantations, payment to labourers, for processing, etc. In early phase of development of tea gardens, foreign capital has been used. In other words, all the tea gardens in Monsoon Asia were owned by colonial financers. But now these gardens are either owned by multinational companies, individuals or by semi-government organizations.

Production and Distribution

The world production of tea in 2006–07 was 3,646 thousand metric tonnes, which was raised to 3,888 thousand metric tonnes in 2007–08.

India and China are the leading tea producing countries and produce about 30 and 24 per cent of the world production respectively. Other important tea producing countries of the world are Kenya, Sri Lanka, Turkey, Vietnam, Indonesia, Japan, Argentina, Iran, Bangladesh, Malawi and Uganda.

The world distribution of tea-growing areas is shown in Figure 26.5

Figure 26.5 *Tea-Growing Areas of the World*

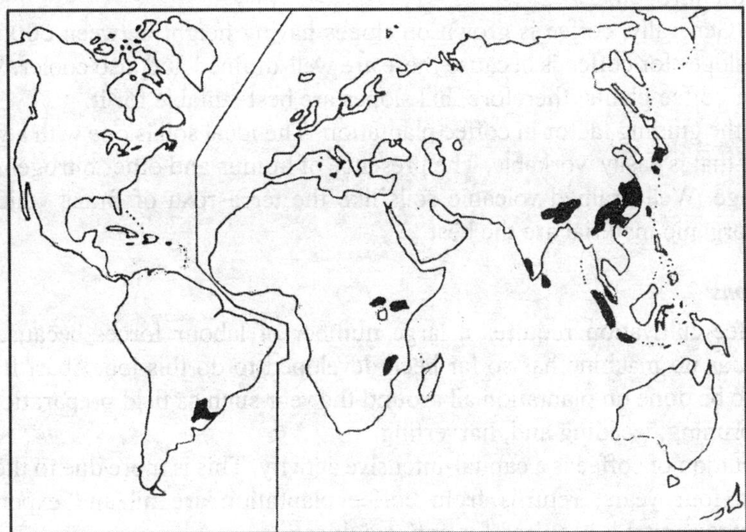

Coffee

Coffee is a popular drink all over the world. The coffee tree is a native of the highlands of southern Ethiopia and the name coffee is derived from that of the highland district of Kaffa where it was found. From Ethiopia, it was taken across the Red Sea to Saudi Arabia.

The status of coffee-producing areas has changed from time to time. Arabian Peninsula, the home of Mocha coffee, at first had a monopoly in its production. It was then introduced in Jamaica and other West Indian islands and the Blue Mountain coffce became world renowned. Dutch then introduced coffee plantation in Indonesia, brought about the emergence of Java Coffee, which is highly flavoured. In Sri Lanka also coffee cultivation was started and finally Brazil then rose to prominence and has become the leading producer of coffee in the world.

There are three types of coffee according to their nature and geographical distribution:

(a) *Coffee Robusta:* Mostly grows on lowlands of African plateaus. It contributes 15–20 per cent of global output.

(b) *Coffee Arabica:* According to volume of production and popularity, it is the most important and contributes 70–80 per cent of global output. It grows in the slopes of the highlands.

(c) *Coffee Liberica:* Inferior grade coffee, contributes only 5–10 per cent of global output and grows in Africa.

Growth Conditions for Cultivation

Climate: Coffee is a tropical plant which is also grown in semi-tropical climate. The coffee tree requires heat, humidity and abundant rainfall. The following temperature and rainfall conditions are necessary for coffee cultivation.

(a) *Temperature:* Coffee requires an average temperature between 20°-27°C. Although it grows in day temperature over 32°C in the Arabian Peninsula. Growth is most rapid during hot rainy season and during cool dry season berries ripen and are ready for picking. Bright sunshine and warm weather are necessary for the harvesting.

(b) *Rainfall:* Coffee needs abundant rainfall, i.e. 100 to 200 cm annually. The hill slopes which receive orographic rainfall are thus best for coffee cultivation.

Shade: Direct sunlight is harmful for coffee plants; therefore, these are planted under shade of taller trees such as bananas. In Brazil, leguminous plants are used which not only provide shade but also enrich soil with nitrogen.

Topography: Generally, coffee is grown on slopes having height between 600 to 1,800 metres. The suitability of slopes for coffee is because these are well-drained and also cooler. Water stagnation is very harmful for coffee plants; therefore, hill slopes are best suitable for it.

Soils: Soil is the guiding factor in coffee plantation. The ideal soil is one with a good sub-surface drainage, and one that is easily workable. The presence of humus and other nitrogenous matter in the soil is an advantage. Well-drained volcanic soils like the terra-roxa of Brazil which contain much potash as well as organic material are the best.

Economic Conditions

(a) *Labour:* Coffee cultivation requires a large number of labour forces because coffee is to be hand-picked, as no machine has so far been developed to do this job. Apart from this, there is much work to be done on plantation all around the year such as field preparation, transplanting, cultivating, pruning, weeding and, harvesting.

(b) *Capital:* Plantation of coffee is a capital-intensive activity. This is more due to the fact that during first three to four years, returns from coffee plantation are nil and expenditure is more. Therefore, huge capital is required for coffee cultivation.

(c) *Transport:* For a successful coffee plantation a good network of both internal and external transport is necessary. The internal rail and road transport help in collection of coffee from production areas, while distribution is done either by sea and air routes or sometimes by land routes.

(d) *Markets:* Coffee is having international market. It is necessary to maintain international market because nowadays there is much competition in the market. Similarly market fluctuations also have a direct impact on the production and price of coffee.

Production and Distribution

Coffee is mainly produced in tropical countries of the world. The total production of coffee in the world in 2007 was 7,743,000 metric tonnes. The distribution of coffee-growing areas has been shown in Figure 26.6.

Figure 26.6 shows that there are four major areas of coffee production in the world. These are:

(a) *South America:* Comprising countries like Brazil, Venezuela, Columbia, Ecuador and Peru. This region contributes more than 70 per cent of global production.

(b) *South-east Asia:* Comprising countries like Indonesia, India, etc.

(c) *Africa:* Comprising countries like Kenya, Uganda, Congo, etc.

(d) *Caribbean Region:* Comprising countries like Mexico, Guatemala, etc.

Figure 26.6 *Coffee-Growing Areas of the World*

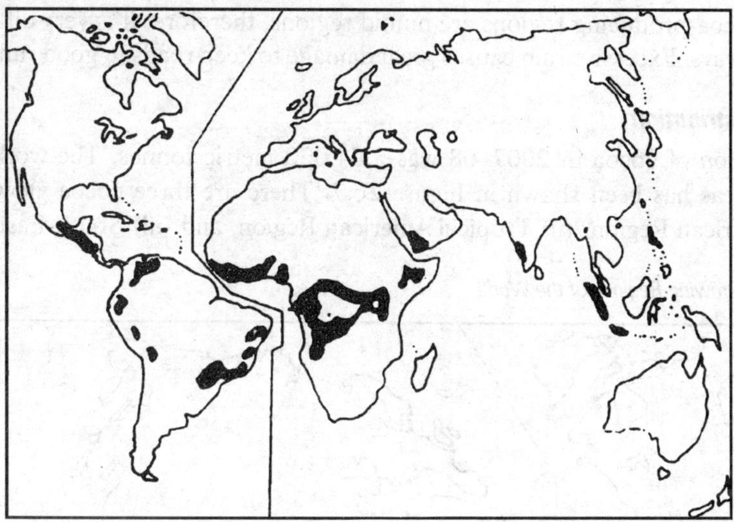

Cocoa

Cocoa is obtained from the beans or seeds of the tropical tree whose botanical name is Theobroma cacao and is used as drinks as well as for chocolate making.

The cocoa tree, indigenous to tropical America, was first found growing wild in lowland Central America, from Panama to the Yucatan Peninsula and in the river basins of the Amazon and the Orinoco. It was used to make chocolate, a national drink of the Aztecs for many centuries, and was regarded as the 'Food of Gods' because of its high nutritional value and its good flavour. But, the drink was not known outside tropical America till the early 16th century when Hernando Cortez, the Spanish conqueror of Mexico, brought some cocoa beans back to Spain. Because of its pleasant taste it soon became a popular drink in many countries of Europe, but in the rest of the world it gained favour much more slowly than either tea or coffee. It was not until the art of chocolate making became known in the industrial west that large-scale cultivation of cocoa began, and only in the 20th century it has become important in the world economy.

Conditions of Growth

Climate: Cocoa is essentially a plant of tropical region. The equatorial environment with high temperatures and heavy rainfall is the best. An annual rainfall of 200 to 350 cm is ideal but it can grow in areas with 150 cm of rain. The cocoa farming is best developed in regions having about 27°C of average annual temperature.

Shade: Although cocoa requires much heat but direct sunlight is harmful to it. Some protection is necessary, especially when trees are young. In cocoa plantations, tall banana plants or other shady trees are usually grown to provide shade.

Soil: Cocoa can be grown on a wide range of soils provided they are fairly deep and well-drained. Loamy soils rich in iron and potassium are ideal but light clays are also suitable. The heaviest crops come from cocoa grown on newly cleared virgin equatorial forest soils with high humus content but manuring in long-established groves enhances the yields.

Labour: Cocoa cultivation and processing are labour-intensive. The cocoa tree bears fruit all the year round but beans are usually harvested during the drier season, i.e. from September to February in West Africa, or April to December in Brazil. Cutting down the pods and splitting them, fermenting, drying and sacking the beans require much manual labour.

Transport: Since much of the cocoa produced is exported, a good network of transport is necessary. The cocoa-producing regions are humid regions, therefore, it is very difficult to maintain the roads and railways. Excessive rain causes great damage to keep roads in good running condition.

Production and Distribution

The total production of cocoa in 2007–08 was 3,64,000 metric tonnes. The world distribution of cocoa-growing areas has been shown in Figure 26.7. There are three cocoa growing areas in the world: (i) West African Region, (ii) Tropical American Region, and (iii) South-East Asian Region.

Figure 26.7 *Cocoa-Growing Regions of the World*

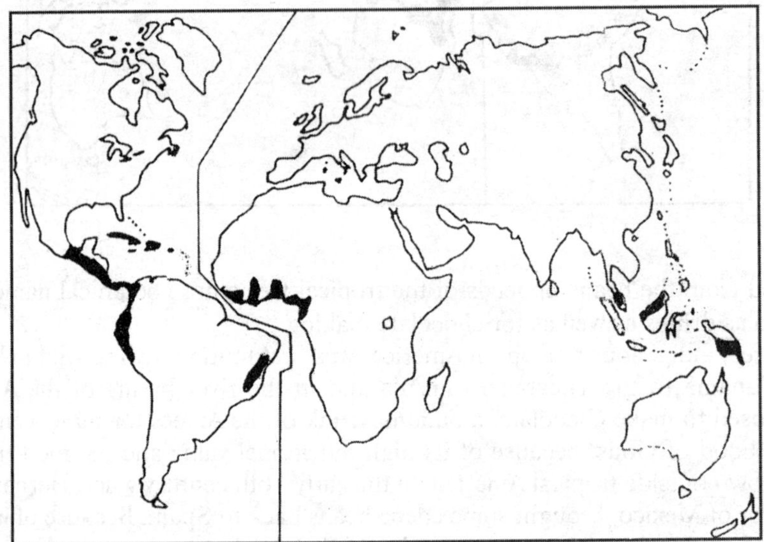

Ivory Coast (Kotede Ivory) is the highest cocoa producing country in the world and produces more than 39 per cent of the world production. It is followed by Ghana with about 20 per cent of the world production. Other important cocoa producing countries are Indonesia, Nigeria, Cameroon, Brazil, Ecuador and Malaysia.

Natural Rubber

Rubber is an important commercial agricultural product, collected from the latex of the tree *Hevea brasiliensis*. It is an indigenous tree of tropical rainforest of Amazon basin of Brazil. Rubber is useful because of its invaluable qualities such as elasticity, resistance to water and non-conduction of electricity. It is thus a useful substance for waterproofing, insulating and for use in making pneumatic tyres. Rubber is obtained from various tropical trees such as balata, funtumia elastic and gutta percha, but Heavea tree has become the almost exclusive provider of natural rubber.

Rubber was used by the native Brazilians since time immemorial but was only introduced into Europe in the 18th century. Its actual value as waterproofing agent was realized by Charles Mackintosh who used it on cloth to make water-proof material. The use of rubber was possible after the discovery of the vulcanizing process in 1839. Charles Goodyear of USA found that, by heating the rubber and mixing it with sulphur a stronger substance could be made. With the development of motor car and use of pneumatic tyres after 1895, the importance of rubber became well-established.

Conditions of Growth

Climate: Rubber is a tropical tree. It requires high temperature throughout the year ranging between 20°-35°C and average monthly mean temperate of 27°C. Less than 20°C temperature is detrimental. Similarly, rubber also requires heavy rainfall. The annual average rainfall of not less than 200 cm is optimum. Rubber tree thrives well when the distribution of rainfall is uniformly high all over the year. The equatorial regions of the world are suitable for rubber cultivation.

Soil: Rubber will do well on a wide variety of soils, some of which would be too poor for many crops. Deep, friable, well-drained soils are ideal as they promote root development, and acidic soils are also suitable.

Land: The plain or gently undulating land is suitable for rubber plantations. This land is generally obtained from clearing of forests. The soil erosion is often a problem in tropical areas, when the original forest cover is removed. Sometimes terraces are used to control soil erosion.

Economic Conditions

Among economic conditions, labour, capital, transport and market are necessary for proper rubber plantation.

(a) *Labour:* It is required for maintenance of plantations, collection of latex, preparation of rubber, nurturing of rubber tree, etc. It is a labour-intensive activity.

(b) *Capital:* For cultivation of rubber, huge capital is required, specially during first 6–7 years, when income is more or less nil. Capital requirement is necessary for labour wages, plantation cost, processing machines, etc.

(c) *Transport:* Both internal and external transport network is necessary for successful rubber cultivation. Internal transport is required for collection of latex and its transportation up to processing centres, while external transportation network is a pre-requisite for export.

(d) *Market:* Although, rubber is in great demand but with the development of synthetic rubber its demand has decreased. Market demand encourages the rubber plantation.

Figure 26.8 *Rubber-Growing Areas of the World*

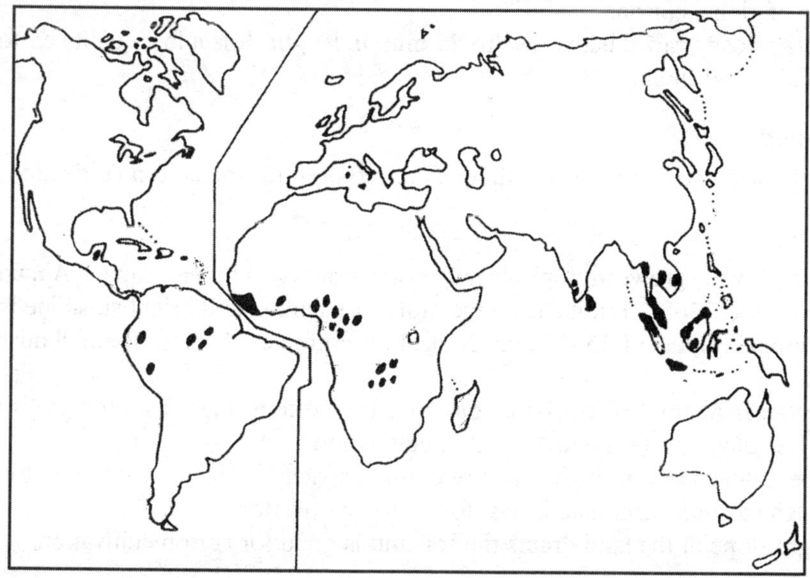

Production and World Distribution

Asia is the largest producer of the world and produces about 91 per cent of the world production, followed by Africa which produces 6 per cent.

In 2008, the total world production was 9,876,000 metric tonnes. Thailand is the highest rubber producer in the world, which produced 31.29 per cent of world production. Indonesia is the second largest producer while Malaysia ranks third in rubber production. There was a time when Malaysia was the highest producer but now its rank is third. The other producers are India, Vietnam, China, Sri Lanka, Cambodia, Nigeria, Ivory Coast, Ghana, Liberia, Kenya, Zaire and Brazil.

Cotton

Cotton is the main natural fibre accounting for about 50 per cent of the world's fibre supply. Besides being the source of fibre, cotton plant is also the basic raw material for the production of edible oil and cottonseed meal. The latter is one of the world's principle oilseeds, one-fifth of the cotton seed is crude protein, yielding 5 to 6 per cent of aggregate world protein supply. The use of cotton fibre to make textiles was known to ancient Egyptians, and cotton has been the major textile in China, India and Central Asia for centuries, but cotton textile remained a luxury in Europe until the 18th century. The cotton textile industry was not only started but grew faster in Europe and in other parts of the world, which resulted in the growth of cotton in various parts of the world.

Types of Cotton

Cotton is derived from the hair surrounding the seeds of plants of the Gossypium family. There are many varieties of the plant, some of which originated in Eurasia and some in the America, while modern varieties have been developed by breeding and hybridization. On the basis of staple length of the fibre, there are three main types of cotton.

(i) *Long Staple:* Staple length of this cotton varies between 25 to 65 mm. The major growing areas are Egypt, Sudan, Haiti, Peru, USA, etc.
(ii) *Medium Staple:* This is also called upland American cotton. Staple length varies between 15 to 35 mm; mostly grown in India, China, Brazil, Argentina, etc. This variety comprises maximum production of global cotton.
(iii) *Short Staple:* Mostly varies between 8 to 25 mm. in length. It is a low quality cotton and mostly found in China and India.

Condition of Growth

The following physical and economic conditions are required for the cotton cultivation:

Physical Conditions

(i) *Temperature:* Cotton grows in tropical and sub-tropical warm humid climate. Annual temperature requirement is 20°-28°C. Equitable temperature distribution and bright sunshine is desirable.
(ii) *Rainfall:* Annual rainfall of 55–100 cm is ideal for cotton cultivation. Rainfall during harvesting is harmful.
(iii) *Frost-Free Days:* About 180 frost-free days are the bottom-line for cotton cultivation. At least 200 frost-free days are desirable for cotton cultivation.
(iv) *Soil:* Fertile, saline soil with high water-retention capacity is ideal for cotton cultivation. Loamy soil with high calcium carbonate is best for cotton cultivation.
(v) *Relief:* Slight slope of the land drains the soil and is good for cotton cultivation.

Economic Conditions

(i) *Labour:* Cotton cultivation is very labour-intensive; large number of cheap labour is required.
(ii) *Capital:* Mechanized cotton cultivation requires huge capital for machinery, pesticides and fertilizers.
(iii) *Transport/Market:* As it is essentially a commercial crop, for export facilities, good transport network and steady demand in international market are pre-requisites.

Production and World Distribution

China is the largest producer of cotton in the world and produces 30.7 per cent of the world's production. India produces 20.4 per cent of the world cotton and ranks second, while USA is the third cotton producing country. Other major cotton producing countries are Pakistan, Brazil, Uzbekistan, Turkey, Greece, Turkistan, Syria, Egypt, Australia, Peru, Bolivia, Sudan, Argentina, Yemen, Nicaragua, Guatemala, Syria and Arab Republic.

The cotton-producing areas of the world are shown in Figure 26.9

Figure 26.9 *Cotton-Producing Areas of the World*

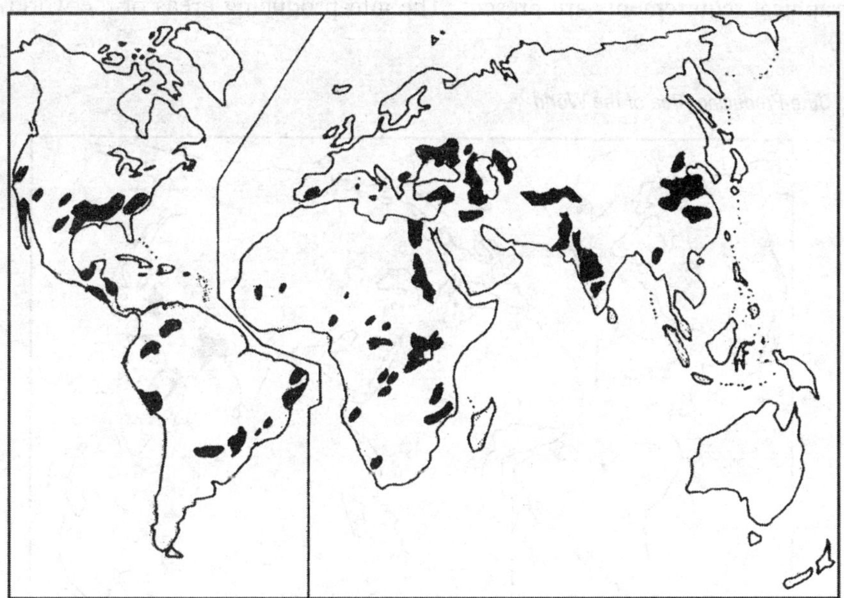

Jute

Cultivation of jute was prevalent in the Indian sub-continent even during the pre-historic period. In spite of its long history, jute cultivation did not develop much until the beginning of the 19th century. The fibre was first spun in England in year 1820 at Oxfordshire. Jute is the fibre obtained from certain species of the Corchorus family, tall reed-like annuals, which grow to a height of 3 to 3.7 metres. Jute is a course fibre and it is demanded for the manufacture of gunny bags, hessian, carpets, ropes, tarpaulins, rugs and coarse cloth.

Conditions of Growth

Jute is essentially a tropical plant and requires high temperature and high precipitation. It needs a well-distributed rainfall throughout with light showers and ample sunshine in the early stages.

Jute, like other fibres, requires rich soils and thrives on river alluvium, specially where annual floods renew the fertility of the soil. The best quality of jute is obtained from loamy soils, whereas the heaviest yield comes from clayey soils. The alluvial soils of the lower Ganga valley are the ideal. Jute requires so much moisture that it is usually grown under flood conditions.

Jute is a labour-intensive crop. It requires large labour force during growing and harvesting season.

There are two main varieties of jute:

(i) *Corchorus Capsularies:* lowland type, and
(ii) *Corchorus Olitorius:* upland type.

Production and Distribution

Jute production is very limited and 66.35 per cent is produced by India and another 24.80 per cent by Bangladesh. Other jute-producing countries are China, Ivory Coast (Kote de Ivory), Thailand, Myanmar, Brazil, Uzbekistan, Nepal and Vietnam. The total world production of jute in 2008 was 3,225 million metric tonnes.

Almost all the production of jute comes from India and Bangladesh. Most of the jute is grown in the fertile, alluvial, annually flooded area at the mouth of the Ganges and Brahmaputra rivers, where all the geographical requirements are present. The jute-producing areas of the world are shown in Figure 26.10

Figure 26.10 *Jute-Producing Area of the World*

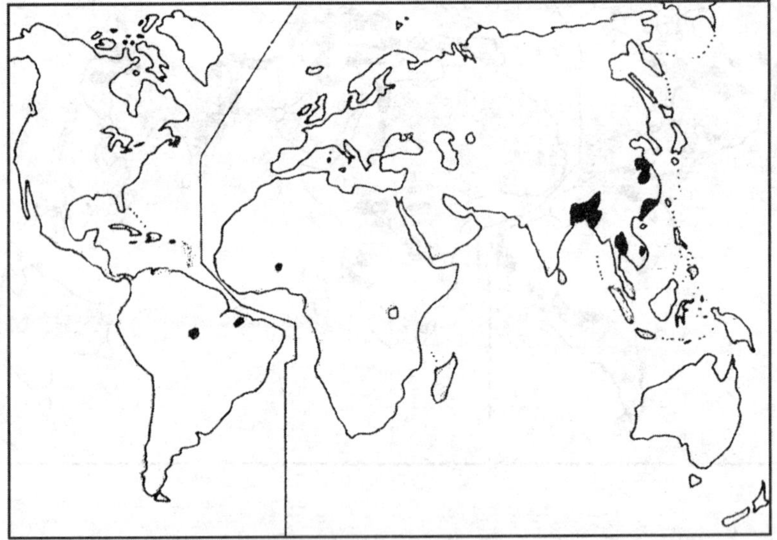

Sugarcane

The most important commercial crop of tropical countries is sugarcane. Sugarcane is a member of the grass family, and is a tall tropical variety, with a hard, thick stem and sugar stored in them. It is believed that the plant originated in Eastern Asia and was taken to the Middle East and Southern Europe by Arab trades in the Middle Ages. The Portuguese realized the profits to be made from growing sugar and introduced it to their colonies in Maderia and the Canary Islands and later to Brazil in the 16th century. From Brazil it spread to the West-Indies and Caribbean countries. It has also been introduced to Africa (especially Natal in South Africa), Australia, Hawaii and Fiji. Now sugarcane is cultivated in most of the tropical countries.

Geographical Conditions for Growth

Sugarcane is a tropical plant, therefore, requires a year of warm weather to reach maturity. The areas having temperature of 20° to 26°C and an average rainfall of 150 cm are suitable for its cultivation. In many areas water is partly supplemented by the canals. Much sunshine is required, particularly at the end of the growing season to produce cane with a high sugar content.

Sugarcane grows well in alluvium and light-clay soil. Moist clay-loam with good drainage is the best soil for cane. As the crop is a valuable one, large amount of fertilizers are used to increase the yield. Sugarcane cultivation also needs sufficient capital and labour.

Production and Distribution

Sugarcane is grown in a very wide range of tropical and sub-tropical countries but the largest producers are from Latin America and Southern and Eastern Asia. In year 2008, the world production of sugarcane was 1, 558 million metric tonnes. Brazil is the largest producer of sugarcane in the world. Its annual production is 5,140 lakh metric tonnes which is 33 per cent of world's production. India and China rank second and third with 22.8 and 6.8 per cent of world's production respectively. Other major sugarcane producers are Thailand, Pakistan, Mexico, USA, Columbia, Australia and Philippines. Cuba, Indonesia, Fiji and Egypt are also producers Sugarcane. The distribution of sugarcane-producing areas is shown in Figure 26.11.

Figure 26.11 *Sugarcane and Sugar Beet Producing Areas of the World*

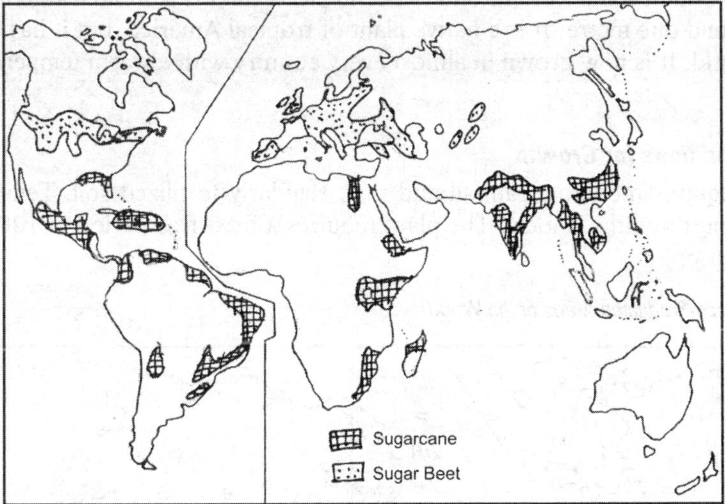

Sugar Beet

Beet is the second most important source of sugar. Sugar beet is a root crop of the beet family. Sugar beet is a large white or yellow root which grows in temperate regions. Its sugar content was first discovered in Germany in the 18th century. But its wide cultivation was started in early 20th century in Europe. Its use has spread to most European countries and to North America as well as in Russia.

Conditions of Growth

Sugar beet is a native of temperate climatic regions with moist and mild winters and moderate summer temperature and thus, in wide contrast to sugarcane it grows in mid-latitude regions. It is very sensitive to annual ranges of rainfall which should necessarily be around 60 to 65 cm. The most suitable geographical conditions for beet-growing are found where the soil is deep, friable, well-drained and contains certain proportion of lime in it. Sugar beet is usually cultivated as a rotation crop on the characteristic mixed farms of Europe and North America. It replaces turnips or other root crops in rotation and is usually alternated with a grain crop. Economic condition is also very important for sugar beet. It is a highly capital-intensive crop, its cultivation is limited only to the affluent countries of the world. It also requires cheap and easy supply of labour to perform the works from planting to harvesting the crop.

Production and Distribution

The total production of sugar beet in the world is 242 million metric tonnes. France is the leading producer of sugar beet and produces about 12 per cent of world's production, while Germany and USA rank second with each producing 10.3 per cent of world production. Other Sugar beet producing countries are Russia, Ukraine, Turkey, Italy, Poland, UK and Spain. Most of the sugar beet producing countries are located in Europe, followed by USA and Russia. In fact, temperate climatic conditions of Europe, Russia and USA are most suitable for sugar beet production.

Tobacco

Tobacco is a commercial crop of considerable importance that provides, directly and indirectly, employment to millions of people. Tobacco has a very wide market as it is consumed by large number of people of the world in some form or other. The tobacco plant is a broad leaved annual growing to a height of the around one metre. It is a native plant of tropical America, but it has been introduced to many parts of world. It is now grown in almost every country with a warm temperate, sub-tropical or tropical climate.

Geographical Conditions for Growth

Tobacco plant requires moderate rainfall and rich, regularly fertilized soil. Temperature exceeding 18°C in the warmest months is ideal. The plant requires a frost free period of 120 to 180 days and a warm growing season.

Figure 26.12 *Tobacco-Producing Areas of the World*

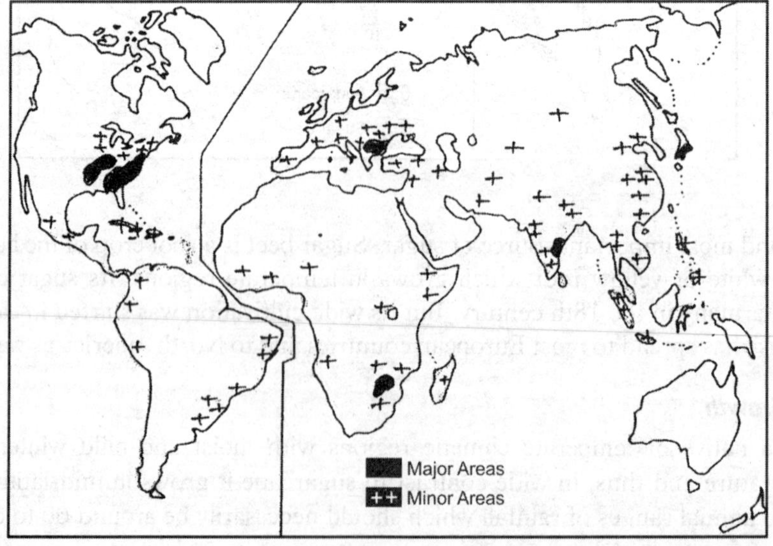

Tobacco is grown mostly wherever the soil is rich, sandy loam. The best kind is well-drained, friable, sandy loam, not too rich in organic matter, but rich in mineral salts like potash, phosphoric acid and iron. Light soil which allows a full development of the roots of tobacco is the best for it.

The tobacco farming requires cheap supply of labour in all stages – field preparation, transplantation, harvesting, processing as well as marketing.

Production and World Distribution

Tobacco is produced in several parts of the world. On the continental basis, Asia produces about 63 per cent tobacco of the world followed by Central America (10%), Europe (8%), Africa and south America (7% each); the total world production of tobacco is about seven million tonnes. China is the largest tobacco producing country; its annual production is 2.64 million tonnes which is 39.6 per cent of the world. India ranks second and produces 9.6 per cent of the world's tobacco. Other major tobacco producing countries are Brazil, USA, European Union, Turkey, Zimbabwe, Indonesia, Malawa, Russia and West Indies. Small quantity of tobacco is also produced in Uganda, Ethiopia and Argentina (Figure 26.12).

27

Livestock and Fisheries

LIVESTOCK

The relationship between men and animals is as old as human civilization. Prior to the domestication of animals, men used to practice hunting, which was the major source of food at that time. Men soon realized the importance of animals in their life and started their domestication probably more than 5000 years ago. From man's point of view, animals have many uses; they can provide food in the form of meat or milk, clothing in the form of hides, furs, wool, hairs, many other products such as horns, bones, feathers, and transport for man and goods, power for ploughing, traction or driving machinery and manure for fertilizing farmland. The importance of livestock raising plays an important role in the economy of many countries of the world. This has become possible only after the commercialization of the pastoralism.

Pastoralism is a worldwide activity prevalent in many forms in different regions of the world. These are:

(i) *Nomadic Herding:* Nomadic herding is still prevalent in arid and semi-arid regions of Africa and Asia. This practice is also prevalent in high altitude areas of Himalaya as well as in Polar Regions. In these areas animals are mostly used as a source of food in form of meat and as a means of transportation.

(ii) *Livestock Ranching:* This practice is performed in areas of hostile climate and where terrain is not suitable for agriculture, use of cattle, goat, sheep, etc. is done for commercial purposes. Dairy products, meat, hide, wool are extracted and sold in the market. Such type of subsistence livestock ranching is done in parts of Argentina, Venezuela and Australia.

(iii) *Mixed Farming:* In many regions of the world pastoralism is performed along with agriculture. In some areas of Africa and Asia it is just in the form of subsistence, which supplements the agricultural activity. But this practice is performed in a commercial way as is done in Western Europe and in North and South America. The grasslands of Prairie in North America, Pampus in South America, Veld in South Africa are suitable areas for mixed farming.

(iv) *Dairy Farming:* Today, in many parts of world, dairy farming is an important commercial activity, in which livestock are reared properly for the production of several milk products. Dairy farming is common in Western Europe, New Zealand, USA, Canada and in some parts of South America. Dairy farming plays vital role in the economy of the countries like New Zealand, Denmark, Sweden, Norway, Spain, France, England, Germany, Canada, USA, etc.

Geographical Factors Controlling Livestock Ranching

The growth and development of livestock ranching, specially commercial livestock farming always depend upon certain geo-economic factors. These factors are as follows:

(i) *Climate:* Most commercial grazing occurs in dry and moderate climate. Dry climate prevails in five of the largest ranching areas – those of North America, Southern South Africa, Australia and the Central Asia. New Zealand and portions of South Africa are having humid climate. Moderate temperature and medium rainfall favour grassland development which is most suitable for livestock ranching.

(ii) *Topography:* The undulation, flat and rolling plains are favourable for the growth of grassland as well as for livestock ranching. The Savannah in Africa, Steppes in Russia, Prairie in North America, Pampus in Argentina, Veld in South Africa and Downs in Australia, all are having rolling plains.

(iii) *Mixed Farming:* In many regions of the world, pastoralism is performed along with agriculture. In some areas of Africa and Asia it is just in the form of subsistence, which supplements the agricultural activity. But this practice is performed in a commercial way as is done in Western Europe and in North and South America.

(iv) *Natural Vegetation:* Livestock ranching occurs in regions of grasslands and desert shrub. The type of grass varies regionally. In the low latitudes are the tropical grasslands generally termed as Savannahs such as the Lanos of Venezuela, the Campos of Brazil, the Gran Chaco of Argentina-Paraguay-Bolivia. The mid latitudes have the best grasslands: Prairies and Steppes. Marginal grasslands are also used, such as desert fringes adjoining Steppes and grassland patches on the edges of forests that adjoin steppes.

World Distribution of Grasslands and Livestock

The whole pastoral economy depends upon grasslands. Therefore, spatial concentration of commercial livestock ranching in the world is controlled by grazing grounds or grasslands. General distribution of livestock in the world and grassland are interdependent on each other. North American grassland provide excellent grazing ground. Vast stretches of land covering parts of Canada and USA are used for livestock rearing. Among all the South American countries, Argentina hold the pride position in commercial livestock ranching. The lushgreen grassland pampas provides excellent grazing facilities to millions of cattle and sheep. The alfalfa grass grows luxuriantly all over the pampas grassland which enables ranchers to improve their livestock health in such a way that Argentina is the largest exporter of dairy product and wool. Apart from Argentina, Uruguay has also improved livestock farming in a big way.

South African grassland, known as Veld is famous for livestock ranching. Low density of population, availability of land, extensive grassland and traditional skill of livestock ranching provided incentives to commercial grazing on a grand scale. New Zealand is famous for its dairy industry in the world and its economy is based on livestock. South African grassland is also famous for sheep rearing. This region produces sizeable quantity of high quality wool.

Tropical Grassland Areas of Livestock

Intervening between the tropical forests and the dry deserts of the trade-wind belts, occur stretches of warm country with a moderate rainfall only. Here there are very few trees but grasses of various kinds which afford abundant food to herds of grazing animals. Such savannahs include the African Sudan, the Venezuelan Lanos of the Orinoco basin, the Brazilian Campos or Selves of Amazon and Shola of India.

These Savannah grasslands generally occupy those areas between the climatic extremes of humid woodlands and dry desert shrub, the transition belts, therefore, between typical grass Prairie and either of these extremes contain more or less tree growth on the one hand, or desert shrub on the other. Soil moisture, rather than total rainfall, usually determines the extent of natural grassland. If the upper layers of soil are moist during part of the year, but the deeper layers remain dry, tree growth cannot compete with grass. The Savannah of the tropics annually receive substantial total rainfall, but a prolonged dry period each year combined with high temperatures and evaporation, prevent the development of forests.

Figure 27.1 *Cattle Rearing Areas of the World*

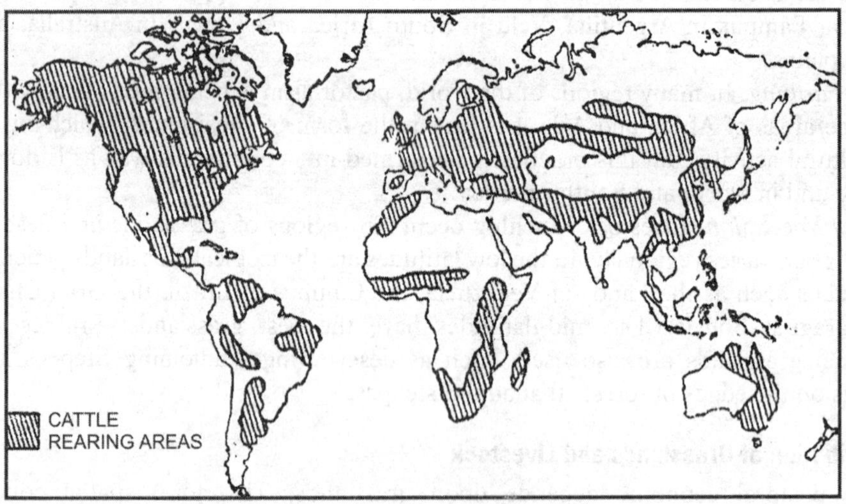

Commercial Livestock Rearing

The most important aspect of this livestock rearing is the commercialization, which is not only responsible for its systematic and scientific development but also a source of income to many countries of the world. The various livestock products and their marketing is a notable feature of international trade. Therefore, it is necessary to discuss in detail these aspects of livestock rearing which are commercially important. These are:

1. Dairy Farming or Dairy Industry
2. Meat Industry
3. Sheep Rearing

Dairy Farming or Dairy Industry

Factors of Growth

Dairy farming has developed only in certain parts of world. This is because of the fact that certain geo-economic conditions are required for its growth and successful development. The factors favourable for the growth of dairy farming are as follows:

(i) *Climate:* Temperate climate is most suitable for dairying. Commercial dairy farming has developed in humid temperate regions. Dairy cattle best thrive in 10°C to 17°C temperatures and 50 cm to 75 cm rainfall. High summer temperature and below freezing winter temperature are harmful for raising cattle. The coastal areas having marine climate and humid continental climate are also suitable for dairying.

(ii) *Fodder:* Dairy cattle require green grass throughout the year. The fodder is also obtained from certain crop such as clover an leguminous crops, which are especially grown for dairy cattle. The good quality of grass available in western Europe, USA, Canada, New Zealand is the main factor of the development of dairy industry in these areas.

(iii) *Water:* On an average 200 litres of water is required for a cow, therefore good quality of water and its availability throughout the year is necessary for dairy industry.

(iv) *Urbanism:* There is a close relationship between dairying and urbanism. The concentration of population in cities create great demand of milk and milk products. The two world's main dairy regions, i.e. Anglo-America and Western Europe are located in highly urbanized areas. The specific type of dairy farming in a dairy region depends to a large degree on the distance of urban markets. The milk consumption in large cities is much more than other regions.

(v) *Capital:* Dairy farming nowadays needs capital for the purchase of cattle, their maintenance, for the purchase of machines as well as for marketing of the products. The processing of milk and their preservation require special types of machines. Similarly, packing and marketing also need capital.

(vi) *Transportation:* Quick and efficient transport network is essential for the dairy industry. The internal transport system helps in collection of milk from production centres, which are generally located away from the production centres. Similarly, external transport system is required for the export. Nowadays air transport is also in use for the export of dairy products to distant regions.

(vii) *Market:* Since dairy products are in great demand, there is no problem of consumption of these products. But when markets are located in nearby places, problem of transportation is lesser and the production is cost-effective. Dairy products are having worldwide market. The dairy products of New Zealand and Denmark are in great demand throughout the world.

Among other factors availability of good quality of milk-producing cattle, skilled labour, government policy, prevention of cattle diseases, etc. are also important. Regular researches and constant experiments with regards to breeds, health, forage, etc. are also necessary.

Figure 27.2 *Dairy Farming Regions of the World*

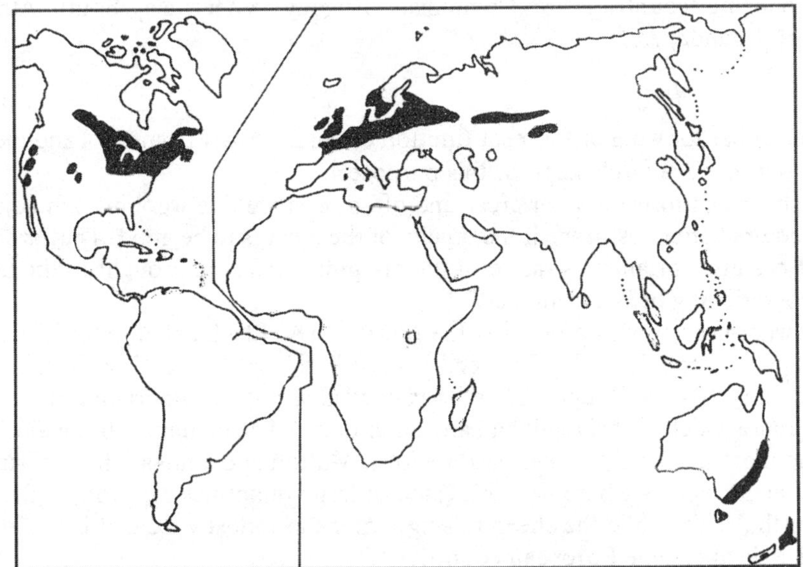

World Distribution

The spatial distributions of dairy farming in the world are widespread. But three main regions of commercial dairy farming are important. The largest is in Western Europe, a belt extending about 3000 km from Atlantic Coast almost to Moscow. The second dairy region, that of Anglo-America, is likewise a belt, this one beginning at Atlantic Coast to the Prairie Provinces of Canada. The third region is the Australia-New Zealand region. The other dairy regions are smaller and located in South Africa and South America. In Asia, dairying has developed in Japan, South Korea, China, India, Pakistan, etc. but most of the productions are consumed locally. Figure 27.2 depicts the world distribution of dairy farming.

Beef Cattle Rearing (Meat Industry)

One of the main objectives of cattle rearing is to obtain meat or beef. Beef cattle are much more widely distributed than dairy cattle because they are harder and can tolerate wider range of physical and climatic conditions. They can survive on sparse pasturage. Beef cattle rearing are also the means of livelihood for thousands of people in many areas of Asia and Africa. While in many other countries like Argentina, Australia, New Zealand it is done for the commercial purpose. Due to the introduction of refrigeration in the 1880s, the international trade in beef expanded rapidly. Chilled meat is transported in cold storage chamber at a temperature just below freezing point, so that it arrives in the consuming countries in fresh state. There is a considerable trade in corned beef (canned, cooked meat), beef extracts or-tongues and other prepared and cooked canned beef products.

There are several breeds of beef cattle. Western Europe is the home of many beef cattle breeds. The Aberdeen Angus breed originated in Abesdeenshire, Scotland. It was first imported into the USA in 1850 and now a new breed, the Red Angus has become important. The Hereford breed now cross-breed called Polled Hereford, developed in US and Canada. The Scotland breed of north-east England, the Galloway of south-west Scotland, the Devon of South-west England, the Highland of Scottish highland are some of the very important breeds of beef cattle, now widespread in many countries. There are several other local and regional breeds of beef cattle found in various countries of the world. The spatial distribution of beef cattle in the world is quite widespread but there is concentration in certain areas. USA, Brazil, China, Argentina, Russia, France, India, Mexico, Germany, Canada, Italy, Ukraine, UK, Columbia and New Zealand are important beef-producing countries. Other beef producing countries are Columbia, Uruguay, Venezuela, South Africa, Ethiopia, Tanzania, Kenya, Uganda, etc.

Sheep Rearing

Sheep are reared either for wool or for meat (mutton or lamb). Apart from wool and meat, sheep may also yield milk, but they are rarely kept for this purpose.

Although sheep can thrive in poor areas and often produce fine wool in such conditions, sheep kept for meat require better pasturage if the quality of the meat is to be good. They are therefore kept in rich lowland pastures, or animals raised in the marginal lands are brought to the better areas for fattening and conditioning before slaughter.

The main wool-bearing sheep breed is the Merino. The Blackface of Scottish Highland yields better meat but poor wool. The Cheviot breed is used both for meat and wool. Leicester, Lincoln, Romany March, Black Welsh, Rough Fell and Swallowtail are other important breeds of sheep.

Sheep are more widely distributed than any other domesticated animal, but majority are kept for wool rather than meat, especially in semi-arid regions. Mutton and lamb are not consumed so widely as beef and the main regions where they are eaten in large quantities are Britain, the drier parts of Eurasia and North Africa, while the sheep raising regions are mostly located in the southern continents, China, Russia and some European countries.

FISHERIES

Fishing is one of the oldest of man's activities, and it is still practiced in all parts of the world where suitable conditions exist. Fishing is practiced in those waters which afford at the same time a suitable land base for fishery and abundance of fish of the species in demand for human consumption. Fish provide an important source of protein in the diet for much of world's population. For many parts of the world, fish is a predominant source of high-quality animal protein. Dependence on fish is heaviest in China, Japan, southeast Asia, Sub-Saharan Africa, and parts of Latin America. Fish is also serving a growing role in the diet of many developed countries, including the United States.

Fisheries are commonly divided into: (i) inland (fresh water) fishery, (ii) coastal fishery, and (iii) open-sea fishery. Fisheries may be commercial or subsistence. Fresh water fish comprise about 16 of the total annual fish catch; while marine fish comprise 84 per cent. Most of the freshwater fisheries are located in lakes and rivers.

Fishing Areas of the World

There are four important fishing areas in the world which are located in the temperate regions. These are:

1. North-West Pacific Region
2. North-West Atlantic Coast
3. North-East Atlantic
4. North-East Pacific
   1. *North-West Pacific Region*: seas bordering central and northern Japan, Northern China, Korea and north-eastern Russia. Japan occupies a strategic position for fishing. The mixing of two currents, Kuroshia warm water current with the cold waters of Oyashio favours the vigorous growth of fish. Japan produces about 8–10 billion pounds fish annually. Besides Japan, China, Korea and Russia are major fish producing countries of this region. China ranks first in fish production in the world and produces about 32 per cent of world's fish production.
   2. *North-West Atlantic Coast:* The banks off the coasts of New England and New Foundland, constitute one of the most important fishing submerged platforms in this area which are used as fishing grounds. This area is a good breeding ground for fish as it is the meeting place of the warm Gulf Stream and cold Labrador current, so it abounds in organic food and plankton.
   3. *North-East Atlantic (The Coasts of North-West Europe):* The seas in this area case important fishing grounds in the world. Dogger Bank and Great Fisher Bank the main fishing grounds of the North Sea. Britain, France, Holland and Norway are the four leading nations participating in large scale fishing. Belgium, Denmark and Spain also take part in fishing in this area.
   4. *North-East Pacific (The Western Coasts of North America):* The area of this region extends from Alaska to California and is noted for salmon, pilchard and tuna fisheries. USA and Canada are the main countries producing fish from this area.

Other Fishing Areas: Tropical Seas are also having varieties of fish species but in these seas commercial fishing is very limited. In India, Pakistan, Indonesia, Pacific Islands and Africa, fish is mostly consumed locally. Similarly in Peru, Chile and Argentina, fishing is done mostly for local consumption.

China, Peru, Japan, USA, Russia, Indonesia, India, Chile, Thailand, Norway and Denmark are the major fishing nations of the world.

28

Mineral and Energy Resources

MINERAL RESOURCES

Minerals are of vital significance to every country because it is the minerals which make possible the power and machine. The tools and machines which modern man uses, are fashioned primarily from minerals taken from the earth. Modern civilization is in many respects, dependent on mineral products. The development of fundamental science, engineering and technology have made it possible to extract and use a great variety of minerals in increasing proportion. Mineral resources are, however, fixed in quantity; they cannot be increased or replaced. Once minerals are extracted from the earth, they are gone forever. It takes probably millions of years for nature to turn earth substance into minerals. From a practical viewpoint, therefore, mineral deposits are exhaustible and mining is considered as a 'robber industry'.

Classification of Minerals

There are several types of minerals, but according to their general characteristics and commercial use, they are classified as follows:

1. *Metallic minerals:* Metallic minerals are being used for a very wide range of purposes. These are sub-divided into the following groups:
   (a) Iron.
   (b) Base metals such as tin, copper, aluminium, lead and zinc.
   (c) Ferro-alloys such as manganese, chromium, nickel, cobalt, tungsten, molybdenum and vanadium can be alloyed with iron to produce better steel.
   (d) Precious metals like gold, silver and platinum.
   (e) Others include uranium, which has specialized uses.
2. *Non-metallic minerals:* These minerals are used for structural purposes, chemicals and various industries. The more common non-metallic minerals are asbestos, mica, graphite, fluorspar, sulphur, diamonds, phosphate rocks, gypsum, etc. Certain precious stones, e.g. diamonds are also non-metallic minerals.
3. *Mineral fuels:* Coal, petroleum and natural gas are the mineral fuels. These three are also known as fossil fuels.
4. *Rocks:* Rocks are also aggregations of minerals. Those commercially exploited include granite and other crystalline igneous and metamorphic rocks.

Iron Ore

Iron is a basic mineral and the most useful of all the metals. The success of almost every industrial enterprise depends upon the extensive and efficient use of machinery and other economic equipment made wholly or in part from iron and its alloys. Iron ore has unique importance in the modern world. It provides steel which is the basis of all industrialization.

There are four types of iron ore, viz. Magnetite, Haematite, Limonite and Siderite.

(i) *Magnetite* (Fe_3O_4): This is the finest type of iron ore with a very high content of iron up to 70 per cent. Magnetite is black in colour and formed in igneous or metamorphic rocks. It has excellent magnetic qualities.

(ii) *Haematite* (Fe_3O_3): The iron content in this ore is between 50 to 65 per cent and is the most important industrial ore in terms of quantity used. Haematites are red ores derived from sedimentary rocks and occur in crystalline or in powdery forms.

(iii) *Limonite* ($2Fe_2O_3H_2O$): The iron content in this ore is less than 50 per cent. This is a brown ore occurring in sedimentary rocks over large areas of the world. It also sometimes occurs in swamps, where it is known as bog iron.

(iv) *Siderite* ($FeCO_3$): This is a carbonate of iron, ash-grey in colour, and is found interbedded with sedimentary rocks. Its iron content is usually between 20 to 30 per cent.

Iron ore-producing areas are widely distributed in the world. There are about 60 countries in the world, which produce iron ore. China, Brazil, Australia, India, Russia, Ukraine, USA, South Africa, Canada, Sweden, etc. are the main producers of the iron ore. China is the leading producer of iron ore and produces more than 32 per cent of the world's iron ore. (Table 28.1)

Table 28.1 *Leading Iron Ore Producers in the World*

Country	*Percentage of Total Production*
China	32.5
Brazil	17.7
Australia	15.2
India	9.6
Russia	5.8

Brazil produces about 17 per cent and Australia 15 per cent of the world's iron ore production, while India ranks fourth in iron ore production and produces 9.6 per cent of the world's iron ore.

The world distribution of iron ore-producing areas is shown in Figure 28.1

Ukraine, Brazil, Russia, China and Australia are having major reserves of the iron ore. The total reserve of iron ore in the world is estimated as 37,000 billion metric tonnes.

Manganese

Manganese is a ferro-alloy metal and essential to the production of steel. Almost all steel contains some manganese from 0.5 per cent to 12 per cent. A little manganese added to iron removes gases, and acts as a 'cleanser' in manufacture of steel. The addition of 2 per cent manganese adds strength and 12 per cent imparts great toughness and resistance to abrasion, as in the making of heavy duty machinery, railway lines, dredge-bucket teeth and jaws by rock crushers, etc.

The main producers of manganese in the world are China, South Africa, Australia, Gabon, Kazakhstan, Brazil, India, Ghana, Ukraine, Georgia and Mexico. The production of manganese in the world is given in Table 28.2.

Figure 28.1 *World Distribution of Iron Ore Producing Areas*

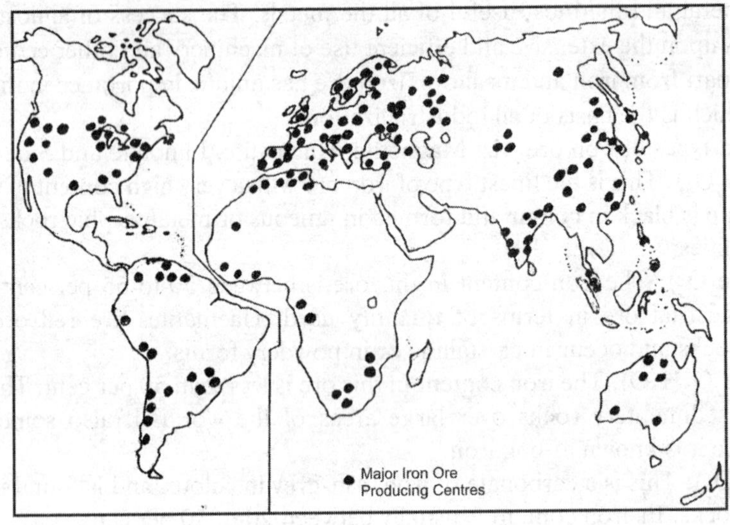

Table 28.2 *Leading Manganese Producers in the World*

Country	*Percentage of World Production*
China	21.6
Kazakhstan	16.2
South Africa	11.0
Gabon	8.9
Ghana	7.7
Australia	7.6
India	4.7

The world distribution of manganese has been depicted in Figure 28.2

Figure 28.2 *World Distribution of Manganese*

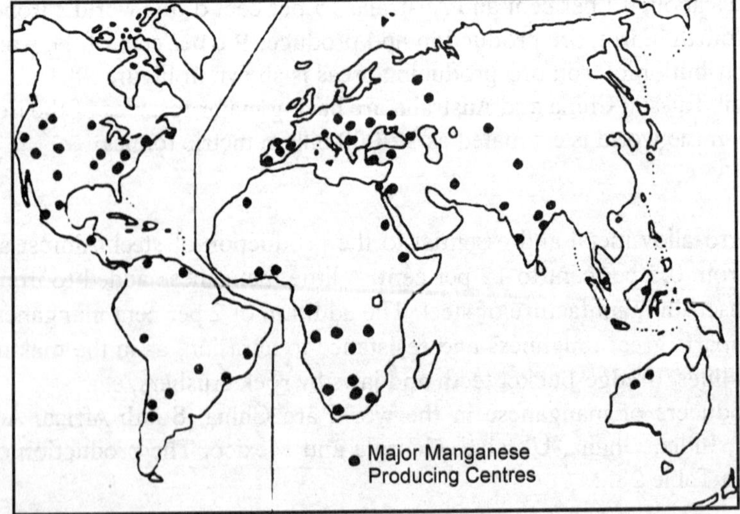

Copper

Copper is a soft brown metal found almost solely in igneous and metamorphic rocks. It was one of the earliest metals used by man because it often occurs in a pure state and because it is very easily worked. The use of copper was recorded in Egypt as early as 5000 BC, and in the far east around 2500 BC. It was also very extensively used in Cyprus giving the island its name. Copper is unfortunately very soft, but by mixing tin with copper, bronze can be obtained, which is harder and tougher than pure copper and more suitable for weapons and tools. The Bronze Age in history thus marked a significant milestone in the progress of human civilization for metals began to be widely used. The present day industrial significance of copper began only with the invention of electricity.

Copper occurs in three forms: (i) as native metal in its pure state, (ii) as oxides, and (iii) as sulphides. Commercial copper ores are not obtained from copper oxides or copper sulphides with a copper content of 2 to 6 per cent. Most copper mines are operated by the open-cast method and are highly mechanized.

Copper is found in every continent and produced in more than 40 countries of the world. The main producers of copper in the world are Chile, Peru, USA, China, Australia, Indonesia, Russia, Canada, Zambia, Poland, Kazakhstan and Mexico.

Table 28.3 *Leading Copper Producers in the World*

Country	*Percentage of World Production*
Chile	34.8
United States of America	8.8
Indonesia	7.9
Australia	6.3
Peru	5.9

Chile is the largest producer of copper in this world. It produces 34.8 per cent of the world's copper, followed by USA with 8.8 per cent production of copper. Indonesia produces 7.9 per cent, Australia, 6.3 and Peru, 5.9 per cent of the world's copper production.

The world distribution of copper is shown in Figure 28.3

Figure 28.3 *World Distribution of Copper*

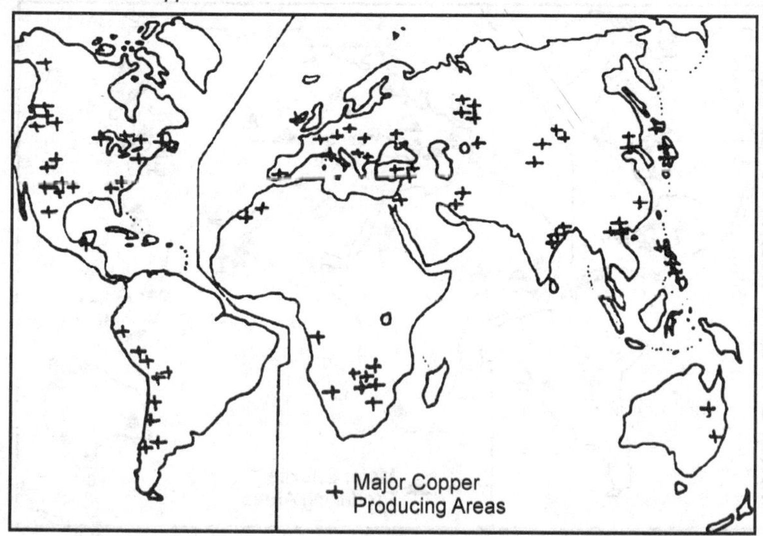

Aluminium (Bauxite)

Aluminium is not found in the native states as a pure metal. It generally occurs as an ingredient of chemical compounds in highly complex minerals. The following four are the natural compounds of aluminium:

(i) *Cryolite* is a double fluoride of aluminium and soda. It was first used as an ore and later in the electrolysis of alumina as a solvent. The west coast of Greenland is the only place where it is found in large quantities. Now, of course, it can be produced synthetically.

(ii) *Corundum* is a native alumina (Al_2O_3), i.e. aluminium oxide. It is used as an abrasive.

(iii) *Kaolin,* the alumina content being too low and silicate content being too high, is not used in aluminium extraction.

(iv) *Bauxite* is the ore containing hydrated aluminium oxide as its important mineral and this is the only source from which aluminium is extracted commercially. Commercial bauxite should contain at least 50 per cent alumina (Al_2O_3) and less than 7 per cent silica. Bauxite is found chiefly in the tropics or in regions of tropical climate. Tropical climates favour the formation of bauxite.

The production of bauxite is done in more than 40 countries of the world. The important bauxite producers are (their percentage in world's production is given in bracket): Australia (31.34%), China (18.41%), Brazil (13.93%), Guinea (8.36%), Jamaica (3.98%), Russia (1.64%), Venezuela (2.39%), Surinam (1.99%), Kazakhstan (2.44%), Greece (1.09%), Guyana (0.60%) and Vietnam (0.01%). The total world production of bauxite is given in Table 28.4

Table 28.4 *Bauxite Production in the World*

Year	*Production in Lakh Metric tonnes*
2003	1,334
2005	1,650
2008	2,050
2009	2,010

The world distribution of bauxite-producing areas is shown in Figure 28.4.

Figure 28.4 *Bauxite Producing Areas of the World*

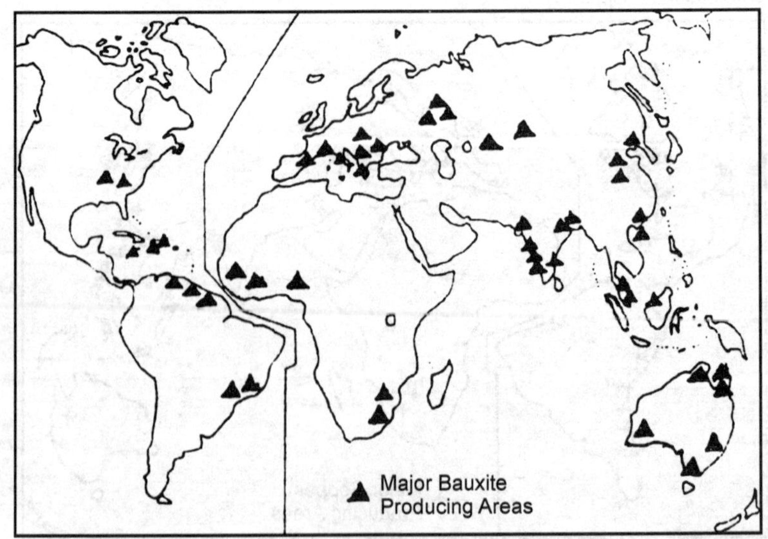

Tin

Tin was found and used in early times but it has only attained its present importance since the invention of tin cans. Tin is extensively used in the canning industry and as an alloy with copper in bronze. It is also used for coating thin sheets by steel as it is corrosion-resisting. Tin plating consumes about 40 per cent of the world's tin output. An electrolytic process is used to coat thin steel plates with a very fine layer of tin.

The tin-producing areas of the world are limited in certain pockets and very unevenly distributed around the world. Nearly 75 per cent of the total output comes from China and other tin producing countries are Peru, Bolivia, Brazil, Congo, Vietnam, Malaysia, Russia and Australia. The major producers and their percentage share in the world is given in the Table 28.5.

Table 28.5 *Important Tin-Producing Countries and Their Percentage in the World*

Country	*Percentage in the World*	*Country*	*Percentage in the World*
China	45.05	Congo	0.90
Indonesia	30.03	Vietnam	1.05
Peru	11.41	Malaysia	0.60
Bolivia	4.80	Russia	0.60
Brazil	3.60	Australia	0.03

Source: US Geological Survey, 2010

Lead

Lead is a useful metal because of its properties of softness, resistance to corrosion and malleability. Lead was one of the earliest metals used by man. It is very easily shaped into pipes or rods. The Romans used considerable quantities of lead to make pipes. Lead is dull grey metal. The ore galena or lead sulphide, with a lead content of 85 per cent is commercially the most important source of lead. Lead seldom occurs in its native form and invariably contains some zinc or silver.

The world's total production of lead in year 2009 was 3,900 thousand metric tonnes. The main producers of lead in the world are China, Australia, USA, Peru, Mexico, Canada, India, Bolivia, Poland, Russia, Sweden, Ireland and South Africa.

The leading lead producer countries of the world are as follows:

Table 28.6 *Leading Lead-Producing Countries and their Percentage in World Production*

Country	*Production Year/Percentage in the World*		
	2005	*2008*	*2009*
China	29.1	39.06	43.33
Australia	23.1	16.8	13.23
USA	13.4	10.38	10.26
Peru	9.3	8.98	7.82
Mexico	4.4	2.63	3.97
Canada	2.3	2.58	2.44
India	1.05	2.27	2.26
Bolivia	—	2.14	2.56

Other lead producing countries are Poland, Russia, Sweden, Ireland, South Africa, Korea, Japan, Myanmar, Iran, Turkey, Thailand, Brazil, Greece, Spain, Romania, Morocco, Namibia, etc.

Zinc

Zinc used to occur in various minerals. The ores of commercial value are chiefly sulphides, of which Sphalerite is the most important. It is used in galvanized roofing and siding sheets, wire fence, pails, etc. It is corrosion-resisting. It is found in association with ores of other metals, especially lead. So zinc-producing districts coincide with lead-producing areas.

Since lead and zinc so often occur together, the major zinc producers are much the same as for lead. The main producers of zinc in the world are China, Peru, Australia, USA, Canada, India, Kazakhstan, Ireland and Mexico.

The distribution of zinc-producing areas along with lead has been depicted in Figure 28.5.

Figure 28.5 *Distribution of Lead and Zinc in the World*

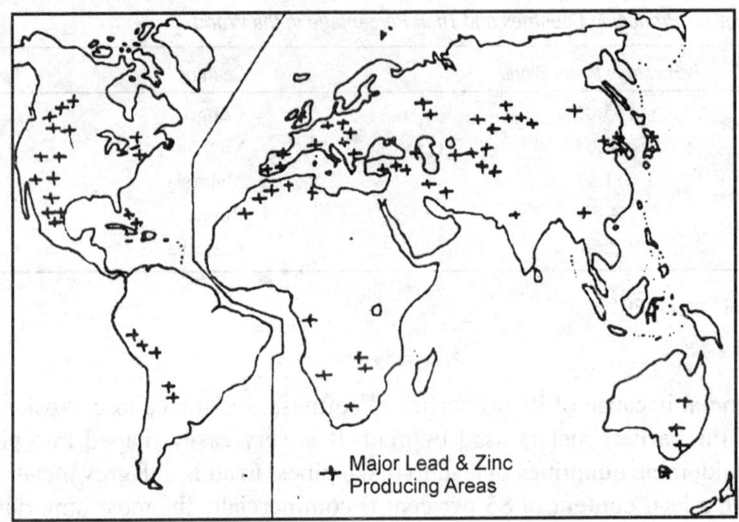

Nickel

Nickel is a hard and silvery metal, used in making stainless steel and for electroplating. It is one of the hardest, most malleable and least fusible metals known. It is less magnetic than iron and is thus useful for constructing metal parts located near compasses. The main ore of nickel is pentlaudite, a complex mixture of nickel, iron and sulphur. It is also often found in association with copper.

The important nickel-producing countries and their percentage in world's production is as follows: Russia – 17.14 per cent, Canada – 15.53 per cent, Indonesia – 13.11 per cent, Australia – 11.18 per cent, New Caledonia – 5.75 per cent, Columbia – 4.65 per cent, China – 5.28 per cent, Philippines – 5.49 per cent, Brazil – 4.70 per cent and Cuba – 4.78 per cent.

Some nickel is also produced in South Africa, Dominican Republic, Norway and Germany. In India, nickel ore is found in Cuttack and Mayurbhanj districts of Odisha.

Chromium

Chromium is a hard, silvery metal, essential in making stainless steel, cutlery, kitchenware, household fittings, surgical apparatus and chemical equipments. Chrome Steel is superior steel used for marine and aircraft parts. Chromium is also used in making nickel-chromium alloys and chromite (chrome ore).

South Africa is the largest producer of chromium in the world. It produces 44.65 per cent of the total world production of chromium. Kazakhstan is the second largest producer, which contributes 17.21 per cent of the total production of chromium in the world.

India is the third largest producer of chromium in the world and its percentage in world production is 15.35. In India deposits of economic significance occur in Bihar, Karnataka, Maharashtra, Odisha and Tamil Nadu.

Other chromium-producing countries are Turkey, Philippines, Albania, Brazil and Finland.

Tungsten

Tungsten is a very heavy metal, with a melting point higher than any other metal. Tungsten is widely used in steel alloys for the manufacture of high-speed cutting tools. Tungsten carbides are very hard than any other metal and employed in the cutting edges of machine tools, razor blades, armour plates, jet engines, rockets and missiles. Pure tungsten is used in making filaments of electric light bulbs. The chief source of tungsten is Wolfarm or Wolframite, often found in association with tin cassiterites in quartz veins.

China has a monopoly in tungsten production. China produces 75 per cent of the world's tungsten. The majority of the deposits occur in the Nanking Mountains of Southern China. Hunan and Kiangsi provinces are also important in tungsten production.

Russia is the second largest producer and produces 5.86 per cent of world's tungsten.

Canada produces 4.76 per cent and Austria, 2.20 per cent of world's tungsten. Other tungsten-producing countries are Bolivia (2.01%), Portugal (1.65%) and North Korea (1.10%). The minor producers of tungsten are Kazakhstan, USA, Brazil, Zaire, Rwanda, Burundi, Australia, Spain, France and Sweden.

Antimony

Antimony is obtained from stibnite and it also occurs in very small quantity along with copper, silver, gold and lead ores. Since antimony has a low melting point, it is used as alloys with lead.

China is a not only the largest producer but it also produces about 84.8 per cent of the world's antimony. Hunan and Kwangsi provinces are having large deposits of antimony.

South Africa produces 4.2 per cent of world's production of antimony. Here antimony occurs as a by-product from gold mine in the Transvaal.

Russia produces 3.8 per cent and Tajikistan produces 1.7 per cent of world's antimony.

Bolivia and Australia also contribute 1.6 and 1.5 per cent of world's antimony respectively. Other antimony-producing countries are Peru, USA, Guatemala, Turkey, Mexico and Kazakhstan.

Vanadium

Vanadium is a hard silvery metal named after a Scandinavian goddess of beauty by its discoverer Sefstorm in 1830. The principal use of vanadium is in making alloy of steel and iron. It acts as a 'scavenging agent' removing the non-metallic impurities. It also increases strength, ductility and resilience. The vanadium compounds and salts are used in making paints and dyes, coloured glasses and glazes and also used in chemical industry.

South Africa is the largest producer of vanadium. It produces nearly 37 per cent vanadium of the world. Russia is another important producer, followed by USA. The other producers of vanadium are Finland, Chile, Namibia and Norway.

Molybdenum

Molybdenum is a silvery metal having many properties similar to those of chromium. Molybdenite (MoS) is the principal ore, the production of which is given in terms of concentrates ranging from 48 to 49 per cent metallic molybdenum. Some molybdenum is also derived as a by-product of copper smelting.

USA is the largest producer of molybdenum. It produces more than 56 per cent of the world's production. In USA, Colorado, Utah, Nevada, Arizona, California and New Mexico states are having molybdenum mining areas. Canada is the second largest producer followed by Russia. The other molybdenum-producing countries in the world are Chile, Peru, China, Mexico, Norway and Japan.

Gold

Gold is regarded as a symbol of prosperity and has been used from the earliest days because of its high metallic luster and its attractive yellow-reddish colour. It is used in jewellery and in the manufacture of luxury articles. Because of its rarity, value and durability, it has always been highly prized as a medium of exchange. It was also used as coin in the past in few areas of the world. Today, the international system is backed by gold.

Although gold is produced in many countries but the important gold-producing countries of the world are South Africa, Australia, USA, China, Peru, Russia, Canada, Uzbekistan, Indonesia and Papua New Guinea.

South Africa is the largest producer of gold in the world. Its annual production is 341.48 thousand kilograms. The Witwersand district in the Transvaal is leading gold-producing area of South Africa. Another area is the Odendabsrust in Orange Free State. Johannesburg, Germinston, Benoni, Booksburg, and Krugerdrop are other gold-producing centres of South Africa.

Australia is the second leading gold producer in the world. Kalgorlie and Koolgardie in Western Australia are the famous gold mining areas. The other gold mining areas of Australia are Bendigo and Bellwort in Victoria, Mount Morgan in Queensland and New South Wales.

USA is the third leading producer of gold in the world. In USA, California, Coloradi, Nevada, South Dakota are the main gold-producing states.

In *Canada*, Prokupine, Kirkland, Larder lake, Pikil lake, Red lake, Yellow Knife and Clondyke are the main gold-producing areas.

China has now emerged as a gold-producing country. Its main gold-producing areas are Yunnan, Manchuria and Tsinling Shan.

In *Russia*, the Yakuts region of north-eastern Siberia is the main gold-producing area.

The other gold-producing countries are Peru, Uzbekistan, Indonesia, Papua New Guinea, Chile, Mali, Tanzania, Philippines, Zimbabwe and Ghana. A little amount of gold is produced in Kolar mines of Karnataka.

Silver

Silver is a lustrous white metal, though much less valuable than gold. Only a small quantity of silver is mined as argenite (silver ore). Most of the silver is derived from ore of lead, zinc, copper and gold. In fact, most of the silver is obtained as a by-product.

Mexico is the largest producer of silver in the world and produces about 16 per cent of the world's production.

China is the second largest silver-producing country in the world, accounting nearly 12 per cent of the world's total output.

Australia, Chile and *Canada* are third, fourth and fifth silver-producing countries of the world respectively.

The other silver-producing countries of the world are Russia, USA, Korea Republic, Kazakhstan, Bolivia, Sweden, Indonesia, Morocco, Turkey, Uzbekistan, Myanmar, Germany and Spain.

Platinum

Platinum is included among the rare metals. It is a white metal having lustrous appearance and a very high melting point. Platinum is found with other rare metals such as osmium, palladium, iridium and rhodium. Apart from its use in jewellery, platinum is also used in industrial applications, especially in the chemical, electrical and metallurgical fields. It is also used as a catalyst in petroleum refining.

Columbia, Canada, Russia, South Africa and USA are the important producers of platinum in the world.

Other Industrial Minerals

Mica

Mica is used in the electrical industry, in condensers, insulators, heating elements, radio components, etc. Mica is a transport material (sheet) and found in white, black and brown colour. Mica occurrs in igneous rocks and its commercial types are Muscovite (white) and Phlogopite (brown). Powdered mica is used in coating and roofing.

India is the largest producer of mica and supplies about 80 per cent of the world's output. Bihar, Jharkhand and Rajasthan are the principal mica-producing states. Other mica-producing states are West Bengal, Odisha, Chhattisgarh, Tamil Nadu, Kerala, Karnataka, Haryana and Himachal Pradesh.

Brazil is also having large mica deposits. In Brazil mica occurs in a 480 km long and 190 km wide belt along the Atlantic coast. Minas Gerais, Sao Paulo, Matto Grasso, Goiaz and Paribas are having mica-producing areas.

USA, Canada, Russia, Czech Republic, Slovakia, France, Sweden, Norway, China, Australia, South Africa, Tanzania, Zimbabwe and Malaysia are other producers of mica.

Asbestos

Asbestos is found in the veins of igneous rocks and is a silky and fibrous mineral. Its main ores are Chrysolite and Actinolite. It is used to make fireproof material, industrial cloths, protective clothing, etc. It is also used for brake-linings, safety sheets, roofing sheets and tiles, insulating boards, fireman's ropes, special papers and asbestos cement. Russia, Ukraine, Canada, South Africa, Zimbabwe, China, Italy and Brazil are the main producers.

Phosphate

Phosphate occurs in sedimentary rocks or as phosphatic nodules. Another source is bird droppings or guano. The phosphate is the most important source of phosphorus and is mainly used in fertilizer. It is also used in chemical, drug and food industry. But its primary use is in fertilizer and world's more than 80 per cent phosphate fertilizer is made by it.

USA is the largest producer of phosphate rock. It is mostly found in western states of Idaho, Montana, Utah, Wyoming and Florida. The other phosphate-producing countries are Russia, Kazakhstan, Morocco, Tunisia, Togo, South Africa, Algeria and Egypt. Some Pacific and Indian Ocean islands also produce phosphate such as Carolina, Mariana, Ruckus, Marshall, Naru, and Society islands. The Guano deposits are found in Peruvian and Chilean desert in South America.

Sulphur

Sulphur is used mainly to manufacture sulphuric acid, which is important in the heavy chemical industry. Its other uses are in manufacture of fertilizers, insecticides, explosives, pulp and paper. There are three sources of sulphur, i.e. native sulphur, sulphide of metals mainly iron (pyrites) and sulphide ores such as copper, zinc and silver bases emitted from volcanoes are highly sulphurous and sulphur thus deposited in volcanic regions. The main sulphur-producing countries are USA, Mexico, Ukraine, Turkmenistan, Kazakhstan, Uzbekistan, Russia, Italy, Norway and Germany.

Potash

Potash or potassium is a salt, similar to common salt and usually occurring within beds or domes. The potassium minerals commonly occur in association with other soluble minerals, such as halite, gypsum and anhydrite, all of which have been precipitated from solution in ancient sea. Potash is used to manufacture soaps, explosives, medicine, glass, paper, dyes, and photographic materials as well as in fertilizers.

In the world, potash-producing countries are Germany, France, USA, Kazakhstan, Turkmenistan, Russia, Spain, Canada and Israel.

ENERGY RESOURCES

Coal

Coal is black or brown rock, consisting mainly of carbon, which is formed by the compressed vegetative remains of past ages. Much of the present-day high quality coals were deposited during carboniferous era, i.e. about 300 million years ago. More recent deposits of Tertiary age are usually composed of lignite or brown coal.

Coal has been classified according to chemical composition, heating value or physical characteristics. The following types of coal have been identified according to their physical properties:

(i) *Peat Coal:* This is the first stage in the change of wood to coal. It is light brown and friable. Producer gas is manufactured from peat. It is also known as brown coal.

(ii) *Lignite Coal:* Lignite burns with much smoke and flame. The first product is lignite coal which ranges in colour from yellow to dark brown, and is still rich in volatile constituents. About half of the world's lignite is mined in Germany. It is also an important raw material for German chemical industries. Lignite coal has the lowest carbon content about 45 per cent and the highest powder content.

(iii) *Bituminous Coal:* Bituminous coals are subdivided into coking and non-coking coals and each one of them in turn is divided into gas coal and steam coal. Coking coal is used for coke-making domestic fuel, the soft coke. It contains a higher percentage of carbon (70–75%) and less water and oxygen than lignite.

(iv) *Anthracite Coal:* Anthracite coal is the last stage in coal formation. It is extremely hard, and has a brilliant black luster. It has the highest carbon content (80–90%) and produces practically no smoke or flame. Among the nations of the world, only China and the United States have appreciable deposits of anthracite.

Coal is produced in many countries of the world but its important producers are China, USA, India, Australia, South Africa, Russia, Indonesia, Poland, Kazakhstan, Columbia, Canada, Germany, Vietnam, UK and Ukraine. The following table indicates the coal production in the world:

Coal Production in the World (in thousand metric tonnes)

Country	Production	Country	Production
China	1,992,324	South Africa	2,42,821
USA	5,10,918	Russia	1,189,758
India	3,82,615	Indonesia	1,19,700
Australia	2,51,136	Poland	1,01,230

The world distribution of coalfields is shown in Figure 28.6.

Figure 28.6 *World Distribution of Coalfields*

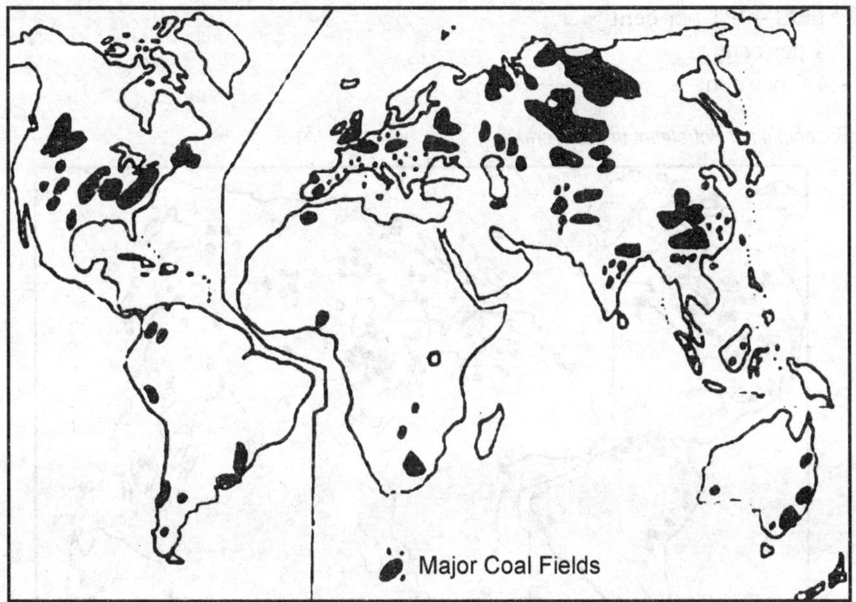

China ranks first in the world in coal production. Nearly 38 per cent of world's output is from China. USA is the second largest producer of coal in the world and produces about 17 per cent coal of the world while India ranks third with 7.5 per cent of world's coal output.

According to an estimate the total coal reserves of the world are 90,906 crore metric tonnes and their largest share exist in Europe and Asia, but USA is the single largest country having more than 27 per cent of the coal reserves of the world followed by Russia and China having 17.3 and 12.6 per cent coal reserves respectively. India is also significant in terms of coal reserves of 10.2 per cent. Other important countries having coal reserves are Australia, South Africa, Ukraine, Kazakhstan, Poland and Brazil.

Petroleum (Mineral Oil)

Petroleum, nowadays, is a main source of energy in the world. This is also due to its multiple usability in different fields of machine civilization. Every aspect of day to day life of man is somehow influenced by the use of petroleum. Transportation, defense, technology, industry, commerce, research and development and many other facts of human activities are directly or indirectly linked with the use of petroleum or its sub-products. It provides fuel for the heat and lighting, lubricants for machinery and raw material for a number of manufacturing industries.

Petroleum Reserves, Production and World Distribution

The West Asia or Middle East is having the largest petroleum reserves of the world, which is about 60 per cent of the world's oil reserves. The total estimated world's oil reserves in 2008 were 1,243 (10^9 bbl). The countries having large petroleum reserves are Saudi Arabia, Canada, Iran, Iraq, Kuwait, UAE, Venezuela, Russia, Libya, Nigeria, Kazakhstan, USA, China, Qatar, Algeria, Brazil and Mexico.

The five leading petroleum producer countries of the world are:

1. Saudi Arabia – 12.3 per cent
2. Russia – 11.7 per cent
3. United States – 8.1 per cent
4. Iran – 5.4 per cent
5. China – 4.9 per cent

Figure 28.7 *Distribution of Petroleum in the World*

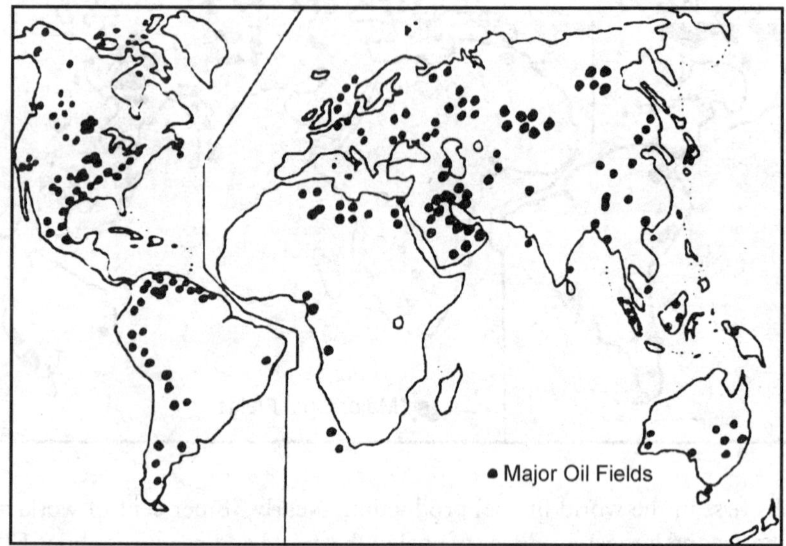

The petroleum-producing countries of the world can be grouped into five geographical regions:

West Asia or Middle East Region

The West Asia region is the largest petroleum-producing region of the world. It also has the largest oil reserves of the world, which is more than 60 per cent of the total world's reserves. The main oil-producing countries of this region are Saudi Arabia, Iran, Iraq, Kuwait, UAE, Qatar, Bahrain, Oman and Abu Dhabi, etc.

Figure 28.8 *Major Oilfields of West Asia*

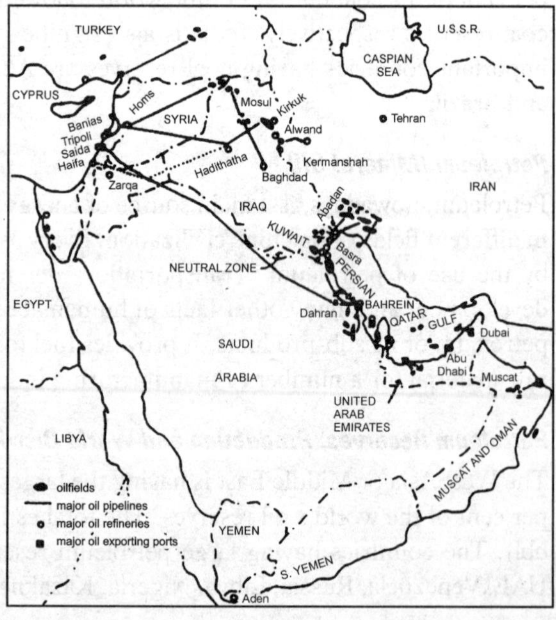

American Region

American petroleum-producing region is having three distinct areas, i.e. (i) North America, (ii) Central America, and (iii) South America.

In North America, USA and Canada are the petroleum-producing countries. On the other hand, Mexico is the major producer of petroleum; while in South America, Venezuela, Brazil, Columbia, Peru and Argentina are the main producers.

Russian Region

The main oil fields of Russia are:

(i) Volga-Caspian Region
(ii) Kamchatka-Sakhalin Region
(iii) Ob-Lena Region
(iv) Pechora-Region

Kazakhstan is having 30 (10^9 bbl) oil reserves and ranks 11th in the world oil reserves countries. Its annual production was 220 m^3d in 2008.

Ukraine is also having petroleum. Much of the oil is extracted from Dnieper basin and Crimean Peninsula.

The Eastern and South Asian Region

China, Japan, Malaysia, Indonesia and India are petroleum producing countries of this region. China ranks 5th in oil production in the world.

African Region

In Africa only northern region or Sahara and Sub-Sahara region are having oil deposits. The main oil-producing countries of Africa are Libya, Algeria, Nigeria and Egypt.

Apart from the above mentioned petroleum regions/countries, the other oil-producing countries of the world are United Kingdom, Romania, Norway, France, Netherlands, Poland and Germany. All these countries are European countries and produce very limited quantity of petroleum.

Natural Gas

Natural gas is the cheapest source of energy, and is generally associated with petroleum. It is a special kind of volatile hydrocarbon which consists of propane, ethane, butane, methane and several other gases. During late 1990s, natural gas has registered the fastest growth as a source of energy and now it contributes about 20 per cent of global commercial energy.

Production and World Distribution

The total production of natural gas in the world was 1,09,789 billion cubic feet in 2008. The following table shows the production of natural gas in the world:

Table 28.8 *Production of Natural Gas in the World (2008)*

Country	*Production (in billion cubic feet)*	*Percentage in the World*
Russia	23,386	21.30
USA	20,377	18.56
Canada	6,037	5.50
Iran	4,107	3.74
Norway	3,503	3.19
Algeria	3,055	2.78
Netherlands	2,991	2.72
Saudi Arabia	2,841	2.59
Qatar	2,719	2.48
China	2,685	2.45

Some other notable natural gas-producing countries are Turkmenistan, Uzbekistan, Kazakhstan, Venezuela, UAE, China, Indonesia, India, Malaysia, South Africa, Gobon, Nigeria, Pakistan, etc.

Hydro-Electric Power

Hydro-electric power is the energy harnessed from running water-streams, rivers or any other artificial or natural water flow. The first hydro-power station was started in France in 1883, but soon this power generation system became popular throughout the world and now hydro-power is a very popular system of power generation in whole of the world. Hydro-electricity now contributes about seven per cent of global electricity production, where only 15.3 per cent of the global exploitable hydro-electric potential is being used. Thus, it has great potentialities of development in future.

Table 28.9 *Potential and Developed Hydro-Power in the World*

Continent	*Potential (in Million Horse Power)*	*Percentage of Total*	*Developed (in Million Horse Power)*	*Percentage of Total*
Africa	272	41.4	0.6	0.6
Asia	151	23.0	13.7	13.6
North America	87	13.2	41.1	40.8
Europe	69	10.5	40.8	40.5
South America	55	8.4	3.1	3.1
Oceania	23	3.5	1.4	1.4

It becomes clear from the above table that Africa possesses the highest potential about 40 per cent of the world's total hydro-power. Most of the water resources of Africa are still utilized in terms of hydro-production. The Congo river has the greatest hydro-power potential in the world. The water of Zambezi, Nile and several other rivers can also be utilized for power generation.

Asia continent is also having large potentialities of the development of hydro-power. It has 23 per cent potential of the total world's potential of hydro-power.

North America, although has developed more than 40 per cent hydro-power but still has only 13.2 per cent of the world's hydel-power potential.

Europe possesses about one-tenth of the world's potential water-power. Its possibilities of development are more in Norway, Sweden, France, Italy and Switzerland. South America is having 8.4 per cent of the total world's potential of hydro-power while Oceania is having the least, only 3.5 per cent of the world's total hydro-power potential.

Nuclear Power

Nuclear power, it is commonly said, holds the key to the future. Atomic or nuclear power is obtained by altering the structure of atoms. When such an alteration is made, much energy is released in the form of heat and this is used to generate electric power. For obtaining atomic power, nuclear fission is required, which takes place in a nuclear reactor containing a heat-producing core and a cooling system. Within the core, atoms of the fuel, usually uranium, enriched uranium or plutonium, are split. The heat released by the reactor is used to generate steam and the steam is used to generate electricity. The first nuclear power station was built in Britain in 1956 at Calder Hall.

The major sources of nuclear power are uranium and thorium. The other minerals like plutonium, barium, zircon, monoxide, etc. are also considered as atomic minerals.

Uranium is the most important mineral widely used for the production of atomic energy. The principal producers of uranium are Canada, Australia, Niger, Russia, USA, Uzbekistan, South Africa, Kazakhstan, Czech Republic, France, China, Gabon and Hungary. India also produces small quantity of uranium.

Canada is the largest producer of uranium contributing about 31 per cent of the world's output. Australia holds the largest reserve of recoverable uranium in the world – over 517 thousand tonnes followed by Canada, USA, Russia, South Africa, Niger, Brazil and Ukraine.

Thorium is another metal used in atomic power production. Thorium is generally extracted from monoxide sand. Australia, Brazil, Nigeria, USA and Canada are principal producers of thorium. Kerala coast in India is also having monoxide sand.

Nuclear energy now has become an important source of energy in many countries of the world. The first nuclear power station was built in Britain in 1956 at Calder Hall and there are now 33 nuclear stations in the UK and nuclear power provides 14 per cent of total electricity output. Britain was rapidly overtaken in this field by the USA, where more than 150 stations of nuclear power provide more than 11 per cent of USA's electricity supplies.

Other countries with a significant investment in nuclear power are China and Japan, West Germany, Russia, Ukraine. France, Sweden, Canada and Spain. Large increases in nuclear output are planned in some countries such as the USA and France which may produce as much 70 per cent of its electricity by this means. Other countries have undergone rapid expansion in the past and are now less keen to develop further plants.

The most recent trend for a number of developing countries is to opt for nuclear power as an alternative to conventional forms of electricity production. Among these countries are Egypt, Brazil, Iran, Iraq and South Korea. India, Pakistan, Taiwan and Argentina already have nuclear plants.

India is developing its nuclear power. The atomic power plants in India are located at Tarapur (Maharashtra), Kalpakkam (Tamil Nadu), Rawatbhata (Rajasthan), Narauru (Uttar Pradesh), Kaiga (Karnataka) and Kakrapara (Gujarat).

29

Industrial Regions and Major Industries of the World

INDUSTRIAL REGIONS

Industrial regions are the areas, where concentration of industries has occurred due to favourable geo-economic conditions. These are the areas within which manufacturing industry is carried out on a relatively large scale and employs a relatively large proportion of population. The spatial distribution of manufacturing units shows a distinct trend of localization towards a few selected areas; these regions are referred to as 'industrial regions'.

The common characteristic features of the industrial regions are:

1. Large population engaged in industrial pursuits,
2. Large industrial complex in hierarchical order,
3. Integration of some main industries with a group of subsidiary industries,
4. Large banking and credit facilities,
5. A network of communication lines, and
6. A large market of labour supply, etc.

The industrial regions of the world are unevenly distributed over the face of the earth. Both natural and cultural factors are involved in any explanation of distribution of manufacturing regions of the world, but the great differentiation in localization of any industrial region can be explained to a large degree by the availability of resources. Since the development of machinery, many industrial centres have sprung up where there is abundance of coal, or coal and iron, or extensive water-power – the mainsprings of modern industry, and all such towns are more or less business centres.

There are several factors affecting the concentration or localization of industry. The main localizing factors are the market, the labour supply, the cost of land resources, the situation of raw material, the nature and situation of the natural and human resources, the value of the commodities produced in relation to the cost of the various items entering into their production, and finally, the supply of capital.

It is obvious that the profitableness of large scale operations must depend on the adequacy of the market, which is governed by various conditions as:

1. The number of people where the industry is carried on.
2. The purchasing power of the people, a great contrast in this respect being presented by developing countries as compared with developed countries and other semi-developed countries, in which the latter's purchasing power is enhanced by the diffusion of education, and probably still more by the extent of the available undeveloped resources.

3. The nature of the commodity for which a market is sought – cheap goods for people of small purchasing power, more valuable commodities for regions in which individual's wealth is greater.
4. Facilities for transport in enlarging the range of the market.

The spatial distribution of the industries shows a trend of concentration or agglomeration of industries in few regions where geo-economic conditions of industrial growth are favourable. These industrial regions have variety of industries and other infrastructures famous for their production of particular or varied types of industrial goods. The major industrial regions/centres of the world are as follows: (Figure 29.1).

Figure 29.1 *Major Industrial Regions of the World*

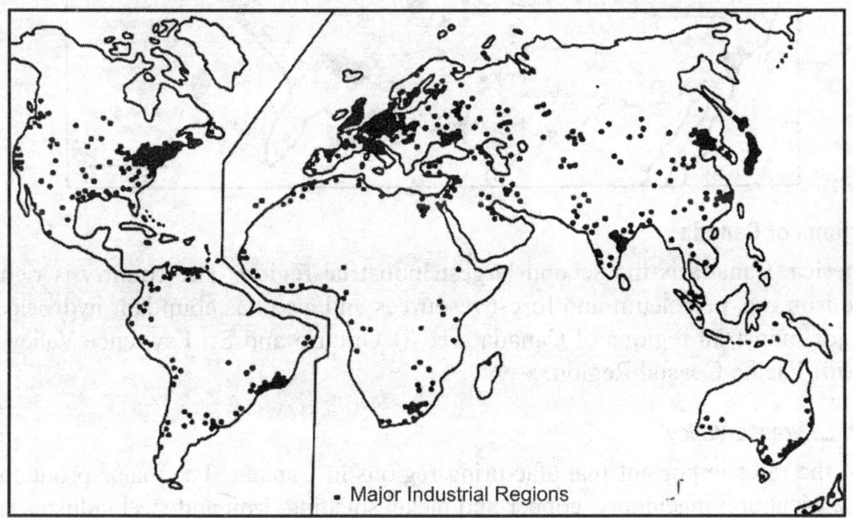

Major Industrial Regions

1. North American region, comprising USA and Canada.
2. European region, consisting of the industrial region of UK, Germany, France, etc.
3. Industrial region of former USSR.
4. Asian region, comprising the industrial regions of Japan, China, India, etc.
5. Other isolated industrial regions.

North American Industrial Region

North American Industrial Region comprising of USA and Canada is a highly developed industrial region of the world. USA is now the wealthiest and most highly developed nation in the world. Southern Canada is also well developed.

The industrial regions of USA are as follows:

(i) Southern New England
(ii) Mid-Atlantic States
(iii) Pittsburgh-Lake Erie Region
(iv) Detroit Industrial Region
(v) Lake Michigan Region
(vi) Southern Appalachian Region
(vii) Eastern Texas
(viii) The Pacific Coastal Region

Figure 29.2 *Major Industrial Areas of USA and Southern Canada*

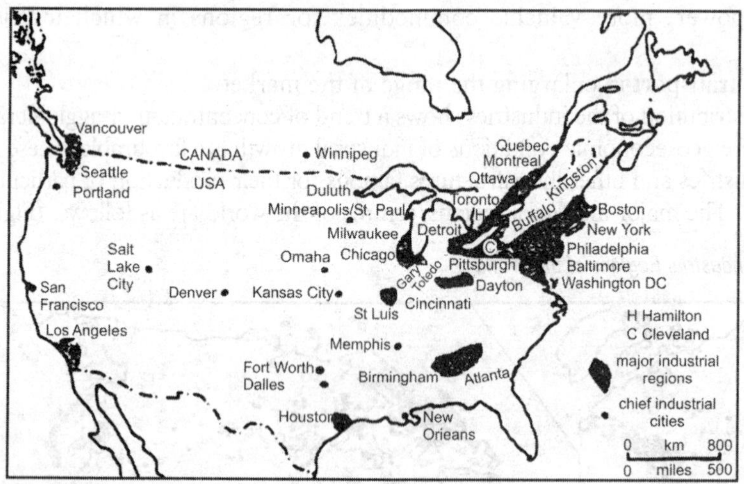

Industrial Regions of Canada

In North America, Canada is the second largest industrial region. The country is rich in mineral resources like iron ore, petroleum and forest resources and also has abundant hydroelectric power. The three major industrial regions of Canada are: (i) Ontario and St. Lawrence Valley, (ii) Prairie Region, and (iii) Pacific Coastal Region.

Ontario and St. Lawrence Valley

This is one of the most important manufacturing regions in Canada. The major products are paper, cheese, flour, agricultural machinery, copper and nickel smelting, iron and steel industry and chemical industry. The major industrial centres are Quebec, Ontario, Ottawa, Toronto, Hamilton, etc.

Prairie Region

As far as manufacturing industry is concerned, this region is not very developed. The major centres of production are Manitoba, Winnipeg, Edmonton, Alberta, etc. Besides agro-based industries, the other noted industries in the region are petroleum refinery and chemical industry.

Pacific Coastal Region

In this region, industries like paper and pulp, furniture, agricultural machinery, etc. have been developed. The major industrial centres of this region are Vancouver and Prince Rupert.

European Industrial Regions

Industrial revolution was first started in Europe, with the result of which, heavy industries on a large scale have developed in many parts of the Europe. In spite of industrial development in many regions of the world, Europe is still a great industrial power. In Europe, industries are distributed in many countries and the manufacturing belt of Europe is not continuous.

The principal manufacturing zone extends from west to east, from Britain through north-eastern France, Belgium, the Rhineland of West Germany, Saxony-Bohemia to Silesia. Other important industrial areas are found on the Swiss Plateau, northern Italy, in central Sweden and in many large towns and cities throughout Europe such as London, Paris, Berlin and Milan (Figure 29.3).

Figure 29.3 *Industrial Belt of North-West Europe*

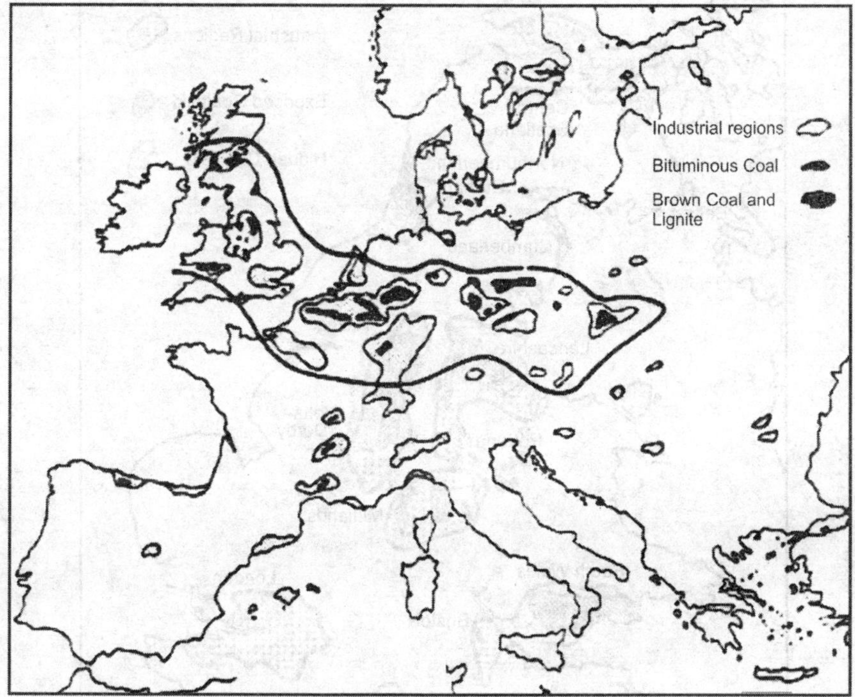

The leading industrial countries of the Europe are United Kingdom, Germany, France and Italy, while several other regions are scattered in other different European countries. A brief description of important industrial regions of Europe is given below:

United Kingdom

UK was the first country in the world to become highly industrialized. Most of its industrial regions are closely associated with the coalfields. At present, the industry is declining due to cheaper and more efficient production overseas, and with old equipment and labour troubles at home. But still the industrial structure of UK is considered important. In United Kingdom there are certain areas which are more specifically industrial. The major industrial regions of UK are shown in Figure 29.4.

(i) The North East Coast
(ii) Cumberland and Westmorland
(iii) The West Riding of Yorkshire
(iv) Nottinghamshire and Derbyshire
(v) Lancashire
(vi) The Birmingham Region (Midlands)
(vii) Greater London
(viii) South Wales
(ix) Scotland

Germany

Germany is considered as a highly industrialized country of Europe. The major industrial regions of Germany are as follows:

Figure 29.4 *Industrial Regions of United Kingdom*

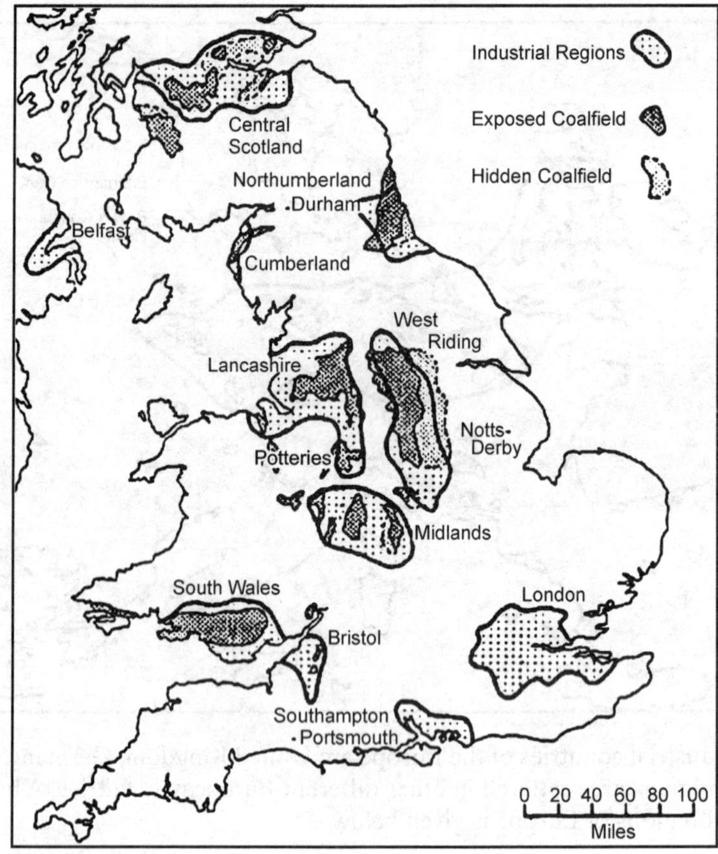

(i) Rhine Industrial Region
(ii) The Middle Rhine Industrial Region
(iii) The Hamburg Industrial Region
(iv) The Berlin Industrial Region
(v) Lipzig Industrial Region

France

France is the third industrial country in Europe, next to UK and Germany. The major industrial regions of the France are as follows:

(i) The North-East Industrial Region
(ii) Lorraine Industrial Region
(iii) The Greater Paris Basin
(iv) Central Plateau Industrial Region
(v) Rhone Sone Valley Region
(vi) Mediterranean Industrial Region

Italy

In Italy industrial development began only after the Second World War. The two distinct industrial regions of Italy are northern and southern regions.

(i) The Northern Region

About four-fifth of the industries is concentrated in Northern Italy. The major industrial regions are Lombardy, Piedmont, Liguria, etc. Most of the manufacturing units are concentrated in urban centres, e.g. Venice, Trieste, Genoa, Savona and in Milan and Turin in great Po river valley. Entire region is highly developed with several types of industries. Important among these are textiles, silk, iron and steel, paper pulp, agricultural machinery, aircraft, machine tools, electrical and automobiles.

(ii) The Southern Region

This region is far less developed in manufacturing than its northern counterpart. Naples is the only major industrial centre, having textiles, machinery and, iron and steel plants.

Other European Industrial Regions

Most of the European countries have important manufacturing activities. These are outlined in brief below:

Belgium has heavy industries, including iron and steel making and the manufacture of armaments. Liege, Brussels, Antwerp, and Ghent, are the major industrial centres of Belgium.

Luxembourg is known for its iron and steel industry. Luxembourg city has iron and steel as well as engineering industries.

The Netherlands has 40 per cent of its population engaged in the industrial sector and many branches of modern industries are well-represented there. Marine engineering and ship building are important in Rotterdam, Schiedam and Dordrecht. Engineering industries are well-developed in Utrecht (light machinery), and Eindhoven (electrical engineering, home of the Philips electrical company). Chemical plants are found in Pernis and Europoort near Rotterdam, and Maastricht. Groningen is the centre of industries based on the extraction of natural gas from the North Sea. There are a number of towns noted for textiles viz. Tilburg (woolen textiles), Eindhoven (linen), Enschede (cotton), Arnhem (rayon) and Breda (synthetic fibres). Amsterdam has long been a centre for diamond cutting and Arnhem for tin smelting. Oil refineries are located in Rotterdam, Pernis and Europoort. Dairying, brewing, sugar refining, food processing and oil milling are also important industries in the larger towns in the agricultural polder-lands.

Sweden is the most industrialized country. The main industrial regions is in central Sweden, the so-called Lake Depression, with its main centre at Stockholm. Sweden has the richest iron ore resources of Europe, has developed much hydroelectric power and has a long history of technological skills. Some of its internationally-known exports include Volvo cars, Bofors guns, Aga beacons, Electrolux refrigerators, Laval cream separators, Ericsson telephones, Nobel dynamites and Johansson's instruments. Eskilstuna is known as 'the Sheffield of Sweden' and produces excellent cutlery and ornamental goods. Other metal and engineering centres are Vasteras, Falun, Dannemora and Oxelosund. Orebro makes footwear, Norrkoping makes textiles; and Karlstad, Norrkoping, Orebro and Trollhattan make paper.

Norway's leading industries are marine engineering, shipbuilding, fish-canning and the pulp and paper industries. There are aluminium smelting plants at Odda and Ardal, and iron and steel is made at Mo-i-Rana. The capital Oslo, located on Oslo Fiord, and its surrounding region has pulp mills, shipyards, chemical plants and fish-canning factories. The towns of Bergen, Trondheim, and Stavanger are noted for their fishing and shipping activities and Stavanger has become a major port servicing the North Sea Oil and gas fields. Electro-chemical industries are well-developed in the Glommen valley and Telemark where cheap hydroelectric power is available. Such plants manufacture fertilizers, plastics, explosives, and smelt ferro-alloys.

Denmark is famous for its dairying and agricultural industries. Other industries developed in Denmark are chemicals, textiles, fishing, vessels, beer, silverware, machinery, electric equipments, etc. Copenhagen, Odense, Flensbug, Esbjerg and Aarthus are the industrial centres.

Switzerland, although a mountainous region, is famous for watch-making, engineering, chemicals and textiles. The Swiss plateau has the greatest concentration of industrial plants. Zurich, Basel, Baden, St. Gallen, Appenzell, La Chaux-de-Fonds, Biel and Le Locle (Jura towns) are the main industrial centres.

Poland is having an important industrial region in south of the country. Iron and steel, chemicals, textiles and zinc/lead refining are important industries developed in this region. The main industrial centres are Wroclaw, Bytom, Czestochowa, Katowice, Chorzow, Krakow, Warsaw and Gdansk.

Elsewhere in Europe there are a few industrial towns having well-developed industries. These include Barcelona, Madrid, Bilbao, Lisbon, Athens, Budapest, Vienna and the capital cities of the other east European countries.

Industrial Regions of the Former USSR

Before the revolution in Russian industry which began in 1928, manufacturing was confined to the western parts of the Soviet Union. Its chief centres were Leningrad, Moscow with its ring of smaller towns, and the Ukraine. Also, the Donetz basin, with its reserves of coal and iron, constituted the foremost centre of the metallurgical industries. The development, initiated by Stalin in the First Five Year Plan was designed to create a series of new industrial regions, with the result of which USSR became a highly developed industrial nation in the world. Although with the disintegration of the country into several new countries, these industrial regions have also been divided, but still their present pattern shows a general industrial scenario of this region. The well-developed industrial regions are as follows:

(i) The Moscow Region
(ii) The Ukraine
(iii) The Leningrad Region
(iv) The Ural Region
(v) The Kuzba Region
(vi) Soviet Central Asia
(vii) The Caucasus Region
(viii) The Soviet Far East is an area rich in forests and minerals. It is, however, remote from other parts of the Soviet Union, with which it is joined by the Trans-Siberian Railway. Its exposure has led to the development here of certain 'strategic' industries as part of the defence plans of the USSR.

Industrial Regions of Asia

In Asia, China, Japan and India have done considerable progress in industrial development. Hong Kong, Korea and Taiwan have become major exporters of textile and other goods.

The South-East Asian Region, though traditionally agricultural in outlook, is also developing industries. Singapore is the most industrialized of the South-East Asian countries. Oil is the major resource of Middle East, therefore, oil refining and petrochemical industry have become important in the region. Iran has developed a wide range of industries.

China

Industrial development in China began only after the beginning of communist rule in 1949, and now China is not only an industrial power of Asia but also of the world. There has been a complete transformation of the industrial system during the last 60 years. Under the new system and policy, China is developing its industrial system in a planned manner. Rapid development has made China a leading producer of iron and steel, textiles, and cheap consumer goods such as toys, household goods and light metal goods. In China, following industrial regions have been identified:

(i) Manchurian Industrial Region
(ii) Tientsin and Beijing Region
(iii) Lower Yangtze Industrial Region
(iv) The Middle Yangtze Industrial Region
(v) Si Kiang Delta Region

Japan

Japan is the highly industrialized country of Asia. Despite its shortage in industrial raw materials and solid fuel, it has been able to develop industries at a very fast rate. The reasons of rapid industrial growth in Japan are:

(i) Availability of hydroelectric power.
(ii) Coastal location and large ports which helps in import of raw material and also in export.
(iii) Proximity to mainland of Asia providing a ready market. Now, Japan has a worldwide market for its products.
(iv) Country's large population provides a ready source of labour.
(v) Technological development.
(vi) Government encouragement, etc.

There are four major industrial regions of Japan as shown below in Figure 29.5:

(i) Tokyo-Yokohama Region
(ii) Osaka, Kobe and Kyoto Region
(iii) The Nagoya Industrial Region
(iv) Northern Kyushu Region

Figure 29.5 *Industrial Regions of Japan*

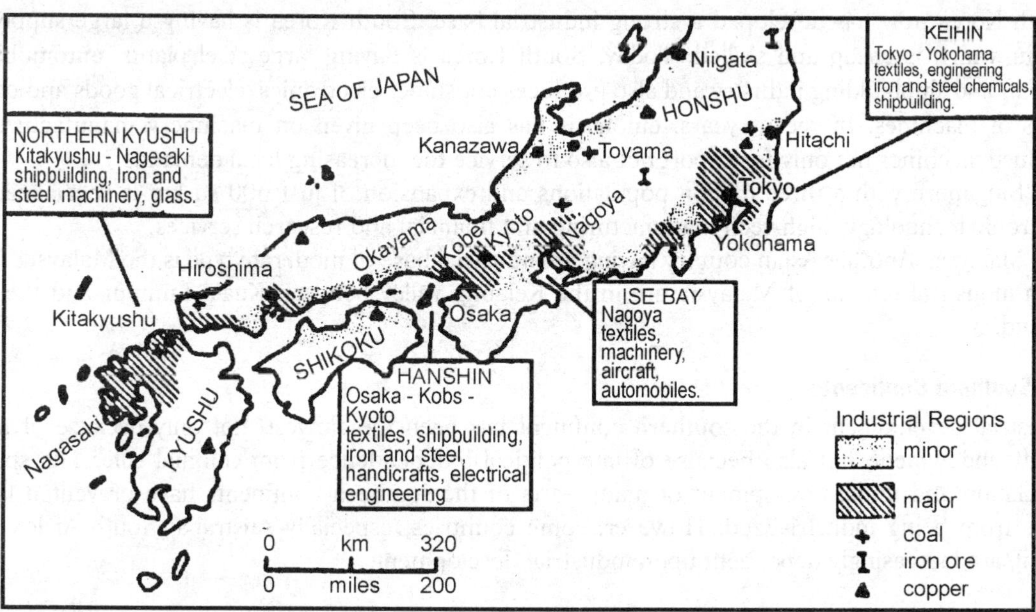

India

Since gaining independence in 1947, India has gradually emerged as an industrialized nation, in spite of the fact that agriculture is still the base of the Indian economy. The industrial development has occurred through government policies adopted during various plan periods. Both private and public sector industries have been developed, with the result India now is considered as an industrial nation of Asia after China and Japan.

The distribution pattern of Indian industries is varied. Among the states, Maharashtra contributes the largest amount of industrial products, followed by Gujarat, Tamil Nadu, West Bengal, Uttar Pradesh, Jharkhand, Bihar, Karnataka, etc.

The following industrial regions are prominent in India:

(i) The Kolkata Region
(ii) Mumbai-Pune Region
(iii) Ahmedabad-Vadodara Region
(iv) Coimbatore-Bengaluru Region
(v) Chotanagpur Plateau Region
(vi) Delhi Capital Region

Other Asian Industrial Regions

The four Asian regions – Hong Kong, Taiwan, South Korea and Singapore are notable for their industrial production. Hong Kong has developed both heavy and light industries. Most of the industries are export-oriented. Hong Kong is famous for toys, plastic goods, domestic appliances, radios, sports goods, handbags and footwear, watches and clocks, photographic goods, jewellery and printed matter. Now, heavier industries like textiles and electrical goods have also been developed. Taiwan is located about 200 km away from the south-east of China. Taiwan's industrial development includes textiles, electronic goods, plastics, toys, small tools, etc. along with iron and steel industry, aluminium and chemical industries.

South Korea is also considered as an important producer of industrial products not only for its own consumption but for export also. With the help from USA, Japan and some European countries, South Korea now has developed a strong industrial base. South Korea is having a large supply of labour which is cheap and skilled. Today, South Korea is having large steel plants, automobiles, aircraft and shipbuilding industry and also produces consumer electronics, electrical goods and other types of machines. In recent years, emphasis has also been given on machinery manufacture to produce machines not only for export but also to service the increasing local demands.

Singapore with a three million populations and expansion of just 600 sq km is specialized in electronic technology, high-tech manufacturing and financial and research services.

Malaysia: Another Asian country having an industrial base of moderate type is the Malaysia. The main industrial regions of Malaysia are in the Kelaung valley between Kuala Lumpur and the sea around.

The Southern Continents

Industrial development in the southern continent has been the slowest, not only because of their scanty endowment, but also because of late political independence from colonial rule. The sparse population and underdevelopment of many parts of the southern continents have prevented large areas from being industrialized. However, some countries, especially Australia, South Africa and Brazil, are increasingly dependent upon industrial development.

South America

Industrial development has been most marked in South America, perhaps the best endowed of all the southern continents. It has had the advantage of a large population, and several minerals are available in some countries. The industries of South America are dotted in many centres, none of them very large, around the coasts of the continent.

Brazil is the most industrialized of Latin American countries. The chief industrial region is the south-east and the city of Sao Paulo has diverse industrial activities including steel mills, chemical plants, motor vehicle assembly plants, and paper, cement and beer industries. Rio de Janeiro has similar industries with shipbuilding and aircraft engineering in addition. Belo Horizonte is the major metallurgical centre with iron and steel and other metal industries based on hydroelectric power and the rich mineral resources of Minas Gerais.

In Argentina and Uruguay the main industrial region is along the shores of the Plate estuary extending inland as far as Rosario. The main industries are shipbuilding, at Buenos Aires, Rosario, Cordoba and General Pacheco. The La Plata urban districts have chemicals, textiles, aircraft engineering and steel industries. Meat-packing, dairying and flour-milling are the dominant industries of the Pampas lands.

Chile's main industries are found in Santiago, Valparaiso and Concepcion. Elsewhere in South America, industries are concentrated in the capital cities of the various states, e.g. Lima, Caracas, and Bogota and in certain other large towns.

Africa

Africa is the least developed of all the southern continents. The large spell of colonial rule, lack of capital and entrepreneurial skills and political instability as well as underdeveloped economic structures are the main cause of industrial backwardness in many countries of Africa.

The main industries are those connected with either mining or agriculture, e.g. smelting and refining of copper in Zambia and Zaire, and processing of rubber, oil palm fruits, etc. in West Africa. Nigeria also has petrochemicals industries. South Africa is, however, an industrially developed country. The main industrial region is the Witwatersrand where there is not only gold but also coal, iron and a range of ferrous and non-ferrous metal deposits, as the basis of iron and steel, engineering, locomotives, and other industries. Chemicals, textiles, cement and light industries are also fairly important. Elsewhere in Africa, the main industrial development has been in capital cities such as Salisbury, Dar-es-Salaam, Nairobi, and so on. Such cities have cement, brewing, food processing and light industries mainly geared to import substitution.

Australia

In Australia, the main industrial area is the south-east where the concentration of early settlement, the favourable climate, the presence of coalfields around Sydney, and of iron ore resources, led to the establishment of the iron and steel industry, especially at Newcastle and Port Kembla. There are also many engineering industries, including the manufacture of cars, locomotives and aircraft, cement works, chemicals and shipbuilding.

Melbourne has also developed many industries similar to those of Sydney, but based on the Gippsland lignite deposits and the hydroelectric power from the Snowy River Scheme. Melbourne has chemicals, shipbuilding, aircraft engineering, railway equipment and motor vehicle industries. Adelaide has well-developed agricultural industries such as fruit-canning, flour-milling, dairy industries and woollen textiles. It also has iron and steel (at Whyalla) and engineering industries. Brisbane is the major industrial city of Queensland and is located near the Ipswich coalfield. Locomotives are made at Brisbane, Ipswhich, Maryborough and Mackay. In western Australia, Perth is the main industrial city dealing with agricultural products.

MAJOR INDUSTRIES

Iron and Steel Industry

The iron and steel industry is the basic manufacturing industry. The economic growth of both developed and developing countries is largely dependent on its steel-making capacity. The establishment, development and concentration of iron and steel industry requires many things. Among the localization factors, the primary factor is, of course, availability of raw material, market, energy supply and labour. While second category of factors are the factors of survival, such as (i) establishment coasts like taxes, duties, rent, etc. and (ii) production cost, e.g. labour, wage, transport charges, sale tax, income tax, etc.

The growth and development of iron and steel industry is a reflection of global economy. The iron and steel industry depicts a changing nature in its growth and production pattern. In the mid-1970s, the relatively developed countries of North America, Western Europe and Japan accounted for nearly two-third of the world's steel production. But gradually the spatial pattern has changed and attention has now shifted to the developing regions. Towards the end of the last century, the growth of steel production in countries like China, South Korea, Brazil and India has changed the entire pattern of steel production in the world. Now main producers of iron and steel in the world are China, Japan, USA, Russia, Germany, South Korea, Brazil, Ukraine, India, France, Italy and Great Britain. The other steel-producing countries are South Africa, Australia, Austria, Netherlands, Czech Republic, Romania, Spain, Belgium, Sweden, etc.

China is the leading producer of iron and steel in the world, which accounts for about 23.9 per cent production of pig iron and 17 per cent of crude steel of the world's production. Japan is the second largest producer with 14.7 per cent pig iron and 13.9 per cent crude steel production in the world. USA, once the highest producer now ranks third in the world followed by Russia. India's position is 9th in the iron and steel production and its production of pig iron and crude steel accounts for 3.9 and 3.6 per cent respectively.

The spatial distribution pattern of iron and steel industry in major countries of the world is as shown in Figure 29.6.

Figure 29.6 *Distribution of Iron and Steel-Producing Areas in the World*

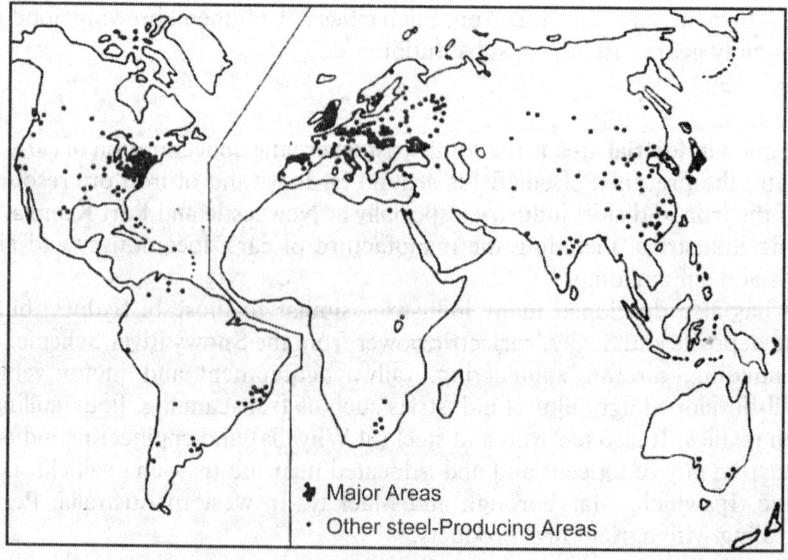

Cotton Textile Industry

Figure 29.7 *Centres of Iron and Steel Industry in Japan*

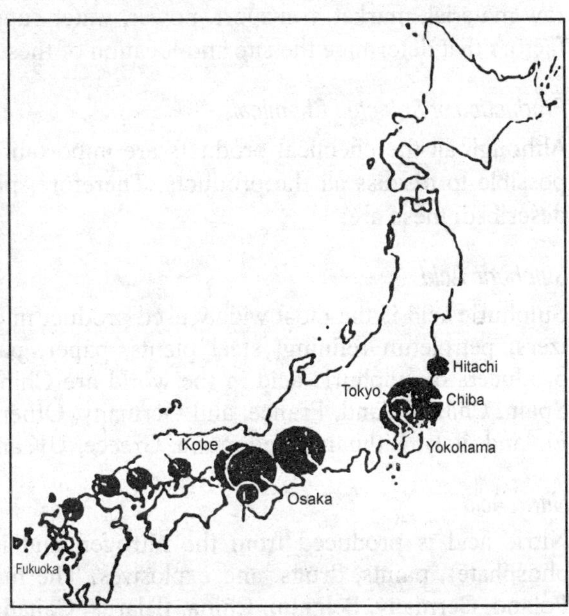

Cotton textile industry is the oldest among all the manufacturing activities. The sign of existence of cotton textile industry was well established in all early civilizations like Indus valley, Egyptian, Vedic, Roman, etc. In fact, the history of textile industry is related with the growth of civilization. Earlier, hand spinning and handlooms were the practices for making the cotton textile. But the real development of cotton textile started in the 18th century after a series of British inventions in the field of ginning, spinning and subsequent inventions of steam engine and cotton gin. All these inventions led to a rapid growth of cotton textile industry in Great Britain followed by USA and Japan in the 19th century. The industry has also developed in former USSR, China, Egypt, Mexico, India, Brazil, Turkey, etc. The cotton textile industry is a combination of ginning, carding, spinning, weaving and dyeing-bleaching activities. Nowadays cotton accounts for about 50 per cent of world's industrial fibre consumption.

The localization of cotton textile industry depends upon many geo-economic factors. The important factors are: (i) climate, (ii) source of power, (iii) raw material, (iv) labour, transportation facilities, and (vi) market.

Cotton textiles industry is quite widespread in the world and as many as 90 countries are producing cotton yarn and/or cloth in varying quantity. But the main concentration of textiles industry is limited to few countries. There are two types of production related with cotton textile, one is the production of cotton yarn and another is the production of cotton cloth. Although many countries produce both the items.

The leading producers of cotton cloth in the world are China, India, Russia, USA, Japan, Italy, Germany, Hong Kong, Egypt, France and Romania. The other producers of cotton cloth in the world are Brazil, Spain, Pakistan, Turkey, Uzbekistan, Bolivia, Vietnam, Korea Republic, Czech Republic, Portugal, Belgium, Poland, South Africa and Syria.

Chemical Industry

The work through chemicals to affect transformation of matter and molecular structures of planned design has been going on for many years, but its impact on the well-being of mankind was never so far reaching as it was during the last century and as it is today. The chemical industries are considered basic for health, industrial and agricultural development as well as defence.

The chemical industries are not only concerned with the manufacture of pure chemicals, but also with many kinds of industrial raw materials that are mixtures of substances. The raw material of chemical industries are derived from various sources such as mines, forests, sea, air, land, oil, gas and brine wells. The chemicals produce serve as essential raw materials for making thousands of articles – medicines, fertilizers, synthetic rubber and fibre, plastics, explosives, etc. Chemicals are also used in many industries to improve or preserve the quality of non-durable and durable goods.

Although factors of localization of chemical industry are similar to that of other industries, e.g. raw material, market, transport, power, water supply, capital, etc. but still there are certain specific factors that determine the site and location of these industries.

Production of Selected Chemicals

Although all the chemical products are important, but they are too much in number and it is not possible to discuss all the products. Therefore, production of a few important products has been described; these are:

Sulphuric Acid

Sulphuric acid is the most widely used product in chemical industries. Several industries, e.g. fertilizers, petroleum refining, steel plants, paper, paints, dyes, glass, drugs, etc. use it. The main producers of sulphuric acid in the world are China, Russia, Japan, India, Brazil, Tunisia, Canada, Spain, Chile, Poland, France, and Germany. Other sulphuric acid-producing countries are Ukraine, Finland, Italy, Lithuania, Indonesia, Greece, UK and USA.

Nitric Acid

Nitric acid is produced from the nitrogen and is used in the manufacture of fertilizers (super phosphate), paints, drugs, and explosives. The main nitric acid producers in the world are USA, Poland, Germany, Belgium, China, Belarus, Canada, Spain, Bulgaria, Japan, Italy, and Brazil. Other nitric acid-producing countries are Ukraine, Turkey, Greece, Croatia and Russia.

Soda Ash

Soda ash is an alkali, used in the manufacture of paper, soda, glass and many other chemical products. The main producers of soda ash in the world are China, USA, Russia, India, Germany, Poland, France, Italy and UK.

Other soda ash-producing countries are Bulgaria, Turkey, Japan and Ukraine.

Caustic Soda

Caustic soda is used in glass, paper, textile, soap and in oil refining. USA is the leading producer of caustic soda and produces about 29 per cent of the world's total yield. China produces 17 per cent and Japan 11 per cent of the world's total production of caustic soda. Other caustic soda-producing countries are Germany, India, Brazil, Russia, Republic of Korea, Canada, UK, Italy, Spain, etc.

Chlorine

Chlorine is used for washing of textiles as well as for purification of potable water. Chlorine is produced by many countries but its major producers are USA, China, Japan, Germany, France, India, Brazil, Republic of Korea, Canada, UK, Russia, Italy, etc.

Ammonia

Ammonia is used as a chemical product in industries. USA is the leading producer of ammonia followed by Russia, Germany, Japan, UK, China, India, etc.

Distribution of Chemical Industry

Since chemical industry is based on science and technology, its development is more important in developed countries. But now, so many developing countries are also important producers of several chemicals. The global chemical industry is mostly controlled by few multinational companies as they have the patent right of the products. The major chemical-producing countries are USA, Germany, China, Russia, Japan, France, India, Brazil, Italy, Poland, Belgium, UK, etc.

Chemical Fertilizer Industry

The growth, development and productivity of agriculture depends upon soil fertility and the fertility of soil depends upon several factors, especially on its mineral components such as phosphorus, potassium, nitrogen, calcium, magnesium, sulphur, etc. The critical problem of modern agriculture is the maintenance of soil fertility. Each crop removes something from the soil. Our ancestors coped with this problem by following a 'gap season' with no crop on the land, so that it could recover naturally something of what it had lost. But now under the pressure of population, land hardly gets this rest, therefore, loss of soil fertility and lower productivity has now become a common problem.

The common chemical fertilizers are:

(i) *Nitrogenous fertilizers* in the form of ammonium sulphate, urea, nitro-limestone and ammonium nitrate.
(ii) *Phosphatic fertilizers* in the form of super-phosphate like nitro-phosphate and ammonium phosphate, and
(iii) *Potassium fertilizers* in the form of potassium.

Nitrogen Fertilizers

Nitrogen fertilizers are the most widely produced and the most commonly used fertilisers. Nearly 75 per cent nitrogen is now obtained from air, 20 per cent is derived as a by-product of coke-oven plant in the form of ammonium gas, four-fifth of which is nitrogen and the rest of 5 per cent is obtained from the (mineral) sodium nitrate.

China is the largest producer of nitrogen fertilizer and contributes 24 per cent of the world's production. India is the second largest producer with 11 per cent of the world's production followed by USA (8.4%), Ukraine (6.4%) and Russia (5.8%). Countries like the Netherlands, UK, Germany, France, Spain, Turkey, Iran, Brazil, South Africa and Uzbekistan are the other producers of nitrogen fertilizers.

Phosphatic Fertilizers

Phosphate fertilizers, generally known as phosphorus have often been called the master key to agriculture, as low crop production more often is due to lack of phosphorus. Phosphorus fertilizers are classified into three types, e.g. water-soluble phosphorous, citrate-soluble phosphorus, and isolable phosphorus.

The fossil remains of animal life in the form of phosphate rock are the largest source of phosphorus today. Most of the rock phosphate reserves of the world are concentrated in North Africa, USA and Russia, as well as in Peru, Egypt, Spain and Nature islands in the Pacific Occan. USA, China, India, Russia, Belgium, Brazil, Tunisia, Poland, South Africa, Morocco, etc. are the major producers of phosphate fertilizers in the world.

Potassium Fertilizers

A large quantity of potassium is derived from the deposits of potash salts. The largest deposits of potash are found in Germany, Belarus, Russia, France, Spain, USA and Israel-Trans Jordan. Playa deposits in desert areas and the brines of some salt lakes are also rich in potassium. Acid areas usually need potassium fertilizers more than neutral and alkaline soils because acid soils have developed in areas of high rainfall that leaches fertilizers out of available potassium. Japan uses potassium fertilizers on a large scale for rice production.

Major producers of potash fertilizers are Belarus, Canada, Russia, China and Germany. Apart from the countries mentioned in the table, Brazil, Poland, Republic of Korea also produce potash fertilizers.

Pulp and Paper Industry

All types of writing work are done on paper, thus, paper is considered as a base of modern civilization. There is evidence that paper was first made in China in 105 AD by Ts'ai lun and this art spread in other countries gradually. Until the mid-19th century, rags remained the chief paper making material. The use of wood pulp with rags for making paper began in Germany in 1840 and in USA in 1880. Thereafter, throughout the 20th century, most of the world's paper has been made from wood pulp and this process is still going on. In fact, paper is a forest-based industry. Both pulp and paper industry is complementary to each other. Paper is usually manufactured from wood fibres. The rounded wood and processed wood is converted into wood pulp, which is considered as a principal raw material of paper industry.

The importance of wood pulp is fast growing. The main uses to which wood pulp is at present put to are the manufacture of paper and cellulose which is then used for the manufacture of artificial silk. The chief woods used for the manufacture of pulp are the species of spruce, fir, and pine for 'mechanical' pulps; and aspen, polar and other deciduous trees for 'chemical' pulp. Hemlock is used in considerable quantity in producing pulp for inferior papers. Mechanical pulps are made by grinding the wood and are used for interior kinds of paper. Chemical pulps are made by dissolving wood into pulp through chemical solutions. These are generally used for better quality paper. At present, wood pulp is largely used in the production of newsprint only, in the manufacture of which two kinds of pulps are mixed together in proportion of 70 mechanical and 30 chemical. The use of wood pulp for paper of other kinds also is, however, growing fast, owing to the improvements being made in pulp making.

The production of wood pulp is an important industry in many countries of the world. Canada is the highest wood pulp producer in the world and produces 32.2 per cent of the world production. USA is the second largest producer followed by Finland and Sweden. The other major producers of wood pulp in the world are Japan, Norway, Germany, Russia, France and New Zealand.

Paper is produced in many countries of the world but its major producers are Canada, USA, Russia, Germany, Norway, Sweden, UK, Japan, China, India, etc. There are several varieties of paper produced in the world, among them newsprint paper is very important because of its large quantity of consumption throughout the world.

30

World Transportation Patterns

Transportation is a measure of the relations between areas and is, therefore, an essential aspect of geography. The economic relations between areas are reflected in the character of transportation facilities and in the flow of traffic. In the reconstruction of a region or a nation, transport system invariably plays a vital role. The growth and development of transportation provides a medium, contributing to the progress of agriculture, industry, commerce, administration, defence, education, health or any other community activity. In fact, transportation pattern is a reflection of regional and/or national development.

Transport systems, which are operated at the global scale, are the expansion of the need for links between both individual nation and trading blocks, and have complex networks. The transportation system can be classified into the following three broad categories:

1. Land transportation
2. Water transportation
3. Air transportation

LAND TRANSPORT

Roads

Roads are the most common means of surface transport, the importance of which is mostly restricted to national limits. The neighbouring countries may allow or not allow international movement. It depends upon their political and strategic interest. But this much is clear that all the countries of the world are having road transport and more a developed country, more efficient and developed is its road transport connectivity. Even developing countries are having good road network.

All the European countries are having well-developed system of road highways and the road network is very close in the more populous and industrialized areas. The countries of North-West Europe like UK, Germany, France, Netherlands, Belgium, etc. and also Spain and Italy have dense road network. Many of the main highways follow age-old routes; in England and France and in many parts of southern Europe, roads are often patterned upon the original Roman network. The French road network is the longest in the Europe, followed by Germany, Belgium, Netherlands, Sweden, Italy and Spain. Eastern Europe is also having good road connectivity, both within the country and with other adjoining countries. The 10,700 km of autobahn in Germany is the largest network in Europe.

The former USSR, with its vast area, has developed road transport system, but in comparison with European countries and USA, its road connectivity is limited. This also is because of the fact that a vast sizeable area is under the cover of ice, and in many other areas of central and eastern part, the topographical conditions are not suitable. In spite of all these conditions, the length of road network has increased more than tenfold since the revolution. During the last twenty-five years, the progress in road transport development is very fast and all the countries of former USSR are developing their internal road transport system.

In North America, the USA has about 34 per cent and Canada around 5 per cent of the roads in the world. USA is now having a well-developed and highly dense road network in the world. Its total length of roads is about 7 million km. Another characteristic feature of the United States is that there are several trans-continental highways. The Trans-Canada Highway links St. John's, Newfoundland, with Victoria on Vancouver Island in British Columbia, a distance of 7,821 km.

Most of the roads in Canada lie in the southern parts of the country. Northern Canadian region is still having limited roads due to adverse geographical conditions.

In Latin American countries, road development is limited and varied. Argentina and Uruguay have reasonable adequate highway networks, but extensive areas are ill-served with roads. Brazil had some 1.2 million km of road, but apart from the network in the south-east, which accounts for nearly 75 per cent of the total road length, most of the areas are still not having road connectivity. The Pan-American Highway is a unique feature of the road transport in Latin America. These highways have been designed to link all the countries of North and Latin America. The Pan-American Highway linking Fairbanks, Alaska with Puerto Mont, Chile and Buenos Aires is a notable highway. Another international highway is the Garretera Marginal Bolivariana de la selva (commonly known as 'La Marginal') running along the eastern foothills of the Andes through Columbia, Peru and Bolivia (Figure 30.1).

Figure 30.1 *Important Highways of South America*

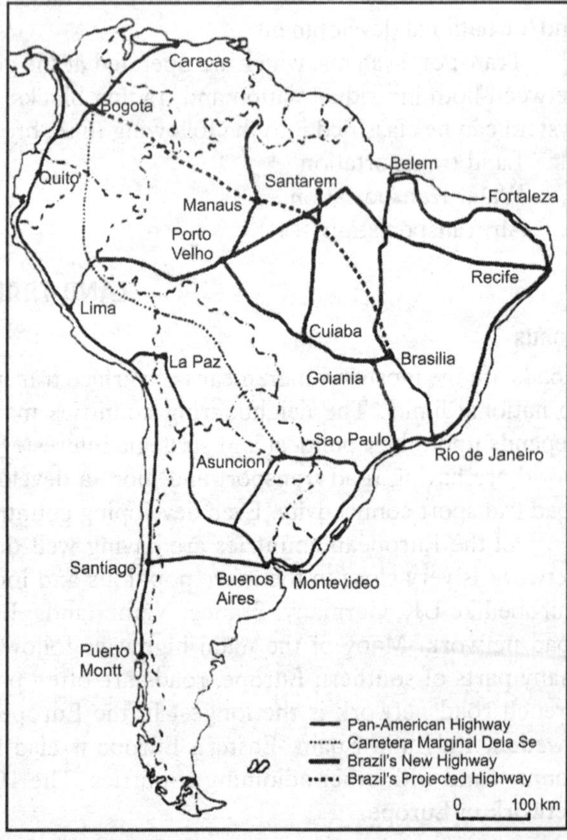

In Asia, countries like China, Japan and India have a developed highway system and a dense road network. In other countries of South-East Asia also, road transport is the main source of transportation. While in West Asia, desertic conditions have restricted the road development, but now gradually these countries are also developing their highway system. In China when communists came into power they began a strenuous programme of highway construction which continues to the present and it is estimated that there are now over 5 lakh km of motorable roads in the country. In Japan, although railways are more important, but the country has also developed a good road network. In Japan, in addition to the 'A' grade expressways that have been constructed, there is a national plan for new arterial highways.

In India, roads are the principal means of transportation. Indian road network is one of the largest in the world. In India, National Highways are the prominent interstate roads constructed and maintained by the central government, while state government maintains state highways, district and village roads. The length of National Highways increased from 19,811 km in 1951 to 70,550 km in 2010. Nearly 40 per cent of road traffic in India is carried on these highways. A comprehensive plan of National Highway development in the form of Super National Highways, Mega Highways, Express Highways, North-South Corridor and Golden Quadrilateral is also ongoing presently. On the whole, India is developing its road network on priority basis but still has a long way to go.

Africa is a vast continent, but because of its physical conditions and economic backwardness, road development is still limited in many of the African countries. The total road length of Africa is estimated around 2 million km, of which perhaps 20 per cent are paved.

The total length of roads in Australia suitable for general traffic is about 8.6 lakh km. Most of the road development in Australia has occurred in south-eastern and south-western part, that too, along the coast.

Railways

Railways are the most important means of inland transport. Railway was, in a real sense, a product of the industrial revolution and afterwards became a predominant mode of inland transport. Railways solved two important needs: (i) the economic carriage by land of (a) materials in bulk, (b) bulky commodities, and (ii) the relatively rapid movement of large number of people as well as goods. The rails always revolve around their fixed tracks. This provides guidance for the wheels and also enables very heavy loads to be carried.

The opening of public traffic on 27 September 1825 of the Stockton and Darlington Railways is among the great events of history. The date marks the inauguration of the railway era in which we now live.

The two categories of railway, i.e. transcontinental and national are important. Transcontinental railways are those which are running across the continent. The railways running from the Atlantic to the Pacific Coast in North America are described as "Transcontinental Railways". Such also are the Canadian National Railway, and the Canadian Pacific Railway. But, in common use, all those railways which reach the Pacific Coast from the Mississippi or from Chicago are described as 'transcontinental', whether there is a through coast-to-coast service or not. Other such railways are the Cape-to-Cairo Railway in Africa and the Trans-Siberian Railway in Asia. The Paris-Berlin-Moscow route, Paris-Milan-Brindisi route; and the Berlin-Vienna-Istanbul route may be called transcontinental railways' systems in Europe.

The Trans-Siberian Railway connects Russia with the far east. It runs from Moscow to Vladivostok on the Pacific, the distance being 9,500 km. The line was constructed by the government for the purpose of facilitating the work of administration in Asiatic Russia. It is a single-track system. From Moscow the line goes to Omsk after crossing the Urals.

The Trans-Caspian Railway connects Central Asia with European Russia. The line runs from Krasnovodsk, on the Caspian Sea, to the heart of the cotton-growing region of Turkestan, throwing off a branch on the Afghan frontier from Merv to Kushk. Kransnovodsk is connected with Moscow via Tashkent.

The Canadian-Pacific Railway was built during the years 1882–86. The length of the line is 4,340 km. This line connects the Atlantic coast of Canada with its Pacific coast. The line runs from Halifax to Vancouver.

The Chile-Argentina Railway or Trans-Andean Railway of South America connects Buenos Aires with Valparaiso; the distance is nearly 1,440 km. This route was opened for traffic in 1910.

Union Pacific Railway passes through the central part of the USA. This railway was built in the year 1869 and is one of the longest railways of USA. It starts from Chicago and reaches up to San Francisco located on the Pacific Coast. Its two branches:

(a) Chicago to Mississippi and (b) Chicago to New York are also important.

In USA, other transcontinental railways are Great Northern Railway, Northern Pacific Railway and Southern Pacific Railway.

Cape to Cairo Railway: In Africa a transcontinental railway was proposed from Cape of Good Hope to Cairo. The scheme, however, did not work out. A railway line from Cape Town goes up to the border of Congo via Bulawayo and Elizabethville, From Elizabethville, the capital of Katanga, a river-cum caravan route proceeds to Lake Victoria, from where a motor road runs to the Nile gorge. From here steamers maintain regular service to Khartoum. From Khartoum a railway line goes to Wadi Haifa, thence by river-transport Shellal is approached. From Shellal a train runs to Cairo.

Perth-Adelaide Railway: In Australia a transcontinental railway runs between Perth located at south-western coast to Sydney located at south-eastern coast of Australia. The main in-between stations on this railway line are Kalgoorlie, Koolkordi, Post Agusta and it also connects Adelaide, Melbourne and Canberra with Sydney.

Orient Express Railway is one of the most important railways which connect Paris with Istanbul, the capital of Turkey.

Apart from trans-continental railway, every country of the world is having its own railway system, which is necessary for the development of a country. But development of railways varies from country to country according to their economic development and resources availability. Some physical factors like snow cover, hilly terrain, desertic and marshy conditions, etc. have also restricted the expansion of railways within a country. On one hand, there are countries like USA, Russia, Canada, China, India, Japan, Germany, France, UK, etc. which have a dense network of railways, while on the other hand, the countries mostly African, Latin American and Asian, are having limited railway facilities.

Pipelines

Another mode of land transport is the pipelines which are used to transport certain items in liquid and gas form. A pipeline may be defined as a line or conduit of pipe of variable diameter and length, but sometimes hundreds of kilometres long, traditionally used for carrying a liquid or gas from a point of supply to a point of consumption. The use of pipe dates back to the Christian era; water supplies are known to have been piped over long distance since very early times. In China, bamboo pipes were being used in circa 5000 BC, and pipes made of clay or stone were widely used in the ancient civilizations of Egypt, Mesopotamia, Greece and Rome.

The discovery and use of petroleum ushered in a new era in pipelines use. The first crude oil pipeline, constructed of wood, was laid down in the United States in 1861 but this proved to be unsuitable, and the first successful pipeline was made of cast iron in Pennsylvania in 1865. In the early days of the oil industry, pipelines served local purposes, only transporting the crude oil to a nearby refinery or to a rail-head, but gradually the value of piping crude oil or the refined products in bulk came to be recognized and before the end of the 19th century a number of trunk lines had been constructed. Today, nearly half a million kilometres of oil pipeline exist in the world, together with a slightly smaller distance of natural gas pipeline. The increasing use of pipelines in recent years represents one of the most notable revolutions in the history of transport, especially in the transport of energy.

The United States was the first country to use pipeline transport on a massive scale. In 1955 it possessed 88 per cent of the total world oil and gas pipeline mileage, excluding that in the Soviet Union of nearly 2, 24,000 km of oil pipeline, the United States had 1,84,000 km and of some 2,50,000 km of pipe carrying natural gas it had approximately 2,24,000 km, since that date there have been phenomenal developments in pipeline usage, especially in other parts of the world. There has been spectacular growth of pipeline system in Western Europe, in former Soviet Union, Middle East, North Africa, Canada and Latin America. Now China, India, Japan and several other countries are using pipelines as means of transportation.

WATER TRANSPORT

Water transport can be divided into two categories: (i) inland water ways, and (ii) oceanic transport.

Inland Waterways

There are three types of inland waterways, namely, rivers, canals and lakes. In earlier times, perhaps, most of the inland carriage of commodities was by water. This was possible when vessels were small, the volume of traffic limited, and the time factor was not particularly pressing. But during the 18th century, ships began to grow in size, trade began greatly to expand, and speed of carriage came to be of greater importance. In order to overcome the limitations of many rivers and, also, to provide many inland towns with water communications, canals began to be built. In England, a pioneer in canal construction, the building of these new man-made waterways became almost a mania. Canal building on the continent came somewhat later, although there are some examples of early canals, e.g. the Canal du Midi in southern France was constructed in 1681. In 19th century, there was some decline in inland water transport due to development of railways and road transport. But after sometime, water transport has tended to make a comeback, this is closely linked with the cheapness and capacity for bulk carriage of water transport – witnessed by the recent and current developments of the Albert Canal, the canal between Zeeburgge and Ghent, the canalization of the Moselle, the Rhone Valley Scheme, etc. Inland waterways have both advantages and disadvantages.

Europe

In Europe, France, Germany, Belgium, Netherlands as well as Russia have very extensive inland waterways including rivers and canals. France is having 5,600 km of navigable rivers and another 4,800 km of canals. The major French rivers, e.g. Loire, Garonne, Seine, Rhone, Meuse and Moselle have been modified and are linked by canal systems so as to travel entirely by river and canals from Mediterranean Sea to English Channel or from Rhine to Atlantic Ocean. On the other hand, Germany is having 7,040 km of inland waterways.

(i) The Rhine Waterway

The Rhine is the busiest navigable river of the world. On both sides of it have developed heavy industries which benefit from cheap water transport. The river is navigable by ocean-going steamers of small size. From the point of economic geography, the rivers prolong the great ocean routes into the interior of lands. Of the "flowing roads" the Rhine is one of the most remarkable waterways. It leads from the ocean into the very heart of the continent. The Rhine is a "Coal river". The lignite of the Aachen basin and the coal of the Ruhr furnish the greater part of the river freight.

(ii) Waterways of the Germanic-Baltic Lowlands

An extensive network of waterways consisting of east-west canals joining the north-south flowing rivers crosses the northern German plain. The Mittelland Canal, also known as the Midland Canal, built in 1938, joins the three major rivers of Ems, Weser and Elbe. Kiel Canal is 96 km long, links the Elbe estuary to the Baltic Sea. The Dortmund-Ems canal runs north-south and links the Rhine with ports of Bremen and Emden.

Waterways of Southern Germany

Danube is the main river, which flows through seven countries – Germany, Austria, Czech Republic, Hungary, Yugoslavia, Romania and Bulgaria and navigable for about 2,400 km. Canals like Ludwing Canal and Rhone-Rhine Canal provide a good inland waterway.

In Belgium, total length of inland waterway is 1,535 km. Albert Canal (built in 1940) and also other canals on coastal plain serve the towns of Ghent, Burges, Zeebrugge and Ostend. Netherlands, at the mouth of Rhine, is criss-crossed by its distributaries, and also has extensive man-made waterways.

The former Soviet Union, has developed a system of navigable waterways totaling 1,44,000 km. Most of these waterways are located in European Russia. The Baltic and White Sea Canal, the Moscow-Volga Canal and Volga-Don shipping canals are nodal. There are several rivers like Volga, Dvina, Don, Dnieper and Dneister, which are navigable in many parts. But many Russian rivers remained blocked during winter season. In spite of such defects Russian rivers are very important for domestic and foreign trade.

North America

In North America the rivers mostly used for navigation are the Mississippi and Missouri and the most important canals are those of St. Lawrence, which unite the Ontario and St. Lawrence; the Sault Sainte Marie canal, between Superior and Huron; the canal which links the Chesapeake to the Ohio; the New York canal; and the canals between North Allegheny and Erie.

The length of navigable waterways in United States is over 36,072 km. The Mississippi river system, largest of all, provides more than 8,000 kilometres of waterways. The Missouri, a tributary of Mississippi, is navigable for 1,216 km to Sioux City, Iowa.

On the recommendation of The Ralph M. Parsons Company, a private engineering and construction enterprise with headquarters in Los Angeles was set up under the project to carry out technical assessment of water and power potential of North America. The project is referred to by the company as NAWAPA – the North American Water and Power Alliance.

The basic idea behind 'NAWAPA' is to capture the surplus waters of the Fraser, Yukon, Peace and Anthabaska river systems in north-western North America and to direct, via an elaborate system of canals, reservoirs and tunnels, the surplus water to deficit areas of Canada and United States.

The St. Lawrence Waterway: The St. Lawrence, with the great lakes, forms a very important commercial route into the heart of North America.

Asia

Asia is not having elaborate system of inland waterways, but rivers in many countries are used as inland waterways.

In China, the rivers have made a significant contribution to the development of commerce. The three great rivers, the Hwang-ho, the Yang-tse-kiang and the Sikiang, cross the country from west to east. China's greatest river is the Yang-tse-kiang, the most important waterway to navigation in the country. It is doubtful whether there is another equally extensive region of wealth in the world where the people depend solely upon a single artery of traffic and upon one entrepot as do the inhabitants of the Yangtse basin. About half of the population of China lies in this fertile area, utilizing the river, its tributaries and its network of canals as their chief means of communication. The Yang-tse-kiang rises in Tibet, and with it tributaries drains the heart of China. It is navigable by steamers up to the port of Hankow. The Sikiang rises in the highlands of Yunan and has a fairly direct course eastward to its mouth. It is navigable for the greater part of its course. The Pei-ho is important for communication and can be navigated up to Tientsin.

Northern India is especially endowed with three large navigable rivers. Three rivers are the Ganga, the Brahmaputra and the Yamuna. The Ganga can be navigated by steamers as far as Kanpur from its mouth. The Lower Ganga is even now very important, and there is traffic all the year around. The Brahmaputra flows through Assam and Bangladesh and is navigable as far as Dibrugarh. The Indus in Pakistan is navigable by steamer up to Dera Ismail Khan in the North Western Frontier Province.

Burma is very fortunate in having a large number of navigable rivers. The Irrawaddy, the most important and the largest, is navigable by steamers for more than 800 km from its mouth and country boats can proceed farther.

Africa

In Africa, some rivers are navigable, that too for a very limited length. The Nile is the most important river in north-east Africa, but its great defect is the succession of cataracts. In its upper course the Nile has rapids and falls; in its middle course there are cataracts. It is navigable in the delta and in its lower course. The rivers of South Africa are of little use for traffic. The Zambezi is navigable for only 350 km, while the Limpopo can be navigated only for a short distance. The Orange is not navigable. In tropical Africa, the Congo provides a magnificent system of waterways. It tries in the highlands between the lakes Tanganyika and Nyasa. But at several places navigation is interrupted by rapids and falls. The Uban, the chief tributary of the Congo, can be navigated almost up to its head. In West Africa the Niger is easily navigable for 500 miles and in the wet season navigation is continued further. The Gambia is navigable for 260 km from its mouth.

South America

South America is having some long rivers but their use as inland waterways is limited. Amazon River is the longest river of the continent. But till now the Amazon system is of relatively little use, because the region through which the river flows is densely forested, scantily populated, undeveloped and largely unexplored. The Orinoco which flows through Venezuela is a long waterway. But the most useful in South America is the Parana system which penetrates the heart of Argentina, Paraguay, Uruguay and South Brazil. In the southern side of South America the river Rio Negro drains the sheep-rearing land of Patagonia.

Australia

Australia is deficient in waterways. Her river-system consists of small streams flowing from the highlands to the coast, thus not suitable for navigation purpose. The two most important rivers are Murray and Darling. Darling river remains almost dry during winter and spring seasons. River Murray is partially used as a waterway.

Ocean Transport

Ocean transport is the most important water transport, because it has certain advantages over land carriage. The sea offers a ready-made carriageway for ships which, unlike the roadway or railway, requires no maintenance. Water surfaces are two-dimensional and, although sea-going vessels frequently keep to shipping lanes, ships can travel, within a limited number of constraints, in any direction. Because of floatability and reduced friction, ocean vessels are capable of carrying far greater loads and far greater weights than can be handled even by the longest railway train, the most powerful lorry and trailer, or the largest aircraft; accordingly, ocean transport is usually the cheapest of all forms of transport. Again, except for fog and floating ice, and occasionally stormy weather which may hinder progress, ocean-going vessels have fewer physical obstacles to surmount than those which so often handicap overland transport. The principal ocean trade routes of the world are as follows (Figure 30.2):

Figure 30.2 *Major Ocean Trade Routes of the World*

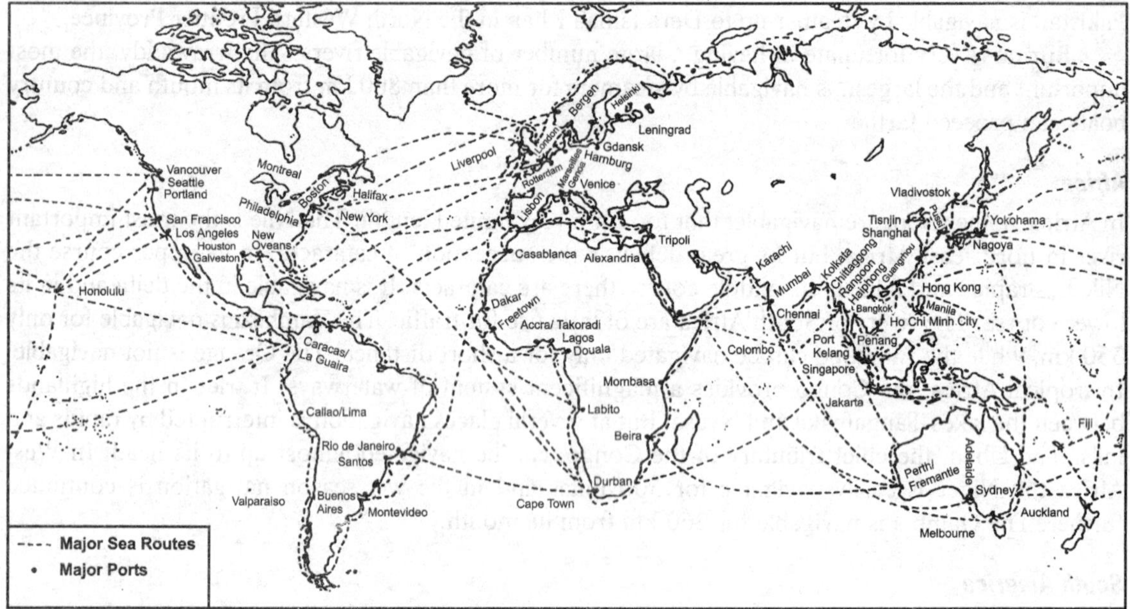

The North Atlantic Ocean Route

The North Atlantic Ocean Route has the greatest traffic of all ocean routes. Nearly one-fourth of the tonnage of the world's merchant vessels serves this route. In volume and variety of cargo, this route far exceeds any other. This route connects the ports of Western Europe with those on the east coast of North America. These two regions are the most populous and highly developed regions in the world. North America and Western Europe are the world's greatest producers of goods of quantity and diversity. Ports on the Western coast of Europe are Glasgow, Liverpool, Manchester, Southampton, London, Rotterdam, Bremen, Bordeaux and Lisbon. Ports on the eastern coast of the USA are Quebec, Montreal, Halifax, St. John, Boston, New York, Baltimore, Charleston Galveston and New Orleans.

This oceanic route is the busiest trade route of the world. Large quantities of manufactured items; textiles, chemicals, machinery, fertilizers, steel, wine, etc. are exported from these ports across the North Atlantic to the United States and Canada.

The exports of Canada and the USA to Europe are timber, fish, wheat, raw cotton, tobacco, oil, machinery and vehicles, metals, paper and chemicals.

The Suez Canal or Mediterranean Asiatic Route

This route is second to the North Atlantic in respect of volume of traffic. It commands the markets of eastern Africa, Iran, Arabia, India, the Far East, Australia and New Zealand. In fact, the route passes through the heart of the world and touches more lands and serves more people than any other route. Throughout its many ports of call, it reaches about three-quarters of the total population of the globe. After crossing the Red Sea, the route follows two directions – one along the eastern coasts of Africa to Durban; another to farther east – to India, Australia, etc. Ports of departure are London, Liverpool, Southampton, Hamburg, Rotterdam, Lisbon, Marseilles, Genoa and Naples. The ports of call are Aden, Mumbai, Kolkata, Rangoon, Penang, Singapore, Manila, Hong Kong, Perth, Adelaide, Melbourne, Sydney, Mombasa, Zanzibar, Mozambique and Durban.

The Cape of Good Hope Route

This route was once the subsidiary alternative to Suez Canal route, but because of its long and circuitous journey, was avoided by most of the shipping companies. During the closure of Suez Canal in 1967 all the ships had no choice but to take this route. Even after Suez Canal reopened in 1975, much trade continues to follow this route because tanker and other vehicles nowadays are much larger.

The Panama Canal: West Indian Central American Route

The construction of Panama Canal was completed in 1913. The Panama Canal is the gateway to the Pacific and eliminated the long and hazardous voyage round the Cape Horn. It has benefited on both Atlantic and Pacific seaboard, facilitating trade in minerals, oil, foodstuff, raw materials, and manufactured products. But the greatest benefits have accrued to traffic between the east and west coasts of USA. The Panama route has also greatly facilitated trade in the West Indian island and the Pacific states of North, Central and South America, especially the Andean steas which are rich in mineral resources and have good markets in North America. The Latin American states import manufactured goods and mining equipment from the USA and the western European countries. Much trade destined for the far east, the Pacific islands and Australasia from North America and Western Europe also goes through the Panama Canal. With the greater economic development of East Asian countries especially China, Japan and the South-East Asian states, the Panama route is assuming a greater role in the exchange of products between the East and West. The distance saved from Auckland to New York via Panama instead of Cape Horn, is more than 4,000 km.

The South Atlantic Route

This route leads to West Indies, Brazil and Argentina. The chief ports of call on the route are Kingston (Jamaica), Havana, Vera Cruz, Tampico, Pernambuco, Bahia, Rio de Janeiro, Stanos, Montevideo, Buenos Aires and Rosario. The exports along the route are sugar, bananas, raw cotton, mahogany, tobacco, coffee, grain, wool and meat, while the imports are manufactured goods.

This route maintains trade connections between Europe on one hand and West Indies, Caribbean seaboard, Brazil, Uruguay and Argentina on the other.

The Trans-Pacific Route

There are several routes in North Pacific which converge at Honolulu for refueling and servicing. The direct route further north is a great circle which links Vancouver and Yokohama without calling the Hawaiian Island, reduces the travelling distance by half. The North Pacific trade include Vancouver, Seattle, Portland, San Francisco and Los Angles on the American side, dealing with wheat, timber, paper and pulp, fish, dairy product and manufactured goods. The destinations across the 7,200 km (4,500 mile) wide Pacific are usually Yokohama, Kobe, Shanghai, Guaznhou (Canton), Hong Kong, Manila and Singapore. The east-bound trade from East Asia to North America consists mainly of manufactured goods, e.g. textiles, electrical equipment, from Japan, Hong Kong, S. Korea, and Taiwan, and tropical raw materials from South-East Asia, e.g. rubber, copra, palm oil, teak, tin and tea. In addition to international links the North Pacific is an important domestic routeway from the US mainland to the isolated states of Alaska, in the north, and Hawaii in the mid-Pacific.

In the South Pacific, the traffic consists mainly of ships travelling via the Panama Canal between West Europe or North America and Australia, New Zealand, and the scattered Pacific islands. Goods transported are mostly wheat, meat, wool, fruits, dairy products and manufactured articles.

The other important routes are eastern North American-east South American (from New York to Cape Sao Roque), North American-western South American (from New York to Punta Arenas via Panama Canal), North American-Australasian (from New York and Vancouver to Sydney and Wellington via Honolulu) etc.

The Suez Canal

The Suez Canal is one of the great international waterways of the world – cuts across the Isthmus of Suez and provides navigational facilities between the Mediterranean Sea and the Indian Ocean. The history of a canal connecting the Mediterranean with the Red Sea dated back to 13th century BC, when Nile-Red Sea canal is known to have been in use until the end of 8th century AD. From 16th century onwards, one or the other European power became interested in the idea of either reopening the old waterway or cutting a new one from the Mediterranean. In 1834, Ferdinand de Lesseps, a member of the French Consular service at Alexandria was interested in the Suez Canal scheme. In 1854, he discussed the project with the Viceroy of Egypt (Khedive) and got his approval. A concession to run for 99 years from the date the canal's opening was granted to de Lesseps, authorizing him to form an International company for the purpose of constructing a waterway. The excavation was started in 1859 under de Lesseps, who took full ten years to construct the canal. It was opened in November, 1869. It is about 160 km long (including distance of lakes) and is 11 to 15 metres deep. The width of the floor is 40 m and varies at the surface; Suez Canal links the Mediterranean Sea with Red Sea. Port Said is located at Mediterranean Sea; while Port Suez is on the Red Sea.

Figure 30.3 *The Suez Canal*

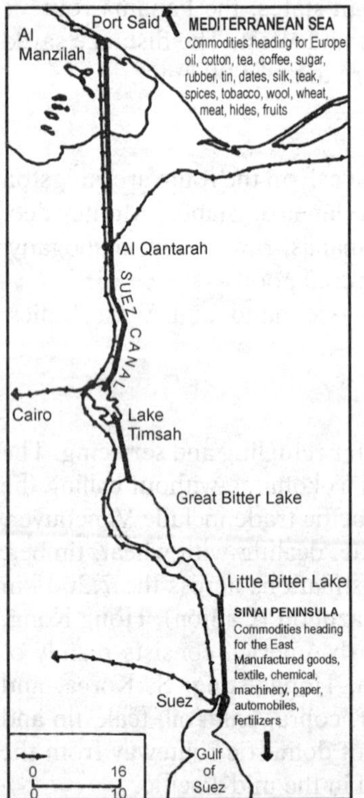

Figure 30.4 *The Panama Canal*

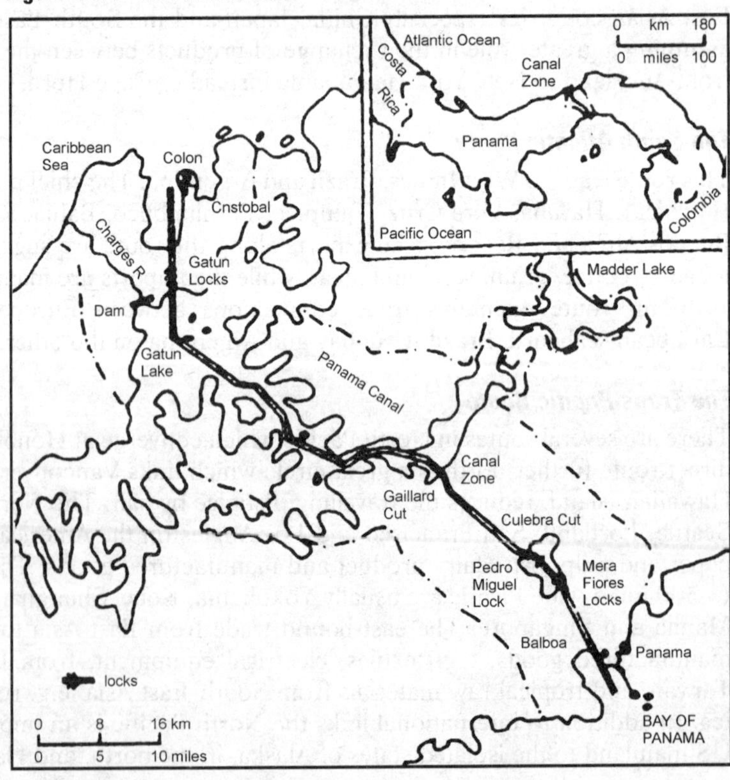

The Panama Canal

The Panama Canal connects the Atlantic and the Pacific oceans by means of two bays, an artificial lake, a natural lake, and three systems of locks. It has been constructed across the narrow Isthmus of Panama where the long Continental Divide dips to one of the lowest points. The canal is 72 km long from deep water to the oceans. It was opened on 15 August 1914.

All the locks are double, so that ships can pass in both directions without any congestion of traffic. The depth of the channel varies from 12 to 26 metres and the width varies from 91 to 305 metres. The time taken to pass through the canal from Panama to Colon is 14 hours.

AIR TRANSPORT

The use of aircraft and development of air transportation belong to the 20th century. The countries which took the lead in developing commercial air routes in the world were France, Belgium, Netherlands and UK. These countries, after the First World War, put to use their experience of flying and the accumulated planes for commercial aviation, primarily for the purpose of maintaining communication with and defence of their colonies. The USA entered the field of commercial aviation after 1927. Lack of colonial interest was one of the factors for delay. The real progress of air transportation, however, started after 1946. The World War II gave a great impetus to the development of aerial transport and the period between 1946 and 1956 was characterized by quick development of new equipment and by a big upsurge in the volume of air travel. The greatest innovation of the period was the application of jet propulsion to transport aircraft.

The world map (Figure 30.5) showing pattern of air routes brings out certain features; these are:

(i) The network shows that there is a worldwide coverage;
(ii) An east-west girdle of routes in the northern hemisphere;
(iii) Longitudinal extensions from the latter into the three southern continents;

Figure 30.5 *Major Air Routes of the World*

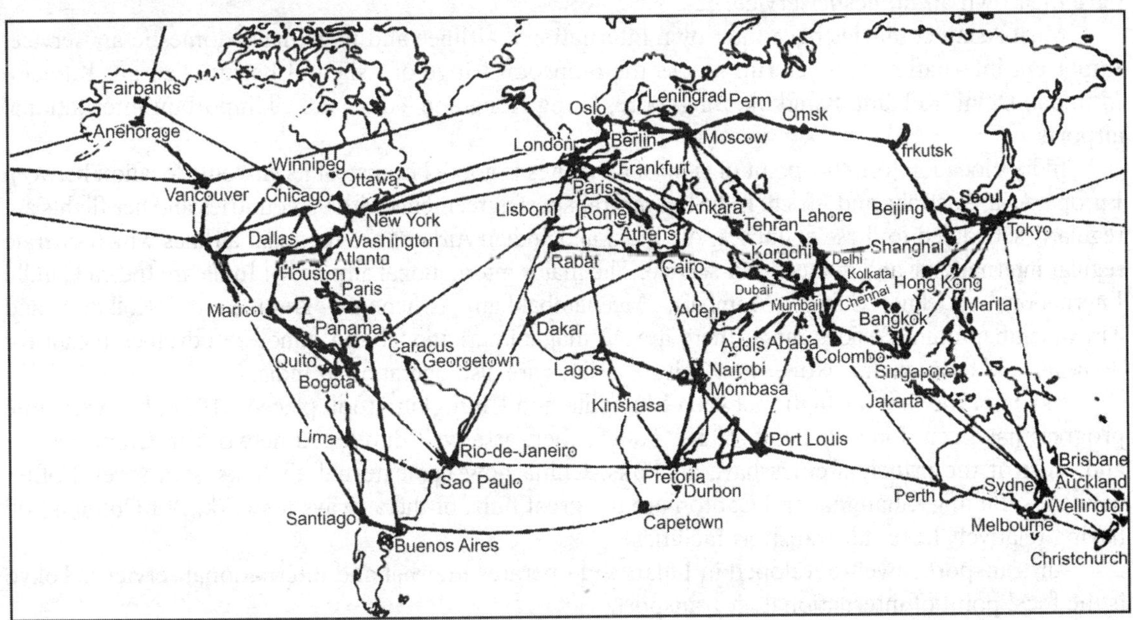

(iv) A great focusing of routes in three particular regions: western Europe. Eastern USA and the Caribbean, and South-East Asia and

(v) Several important foci from which numerous routes radiate, e.g. New York, Chicago, Los Angeles, Washington, Seattle, San Francisco, London, Paris, Berlin, Frankfurt, Rome, Moscow, Cairo, Delhi, Mumbai, Kolkata, Bangkok, Singapore, Hong Kong, Tokyo, Sydney, etc.

The greatest development in air transport has taken place in United States of America. USA is country of long distances so that there are no political frontiers with all the customs formalities, to be crossed by aeroplanes, The Atlantic and the pacific coasts of the United States are among the most developed parts of the country, and the quickest way to connect these areas is by air. Because of this the number of people wishing to use this quick method of contact between the two far-flung borders of USA is very large.

United States alone is served by about 9,000 air terminals, has over 1,50,000 registered civil aircrafts and accounts for almost half of the world's air passenger traffic.

In Canada, the largest airline company is the Trans-Canada Airline, with Montreal, Toronto and Vancouver as the busiest air terminals.

In Europe, most of the countries have their own national airlines. After US, the greatest volume of air traffic is found in Europe. The major air terminals of Europe are London, Paris, Rome, Madrid, Berlin, Warsaw, Vienna, Geneva, Moscow, etc. England-Australia air route is one of the busiest and important air routes. This route starts from London and follows Marseilles, Athens, Alexandria, Cairo, Gaza, Baghdad, Bahrein, Sharjah, Karachi, Delhi, Kolkata, Rangoon, Bangkok, Penang, Singapore, Batavia, Darwin, Brisbane and finally reaches Sydney.

Russian international air routes are mostly controlled by Moscow, Leningrad, Novosibirsk, Irkutsk, and Vladivostok.

Africa is served by international airlines following the East African route through London, Rome, Cairo, Khartoum, Addis Ababa, Nairobi, and Johannesburg with Mauritius, Sri Lanka and South-East Asia. The Central African route is through Paris, Algiers, Kano, Lagos, Kinshasa, Johannesburg, Durban or Cape Town whereas West African route goes through London, Madrid, Casablanca, Dakar, Freetown, Monrovia, Accra and Luanda. In addition, many African countries have their own air domestic service.

Most Asian countries have their own international airlines and many have domestic air service. Important international routes run across the monsoon fringe of Asia and Beirut, Tehran, Karachi, Mumbai, Delhi, Kolkata. Bangkok, Singapore, Hong Kong and Tokyo are all important international airports.

India's location from the point of view of air routes is good, because of its en-route location between Europe, Asia, Australia and Africa. India has air transport agreements with 93 countries and her flights are regularly scheduled to these countries. Air India and Indian Airlines are the main airlines which operate regular international and domestic air service. The major international airports of India are Indira Gandhi International (Delhi), Sahar (Mumbai), Meenambakkam (Chennai), Dum Dum (Kolkata) and Tiruvananthapuram. Besides these, there are 92 major international and minor aerodromes meant for domestic and other flights. Now, some private airlines are also operating in India.

In China the first air flight between Shanghai and Chungking took place in 1928, but very little progress has been done till 1950–1958. Today, there is a well-developed network in China proper and most of the principal cities have air links. China now has external air links with several other countries. Beijing, Shanghai and Canton are the great hubs of the airways network. But China is still having relatively lesser air transport facilities.

Air transport is well-developed in Japan and operates internal and international services. Tokyo is the focal point of international air transport.

The other countries of South-East Asia are also having international aviation. Similarly, countries like South and North Korea, Pakistan, Bangladesh, Sri Lanka, etc. are well-connected with air transport.

The Central and South American countries have developed air transport and all the capital cities are connected with international flights; the main focal points of air transport are Caracas, Bogota, Lima, Santiago, Buenos Aires, Sao Paulo and Rio de Janeiro.

In a large country like Australia where population is less and widely scattered and half of the people live in or near six major cities, the development of internal air service has been of enormous value. Thus, Australia has a well-developed internal air service and is also well-connected with other parts of the world. Sydney is its most important international airport. New Zealand also has well-developed internal and external air transport facilities. Auckland, Wellington, Christchurch and Dunedin are the principal airports of New Zealand.

Multiple Choice Questions

SET – 1
WORLD: Physical Geography

1. What is the average distance (approximate) between the Sun and the Earth?
(a) 70×10^5 Km
(b) 100×10^5 Km
(c) 110×10^6 Km
(d) 150×10^6 Km

2. Which one of the following planets has the largest number of Natural Satellites or moons?
(a) Jupiter (b) Mars
(c) Saturn (d) Venus

3. In order of their distances from the Sun which of the following planets lie between Mars and Uranus?
(a) Earth and Jupiter
(b) Jupiter and Saturn
(c) Saturn and Earth
(d) Saturn and Neptune

4. What is the unit of measurement of distances of stars from the earth?
(a) Light year (b) Fathoms
(c) Nautical miles (d) Kilometres

5. Light from the nearest star reaches the earth in:
(a) 4.3 seconds (b) 4.3 minutes
(c) 43 minutes (d) 4.3 years

6. Which of the following will never get the vertical rays of the Sun?
(a) Srinagar
(b) Mumbai
(c) Chennai
(d) Thiruvanathapuram

7. Which comet appears every 76 years?
(a) Hailey's (b) Halwe's
(c) Donat's (d) Aplha Centauri

8. Which one of the following scholars suggests the Earth's origin from gases and dust particles?
(a) James Jeans (b) H. Alfven
(c) F. Hoyle (d) O. Schmidt

9. Consider the following statements:
(i) Most magmas are a combination of liquid, solid and gas.
(ii) Water vapous and carbon dioxide are the principal gases dissolved in magma.
(iii) Basaltic magma is hotter than the silicic magma.
(iv) The magma solidified between sedimentary rocks in a horizontal position is known as dike.
Which of these statements are correct?
(a) 1, 2 and 3 (b) 2, 3 and 4
(c) 1 and 4 (d) 1, 2 and 4

10. The origin of the universe is about:
(a) 15 billion years ago
(b) 16 billion years ago
(c) 17 billion years ago
(d) 18 billion years ago

11. Asteroids are found between:
(a) Mars and Jupiter
(b) Jupiter and Saturn
(c) Saturn and Uranus
(d) Uranus and Neptune

12. Examine the following statements and select the correct answer using the code given below:
(i) The outer most layer of the Erath is known as crust.
(ii) The Earth crust is above Mohorovicic (Moho) discontinuity.
(iii) Earth crust represents less than one per cent of total volume of the earth.
Code:
(a) 1 and 2 are correct
(b) 2 and 3 are correct
(c) 1 and 3 are correct
(d) 1, 2 and 3 are correct

13. Consider the following geological phenomena:
(i) Development of a fault.
(ii) Movement along a fault.
(iii) Impact produced by a volcanic eruption.
(iv) Folding of rocks.
Which of the following cause earthquakes?
(a) 1, 2 and 3 (b) 2 and 4
(c) 1, 3 and 4 (d) All of these

14. In the structure of planet Earth below the mantle, the core is mainly made up of which one of the following?
(a) Alluminium (b) Chromium
(c) Iron (d) Silicon

15. Which of the following pairs is/are correctly matched?

Theory/Law	**Associated Scientist**
1. Continental Drift	Edwin Hubble
2. Expansion of Universe	Aflired Wegenes
3. Photoelectric Effect	Albert Einstein

Select the correct answer using codes given below:
(a) 2 and 3 (b) Only 3
(c) Only 2 (d) Only 1

16. What is the approximate mean velocity with which the Earth moves round the Sun in its orbit?
(a) 20 km/s (b) 30 km/s
(c) 40 km/s (d) 50 km/s

17. The South Pole experiences continuous light at the time of:
(a) Winter solstice
(b) Summer solstice
(c) Vernal equinox
(d) Autumnal equinox

18. Which one of the following is the most common variety of igneous rocks?
(a) Basalt (b) Diorite
(c) Granite (d) Periodite

19. Which planet takes the longest time in completing one revolution around the Sun?
(a) Neptune (b) Jupiter
(c) Saturn (d) Uranus

20. The equatorial diameter of the earth is:
(a) 12,737 km (b) 12,797 km
(c) 12,717 km (d) 12,757 km

21. The primary cause of day and night is:
(a) Earth's annual motion
(b) Earth's daily motion
(c) Inclination of the earth's axis and its rotation
(d) Inclination the earth's axis and its revolution

22. The latitude of a place is determined by:
(a) Measuring the earth
(b) By celestial observation
(c) By establishing the meridian of a place
(d) East or West of Greenwich

23. When is the Sun vertical over the Tropic of Cancer?
(a) 21 March (b) 21 June
(c) 23 September (d) 22 December

24. Australia experiences the longest day on:
(a) 22 December (b) 23 September
(c) 20 June (d) 15 March

25. What is the earliest period of geological history?
(a) Permian (b) Pre-Cambrian
(c) Cambrian (d) Tertiary

26. A narrow stretch of land connecting two large masses is called:
(a) a strait (b) an isthumus
(c) a continent (d) a peninsula

27. Which is the most recent geological era?
(a) Cenozoic (b) Mesozoic
(c) Paleozoic (d) Cretaceous

28. The longitude of a place is its:
(a) angular distance north or south of equator
(b) angular distance east or west of the prime meridian
(c) angular distance with respect to the pole stars
(d) angular distance east or west of International date line

29. When the position of the earth is at its minimum from the Sun, it is called:
(a) Apogee (b) Perihelion
(c) Pesigee (d) Aphalion

30. Match the List–I and List–II and select your answer with codes given below the lists:

LIST–I	LIST–II
A. Prime Meridian	1. 180° longitude
B. Tropic of Cancer	2. 23½°N latitude
C. International date line	3. 0° longitude
	4. 23½°S latitude
D. Arctic circle	5. 66½°N latitude
E. Tropic of Capricorn	

Code:

	A	B	C	D	E
(a)	2	4	5	1	3
(b)	3	2	1	5	4
(c)	3	1	4	5	2
(d)	3	4	5	2	1

31. The poleward limit of the vertical rays of the Sun is:
(a) 15°N and S (b) 5°N and S
(c) 23½°N and S (d) 35°N and S

32. The Continental Drift Theory was given by:
(a) Laplace (b) Jeans and Jeffrey's
(c) Chamberlain (d) Wegner

33. What name has been given to the single landmass of the earth before continental drift?
(a) Massif (b) Angaraland
(c) Pangaea (d) Gondwana

34. The ocean surrounding Pangaea was known as:
(a) Thethis Sea (b) Panthalassa
(c) Great Sea (d) Mid Ocean

35. Why has Earth Grid been evolved?
(a) To measure directions and locations.
(b) To measure elevations on the earth.
(c) To measure movements on earth.
(d) To measure depth of oceans.

36. The sphere which supports life on the earth is known as:
(a) Lithosphere
(b) Hydrosphere
(c) Atmosphere
(d) Biosphere

37. **Assertion (A)** To orbit around the Sun, the planet Mars takes lesser time than the time taken by the Earth.

Reason (R) The diameter of the planet Mars is less than that of the Earth.

Code:
(a) Both **A** and **R** are true and **R** is the correct explanation of **A**.
(b) Both A and **R** are true, but **R** is not the correct explanation of **A**.
(c) **A** is true, but **R** is false.
(d) **A** is false, but **R** is true.

38. Consider the following statements:
1. Total land area of Earth is approximately 1475 lakh sq km.
2. Ratio of land area to water area of Earth is approximately 1: 4.
3. Maximum percentage of Earth's water is in the Pacific ocean.

Which of the statements given above is/are correct?
(a) 1 and 3 (b) 2 and 3
(c) Only 1 (d) Only 3

39. Consider the following statements:
    1. The axis of the Earth's magnetic field is inclined at 23½° to the geographic axis of the Earth.
    2. The Earth's magnetic pole in the Northern Hemisphere is located on a Peninsula in Northern Canada.
    3. Earth's magnetic equator passes through Thumba in South India.

    Which of the statements given above is/are correct?
    (a) Only 3 (b) 2 and 3
    (c) Only 2 (d) All the above

40. A day is added when one crosses:
    (a) 180° longitude from west to east
    (b) The equator from north to south
    (c) 180° longitude from east to west
    (d) 0° longitude from east to west

41. Where is the doldrums belt located?
    (a) Near the equator
    (b) Near the polar axes
    (c) On the Tropic of Cancer
    (d) On the Tropic of capricon

42. The second largest continent of the world is:
    (a) Europe (b) North America
    (c) Africa (d) South America

43. What is the percentage of land area of Antarctica in total world land area?
    (a) 10.4 (b) 9.3
    (c) 7.0 (d) 11.9

44. What is the percentage of land area of Asia continent in the total world land area?
    (a) 29.4 (b) 32.6
    (c) 20.2 (d) 28.5

45. Select the correct descending order of the following continents:
    (a) Africa, Asia, Australia
    (b) North America, Africa, South America
    (c) Antarctica, Europe, Australia
    (d) South America, Australia, Antarctica

46. Consider the following statements and select the correct code for the answer:
    1. Australia is having the lowest population among the continents of the world.
    2. Africa is the second largest continent both in terms of area and population.
    3. North America ranks next to Europe both in terms of area and population.

    Code:
    (a) Only 1 is correct
    (b) Only 2 is correct
    (c) Both 1 and 2 are correct
    (d) Both 2 and 3 are correct

47. The isthumus that joins north and south America is:
    (a) Mexico (b) Bolavia
    (c) Peru (d) Panama

48. Asia is having 29.4 per cent of the total land area of the world. What is its percentage of population in world's population?
    (a) 58 (b) 66
    (c) 61 (d) 63

49. Which of the following forms the boundary between Asia and Africa?
    (a) Isreal
    (b) Suez canal
    (c) Saudia Arabia
    (d) Egypt

50. Asia is also known as:
    (a) Cradle of civilization
    (b) Cradle of mankind
    (c) Orient culture area
    (d) All the above

51. Examine the following statements and select the correct answer using the code given below:
    1. Asia occupies one-third of the total land surface of the earth.
    2. Asia is the birth place of major religions.
    3. Singapore is the smallest nation of Asia.

    Code:
    (a) 1 and 2 are correct
    (b) 2 and 3 are correct
    (c) Only 2 is correct
    (d) 1, 2 and 3 are correct

52. Which one of the following nations has the highest number of volcanoes?
    (a) South Korea (b) Indonesia
    (c) New Zealand (d) Philippines

53. What is the depth of the Dead Sea?
(a) –392 m (b) –284 m
(c) –352 m (d) –278 m

54. Which one of the following statements is not correct?
(a) Anatolia forms Asia's western-most edge.
(b) Bosporus strait joins Blacksea with sea of Marmara.
(c) Anatolia is the eastern plateau of Turkey.
(d) Cyprus means copper in Greek language.

55. Jerusalem is a holy place for the:
(a) Jews
(b) Christians and Jews
(c) Jews and Muslims
(d) Christians, Jews and Muslims

56. Bab-al-Mandab lies between:
(a) Eritrea and Yemen
(b) Djibouti and Yemen
(c) Djibouti and Eritrea
(d) Somalia and Yemen

57. Match List–I with List–II and select the correct answer using the code given below:

LIST–I (Country)	**LIST–II (No. of Islands)**
(A) Indonesia	(1) 14,000
(B) Japan	(2) 10,000
(C) Philippines	(3) 7000
(D) Oceania	(4) 3000

Code:

	A	**B**	**C**	**D**
(a)	1	4	3	2
(b)	1	3	4	2
(c)	2	3	4	1
(d)	2	4	3	1

58. Which one of the following is more sparsely populated?
(a) Arabian Peninsula
(b) Deserts of Central Asia
(c) Plateau of Tibet
(d) Arctic Tundra

59. Which of the following belongs to the Gondwana land?
(a) Chinese Massif (b) Anatolia
(c) Arabian Peninsula (d) Karakoran

60. Which part of Asia is known as roof of the world?
(a) Deccan plateau (b) Tibet Plateau
(c) Anatolia (d) Aldan plateau

61. Which is not a plateau in Asia?
(a) Arabian (b) Yakut
(c) Kirghiz (d) Patom

62. Which is not an important plain of Asia?
(a) Mikang (b) Chang Jiang
(c) Irrawaddy (d) Kolyma

63. In which country is Dasht-e-lat desert located?
(a) Iran (b) Turkey
(c) Egypt (d) Mongolia

64. In which country is Registan Desert located?
(a) Mongolia (b) Afganistan
(c) Pakistan (d) Turkmenistan

65. Where is Dasht-e-Kavir desert located?
(a) Afganistan (b) Turkey
(c) Iran (d) Pakistan

66. Which is not a desert of Asia?
(a) Gobi (b) Taklamakan
(c) Karakum (d) Tigris

67. Which river is known as 'Yellow River"?
(a) Salween (b) Syr-Darya
(c) Hwang He (d) Lena

68. Asia's coniferous forests are known as:
(a) Taiga (b) Selva
(c) Steppe (d) None of the above

69. Which of the following is/are a type of shifting cultivation in Asia?
(a) Ladaug (b) Caingin
(c) Tamori (d) All the above

70. In Malasia and Indonesia shifting cultivation is known as:
(a) Chena (b) Ladaug
(c) Taungya (d) Rai

71. Consider the following statements and select the correct answer code given below:
1. Intensive subsistence agriculture is a characteristic feature of Monsoon Asia.

2. Plantation agriculture is performed only in South-east Asia.
3. In Sri Lanka shifting cultivation is known as 'Chena'

Code:
(a) Only 1 is correct
(b) 2 and 3 are correct
(c) Only 3 is correct
(d) 1 and 3 are correct

72. Examine the following statements and select the correct answer using the code given below:
1. The south region of the Aral Lake is known as Turan.
2. The Pamir and Tienshan lie to the east of Turan.
3. Central Asia is predominantly urban.
4. The eastern coast of the Caspian Sea is rich in natural gas.

Code:
(a) 1, 2 and 3 are correct
(b) 1, 2 and 4 are correct
(c) 1, 3 and 4 are correct
(d) 1, 2, 3 and 4 are correct

73. Cardamom Range lies in:
(a) Cambodia (b) Laos
(c) Thailand (d) Vietnam

74. Which one of the following statements is not correct?
(a) The Arabian peninsula holds the world's largest petroleum reserves.
(b) Culturally, the peninsula of Arabia is homogenous.
(c) About 50 per cent of Yemen's population is urban.
(d) Arabian Peninsula is poor in metallic minerals.

75. Which one of the following statements is correct?
(a) The area where Myammar, Thailand and Laos meet is known as the 'Golden Triangle'.
(b) Phnom Penh is the capital of Laos.
(c) Phuket is a sea resort of Malaysia.
(d) Mekong river makes its delta in Thailand.

76. Which one of the following countries has never been ruled by the Europeans?
(a) Indonesia (b) Myammar
(c) Thailand (d) Vietnam

77. Which one of the following statements is correct?
(a) The Jewish state (Isreal) was created in 1950.
(b) The Tigris, Euphrates valleys of Iraq to Persian Gulf is known as 'Fertile crescent'.
(c) Dates are Iran's most important export item after oil.
(d) The Mediterranean coast of Asia is sparsely populated.

78. Which one of the following is the largest city?
(a) Jakarta (b) Kuala Lumpur
(c) Singapore (d) Yangon

79. Match the List–I and List–II and select the correct answer using the code given below:

LIST–I (Country)	**LIST–II (Capital)**
A. Kazakhstan	1. Ashquabad
B. Tajikastan	2. Astana
C. Turkmenistan	3. Dushanbe
D. Uzbekistan	4. Tashkent

Code:

	A	B	C	D
(a)	4	3	1	2
(b)	2	3	1	4
(c)	2	3	4	1
(d)	4	2	3	1

80. Which one of the following is not correctly matched?
(a) Aden – Bab-al Mandab
(b) Bandar Abbas – Strait of Hormuz
(c) Houstan – Strait of Florida
(d) Istanbul – Strait of Bosporus

81. Which one of the following is the busiest navigable river?
(a) Danube (b) Rhine
(c) Thames (d) Volga

82. Which one of the following is the shallowest sea?
(a) Adriatic Sea (b) Baltic Sea
(c) Mediterranean Sea (d) North Sea

83. Which one of the following is the largest city of Europe?
(a) Berlin (b) London
(c) Moscow (d) Paris

84. Which one of the following is not located along a river?
(a) Edinburg (b) Moscow
(c) Rome (d) Rotterdom

85. The biggest and highest mountain range in Europe is:
(a) Ural Mountains (b) Alps
(c) Pyrenees (d) Caucasus

86. Which one of the following statements is not correct?
(a) Mt. Blanc is the highest peak of Alps.
(b) The main sources of revenue of Monaco is casino.
(c) About 40 per cent of France is productive farmland.
(d) Wheat, sugarbeat and wine are the main agricultural products of France.

87. Which one of the following statements is not correct?
(a) Polar winds bring cold winter weather to northern Europe.
(b) Gulf stream creates more temperate conditions in Western Europe.
(c) In winter, polar winds penetrate countries of south Europe.
(d) Southern Europe has warm, dry summer, and mild, wet winters.

88. The Strait of Denmark lies between:
(a) Denmark and Norway
(b) Iceland and Greenland
(c) Denmark and Sweden
(d) Scotland and Iceland

89. Match List–I with List–II and select the correct answer using the code given below:

List–I (Country)	List–II (Capital)
A. Austria	1. Budapest
B. Albania	2. Ljubljana
C. Hungary	3. Vienna
D. Sloveria	4. Tiraná

Code:

	A	B	C	D
(a)	3	4	1	2
(b)	3	4	2	1
(c)	4	3	1	2
(d)	4	2	3	1

90. Which one of the following is the largest port of Europe?
(a) Amsterdam (b) Le Harvre
(c) Rome (d) Rotterdam

91. Which one of the following is the leading producer of grapes?
(a) France (b) Spain
(c) Italy (d) Germany

92. Which one of the following statements is not correct?
(a) Europe is rich in coal and iron ore.
(b) Petroleum has been explored in the North sea.
(c) Scandinavian countries are rich in coal deposits.
(d) The Ruhr Basin of Germany has the best quality of coal in Europe.

93. Which area of Europe is known as store house of fish?
(a) Coast of Norway
(b) Southern England
(c) Coast of France
(d) Coast of Netherland

94. Which one of the following is a landlocked state?
(a) Belarus (b) Georgia
(c) Romania (d) Yugoslavia

95. Consider the following statements and select the correct answer using the code given below:
1. Geographical location of Europe is suitable for oceanic transportation.
2. European economy is largely dependent on industries and tertiary activities.
3. European countries are rich in power resources.
4. European culture is known as western culture.

Code:
(a) Only 1 and 2 are correct
(b) Only 2 and 3 are correct
(c) Only 3 is correct
(d) 1, 2 and 4 are correct

96. Arrange the following rivers from west to east and select the correct answer using the code given below:

1. Elbe
2. Order
3. Rhine
4. Vistula

Code:

(a)	3	1	2	4
(b)	3	1	4	2
(c)	1	3	2	4
(d)	1	3	4	2

97. The Miami Beach lies in the state of:
(a) Alabama (b) Florida
(c) Georgia (d) Louisiana

98. James Bay is a part of the:
(a) Gulf of Mexico
(b) Gulf of California
(c) Bay of Funday
(d) Bay of Hudson

99. The twin cities of Minneapolis and St. Paul lies in the state of:
(a) Illinois (b) Indiana
(c) Olio (d) Minnesota

100. Match List–I with List–II and select the correct answer using the code given below:

List–I (City)	**List–II (River)**
A. Balimore	1. Hudson
B. Minneapolis	2. Colorado
C. Las Vegas	3. Patasco
D. New York	4. Mississippi

Code:

	A	B	C	D
(a)	3	4	2	1
(b)	3	4	1	2
(c)	4	3	2	1
(d)	4	3	1	2

101. Grand Canyon of America is formed by which river?
(a) Mcanzi (b) St. Lawrence
(c) Colorodo (d) Missouri

102. Which one of the following is not a lake of North America?
(a) Victoria (b) Ontario
(c) Erie (d) Huron

103. Which of the following river meets the Arctic Ocean?
(a) Ucran (b) Maekcnzie
(c) Columbia (d) Colorodo

104. Which of the following is included in Eastern Highlands of North America?
(a) Rockey Mts.
(b) Mexican Plateau
(c) Appalachian Mts.
(d) Labrador Plateau

105. Which of the following rivers does not flow in Atlantic Ocean?
(a) St. Lawrence (b) Columbia
(c) Mississipi (d) Rio-Grande

106. Which one of the following is the smallest lake?
(a) Biakal (b) Hurou
(c) Michigan (d) Tanganyika

107. Which region of North America is having Mediterranean climate?
(a) Eastern coast of Canada
(b) Southern parts of USA
(c) Central and Southern California
(d) Both sides of Hudson Bay

108. Which one of the following is not correct about the north-eastern USA?
(a) The north-east is America's commercial and industrial hub.
(b) The Hudson river flows between Taconic Range and Catskill Mountain.
(c) Los Angles is the largest seaport of USA.
(d) Pennsylvania is the largest iron and steel producing state of USA.

109. Which one of the following is known as the 'Golden Horse Shoe'?
(a) The Ottawa – Montreal region
(b) Great Lakes – St. Lawrence region
(c) Lake Erie – Ontario region
(d) Lake Ontario – Nova-Scotia region

110. Match List–I with List–II and select the correct answer using the code given below:

List–I (Industry)	**List–II (City)**
(A) Iron and Steel	1. Houston
(B) Automobile	2. Detroit
(C) Ship-building	3. Pittsburg
(D) Petrochemical	4. San Diego

Code:

	A	B	C	D
(a)	3	2	4	1
(b)	3	2	1	4
(c)	2	3	4	1
(d)	2	3	1	4

111. Examine the following statements and select the correct answer using the code given below:
1. The Appalachian mountains are low and wide.
2. Applachians are rich in coal.
3. The Appalachain highlands have a large population.
4. The economic importance of the Appalachian region is gradually declining.

Code:
(a) 1, 2 and 3 are correct
(b) 1, 2 and 4 are correct
(c) 1, 3 and 4 are correct
(d) 1, 2, 3 and 4 are correct

112. Which one of the following is not correctly matched?
(a) Seattle – Aircraft
(b) Orlando – Textile
(c) Buffalo – Food processing
(d) Phoenix – Precision instruments

113. Examine the following statements and select the correct answer using the code given below:
1. Canada is the second largest country in area in the world.
2. Ottawa is the capital and the largest city of Canada.
3. Canada extends beyond the Arctic circle.
4. The Middle island (Lake Erie) is the southern most point of Canada.

Code:
(a) 1, 2 and 3 are correct
(b) 1, 3 and 4 are correct
(c) 2, 3 and 4 are correct
(d) 1,2,3 and 4 are correct

114. The total forest area of Canada is about:
(a) 30 per cent (b) 35 per cent
(c) 45 per cent (d) 50 per cent

115. Examine the following statements and select the correct answer using the code given below:
1. The continent of South America has highest percentage of forest area in the world.
2. South America's population is concentrated along the coastal areas.
3. There is cultural pluralism is most of the countries of South America.
4. The western states of South America are dominated by the Portuguese.

Code:
(a) 1, 2 and 3 are correct
(b) 1, 3 and 4 are correct
(c) 1, 2 and 4 are correct
(d) 1, 2, 3 and 4 are correct

116. Which one of the following is not correctly matched?
(a) Compos – Brazil
(b) Granchaco – Nothern Argentina
(c) Llanos – Venezuela
(d) Selvas – Paraguay

117. Brazil has its border with all the mainland countries of South America except:
(a) Ecquador and Bolivia
(b) Ecquador and Chile
(c) Argentina and Chile
(d) Paraguay and Peru

118. Match List–I with List–II and select the correct answer using the code given below:

List–I (Capital)	**List–II (Country)**
(A) Concepeion	1. Ecuador
(B) Lapaz	2. Paraguay
(C) Bogota	3. Bolivia
(D) Quito	4. Colmbia

Code:

	A	B	C	D
(a)	2	3	4	1
(b)	2	3	1	4
(c)	3	4	2	1
(d)	3	4	1	2

119. Which one of the following is not correct:
(a) Brazil is rich in metallic and non-metallic minerals.
(b) Brazil has about 15 per cent of world's export of iron ore.

(c) The Itabara iron are is of high grade.
(d) Brazil is rich in coal deposits.

120. Which one of the following is a leading producer of natural rubber?
(a) Brazil (b) Indonesia
(c) Malaysia (d) Thailand

121. Which one of the following is the longest mountain chain in the world?
(a) Andes (b) Himalayas
(c) Apls (d) Rockies

122. Which one of the following is not a seaport?
(a) Buenos Aires (b) Caracas
(c) Rio de Jenerio (d) Santiago

123. La Palata is a:
(a) Lake (b) River
(c) Volcano (d) Highland

124. Which region of South America is not suitable for economic development?
(a) Pampa (b) Central Chile
(c) Amazon (d) Coaltal plains

125. Equatorial forests of Amazon are known as:
(a) Selves (b) Pampas
(c) Welds (d) Granchaco

126. Which one of the following continents is a leading producer of copper?
(a) Asia (b) Australia
(c) North America (d) South America

127. Examine the following statements and select the correct answer using the code given below:
1. Most of the rivers of Africa are seasonal.
2. Most of the rivers of Africa have water falls.
3. Most of the rivers of Africa are navigable.
4. Nile is the longest river of the world.

Code:
(a) 1 and 2 are correct
(b) 2 and 3 are correct
(c) 2 and 4 are correct
(d) 3 and 4 are correct

128. The Tropic of Capricorn passes through:
(a) Botswana (b) Swaziland
(c) Zambia (d) Zimbabwe

129. The Jabal Akhdar (Green Mountain) lies in:
(a) Algeria (b) Libya
(c) Morocco (d) Tunisia

130. Match List–I with List–II and select the correct answer using the code given below:

List–I (Country)	**List–II (Capital)**
(A) Ethiopia	1. Addis Ababa
(B) Chad	2. Mogadishu
(C) Somalia	3. Khartoum
(D) Sudan	4. N'dyammera

Code:

	A	B	C	D
(a)	2	3	4	1
(b)	1	4	2	3
(c)	3	4	1	2
(d)	3	1	2	4

131. Which one of the following is not a landlocked country?
(a) Lesotho (b) Botswana
(c) Zambia (d) Namibia

132. The highest peak of Africa is:
(a) Mt. Kenya (b) Mt. Ruenzori
(c) Mt. Kilimanjaro (d) Mt. Jabelta

133. Which mountain range is located in north-west Africa?
(a) Atlas (b) Cameroon
(c) Darkensbreng (d) Kenya

134. The unique feature of north-east Africa is the:
(a) Nile valley (b) Ethopia plateau
(c) Lake Victoria (d) Great Rift Valley

135. In which river do floods occurr twice a year?
(a) Zambrezi (b) Nile
(c) Niger (d) Gambia

136. On which river of Africa are Stanley and Livingstone falls located?
(a) Congo (b) Niger
(c) Nile (d) Zambezi

137. The famous Victoria fall is formed by the river:
(a) Congo (b) Zambezi
(c) Orange (d) Limpopo

138. Which of the following regions of Africa receive conventional type of rainfall?
(a) Eastern Africa
(b) North-west coastal region
(c) Congo Basin
(d) Southermost Africa

139. Which one of the following is the leading producer of clove?
(a) Gambia (b) Zambia
(c) Kenya (d) Tanzania

140. Examine the following statements and select the correct answer using the code given below:
1. About 95 per cent of total population of Egypt lives within a belt of 20 km along the banks of Nile.
2. The Nile river discharges maximum water in May and June.
3. The Aswan Dam was constructed on Nile to generate electricity.
4. Lake Nasser is the biggest artificial lake in Africa.

Code:
(a) 1, 2 and 4 are correct
(b) 2 and 3 are correct
(c) 2 and 4 are correct
(d) 1 and 4 are correct

141. Which one of the following is correctly matched?
(a) Copper – Zaire
(b) Gold – South Africa
(c) Bauxite – Nigeria
(d) Phosphate – Morocco

142. In which country of Africa is the best quality of cotton grown?
(a) Sudan (b) Egypt
(c) Kenya (d) Uganda

143. Which one of the following statements is not correct?
(a) The Great Barrier Reef lies to the east of Queensland.
(b) The Great Barrier Reef is about 1200 miles in length.
(c) Wellington is the capital of New Zealand.
(d) Wellington lies at the northern tip of the south island.

144. The strait of Malacca lies between:
(a) Java and Sumatra
(b) Sumatra and Malaysia
(c) Malaysia and Borneo
(d) Borneo and Java

143. Match List–I with List–II and select correct answer using the code given below:

List–I (Capital)	List–II (Country)
(A) Muscat	1. Jordan
(B) Amman	2. Oman
(C) Sana	3. Saudi Arabia
(D) Riyadh	4. Yemen

Code:

	A	B	C	D
(a)	4	3	2	1
(b)	4	3	1	2
(c)	2	1	3	4
(d)	2	1	4	3

146. In which part of Australia are the Blue Mountains located?
(a) New South Wales (b) Victoria
(c) Queensland (d) Tasmania

147. The largest salt water lake of Australia is:
(a) Torens (b) Hores
(c) Eyre (d) Edward

148. Examine the following statements and select the correct answer using the code given below:
1. Australia's major exports are coal, crude oil, iron ore and wheat.
2. New Zealand relies heavily on sales of dairy products, meat and timber.
3. New Zealand is rich in iron-ore and coal.
4. Australia and New Zealand have developed economics with strong and diverse manufacturing sectors.

Code:
(a) 1, 2 and 3 are correct
(b) 1, 2 and 4 are correct
(c) 1, 3 and 4 are correct
(d) 1, 2, 3 and 4 are correct

149. The site of the first major settlement of Europeans in Australia was:
(a) Melbourne (b) Adelaide
(c) Sydney (d) Brisbane

150. Which one of the following statements is not correct?
   (a) Cook strait lies between North and South Island of New Zealand.
   (b) Bass strait lies between Australia and New Zealand.
   (c) Foveaux strait lies between Steward Island and North Island of New Zealand.
   (d) Torres strait lies between New Guinea and Australia.

151. Which of the following is/are India's research stations on Antarcitica?
   (a) Maitri (b) Dakshin Gangotri
   (c) Mount Indra (d) All the above

152. The largest island of the world is:
   (a) Madagascar (b) Grrenland
   (c) Borneo (d) New Guinea

153. Which one of the following islands is an independent country?
   (a) Sumatra (b) Newfoundland
   (c) Luzon (d) Iceland

154. Mindanao island is a part of:
   (a) Philippines (b) Indonesia
   (c) New Zealand (d) Sri Lanka

155. Hispaniola island is a part of:
   (a) Japan
   (b) Dominican Republic
   (c) Philippines
   (d) Cuba

156. In which ocean/sea is Cuba located?
   (a) Pacific ocean (b) Indian Ocean
   (c) Caribbean sea (d) North Sea

157. Which one of the following city-river pairs is not correctly matched?

	City		River
(a)	Berlin	–	Rhine
(b)	London	–	Thames
(c)	New York	–	Hudson
(d)	Vienna	–	Danule

158. Which of the following countries are landlocked countries?
   1. Afganistan 2. Hungary
   3. Malaysia 4. Switzerland

   Select the correct answer using the codes given below:
   (a) 1 and 2 (b) 1, 2 and 3
   (c) 2, 3 and 4 (d) 1, 2 and 4

159. Cape Canaveral, the site from which space shuttle are launched is located on the coast of:
   (a) Florida (b) Virginia
   (c) North Carolina (d) South Carolina

160. Bermuda triangle extends upto:
   1. Southern Florida 2. Puerto Rico
   3. Hawai Islands
   Select the correct answer using the codes given below:
   (a) All of these (b) 1and 2
   (c) 2 and 3 (d) 1 and 3

161. Which of the following oceans have well developed mid-oceanic ridges?
   (a) Pacific and Atlantic
   (b) Atlantic and Indian
   (c) Indian and Pacific
   (d) Arctic and Indian

162. Which Ocean has a mid oceanic ridge resembling letter's'?
   (a) Atlantic Ocean (b) Pacific Ocean
   (c) Indian Ocean (d) Arctic Ocean

163. In which Ocean is Mindanao Island located?
   (a) Indian (b) Atlantic
   (c) Pacific (d) Arctic

164. Azove Sea is a marginal sea of:
   (a) Atlantic Ocean (b) Pacific Ocean
   (c) Indian Ocean (d) Arctic Ocean

165. The Telegraph plateau is located in the:
   (a) North Atlantic (b) South Atlantic
   (c) North Pacific (d) South Pacific

166. Match the List–I with List–II and select the correct answer using code given below:

List–I	List–II
A. Baikal Lake	1. Deepest point of Asia
B. Aral Sea	2. Deepest point on earth
C. Dead Sea	3. Between Kazakistan and Uzbekistan
D. Lake Turnool	4. Turkey

Code:

	A	B	C	D
(a)	2	3	1	4
(b)	3	4	2	1
(c)	2	1	4	3
(d)	3	4	1	2

167. Which of the following is not the marginal sea of western Pacific?
(a) Banda Sea (b) Arafura Sea
(c) Baltic Sea (d) Yellow Sea

168. Kurile Trench is located in:
(a) Atlantic Ocean (b) Pacific Ocean
(c) Arctic Ocean (d) Indian Ocean

169. In which Ocean is Falklands island located?
(a) Pacific (b) Arctic
(c) Indian (d) Atlantic

170. Which one of the following islands is not located in Atlantic Ocean?
(a) Soctra (b) Georgia
(c) Sandwich (d) St. Helena

171. Which one of the following islands is not located in Indian Ocean?
(a) Reunion (b) Zanzibar
(c) Cape Verde (d) Mauritius

172. Which of the following is/are marginal sea(s) of Mediterranean sea?
(a) Sargasso (b) Agean
(c) Adriatic (d) All of these

173. Which is the largest lake among the following?
(a) Michigan (b) Tanganyika
(c) Superior (d) Baikal

174. Which is the smaller lake among the following?
(a) Titicaca (b) Chad
(c) Eyre (d) Ontario

175. In which country is Balkash lake located?
(a) Uzbikistan (b) Kazakhastan
(c) Nigeria (d) Afghanistan

Answer Key – SET 1

Que. No.	Answer	Que. No.	Answer	Que. No.	Answer	Que. No.	Answer	Que. No.	Answer
1	d	36	d	71	d	106	b	141	c
2	c	37	d	72	b	107	c	142	b
3	b	38	a	73	a	108	c	143	d
4	a	39	b	74	c	109	b	144	b
5	d	40	c	75	a	110	a	145	d
6	a	41	a	76	c	111	d	146	a
7	a	42	c	77	a	112	c	147	c
8	d	43	b	78	a	113	d	148	b
9	a	44	a	79	b	114	d	149	c
10	a	45	c	80	c	115	a	150	b
11	b	46	b	81	b	116	d	151	d
12	d	47	d	82	d	117	b	152	b
13	d	48	c	83	c	118	a	153	d
14	c	49	b	84	a	119	c	154	a
15	b	50	d	85	b	120	d	155	b
16	b	51	d	86	c	121	a	156	c
17	a	52	b	87	c	122	d	157	a
18	c	53	a	88	b	123	b	158	d
19	a	54	c	89	a	124	c	159	a
20	d	55	d	90	d	125	a	160	b
21	b	56	b	91	c	126	d	161	b
22	b	57	a	92	c	127	d	162	a
23	b	58	d	93	a	128	a	163	c
24	a	59	c	94	a	129	b	164	a
25	b	60	b	95	d	130	b	165	a
26	d	61	c	96	a	131	a	166	a
27	a	62	d	97	b	132	c	167	c
28	b	63	a	98	d	133	a	168	b
29	b	64	b	99	d	134	d	169	d
30	b	65	c	100	a	135	b	170	a
31	c	66	d	101	c	136	a	171	c
32	d	67	c	102	a	137	b	172	d
33	c	68	a	103	b	138	c	173	c
34	b	69	d	104	c	139	d	174	a
35	a	70	b	105	b	140	d	175	b

SET – 2
WORLD: Physical Geography

1. Which one of the following is a mountain system?
   (a) Blue Ridge (b) Appalachian
   (c) Atlas (d) Sierra Nivada

2. Examine the following statements and select the correct answer using code given below:
   1. Mountains have come out of geosynclines.
   2. All the mountains of the world are fold mountains.
   3. The formation of mountains is due to compression or tension.
   4. The sial layer beneath the main mountain is very deep.

   Code:
   (a) 1 and 2 are correct
   (b) 2 and 3 are correct
   (c) 1, 2 and 3 arte correct
   (d) 1, 3 and 4 are correct

3. Match the List–I with List–II and select the correct answer using code given below:

List–I (Mountain type)	**List–II (Mountains)**
A. Block mountain	1. Appalachian
B. Oldfold mountain	2. Rocky
C. Young fold mountain	3. Black Forest
D. Volcanic mountain	4. Scottish Highland
E. Relict mountain	5. Vesuvious

Code:

	A	B	C	D	E
(a)	1	3	2	5	4
(b)	3	1	4	2	5
(c)	3	1	2	5	4
(d)	4	1	2	3	5

4. What are the typical features of the young fold mountains?
   1. Rugged peaks 2. Broad Valleys
   3. Reduced relief 4. Sharp precipice

   Select your answer using code given below:
   Code:
   (a) 1 and 3 (b) 1, 2 and 4
   (c) 2 and 4 (d) 1 and 4

5. Black forest Mountain is an example of:
   (a) Old Fold Mountains
   (b) Black Mountains
   (c) Relict Mountains
   (d) Young Fold Mountains

6. Mountain Kea is an example of:
   (a) Volcanic Mountain
   (b) Block Mountain
   (c) Relict Mountain
   (d) Old Fold Mountain

7. Aravalli mountain is an example of:
   (a) Mature Fold Mountain
   (b) Relict Mountain
   (c) Block Mountain
   (d) Volcanic Mountain

8. Which of the following is an example of Pre-cambrian mountains?
   (a) Sierra nevada (b) Hawai
   (c) Aravalli (d) Rocky

9. The mountain building theory based on radio-activity was proposed by:
   (a) Kober (b) Joly
   (c) Holmes (d) Wegener

10. The Convection current theory was proposed by:
    (a) Holmes (b) Wegener
    (c) Mackeij (d) Joly

11. Application mountain is the example of:
    (a) Alpine mountains
    (b) Hercynian mountains
    (c) Pre-Combasion mountains
    (d) Caledonian mountains

12. Which is the longest mountain range of the world?
    (a) Himalayas (b) Andes
    (c) Rockey (d) Urals

13. Which among the following is the longer mountain range?
    (a) Altai (b) Urals
    (c) Trans-Antarctic (d) Altai

14. The second highest peak of Himalayas is:
(a) Kanchenjunga (b) Lhotse
(c) Makalu (d) K2(Godwin Austen)

15. Which is the highest peak of Andes mountains?
(a) Mt. Elbert (b) Mt. Aconcauge
(c) Mt. Elbrus (d) Mt. Blance

16. Which is the highest peak of Rocky mountains?
(a) Mt. Elbrus (b) Mt. Mckinley
(c) Mt. Elbert (d) Mt. Blance

17. The place where several mountain ranges come together is known as:
(a) Pamir 'knot' (b) Bering strait
(c) East Indies (d) Bosporus

18. What was the period of formation of the young fold mountains?
(a) Carboniferous (b) Tertiary
(c) Cretaceous (d) Trassic

19. Which mountains form the group called Alpine?
(a) Vosges, Himalayas, Scottish Highlands.
(b) Himalayas, Alps, Scottish Highlands.
(c) Alps, Himalayas, Rockey, Andes.
(d) Scottish Highland, Alps, Andes.

20. The mountain that belongs to Hercynian group is the:
(a) Harz
(b) Scottish Highlands
(c) Alps
(d) Andes

21. Which of the following is considered an inter-montane plateau?
(a) Brazilian Plateau (b) Deccan Plateau
(c) Arabian Plateau (d) Tibetan Plateau

22. Which of the following is an example of volcanic plateau?
(a) Columbia (b) Brazilian
(c) Tibetan (d) Arabian

23. The largest plateau of North America is:
(a) Columbia (b) Colorado
(c) Arizona (d) Utah

24. Which one of the following is a plain of emergence?
(a) Russian plain (b) Paraguay plain
(c) Congo plain (d) Gulf Coast of USA

25. Guinea plain is an example of:
(a) Structural plain (b) Peneplain
(c) Deposional plain (d) Non of the above

26. The plains produced by wind action in Sahara desert are known as:
(a) Reg (b) Desert plains
(c) Flood plain (d) Alluvial plains

27. Which one of the following is not a river plain?
(a) Amazon Basin (b) Parara Plain
(c) Volga Plain (d) N.W. European plain

28. A geographic region has the following distinct characteristics:
(a) Warm and dry climate
(b) Mild and wet winter
(c) Evergreen oak trees

The above features are the distinct characteristics of which one of the following regions?
(a) Mediterranean (b) Eastern China
(c) Central Asia (d) Eastern China

29. What causes wind to deflect towards left in the Southern hemisphere?
(a) Temperature
(b) Pressure
(c) Rotation of the earth
(d) Magnetic field

30. **Assertion (A)** Areas near the equator receive rainfall throughout the year.

Reason (R) High temperatures and high humidity cause conventional rain in afternoons near the equator.

Code:
(a) Both **A** and **R** are true, but **R** is the correct explanation of **A**.
(b) Both **A** and **R** are true, but **R** is not the correct explanation of **A**.
(c) **A** is true, but **R** is false
(d) **A** is false, but **R** is true

31. Which one of the following is a climate characteristic of the Tropical Savanna region?
(a) Rainfall in winter only
(b) Rainfall throughout the year
(c) An extremely short dry season
(d) A definite dry and wet season

32. The climatic type almost absent in the southern hemisphere is:
(a) Equatorial (b) Mediterranean
(c) Hot desert (d) Taiga

33. **Assertion (A)** The Mediterranean region receives rainfall during winters.
Reason (R) The westerlies associated with cyclones bring winter rainfall.
(a) Both **A** and **R** are true, but **R** is the correct explanation of **A**.
(b) Both **A** and **R** are true, but **R** is not the correct explanation of **A**.
(c) **A** is true, but **R** is false
(d) **A** is false, but **R** is true

34. In Koppen's classification, the climate represented by Bwh is:
(a) Low latitude hot desert
(b) Mediterranean
(c) Equatorial
(d) West European type

35. Which of the following regions is not included in Mediterranean type of climate?
(a) Central Chile
(b) California
(c) South Korea
(d) Southern South Africa

36. Match List–I with List–II and select the correct answer using the code given below:

List–I (Climate type)	**List–II (Region)**
A. Tropical rain forest	1. Just at the fringe of Sahara
B. Tropical Savana Climate	2. Western Africa near Equator
C. Mediterranean climate	3. North-west USA
D. Humid Sub-tropical climate	4. South Africa

Code:

	A	B	C	D
(a)	1	2	4	3
(b)	2	1	4	3
(c)	2	1	3	4
(d)	1	2	3	4

37. Which of the following conditions characterize equatorial climates of the world?
1. Cold winter
2. Hot summer
3. Afternoon thunder shower
4. Small diurnal range of temperature

Select the correct answer using the codes given below:
Code:
(a) 1 and 3 (b) 2 and 4
(c) 1 and 2 (d) 3 and 4

38. The hot dust laden wind blowing from Sahara to Mediterranean region is called:
(a) Sirocco (b) Fohn
(c) Chinook (d) Mistral

39. Match List–I with List–II and select the correct answer using the code given below:

List–I (Climate)	**List–II (Code in Koppen's Classification**
A. Savannah climate	1. Aw
B. Steppe climate	2. Bs
C. Humid sub-tropical climate	3. Cs
D. Dry sub-tropical climate	4. Cfa

Code:

	A	B	C	D
(a)	1	2	4	3
(b)	1	4	2	3
(c)	3	2	4	1
(d)	3	4	2	1

40. Which of the following is not a broad climatic zone of the world?
(a) Tropical humid climate
(b) Marine climate
(c) Dry climate
(d) Polar climate

41. Match List–I with List–II and select the correct answer using the code given below:

List–I (Climate region)	List–II (Location and Area)
A. Tundra	1. Continental western margins between 30° to 40°N and S.
B. Monsoon	2. Very high latitudes
C. Mediterranean sub-tropical	3. Large areas of the tropics
D. Equatorial	4. Continental low to moderate interiors of elevation along the equator

Code:

	A	B	C	D
(a)	2	3	1	4
(b)	3	1	2	4
(c)	4	3	1	2
(d)	2	4	3	1

42. Who has been a pioneer in using a letter code for climate classification?
(a) Trewartha (b) Koppen
(c) Thornhwaite (d) Miller

43. 'D' in Koppen's climate classification indicates:
(a) Tropical climates
(b) Mesothermal climates
(c) Micothermal climates
(d) Dry climates

44. The second letter 'S' in Koppen's climate code stands for:
(b) Steppe type climate
(b) Desert type climate
(c) Warm-wet climate
(d) Cool-humid climate

45. The primary divisions in Koppen's climatic classification are based on:
1. Mean monthly temperature
2. Mean annual temperature
3. Seasonality of rainfall
4. Mean precipitation and evaporation

Code:
(a) 1 and 2
(b) 1, 2 and 3
(c) 1, 2 and 4
(d) 1 and 4

46. What is the code for Mediterranean type of climate in Koppen's climatic classification?
(a) Bs (b) Am
(c) Cf (d) Bw

47. In 1948 classification, Thornthwaite has considered the ponts as:
(a) Indicatiors of climate
(b) Makers of soils
(c) Makers of climate
(d) All the above

48. What is the lower critical limit of warm temperature climate according to Koppen's classification?
(a) –3°c (b) 10°c
(c) –10°c (d) 3°c

49. The subdivisions in Koppen's classification are made on the basis of:
(a) Seasonality of rainfall
(b) Annual precipitation
(c) Mean monthly temperature
(d) Annual temperature range

50. What is true about classification of Koppen and Thornthwaite?
(a) Koppen has used a letter code to describe climate types while Thornthwaite has not.
(b) Thornthwaite has not given description names to climate types as Koppen has done.
(c) Both have based their classification on the concept of vegetation as a maker of climate.
(d) Thornthwaite's climatic regions are usually called vegetation regions given climatic names.

51. Which of the following is a characteristic of evergreen forest?
(a) The trees never shed their leaves.
(b) All trees do not shed their leaves simnetaneously.

(c) When the trees are leafless, grasses remain green.
(d) The name has nothing to do with greenery as such.

52. In which areas are broad leaf trees most common?
(a) Hot and humid areas
(b) Cool and humid areas
(c) Hot and dry areas
(d) Hilly areas

53. Plants growing in areas of high temperature are called:
(a) microtherms
(b) megatherms
(c) mesotherms
(d) endotherms

54. Select the correct statement:
(a) Vegetation is determined by climate
(b) Climate and vegetation have a two way relations
(c) Soils of an area is determined only by the climate.
(d) Climate, soil and vegetation all are inter-related.

55. A Savannah is a:
(a) forest vegetation of tropical region.
(b) tropical mixed vegetation dominated by trees.
(c) temperate type of grassland having only grasses.
(d) type of vegetation adopted to arid climate.

56. A large unit of vegetation having uniformity of not only the life form but also of adaptation to environment is called:
(a) a biome
(b) a biochore
(c) a forest
(d) an ecological niche

57. Consider the following statements about the Mediterranean vegetation and select the correct answer using code given below:
1. The vegetation is called scherophylleous.
2. It has leathery leaves.
3. Figs and olive are its typical examples.

Code:
(a) 1 and 2 are correct
(b) 1 and 3 are correct
(c) 2 and 3 are correct
(d) 1, 2 and 3 are correct

58. Match List–I with List–II and select correct answers using the code given below:

List–I (Mediterranean Vegetation)	List–II (Country)
(A) Chaparrel	1. Australia
(B) Mallee	2. California (USA)
(C) Maquis	3. Chile
(D) Mattoral	4. France

Code:

	A	B	C	D
(a)	4	3	1	2
(b)	4	3	2	1
(c)	2	1	4	3
(d)	2	1	3	4

59. Consider the following statements and select the correct answer using the code given below:
1. Temperate grasslands are known as the 'bread-basket'.
2. Temperate grasslands are among the most modified regions of the world.
3. Temperate grasslands are conifined to the Northern Hemisphere.

Code:
(a) 1 and 2 are correct (b) 1 and 3 are correct
(c) 2 and 3 are correct (d) 1,2 and 3 are correct

60. Xerophytes are:
(a) Plants that grow in humid areas.
(b) Plants which sustain high degree of salinity.
(c) Plants which grow in areas of moderate moisture supply.
(d) Plants which can withstand extreme aridity.

61. Which types of forests are known as 'taiga'?
(a) Tropical rainforests (b) Subarctic coniferous forests
(c) Evergreen scrub forests (d) Arid forests

62. Which one of the following is the tropical grassland?
(a) Steppe (b) Veld
(c) Savannah (d) All the above

63. Which major biochore has a dominance of thorny bushes?
(a) Savannah (b) Forests
(c) Tundra (d) Desert

64. The tallest trees in the world are found in the:
(a) Equatoral region
(b) Monsoon region
(c) Mediterranean region
(d) Temperate rain forest region

65. Which one of the following sets represents the deciduous varieties of mid-latitude broadleaf forest?
(a) Hemlock, pines, oak, hickory
(b) Beech, elm, oak, maple
(c) Beech, elm, hickory, maple
(d) Hemlock, spruce, fir, larch

66. Which one of the following statements is not correct?
(a) The northern needle leaf forest is known as boreal forest.
(b) A more dense boreal forest is known as taiga forest.
(c) The Southern Hemisphere has appreciable area of boreal forest.
(d) Moutane forest of needle leaf trees exist world wide at high elevations.

67. Which one of the following is correct?
(a) Spruce and fir are deciduous trees.
(b) Epiphytes flourish in selvas.
(c) Spanish moss is found in Siberia.
(d) Figs are grown in Germany.

68. Which one of the following is not correct?
(a) Congo is the largest area of selvas.
(b) The cloud forests lie in Venezuela.
(c) The tree dwellers of selvas are known as arboreals.
(d) The cloud forests are perpetuated by high humidity and cloud cover.

69. Which one of the following is a Mediterranean vegetation?
(a) Baobab (b) Birch
(c) Sumac (d) Cork

70. Match List–I with List–II and select the correct answer using the code given below:

List–I (Tree)	**List–II (Region/climate)**
A. Elm	1. Temperate rain forest
B. Lianas	2. Desert/Semi-desert
C. Redwood	3. Mid-latitude broad leaf
D. Saguaro	4. Equatorial region

Code:

	A	**B**	**C**	**D**
(a)	2	4	1	3
(b)	3	4	1	2
(c)	3	1	2	4
(d)	2	1	4	3

71. Which one of the following is not correctly matched?
(a) Elephant grass – Herb
(b) Sumac – Shrub
(c) Baobab – Grass
(d) Elm – Tree

72. Match List–I with List–II and select the correct answer using the code given below:

List–I (Biome/Ecosystem)	**List–II (Vegetation Characteristics)**
A. Needle leaf conifers	1. Maple/ Redwood
B. Mid-latitude mixed forest	2. Teak/Sal
C. Tropical monsoon	3. Spruce/Fir
D. Tropical Savannah	4. Trees with flattened crowns

Code:

	A	**B**	**C**	**D**
(a)	3	1	2	4
(b)	3	1	4	2
(c)	4	3	1	2
(d)	4	3	2	1

73. Which one of the following is not correctly matched?
(b) Maquis vegetation grows in southern Europe.
(c) Chaparral vegetation is found in Australia.

(d) Sclerophyllous vegetation is found in the Mediterranean climate.
(e) Chapparal is called as mattoral in Chile.

74. Examine the following characteristics about the 'mid-latitude grassland's and select the correct answer using the code given below:
1. World's bread-baskets
2. Intensive agriculture
3. Large ranches
4. Well known for grain and livestock production.

Code:
(a) 1, 2 and 3 are correct
(b) 1, 2 and 4 are correct
(c) 2, 3 and 4 are correct
(d) 1, 2, 3 and 4 are correct

75. Cork oak grows in the:
(a) Taiga region
(b) Monsoon region
(c) Equatorial region
(d) Mediterranean region

76. Which one of the following is not correct about the tundra biome?
(a) Growing season 100 to 120 days.
(b) Stable high pressure anticyclone.
(c) Soils are poorly developed.
(d) Frost can occur at any time of the year.

Question no. 77 to 80 consist of two statements, i.e.
Assertion (A) and **Reason (R)**.
You are to examine both the statements and select the correct answer using the following codes.

Code:
(a) Both **A** and **R** are true and **R** is the correct explanation of **A**.
(b) Both **A** and **R** are true and **R** is not the correct explanation of A.
(c) **A** is true **R** is false
(d) **A** is false **R** is true.

77. **Assertion (A)**: The equational rain forest canopy forms three leaves.

Reason (R): The equatorial forest has the tallest trees of the world.

78. **Assertion (A)**: The steppe grasslands are characterized by tall grasses.

Reason (R): The steppe grasses are leguminous.

79. **Assertion (A)**: Cold desert vegetation includes short grass and dry shrubs.

Reason (R): The cold deserts have scanty rainfall and cold winters.

80. **Assertion (A)**: The Tropical savannah is a transitional biome between the tropical forests and semi-arid tropical regions.

Reason (R): The savannah grasses are soil exhaustive.

81. Which of the following soils occurs in equatorial regions?
(a) Podzol (b) Alluvial
(c) Chernozem (d) Laterite

82. Soil formation includes the:
1. Weathering 2. Organic activity
3. Erosion 4. Leaching
(a) 1 and 2 (b) 1 and 3
(c) 1, 2 and 4 (d) All the four

83. The pedogeric refime of podzilization is typical in areas of:
(a) Coniferous forests
(b) High temperature and humidty
(c) Dry climates
(d) Tropical monsoon regions

84. The horizontal layers of soils are called?
(a) Soil layers (b) Soil horiznoes
(c) Soil beds (d) Bedding planes

85. Terra rossa is a typical soil of the:
(a) Equatorial regions
(b) Mediterranean regions
(c) Tropical regions
(d) Temperate regions

86. Alluvial soils are formed in areas of:
(a) High temperature and low humidity.
(b) Wind erosion and deposition.
(c) Calcareous rocks easily soluble in water.
(d) Deposition by streams.

87. Alluvial soils are usually productive because they:
1. receive fresh deposits regularly
2. are usually rich in organic matter
3. occupy lowlands rich in water resources
4. are perfectly leached
(a) 1, 2 and 3 (b) 1, 3 and 4
(c) 1 and 3 (d) All the four

88. What is indicated by PH of soil?
(a) Its fertility
(b) Its maturity
(c) Its age
(d) Its chemical reaction

89. What would be the PH value of highly alkaline soil?
(a) About 7 (b) about 13
(c) about 3 (d) It may be nothing

90. What is humus?
(a) Leaves and twigs of plants
(b) Fully decomposed organic matter
(c) Partly decomposed organic matter
(d) Living organic matter

91. Which one of the following is the correct sequence of soil horizons?
(a) A, B, C, E (b) O, A, E, B
(c) A, B, C, O (d) O, A, B, R

92. Which one of the following is not correct?
(a) Humus is a mixture of decomposed organic materials.
(b) Humus is usually dark in colour.
(c) At the bottom of the soil profile is the E horizon.
(d) The middle layers of a soil profile are composed of small silt and clay, etc.

93. Which one of the following is not correct?
(a) Climate types correlate closely with soil types worldwide.
(b) The soils of the equatorial region have more humus content than that of Savannah.
(c) The soils are affected by the present climates.
(d) In the Nothern hemisphere the soils of the leeward slope has more moisture for plants than is available on south facing slopes.

94. Which one of the following is not a dynamic soil forming factor?
(a) Climate (b) Parent material
(c) Relief (d) Time

95. Which one of the following is not correctly matched?
(a) Aridisols – Desert soils
(b) Alfisols – Humid Temperate soils
(c) Mollisols – Taiga soils
(d) Oxisols – Tropical soils

96. Chernozems are:
(a) Silica rich temperate grassland soils.
(b) Iron rich humid upland soils.
(c) Humus rich dark brown colour soils in semi-arid region.
(d) Humus rich tropical grassland soil.

97. Hardpan and clay pan in soil are found in:
(a) A – horizon (b) B – horizon
(c) C – horizon (d) D – horizon

98. Match List–I with List–II and select the correct answer using the code given below:

List–I (Soil order)	**List–II (Location/climate)**
A. Aridisols	1. Tropical soils
B. Alfisols	2. Hot deserts
C. Mollisols	3. Temperate forests
D. Oxisols	4. Grassland (Sub-humid-semi-arid)

Code:

	A	**B**	**C**	**D**
(a)	2	3	1	4
(b)	2	3	4	1
(c)	4	2	3	1
(d)	4	1	3	2

99. Which one of the following soils has the largest percentage of land area of the world?
(a) Aridisols (b) Alfisols
(c) Entisols (d) Inceptisols

100. Which one of the following is the most permanent attribute of soils?
(a) Colour (b) Consistency
(c) Structure (d) Texture

101. Which one of the following is not an area of chernozemic soil?
(a) United States (b) Russia
(c) South Africa (d) South America

102. Which of the following soil is having the lowest percentage of land area in the world?
(a) Spodosol (b) Histosol
(c) Oxisol (d) Ultisol

103. Which of the pedogenic regimes is associated with hot, dry climates?
(a) Podzolization (b) Gleization
(c) Laterization (d) Salinization

104. Laterization occurs in:
(a) Warm, humid areas
(b) Poorly, drained areas
(c) Cool temperate areas
(d) Riverine tracts

105. The pedogenic regime of calcification is commonly associated with:
(a) Hot, humid areas
(b) Cool temperate areas
(c) Mid-latitude steppe lands
(d) Coastal areas

106. Match List–I with List–II and select the correct answer using the code given below:

List–I (Soil order)	**List–II (Location/climate)**
A. Andisols	1. Organic soils, wet places
B. Histosols	2. Areas affected by volcanic activity
C. Entisols	3. Confer forests, cool humid forests
D. Spodosols	4. Recent soils, profile undeveloped, all climate

Code:

	A	B	C	D
(a)	2	1	4	3
(b)	2	1	3	4
(c)	3	2	1	4
(d)	3	1	4	2

107. Which one of the following is not correctly matched?
(a) Entisols – young soils lacking horizons
(b) Histosols – Over 20 per cent organic matter
(c) Andisole – Volcanic parent material
(d) Oxisols – Low climatic activity

108. Examine the following statements and select the correct answer using the code given below:
1. Soil fertility is the ability of soil to sustain plants.
2. Soil has fertility when it has organic substances and clay minerals.
3. Soil can be made sustainable with the half of chemical fertilizers.
4. Soil characteristics vary from place to place.

Code:
(a) 1, 2 and 3 are correct
(b) 1, 3 and 4 are correct
(c) 1, 2 and 4 are correct
(d) 1, 2, 3 and 4 are correct

109. Consider the following statements:
1. Most magmas are a combination of liquid, solid and gas.
2. Water vapour and carbon dioxide are the principal gases dissolved in a magma.
3. Basaltic magma is hotter than the silica magma.
4. The magma solidified between sedimentary rocks in a horizontal position is known as dyke.

Select the correct answer from the codes given below:
(a) 1, 2 and 3 (b) 2, 3 and 4
(c) 1 and 4 (d) 1, 2 and 4

110. Consider the following statements regarding the earthquakes:
1. The intensity of earthquake is measured on Mercalliscale.
2. The magnitude of an earthquake is a measure of energy release.
3. Earthquake magnitudes are based on direct measurements of the amplitude of seismic waves.

4. In the Richter scale each whole number demonstrates a hundred fold increase in the amount of energy released.

Which of the statements given above are correct:

(a) 1, 2 and 3 (b) 2, 3 and 4
(c) 1 and 4 (d) 1 and 3

111. Which one of the following sets represents the most active volcano?
(a) Etna and Mona Lova
(b) Yellowstgone and Mazama
(c) Saint Helenes and Fujyoma
(d) Chimbarazo and Aconcagua

112. Which one of the following sets is an example of extinct volcano?
(a) Etna and Mona Lova
(b) Pinatubo and Mayon
(c) Mazama and Mona Lova
(d) Demavand and Koh Sultan

113. Which of the following regions has the largest number of active volcanoes?
(a) Alaska (b) Canary Islands
(c) Hawai Islands (d) Iceland

114. Which one of the following is the most violent volcano:
(a) Vulcania (b) Hawai
(c) Peleau (d) Strambolian

115. **Assertion (A)** The circum-pacific belt is the most vulnerable zone of Tsunamis.

Reason (R) It is tectonically the most active area and generates world's major earthquakes.

Code:
(a) Both **A** and **R** are true and **R** is the correct explanation of **A**.
(b) Both **A** and **R** are true but **R** is not the correct explanation of **A**.
(c) **A** is true but **R** is false
(d) **A** is false but **R** is true

116. What is the cause of the tectonic earthquakes?
(a) Volcanic eruption
(b) Lava in transition in rock strata
(c) Faulting and folding
(d) Collapse of subsurface caves

117. Where is the focus of the plutonic earthquakes?
(a) On the surface of the earth
(b) At the great depth in the interior of the earth
(c) Just below the crust of the earth
(d) It can be anywhere in basaltic rocks

118. Where is the epicenter of an earthquake?
(a) It is the point of origin.
(b) Below the point of origin
(c) Diametrically opposite the point of origin
(d) Above the point of origin

119. The 'Ring of Fire' lies in the:
(a) Atlantic Ocean (b) Pacific Ocean
(c) Indian Ocean (d) Arctic Ocean

120. Match List–I with List–II and select the correct answer using the code given below:

List–I (Volcanic Eruption)	**List–II (Characteristics)**
A. Effusive eruption	1. Mount Pinatubo
B. Explosive volcano	2. Martinique (Caribbean islands)
C. Glowing cloud	3. Mt. Orizba (Mexico)
D. Composite Cone	4. Kilauea (Hawaii)

Code:

	A	B	C	D
(a)	4	1	2	3
(b)	4	1	3	2
(c)	3	4	1	2
(d)	3	4	2	1

121. **Assertion (A)** The circum-pacific belt is highly prone to earthquakes.

Reason (R) It has folded mountains.

Code:
(a) Both **A** and **R** are true and **R** is the correct explanation of **A**.
(b) Both **A** and **R** are true and **R** is not the correct explanation of **A**.
(c) **A** is true **R** is false.
(d) **A** is false **R** is true

122. **Assertion (A)** Most of the earthquake activity occurs at the divergent plates.

Reason (R) Most of the earthquakes of the Pacific Ocean have their focus more than 25 km deep.

Code:

(a) Both **A** and **R** are true and **R** is the correct explanation of **A**.
(b) Both **A** and **R** are true and **R** is not the correct explanation of **A**.
(c) **A** is true **R** is false.
(d) **A** is false **R** is true

123. **Assertion (A)** The Islands along the mod-Atlantic Ridge are highly prone to volcanic activity.

Reason (R) The Sea-floor spreading is the main cause of earthquakes in the Atlantic Ocean.

Code:

(a) Both **A** and **R** are true and **R** is the correct explanation of **A**.
(b) Both **A** and **R** are true and **R** is not the correct explanation of **A**.
(c) **A** is true **R** is false.
(d) **A** is false **R** is true

124. Match List–I with List–II and select the correct answer using the code given below:

List–I (Volcanic activity)	**List–II (Example)**
A. Oceanic plate convergence	1. Hawaii
B. Seafloor spreading	2. Mount Saint helous Philippines.
C. Land plates convergence	3. Iceland, East Africa
D. Hot spots	4. Himalayan region

Code:

	A	**B**	**C**	**D**
(a)	2	3	4	1
(b)	2	3	1	4
(c)	3	4	1	2
(d)	3	4	2	1

Question no. 125 to 128 consist of two statements, i.e. **Assertion (A)** and **Reason (R)** Examine these two statements and select your answer using the codes given below:

(a) Both **A** and **R** are true and **R** is the correct explanation of **A**.
(b) Both **A** and **R** are true and **R** is not the correct explanation of **A**.
(c) **A** is true **R** is false.
(d) **A** is false **R** is true

125. **Assertion (A)** Laterite soils are porous in structure.

Reason (R) High amount of rainfall results and silica.

126. **Assertion (A)** The profile of podzol soils is best developed.

Reason (R) They develop in coniferous forest belt.

127. **Assertion (A)** Laterites have low fertility.

Reason (R) High rainfall helps in leaching out minerals.

128. **Assertion (A)** Soils of tropical regions are rich in insoluble iron and aluminium compounds.

Reason (R) High temperature and humidity are favourabe to bacterial activity and the soils thus lack humus.

Answer Key – SET 2

Que. No.	*Answer*	*Que. No.*	*Answer*	*Que. No.*	*Answer*	*Que. No.*	*Answer*
1	b	36	c	71	c	106	a
2	d	37	d	72	a	107	d
3	c	38	a	73	b	108	c
4	d	39	a	74	b	109	a
5	b	40	b	75	d	110	a
6	a	41	a	76	a	111	a
7	b	42	b	77	c	112	d
8	c	43	c	78	d	113	a
9	b	44	a	79	a	114	c
10	a	45	d	80	c	115	a
11	d	46	b	81	d	116	c
12	b	47	c	82	c	117	b
13	c	48	a	83	a	118	d
14	d	49	a	84	b	119	b
15	b	50	c	85	b	120	a
16	c	51	b	86	d	121	c
17	a	52	a	87	c	122	d
18	b	53	b	88	d	123	a
19	c	54	d	89	b	124	a
20	b	55	b	90	c	125	a
21	d	56	a	91	b	126	b
22	a	57	d	92	c	127	a
23	b	58	a	93	b	128	b
24	d	59	a	94	a		
25	b	60	d	95	c		
26	a	61	b	96	a		
27	d	62	c	97	b		
28	a	63	d	98	b		
29	c	64	d	99	a		
30	a	65	d	100	d		
31	d	66	c	101	c		
32	d	67	b	102	b		
33	a	68	a	103	d		
34	a	69	d	104	a		
35	c	70	b	105	c		

SET – 3
WORLD: Social Geography

1. The prime component(s) of the world social geography is (are):
(a) Language (b) Religion
(c) Tribes (d) All the above

2. Consider the following statements and select the correct answer using the code given below:
1. Language is a cultural variable.
2. Language is a means of communication between people.
3. Language is a part of physical environment.
4. Language is useful in delimting the social regions.

Code:
(a) 1, 2 and 3 are correct
(b) 2 and 3 are correct
(c) 1, 2 and 4 are correct
(d) 1, 2, 3 and 4 are correct

3. The approximate number of languages that existed four hundred years ago in the world was:
(a) 7000 (b) 6000
(c) 5000 (d) 5500

4. The approximate number of languages existing presently is:
(a) 6500 (b) 6000
(c) 5000 (d) 4500

5. How many languages are included in Indo-European language family?
(a) 200 (b) 185
(c) 144 (d) 125

6. In which language family are the highest number of languages included?
(a) Niger-Kordofian (b) Sino-Tibetan
(c) Amerind (d) Indo-European

7. How many languages are included in Niger-Kordofian language family?
(a) 959 (b) 583
(c) 731 (d) 1064

8. How many languages are included in Indo-Pacific language family?
(a) 583 (b) 731
(c) 258 (d) 144

9. Which of the following language families is having 731 languages?
(a) Indo-European (b) Indo-Pacific
(c) Austronesian (d) Australian

10. Which language family in the world is having the least number of languages?
(a) Austric
(b) Inuit-Aleut
(c) Chukchi-Kamchatkan
(d) Uralic-yukashir

11. Which language is having the highest speakers in the world?
(a) English (b) Hindi
(c) Mandarin (d) Spanish

12. The approximate number of the speakers of English language in the world is:
(a) 800 Million (b) 508 Million
(c) 487 Million (d) 417 Million

13. Match List–I with List–II and select the correct answer using the code given below:

List–I (Language)	**List–II (No. of speakers in millions)**
A. English	1. 1052
B. Mandarin	2. 508
C. Hindi	3. 417
D. Spanish	4. 487

Code:

	A	**B**	**C**	**D**
(a)	2	3	4	1
(b)	2	1	4	3
(c)	4	3	1	2
(d)	3	4	2	3

14. Mandarin language is spoken only in:
(a) South-East Asia (b) Japan
(c) Central Asia (d) China

15. Which one of the following European countries has four official languages?
(a) Belgium (b) Germany
(c) Switzerland (d) Italy

16. Which statement is not correct regarding languages of Europe?
1. Except Germany and Austria, no two countries of Europe have same language structure.

2. English is the common European language spoken in Europe.
3. The three main language groups of Europe are – Germanic, Romanic and salvic.

Code:
(a) 1 and 2 are correct
(b) 2 and 3 are correct
(c) 1 and 3 are correct
(d) 1, 2 and 3 are correct

17. Which is (are) the chief language(s) spoken by Americans?
(a) English (b) Spanish
(c) Portuguese (d) All the above

18. Which country is considered as *lingua-franca*?
(a) Malayasia (b) Philippines
(c) Sri Lanka (d) Vietnam

19. In Philippines, how many native languages and dialects are spoken?
(a) 74 (b) 64
(c) 54 (d) 44

20. In which country, is the writing script common but spoken differently in various regions?
(a) Japan (b) Indonesia
(c) China (d) Sri Lanka

21. Consider the following statements and select the correct answer using the code given below:
1. Language plays a key role in landscape in the form of place names.
2. Language is in landscape and also landscape is in language.
3. Language reflects the physical environment in which its speakers live.
4. When language moves to new environments adds new words.

Code:
(a) Only 1 and 2 are correct
(b) Only 2 and 3 are correct
(c) Only 3 and 4 are correct
(d) 1, 2, 3, and 4 are correct

22. In which of the following countries is found great uniformity of languages?
(a) Nigeria (b) Japan
(c) Indonesia Guinea (d) Papua New

23. Which of the following language (s) belongs to Italic group of language?
(a) French (b) Spanish
(c) Portuguese (d) All the above

24. Modern English is the outcome of:
(a) Old Frision (b) Old Saxon
(c) Anglo Saxon (d) Dutch Flemish

25. The Latin linguistic group is also known as:
(a) Romance languages
(b) Balto-Slavic languages
(c) Germanic languages
(d) Indo-Iranian Languages

26. Which is not a correct match:
(a) Pashto – Afghanistan
(b) Kurdish – Iraq and Turkey
(c) Farsi – Iran
(d) Urdu – Bangladesh

27. Which is the spoken language of Sri Lanka?
(a) Sinhali (b) Tamil
(c) Telugu (d) All the above

28. In South-West Asia, the official language of about 20 countries is:
(a) Farsi (b) Arabic
(c) Urdu (d) Kurdish

29. Which is the official language of Tanzania?
(a) Mausa (b) Yoruba
(c) Swahili (d) Ibo

30. The main religions of the world in descending order according to number of followers are:
(a) Islam, Hinduism, Christianity, Buddhism
(b) Christianity, Islam, Hinduism and Buddhism
(c) Hinduism, Christianity, Islam and Buddhism
(d) Christianity, Buddhism, Hinduism and Islam

31. The percentage of people following no religion in the world is:
(a) 13.4 (b) 12.6
(c) 6.3 (d) 5.9

32. Which religion is having the highest number of followers?
(a) Islam (b) Hinduism
(c) Christianity (d) Buddhism

33. In Asia, which religion is having the highest number of followers?
(a) Islam (b) Hinduism
(c) Buddhism (d) Confucianism

34. Which of the following regions/areas is not having Christianity in dominance?
(a) Europe (b) Eastern Asia
(c) Australia (d) The Americans

35. Which one of the following countries is not a follower of Islam?
(a) Indonesia (b) Pakistan
(c) Turkey (d) Nigeria

36. Which of the following are the ethnic religions?
(a) Judaism (b) Shinto
(c) Taoisma (d) All the above

37. Which one of the following countries is not having Buddhism?
(a) Indonesia (b) Thailand
(c) Sri Lanka (d) Japan

38. Which is/are the universalizing religions in the world?
(a) Christianity (b) Islam
(c) Buddhism (d) All the above

39. Which one of the following is not a ethnic religion?
(a) Judaism (b) Buddhism
(c) Hinduism (d) Shintoism

40. In which country/countries is Buddhism a state religion?
(a) Myanmar (Burma) (b) Laos
(c) Thailand (d) All the above

41. Consider the following statements:
1. Geographers are concerned with the interaction between religion and the landscape.
2. Religion may be studied as a geographic process.
3. Religion also determines the language pattern.
4. Impact of religion on social and cultural practices is also an important area of social and cultural geography.

Code:
(a) Only 1 and 2 are correct
(b) Only 2 and 3 are correct
(c) 1, 2 and 4 are correct
(d) Only 3 and 4 are correct

42. Which one of the following tribes is from South America?
(a) Arawak (b) Crow
(c) Mapuche (d) Boro

43. Which one of the following is not a Malaysian tribe?
(a) Koren (b) Iban
(c) Sakai (d) Semonag

44. Which one of the following tribes is not correctly matched?
(a) Chakma – Bangldesh
(b) Bantu – Nigeria
(c) Cola – Alaska
(d) Toba – Brazil

45. Which one of the following is not an Indonesian tribe?
(a) Dayak (b) Dani
(c) Dieri (d) Asmat

46. Match List–I with List–II and select the correct answer using the code given below:

List–I (Tribe)	**List–II (Country)**
A. Chin	1. Indonesia
B. Dayak	2. Malaysia
C. Iban	3. Myanmar
D. Lisu	4. Thailand

Code:

	A	**B**	**C**	**D**
(a)	4	2	3	1
(b)	3	1	4	2
(c)	3	1	2	4
(d)	4	2	1	3

47. Which one of the following is not an Australian tribe?
(a) Tiwi (b) Arunta
(c) Kadaku (d) Lisu

48. Which one of the following is not a tribe from the Sahara desert?
(a) San (b) Berbar
(c) Fulani (d) Tuareg

49. Which one of the following tribes is not correctly matched?
(a) Rawala – Uzbikistan
(b) Boro – Amazon Basin
(c) Fulani – Sahara Desert
(d) Mabut – Congo Basin

50. Which one of the following is not a hunting tribe?
(a) Blackfoot (b) Kazak
(c) San (d) Taw

51. Match List–I with List–II and select the correct answer using the code given below:

List–I (Tribe)	**List–II (Country)**
A. Apache	1. Chile
B. Lacandon	2. Mexico
C. Mapuche	3. New Zealand
D. Maori	4. USA

Code:

	A	B	C	D
(a)	4	2	1	3
(b)	4	1	2	3
(c)	2	4	1	3
(d)	2	1	4	3

52. Dayak tribe is inhabited in:
(a) Sumatra (b) Vietnam
(c) Indonesia (d) Cambodia

53. Which one of the following is not a hunter tribe?
(a) Crow (b) Semang
(c) Bushman (d) Kurd

54. Which one of the following is not correctly matched?

	Tribe	**Continent**
(a)	Iban	– Asia
(b)	Inuit	– North America
(c)	Saami	– Europe
(d)	Kadaku	– South America

55. Which one of the following tribe is not dependent on shifting cultivation?
(a) Iban (b) Lotha
(c) Chakma (d) Vedda

56. Which is not a tribe of Central Asia?
(a) Khoid (b) Tao
(c) Tajik (d) Kyugyz

57. Which is not a correct match?
(a) Lepcha – Sikkim
(b) Vedda – Sri Lanka
(c) Chakma – Maldives
(d) Yamato – Japan

58. Which one of the following is not a South-East Asian tribe?
(b) Lise (b) Hamong
(c) Karen (d) Shah

59. Match List–I with List–II and select correct answer using code given below:

List–I (Tribe)	**List–II (Region)**
A. Pygmis	1. Malaysia
B. Sakais	2. Kalahari desert
C. Bushman	3. Congo Basin
D. Yukaghirs	4. N.E. Siberia

Code:

	A	B	C	D
(a)	2	3	1	4
(b)	4	2	3	1
(c)	3	1	2	4
(e)	3	2	4	1

60. Consider the following statements and select the correct answer using the code given below:
1. The major tribes of the world are known as the 'First people'.
2. The tribal communities has a close relationship with land and environment.
3. They are the victims of the past colonialism as well as present set up.

Code:
(a) 1 and 2 are correct
(b) 1 and 3 are correct
(c) 2 and 3 are correct
(d) 1,2 and 3 are correct

61. Which one of the following is a tribe of Russia?
(a) Busques (b) Chukchi
(c) Nenets (d) Yugra

62. Berber speaking community frequently in the news, lives in:
(a) Afghanistan (b) Algesia
(c) Argentina (d) Australia

63. Consider the following international languages:
1. Arabic 2. French
3. Spanish

The correct sequence of the languages given above in the decreasing order of the number of their speakers is:
(a) 3, 1, 2 (b) 1, 3, 2
(c) 3, 2, 1 (d) 1, 2, 3

64. Which is not a tribe of North America and Canada?
(a) Blackfoot (b) Haida
(c) Apache (d) Lakona

65. Which is not a tribe of South America?
(a) Arunta (b) Achi
(c) Akrosi (d) Arawak

66. The estimated world population in 2011–12 was:
(a) 6.82 Billion
(b) 7.129 Billion
(c) 7.92 Billion
(d) 7.531 Billion

67. The growth rate of population was 2.2 (highest) in:
(a) 1961 (b) 1962
(c) 1963 (d) 1964

68. Consider the following statements regarding world population projections:
1. A continued increase in population in near future.
2. A steady decline in the population growth rate.
3. By 2050 global population expected to reach between 8.3 to 10.9 billion.

Code:
(a) 1 and 2 are correct
(b) 2 and 3 are correct
(c) 1, 2 and 3 are correct
(d) 1 and 3 arc correct

69. The descending order of the following continents in terms of total population is:
(a) Asia, N. America, Europe and Africa
(b) Asia, Europe, Africa, N. America
(c) Asia, Africa, South America, Europe
(d) Asia, Africa, Europe, North America

70. Which of the following continents is having the highest density of population?
(a) Africa (b) Europe
(c) North America (d) South America

71. Which is the most extensive area of high population density in the world?
(a) Southern Europe
(b) Western Europe
(c) East Asia
(d) Nile Valley and its delta

72. Which is the second major population cluster in the world?
(a) South Asia
(b) Western Europe
(c) Eastern North America
(d) Eastern Europe

73. The third largest concentration of population occurs in:
(a) Africa (b) North America
(b) South America (d) Eurasia

74. The three largest population clusters of the world together have:
(a) More than half population
(b) Almost half population
(c) About one-third population
(d) More than two-third population

75. Which of the following group of countries is in the descending order according to their total population:
(a) Indonesia, Brazil, Pakistan and USA
(b) USA, Indonesia, Brazil, Pakistan
(c) USA, Brazil, Pakistan, Indonesia
(d) Brazil, USA, Indonesia, Pakistan

76. Which of the following countries is having the highest population?
(a) Japan (b) Nigeria
(c) Germany (d) Bangladesh

77. The tendency of the population growth rate during 20th century is that:
(a) It has been rising
(b) It has been falling
(c) It has been stable
(d) It has been fluctuating

78. Which countries have the higher growth rate of population?
(a) Highly developed
(b) Moderately developed
(c) Developing
(d) Backward

79. The country on the verge of attaining a zero growth rate of population is:
(a) USA (b) Russia
(c) France (d) China

80. In which region of the world are the birth rates the highest?
(a) Southeast Asia (b) Latin America
(c) Central Africa (d) East Asia

81. Consider the following statements:
1. Russia has more population than Brazil.
2. Bangladesh has more population than Nigeria.
3. Nigeria is having the highest population among Africa countries.

Select the correct answer from the codes given below:
(a) Only 3 (b) 1 and 2
(c) 2 and 3 (d) All the three

82. Which among the following countries has the largest population?
(a) Indonesia (b) Japan
(c) Pakistan (d) Sudan

83. Consider the following stages of Demographic Transition:
1. Low increase in National population (total)
2. Low birth and death rate
3. High birth rate and low death rate
4. Falling birth rate and low death rate

Arrange the above stages and select the correct answer from the codes given below:
(a) 1, 2, 3, 4 (b) 2, 3, 1, 4
(c) 1, 3, 4, 2 (d) 4, 3, 2, 1

84. Which is not a sparsely populated region of the world?
(a) Tropical rainforests region
(b) Mountainous regions
(c) River delta regions
(d) Desert and acid regions

85. **Assertion (A)** China is the most populous country in the world.

Reason (R) It's population density is much lower than that of India.

Code:
(a) Both **A** and **R** are true and **R** is the correct explanation of **A**.
(b) Both **A** and **R** are true but **R** is not the correct explanation of **A**.
(c) **A** is true but **R** is false
(d) **A** is false but **R** is true

86. What has been the major cause of the population explosion?
(a) High fertility
(b) Improved medical facilities
(c) Low fertility
(d) High mortality

87. India is in which stage of the Demographic Transition model?
(a) Stage–I (b) Stage–II
(c) Stage–III (d) Stage–IV

88. Most of the European countries are in which stage of the Demographic Transition model?
(a) Stage–I (b) Stage–II
(c) Stage–III (d) Stage–IV

89. What is shown with the help of a population pyramid?
(a) Age and sex composition
(b) Growth rate
(c) Natural increase
(d) Occupational structure

90. Net change in the population due to balance between birth and death rate is called:
(a) Growth rate (b) Fertility
(c) Natural change (d) Migration

91. Optimum population is calculated as per man and:
(a) growth rate of population
(b) resources
(c) his economic potential
(d) his working ability

92. The daily movement of people for work is called:
(a) Temporary movement
(b) Trans humance

(c) Seasonal migration
(d) Commuting

93. The theory of demographic transition deals with change in the:
(a) Population increase with time
(b) Vital rates with time
(c) Age and sex composition with time
(d) Migration rates with time

94. Who were the earliest migrants from Europe to South America?
(a) French (b) Britishers
(c) Spaniasds (d) Portuguese

95. Which migration involved maximum number from Europe?
(a) Britain to North America
(b) Portugal to South America
(c) Spain to South America
(d) Britain to South America

96. Which of the following has majority of the Chinese migrants?
(a) Indonesia (b) New Zealand
(c) Singapore (d) Malaysia

97. Which of the following is a politico religious migration?
(a) Spaniards to Brazil
(b) Jews to Israel
(c) Portuguese to S. America
(d) All the above

98. Who first suggested the laws of migration?
(a) Ravenstein (b) Trewastha
(c) Lee (d) Zelinksy

99. The migration theory was given by:
(a) Thompson (b) Notestien
(c) Doubleday (d) E.S. Lee

100. Which of the following areas is having transhumane activity?
(b) Nothern Rockeys Zealand
(b) Southern New
(c) Southern Aples
(d) South Africa

101. The concept of positive checks on population was introduced by:
(a) Sadler (b) Ginni
(c) Doubleday (d) Malthus

102. Which factor makes the Amazon Basin thinly populated?
1. High temperature 2. High rainfall
3. Fertile alluvial soils 4. High incidence of disease

Code:
(a) 1, 2 and 3 are correct
(b) 2,3 and 4 are correct
(c) 1, 3 and 4 are correct
(d) 1, 2 and 4 are correct

103. Which cultural realm is characterized by the highest degree of urbanization?
(a) North American (b) European
(c) South American (d) Asian

104. Which among the following is a tribe of Africa?
(a) Ainu (b) Fula
(c) Aborigines (d) Tao

105. Which one of the following is not a mega city?
(a) Moscow (b) Cairo
(c) London (d) Paris

106. Which one of the following is the largest city?
(a) Sao Paulo (b) New York
(c) Rio de Janerio (d) Mexico city

107. The mega cities of Africa are:
(a) Cairo and Lagos
(b) Cairo and Rabat
(c) Cairo and Nairobi
(d) Cairo and Johannesburg

Answer Key – SET 3

Que. No.	Answer	Que. No.	Answer	Que. No.	Answer	Que. No.	Answer	Que. No.	Answer
1	d	23	d	45	c	67	c	89	a
2	c	24	c	46	c	68	c	90	c
3	a	25	a	47	d	69	d	91	b
4	b	26	d	48	a	70	b	92	d
5	c	27	a	49	a	71	c	93	b
6	a	28	b	50	b	72	a	94	d
7	d	29	c	51	a	73	d	95	a
8	b	30	b	52	c	74	d	96	c
9	b	31	b	53	d	75	b	97	b
10	a	32	c	54	d	76	d	98	a
11	c	33	a	55	b	77	a	99	d
12	b	34	b	56	b	78	c	100	c
13	b	35	d	57	c	79	c	101	d
14	d	36	d	58	a	80	c	102	d
15	c	37	a	59	c	81	d	103	a
16	c	38	d	60	d	82	a	104	b
17	d	39	d	61	a	83	b	105	c
18	a	40	d	62	c	84	c	106	d
19	b	41	c	63	a	85	b	107	a
20	c	42	b	64	d	86	b		
21	d	43	a	65	a	87	b		
22	b	44	b	66	b	88	d		

SET – 4

WORLD: Economic Geography

1. Which of the following is/are primary activity/activities? Select answer using code given below:
   1. Transport 2. Communications
   3. Mining 4. Manufacturing
   5. Agriculture

   Code:
   (a) 1, 3 and 5 (b) 2, 3 and 4
   (c) 2 and 3 (d) 3 and 5

2. Which one of the following is not a type of farming on the basis of the availability of water?
   (a) Humid farming
   (b) Commercial farming
   (c) Irrigated farming
   (d) Dry farming

3. Which one of the following is not a regional type of agriculture?
   (a) Intensive agriculture
   (b) Monsoon type agriculture
   (c) Tropical plantation agriculture
   (d) Mediterranean type of agriculture

4. Which one of the following is not a type of farming on the basis of ownership of land?
   (a) Collective state farming
   (b) Coperative farming
   (c) Commercial grain farming
   (d) Individual farming

5. Which is not a characteristic of shifting cultivation?
   (a) High dependence on manual labour
   (b) Low levels of technology
   (c) Use of chemical fertilizers
   (d) Utilization of poor soils through falowing

6. Match the following and select answer using codes given below:

Shifting agriculture	**Location**
A. Ladang	1. Sri Lanka
B. Chena	2. Indonesia
C. Conuco	3. Malaysia
D. Humah	4. Venezula

Code:

	A	**B**	**C**	**D**
(a)	3	4	2	1
(b)	3	1	4	2
(c)	4	3	1	2
(d)	1	2	3	4

7. Which is not a characteristic of primitive subsistence agriculture?
   (a) Intensive use of land
   (b) Hill slopes are preferred
   (c) The forests are usually cleared by fire
   (d) Cultivation is done with primitive tools

8. In which areas of the world is primitive subsistence agriculture done?
   (a) Tropical South and Central America
   (b) Congo Basin
   (c) Central Australia
   (d) South-East Asia

9. Which one of the following countries is not having intensive subsistence farming?
   (a) Japan (b) Pakistan
   (c) China (d) New Zealand

10. Which is/are the characteristic(s) of intensive subsistence agriculture? Select your answer using code given below?
    (1) Very small holdings (2) Mechanization
    (3) Farming is intensive (4) Dominance of food crops

    Code:
    (a) 1 and 2 (b) 1, 3 and 4
    (c) 1, 3 and (d) 2 and 4

11. Humid farming implies:
    (a) Farming with irrigation
    (b) Farming in rainy areas
    (c) Farming in semi-arid regions
    (d) Counter farming

12. Consider the following statements about plantation agriculture and select the correct answer using the code given below?
    1. The size of forms is large.
    2. Heavy application of fertilizers and insecticides.
    3. It is based on exploitation of labour.

4. In most plantation farms, there are factories to process the product.

Code:
(a) 1, 2 and 3 are correct
(b) 1, 3 and 4 are correct
(c) 1, 2 and 4 are correct
(d) 1, 2, 3 and 4 are correct

13. Intensive subsistence farming lays a greater emphasis on:
(a) Grain crops
(b) Mixed crops and animals
(c) Grain and fiber crops
(d) Commercial crops

14. The commercial mixed farming implies:
(a) Grain crops cultivation at a large scale.
(b) Plantation agriculture with silk worms rearing.
(c) Cultivation of grain crops and rearing of animals.
(d) Animal farming with poultry.

15. The extensive farming is practical only in areas:
(a) of high pressure of population
(b) of availability of large cultivable area.
(c) in areas of high rainfall.
(d) where transport connectivity is very good

16. Consider the following statements about the intensive subsistence agriculture and select the correct answer using the code given below:
1. Small size of holdings
2. Small and scattered fields
3. Use of domestic labour
4. Dominance of cash crops

Code:
(a) 1, 2 and 3 are correct
(b) 1, 3 and 4 are correct
(c) 1. 2 and 4 are correct
(d) 1, 2, 3 and 4 are correct

17. Which one of the following statements is not correct about the extensive agriculture:
(a) It is highly mechanical
(b) The per hectare yield is low
(c) The per worker production is high
(d) It is practiced in the densely populated regions of the world

18. Which one of the following areas is having commercial grain farming?
(a) African Savannas (b) Southern Asia
(c) North Indian Plains (d) North America

19. Which is not a correct match:

	Area	Plantation Product
(a)	Sri Lanka	– Rubber
(b)	N.E. India	– Tea
(c)	Brazil	– Coffee
(d)	Cuba	– Sugar

20. Which one of the following is not a region of Mediterranean type of agriculture?
(a) Central Chile
(b) Eastern Coast of USA
(c) Southern South Africa
(d) Southernmost Europe

21. Which are the main aspects of Mediterranean agriculture? Select your answer using code given below:
1. Orchard farming
2. Viticulture
3. Cereal and vegetable cultivation
4. Limited animal husbendary

Code:
(a) 1, 2 and 3 are correct
(b) 2, 3 and 4 are correct
(c) 3 and 4 are correct
(d) 1, 2, 3 and 4 are correct

22. Viticulture is related to:
(a) Cultivation of tea
(b) Cultivation of Vegetables
(c) Grape cultivation
(d) Animal husbandary

23. Which of the following is/are the regions of mixed farming:
(a) Eurasia (b) United States
(c) Southern Africa (d) All the above

24. Which one of the following is a region of commercial grain farming?
(a) 'Pampa' of argentina
(b) Central Africa
(c) Eastern Asia
(d) Eastern Australia

25. Which one of the following agricultural practices is eco-friendly?
    (a) Shifting cultivation
    (b) Organic farming
    (c) Cultivation of high yielding varities
    (d) Growing plants in glass-houses

26. The South-western part of California is known for:
    (a) Subsistence agriculture
    (b) Forestry
    (c) Mediterranean agriculture
    (d) Commercial grain farming

27. Consider the following statements:
    1. Dry farming means farming without water.
    2. Commercial grain farming specialized in single crops.
    3. Plantation agriculture has developed mostly in tropical areas.
    4. Mediterranean agriculture is unique because it is a mixture of diverse biocultural activities.

    Select answer using the code given below:
    Code:
    (a) 1 and 2 are correct
    (b) 2, 3 and 4 are correct
    (c) 1, 2 and 3 are correct
    (d) 1, 3 and 4 are correct

28. In which region is rice a dominant crop?
    (a) Semi-arid region (b) European region
    (c) Temperate region (d) Monsoon region

29. Which one of the following is not correctly matched?
    (a) Sugar – Cuba
    (b) Sisal – Tanzania
    (c) Qaat – Oman
    (d) Oil-Palm – Malaysia

30. The leading producer of rice in the world is:
    (a) India (b) China
    (c) Bangladesh (d) Indonesia

31. Arrange the following countries in descending order in respect of rice production and select the correct answer using the code given below:
    1. Bangladesh 2. China
    3. India 4. Indonesia

    Code:
    (a) 3, 2, 4 and 1 (b) 3, 2, 1 and 4
    (c) 2, 3, 1 and 4 (d) 2, 3, 4 and 1

32. Which one of the following is the leading producer of wheat in the world?
    (a) USA (b) India
    (c) China (d) Russia

33. Arrange the following countries in descending order in respect of production of wheat and select the correct answer using the code given below:
    1. India 2. China
    3. Russia 4. USA

    Code:
    (a) 1, 2, 3 and 4
    (b) 2, 1, 3 and 4
    (c) 2, 1, 4 and 3
    (d) 1, 2, 3 and 4

34. Spring wheat is mostly grown in:
    (a) Canadian Prairies (b) Argentina (Pampa)
    (c) Southern Australia (d) Northern India

35. Arrange the following countries in descending order in respect of production of barley and select correct answer using the code given below:
    1. Canada 2. Spain
    3. Russia 4. Germany

    Code:
    (a) 3, 2, 1 and 4 (b) 3, 2, 4 and 1
    (c) 2, 3, 4 and 1 (d) 2, 3, 1 and 4

36. The leading maize producing country in the world is:
    (a) Brazil (b) China
    (c) USA (d) Mexico

37. Which country is the largest producer of raw jute?
    (a) India (b) Bangladesh
    (c) Myanmar (d) Pakistan

38. Which of the following is an important paddy-growing region?
    (a) Southwest Asia (b) Southeast Asia
    (c) South Africa (d) Western Europe

39. Which country produces more of long staple cotton?
    (b) India (b) Nigeria
    (c) USA (d) Kenya

40. Which of the following conditions are favourable for tea cultivation?
    1. High rainfall   2. Warm temperature
    3. High altitude   4. Sloping land

    Code:
    (a) 1, 2 and 3   (b) 2, 3 and 4
    (c) 1, 2 and 4   (d) All the four

41. The best quality tea comes from plantations at an altitude of:
    (a) About 1000 metres
    (b) Between 1000 to 1500 metres
    (c) About 2000 metres
    (d) Above 2000 metres

42. What defines the poleward limit of wheat cultivation?
    (a) 120 frost free days
    (b) 100 frost free days
    (c) 150 frost free days
    (d) 180 frost free days

43. Which one of the following countries is the leading producer of millet?
    (a) China   (b) India
    (c) Nigeria   (d) USA

44. Which one of the following is the leading producer of soyabean?
    (a) Brazil   (b) Argentina
    (c) USA   (d) China

45. Arrange the following countries in descending order in respect of production of sugarcane and select the correct answer using the code given below:
    1. Brazil   2. China
    3. India   4. Pakistan

    Code:
    (a) 2, 3, 1 and 4   (b) 1, 3, 2 and 4
    (c) 4, 3, 2 and 1   (d) 2, 1, 3 and 4

46. The leading producer of sugarbeet in the world is:
    (a) China   (b) Russia
    (c) France   (d) USA

47. Which is/are necessary condition(s) for Jute cultivation?
    1. High temperature
    2. High precipication
    3. High capital investment
    4. Mechanization

    Code:
    (a) 1 and 2 are correct
    (b) 2 and 3 are correct
    (c) 1, 2 and 3 are correct
    (d) 2, 3 and 4 are correct

48. Which one of the following is a leading producer of cotton in the world?
    (a) USA   (b) China
    (c) Brazil   (d) India

49. The leading exporter of cotton in the world is:
    (a) India   (b) China
    (c) Egypt   (d) USA

50. Arrange the following countries in the descending order in respect of production of natural rubber and select the correct answer using the code given below:
    1. Malaysia   2. Indonesia
    3. Thailand   4. India

    Code:
    (a) 1, 3, 2 and 4   (b) 2, 3, 4 and 1
    (c) 2,1, 4 and 3   (d) 3, 2, 1 and 4

51. Which of the following conditions is/are suitable for cotton cultivation?
    (a) Annual rainfall 55 to 100 cms.
    (b) 20° to 80°C temperature
    (c) 180 frost free days
    (d) All the above

52. Which is not a coffee producing country of South America?
    (a) Columbia   (b) Brazil
    (c) Chile   (d) Venezuela

53. The largest coffee producing country in the world is:
    (a) Indonesia   (b) India
    (c) Brazil   (d) Ethiopia

54. Which of the following statements are correct? Select answer from the code given below:
    1. Brazil produces best quality of coffee in the world.
    2. Sao Paulo is the largest coffee producing province of Brazil.
    3. Terra-roxa soil is suitable for coffee production.

    Code:
    (a) 1, 2 and 3 are correct

(b) 2 and 3 are correct
(c) 1 and 2 are correct
(d) 1 and 3 are correct

55. Which of the following is the highest cocoa producing country in the world?
(a) Ivory coast (b) Ghana
(c) Nigeria (d) Cameron

56. Which is not a cocoa growing region of the world?
(a) West Africa region
(b) South African region
(c) Tropical American region
(d) Southeast Asian region

57. Which of the following is not a cotton producing country of the world?
(a) USA (b) Brazil
(c) Canada (d) Pakistan

58. Which of the following is not a sugarcane producing country in the world?
(a) Thailand (b) Pakistan
(c) Cuba (d) Ghana

59. The leading Tobacco producing country in the world is:
(a) India (b) China
(c) Turkey (d) Zimbabwe

60. Arrange the following countries in descending order of production of Cocoa and select the correct answer using the code given below:
1. Brazil 2. Ghana
3. Indonesia 4. Ivory Coast

Code:
(a) 1, 2, 3 and 4 (b) 1, 3, 2 and 4
(c) 4, 2, 3 and 1 (d) 4, 3, 2 and 1

61. Which one of the following is the leading producer of groundnut?
(a) USA (b) China
(c) Sudan (d) India

62. Consider the following statements and select the correct answer using code given below:
1. China is the leading producer of rice in the world.
2. Australia is the leading producer of wheat in the world.
3. France is the leading producer of sugrabeat in the world.

Code:
(a) 1 and 2 are correct
(b) 1 and 3 are correct
(c) 2 and 3 are correct
(d) 1, 2 and 3 are correct

63. Which one of the following is the leading producer of coconut in the world?
(a) India (b) Indonesia
(c) Philippines (d) Sri Lanka

64. Match List–I with List–II and select correct answer using the code given below:

List–I (Crops)	**List–II (Geographical conditions)**
A. Barley	1. Hot dry climate with poor soil
B. Rice	2. Cool climate with poorer soil
C. Millets	3. Warm and moist climate with high altitude
D. Tea	4. Hot and moist climate with rich soil

Code:

	A	**B**	**C**	**D**
(a)	2	4	1	3
(b)	3	4	1	2
(c)	2	1	4	3
(d)	3	2	4	1

65. Match List–I with List–II and select the correct answer using the codes given below:

List–I (Crops)	List–II (Producer)
A. Banana	1. Columbia
B. Cocoa	2. Ghana
C. Coffee	3. Jamaica
D. Tea	4. Kenya

Code:

	A	**B**	**C**	**D**
(a)	2	3	1	4
(b)	3	2	1	4
(c)	3	2	4	1
(d)	2	3	4	1

66. **Assertion (A)** Sri Lanka produces more tea in central and eastern regions.

Reason (R) The seasonally occurring heavy rainfall is responsible for much higher production.

Code:

(a) Both **A** and **R** is/are true and **R** is the correct explanation of **A**.
(b) Both **A** and **R** is/are true and **R** is not the correct explanation of **A**.
(c) **A** is true but **R** is false.
(d) **A** is false but **R** is true.

67. Match List–I with List–II and select correct answer using the codes given below:

List–I (Regions)	**List–II (Activities)**
A. Humid Tropics	1. Nomadic herding
B. Hot Deserts	2. Shifting cultivation
C. Monsoon Asia	3. Commercial livestock farming
D. Temperate Grass land	4. Intensive subsistence agriculture

Code:

	A	**B**	**C**	**D**
(a)	1	2	3	4
(b)	2	1	3	4
(c)	2	1	4	3
(d)	1	2	4	3

68. Which of the following statements is/are true regarding primitive subsistence agriculture?
   1. Primitive subsistence agriculture is also known as shifting cultivation.
   2. Shifting cultivation is widely practiced by many tribes in Africa and South America as well as South Central America.
   3. This type of agriculture is known as Milpa in Central America.
   4. The farmer does not return to the starting point or the previous place, where he was habituated.

   Select the correct answer using the codes given below:

   (a) 1 and 2 (b) 2 and 3
   (c) 2, 3 and 4 (d) 1, 2 and 3

69. Which of the following is not included in pastoralism?
   (a) Nomadic herding
   (b) Commercial farming
   (c) Livestock ranching
   (d) Dairy farming

70. Commercial livestock ranching is controlled by:
   (a) Grasslands (b) Soil fertility
   (c) Agriculture (d) All the above

71. Which one of the following is not associated with commercial; live stock rearing?
   (a) Dairy farming (b) Meat industry
   (b) Dry farming (d) Sheep rearing

72. Consider the following factors regarding dairy farming and select the correct answer using the code given below:
   1. Temperate climate is most suitable for dairying.
   2. Dairy farming is concentrated only in cities.
   3. Dairy industry plays a vital role in the economy of New Zealand.
   4. Dairy cattle requires green grass throughout the year.

   Code:

   (a) 1, 2 and 3 are correct
   (b) 1, 3 and 4 are correct
   (c) 2, 3 and 4 are correct
   (d) 1 and 2 are correct

73. The largest belt of commercial dairy farming in the world is:
   (a) South-east Asia
   (b) Eastern USA
   (c) Australia-NewZealand
   (d) Western Europe

74. Which one of the following is not a major commercial beef cattle rearing country?
   (a) South Africa
   (b) Argentina
   (c) Australia
   (d) New Zealand

75. The most common areas of nomadic herding are:
   (a) Hot humid regions (b) Dry areas
   (c) Cool humid regions (d) Rainy areas

76. Which is the leading country in production of beef?
   (a) China (b) Australia
   (c) USA (d) United Kingdom

77. Which country is the leading pork importer in the world?
(a) USA (b) United Kingdom
(c) France (d) Russia

78. Which is the leading producer country of the quality merino wool?
(a) New Zealand (b) South Africa
(c) Turkey (d) Australia

79. The largest importer of wool in the world is:
(a) Germany (b) France
(c) United kingdom (d) USA

80. Which type of climate is most suitable for dairy cattle?
(a) Temperate (b) Tropical
(c) Arid and semi-arid (d) Monsoon

81. Which one of the following country has well developed mutton industry?
(a) South Africa
(b) Brazil
(c) Russia
(d) Australia

82. Animal ranching has well developed in:
(a) Savannah regions
(b) Prairies and Steppes
(c) Monsoon regions
(d) Semi-arid regions

83. 'Transhumace' refers to:
(a) Migration of people in search of jobs.
(b) Migration of animals in ranches.
(c) Seasonal movement of people and their herds from plain areas to mountain and vice-versa.
(d) Migration of nomads.

84. Which one of the following natural regions is not correctly matched with its economic base?
(a) Canadian Prairies – Spring wheat cultivation
(b) Australian Downs – Nomadic Herding
(c) Argentine Pampas – Beef cattle ranching
(d) South African Veld – Maize growing

85. **Assertion (A)** Dairy and meat products are the most important exports of New Zealand.
Reason (R) New Zealand has cool moist climate suitable to support live stock.
Code:
(a) Both **A** and **R** are true and **R** is the correct explanation of **A**.
(b) Both **A** and **R** are true and **R** is not the correct explanation of **A**.
(c) **A** is true but **R** is false.
(d) **A** is false but **R** is true.

86. Which one of the following countries does not practice live stock ranching?
(a) Argentina (b) Kenya
(c) New Zealand (d) Australia

87. Consider the following statements about the nomadic herding and select the correct answer using code given below:
1. Subsistence nomadic hearding is done in the semi-arid regions of the world.
2. It is an ecological system of agriculture.
3. It is most simple form of pastoralism.
Code:
(a) 1 and 2 are correct
(b) 1 and 3 are correct
(c) 2 and 3 are correct
(d) 1, 2 and 3 are correct

88. Consider the following statements about the dairy farming and select the correct answer using the code given below:
1. It is highly mechanized.
2. The farmers are well-off.
3. It is labour intensive.
4. There is fixed ratio between agricultural land and number of cows to be kept in the dairy.
Code:
(a) 1, 2 and 3 are correct
(b) 1, 2 and 4 are correct
(c) 2, 3 and 4 are correct
(d) 1, 2, 3 and 4 are correct

89. Which type of farming is common in areas of low rainfall?
(a) Livestock ranching
(b) Dairy farming
(c) Paddy cultivation
(d) Intensive subsistence farming

90. The essential feature of shifting cultivation is the:
(a) Rotation of crops (b) Mono-cropping
(c) Intensive cropping (d) Rotation of fields

91. Which one of the following is the largest wool producing country in the world?
(a) New Zealand (b) Argentina
(b) Australia (d) South Africa

92. Which one of the following is the most important fishing ground of the world?
(a) Grand Bank (b) Doggar Bank
(c) South Atlantic (d) South Pacific

93. Arrange the following fishing grounds of the world in descending order of production and select the correct answer using the code given below:
1. North Atlantic 2. North Pacific
2. South Atlantic 4. South Pacific
Code:
(a) 2, 1, 3 and 4 (b) 1, 2, 4 and 3
(c) 1, 2, 3 and 4 (d) 2, 1, 4 and 3

94. Which one of the following is the leading fish catching country of the world?
(a) USA (b) Russia
(c) Japan (d) China

95. Which one of the following is not a major fishing area of the world?
(a) Southern Indian Ocean
(b) North-west Atlantic coast
(c) North-east Atlantic
(d) North – East Pacific

96. Which of the following is/are type (s) of fisheries?
(a) Inland fishery (b) Coastal fishery
(c) Open-sea fishery (d) All the above

97. The leading importer of fish in the world is:
(a) USA (b) Russia
(c) United kingdom (d) France

98. Which are the ideal conditions for fish growth?
(a) Temperate continental shelfs
(b) Cool deep ocean
(c) Warm open oceans
(d) All seas are equally good

100. The famous fishing ground Grand Banks are situated in the regions of:
(a) North-west Europe
(b) East Asia
(c) Northeastern North America
(d) East Pacific

101. In which of the following regions are Dogger banks located?
(a) Northeastern North American region
(b) Northwest European region
(c) East Asiatic region
(d) East pacific region

102. Which of the following produces almost 90 per cent of the world's annual fish supply?
(a) The continental shelf
(b) The open sea
(c) The inland water
(d) The coast of Peru

103. Japan, Indonesia, Malaysia and Philippines have developed fishing industry in the world due to:
(a) Coast line (b) Location
(c) Both a and b (d) None of these

104. Which one of the following countries is the largest producer of marine fish?
(a) Japan (b) China
(c) New Zealand (d) Chile

105. **Assestion (A)** Herring and sasdine constitute the largest percentage of commercial catch of marine fish.
Reason (R) These fish are largest found in the ocean of mid-latitudes.
Code:
(a) **A** and **R** are true and **R** is the correct explanation of **A**.
(b) **A** and **R** are true and **R** is not the correct explanation of **A**.
(c) **A** is true **R** is false
(d) **A** is false **R** is true

Answer Key – SET 4

Que. No.	Answer	Que. No.	Answer	Que. No.	Answer	Que. No.	Answer
1	d	31	b	61	c	91	d
2	b	32	d	62	d	92	c
3	a	33	c	63	b	93	b
4	c	34	c	64	c	94	b
5	c	35	a	65	a	95	d
6	b	36	c	66	b	96	a
7	a	37	c	67	c	97	d
8	c	38	a	68	c	98	c
9	d	39	b	69	d	99	a
10	b	40	c	70	b	100	c
11	b	41	c	71	a	101	b
12	d	42	b	72	c	102	a
13	a	43	a	73	b	103	c
14	c	44	a	74	d	104	b
15	b	45	d	75	a	105	b
16	b	46	b	76	b		
17	a	47	c	77	c		
18	d	48	a	78	b		
19	c	49	b	79	d		
20	a	50	b	80	c		
21	b	51	d	81	a		
22	d	52	d	82	d		
23	c	53	c	83	b		
24	d	54	c	84	c		
25	a	55	b	85	b		
26	b	56	a	86	a		
27	c	57	b	87	b		
28	b	58	c	88	d		
29	d	59	d	89	b		
30	c	60	b	90	a		

SET – 5
WORLD: Economic Geography

1. In which category are mineral resources included?
(a) Renewable (b) Non-renewable
(c) Misusable (d) Immutable

2. Which one of the following is not a localized mineral resource?
(a) Iron ore (b) Coal
(c) Water (d) All the above

3. Under which category, are mineral resources included?
(a) Inorganic resources
(b) Organic resources
(c) Flow resources
(d) Inexhaustible resources

4. Match the following and select the answer using the code given below:

A. Lignite	1. Copper
B. Bauxite	2. Iron
C. Magnetite	3. Aluminium
D. Pyrite	4. Coal

Code:

	A	B	C	D
(a)	4	3	1	2
(b)	2	3	1	4
(c)	4	3	2	1
(d)	4	1	3	2

5. Iron is obtained mostly from:
(a) Haematite (b) Magnitite
(c) Siderite (d) Iron Pyrites

6. The largest producer of mercury in the world is:
(a) Spain (b) Italy
(c) USA (d) China

7. Which of the following is the leading producer of Uranium in the world?
(a) Russia (b) Canada
(c) India (d) France

8. The leading producers of rock phosphate are:
(a) Russia and France (b) USA and Mexico
(c) USA and Russia (d) Spain and USA

9. Republic of Congo is the leading producer of:
(a) Diamond (b) Bauxite
(c) Coal (d) Lead

10. Which area is most important for uranium mining?
(a) Mesabi range
(b) New Mexico
(c) Katanga
(d) Urals

11. The mineral which occurs in veines is:
(a) Bauxite (b) Coal
(c) Iron (d) Gold

12. Manganese is used for:
(a) Copper smelting
(b) Storage batteries
(c) Steel making
(d) Petroleum refining

13. Match the List–I with List–II and select answer using code given below:

List–I (Ore)	**List–II (Metal)**
A. Limonite	1. Copper
B. Pyrite	2. Uranium
C. Bauxite	3. Iron
D. Monazite	4. Aluminium

Code:

	A	B	C	D
(a)	1	3	2	4
(b)	3	1	4	2
(c)	3	4	1	2
(d)	1	2	4	3

14. Which mineral occurs most often in association with zinc?
(a) Copper (b) Lead
(c) Iron (d) Gold

15. The anthracite coal has:
(a) Above 70 per cent carbon
(b) 20 to 60 per cent carbon
(c) Low smoke but high moisture
(d) Low carbon and smoke

16. The chief iron mining centres of the Lake Region of USA are:
(a) Los Angeles, Putelo, Pittsburgh
(b) Brimingham, Pittsburgh, Chicago
(c) Duluth, Mesabi, Vermillion
(d) Dallas, Pittsburgh and Mesab

17. Which country has the iron ores with highest content?
(a) United Kingdom (b) Japan
(c) USA (d) Sweden

18. Match the following and select the answer using the code given below:

A. Mica 1. Metallic mineral
B. Nickel 2. Rock
C. Petroleum 3. Non-metallic mineral
D. Granite 4. Fossil fuels

Code:

	A	B	C	D
(a)	2	4	3	1
(b)	3	1	4	2
(c)	3	2	1	4
(d)	4	3	2	1

19. Which one of the following is a type of iron-ore?
(a) Magnetite (b) Silicotite
(c) Haematite (d) Limonite

20. Which one of the following is the leading iron-ore producer country in the world?
(a) China (b) Brazil
(c) India (d) Australia

21. Arrange the producers of iron ore in descending order and select the correct answer using the code given below:
1. China 2. India
3. Brazil 4. Australia
Code:
(a) 1, 3, 2 and 4 (b) 1, 2, 3 and 4
(c) 3, 2, 4 and 1 (d) 3, 4, 2 and 1

22. Which one of the following is not an iron-ore producing country in the world?
(a) Russia (b) USA
(c) Mexico (d) Sweden

23. In which country does the highest known coal reserves exist?
(a) Australia (b) Ukraine
(c) Kazakhstan (d) USA

24. The leading manganese producer country in this world is:
(b) South Africa (b) Russia
(c) Kazkhstan (d) China

25. The second leading manganese producing country in Africa is:
(a) Gabon (b) Ghana
(c) Nigeria (d) Zambia

26. The leading copper producing country in the world is:
(b) USA (b) Indonesia
(c) Chile (d) Australia

27. Arrange the copper producing countries in descending order and select the correct answer using the code given below:
1. USA 2. Australia
3. Chile 4. Indonesia
Code:
(a) 3, 2, 1 and 4 (b) 3, 1, 4 and 2
(c) 1, 3, 2 and 4 (d) 2, 3, 4 and 1

28. Apart from Chile, which is the leading copper producing country in South America?
(a) Brazil (b) Argentina
(c) Peru (d) Colombia

29. The leading copper producing country in Asia is:
(a) Indonesia (b) India
(c) China (d) South Korea

30. Which of the following is/are the natural compound(s) of aluminium?
(a) Bauxite (b) Cryolite
(c) Corundum (d) All the above

31. At present aluminium is extracted commercially from:
(a) Cryolite (b) Bauxite
(c) Kaolin (d) Corundum

32. The leading bauxite producing country in the world is:
(b) Brazil (b) Guinea
(c) Australia (d) Jamaica

33. The top two leading tin producing countries in the world are:
(a) Malaysia and Russia
(b) China and Indonesia
(c) Peru and Bolivia
(d) Bolivia and Brazil

34. The leading lead producing country in the world is:
(b) USA (b) Australia
(c) China (d) Canada

35. Which South American country is among the five leading lead producing countries in the world?
(a) Peru (b) Bolivia
(c) Chile (d) Colombia

36. The largest producer of chromium is:
(a) India (b) South Africa
(c) Kazakhstan (d) Turkey

37. Which one of the following country produces 75 per cent of world's tungsten?
(a) Russia (b) USA
(c) South Africa (d) China

38. Which one of the following is not a tungsten producing country in the world?
(a) Russia (b) Canada
(c) South Africa (d) Bolivia

39. The largest antimony producing country in the world is:
(a) China (b) South Africa
(c) Russia (d) Australia

40. The leading vanadium producing country in the world is:
(a) China (b) South Africa
(c) USA (d) Russia

41. Which country is the largest producer of molybdenum?
(a) China (b) Russia
(c) USA (d) Japan

42. The largest producer of gold in the world is:
(a) Australia (b) South Africa
(c) USA (d) Canada

43. Which one of the following is the largest producer of silver in the world?
(a) Mexico (b) Chile
(c) Australia (d) USA

44. Which one of the following is not a platinum producer country in the world?
(a) Columbia (b) Germany
(c) Canada (d) South Africa

45. Which one of the following is not correct?
(a) The north-west African countries are affluent as compared to the rest of Africa.
(b) Algeria is rich in iron-ore.
(c) Morocco is rich in phosphate.
(d) Egypt is rich in petroleum.

46. Which one of the following statements is not correct?
(a) Arkansas region is rich in petroleum deposits.
(b) The Lake superior region is rich in iron ore.
(c) Vermilion region is rich in natural gas.
(d) Utah region is rich in copper deposits.

47. Which one of the following statements is not correct?
(a) Scandinavian countries are rich in coal deposits.
(b) Europe is rich in coal and iron-ore.
(c) Petroleum has been explored in the North Sea.
(d) The Ruhr Basin of Germany has the best quality of coal in Europe.

48. Which one of the following is the main mineral of France?
(a) Coal (b) Iron ore
(c) Copper (d) Manganese

49. Which one of the following is not correct?
(a) Brazil has about 15 per cent of world's iron ore.
(b) Brazil is rich in metallic and non-metallic minerals.
(c) The Itabasa iron-ore is of high grade.
(d) Brazil is rich in coal deposits.

50. The main minerals of Philippines are:
(a) Iron ore and manganese
(b) Nickel and copper
(c) Copper and gold
(d) Natural gas and coal

51. The Brazilian plateau is rich in:
(a) Petroleum (b) Copper
(c) Iron-ore (d) Nickel

52. The largest producer of mica in the world is:
(a) India (b) Brazil
(c) USA (d) Czech Republic

53. Which is not a rock phosphate producing country?
(a) Canada (b) Kazakhstan
(c) USA (d) Morocco

54. Which is not a sulphate producing country?
(a) Mexico (b) Uzbekistan
(c) Turkmenistan (d) Indonesia

55. Which is not a potash producing country?
(a) Germany (b) Norway
(c) Spain (d) Israel

56. Match List–I with List–II and select the correct answer using the codes given below:

List–I (Minerals)	**List–II (Major Producers)**
A. Mineral oil	1. Zambia
B. Copper	2. Guyana
C. Manganese	3. Venezuela
D. Bauxite	4. Gabon

Code:

	A	**B**	**C**	**D**
(a)	3	1	4	2
(b)	3	1	2	4
(c)	1	3	2	4
(d)	1	3	4	2

57. Consider the following pairs:

	Produce	**Region**
1.	Coal	– Donbas
2.	Iron-ore	– Mognit Ogorsk
3.	Mica	– Kodarma

Which of the pairs given above is/are correctly matched?
(a) Only 1 (b) 1 and 2
(c) 2 and 3 (d) All of these

58. Commercial sources of energy purely consist of:
(a) Power, coal, oil, gas, hydroelectricity and uranium.
(b) Coal, oil, firewood, vegetable waste and agricultural waste.
(c) Power, coal, animal and fire wood.
(d) Coal, gas, oil and firewood

59. The most abundant source of energy is:
(a) Coal (b) Oil
(c) Sun (d) Biomass

60. Most of the coal fields are located in areas of:
(a) Crystalline rocks
(b) Sedimantory rocks
(c) Igneous rocks
(d) Limestone rocks

61. Which type of coal is mainly used in smelting of iron ore?
(a) Anthracite (b) Bituminous
(c) Lignite (d) Peat

62. Consider the following statements and select the correct answer using the code given below:
1. Water, wind solar and tidal are renewable sources of energy.
2. Coal, petroleum and natural gas are the non-renewable sources of energy.
3. Coal is abundant but exhaustible.
4. USA is the leading producer of coal in the world.

Code:
(a) 1. 2 and 3 are correct
(b) 1, 2 and 4 are correct
(c) 1, 3 and 4 are correct
(d) 1, 2, 3 and 4 are correct

63. The leading coal producing country in the world is:
(a) USA (b) Russia
(c) China (d) India

64. Which one of the following is the leading consumer of coal in the world?
(a) Germany (b) India
(c) USA (d) China

65. Arrange the coal producing countries in descending order and select the correct answer using the code given below:
1. China 2. Asutralia
3. USA 4. India

Code:
(a) 3, 1, 2 and 4 (b) 1, 3, 4 and 2
(c) 1, 4, 3 and 2 (d) 3, 4, 2 and 1

66. The commonly found coal in China is:
(a) Anthracite (b) Bituminous
(c) Lignite (d) Peat

67. The British coal is mainly:
(a) Anthracite (b) Bituminous
(c) Lignite (d) Peat

68. The approximately China's percentage in world coal production is:
(a) 42 per cent (b) 40 per cent
(c) 38 per cent (d) 32 per cent

69. Which is the most important coal producing region of China?
(a) The Shansi-Shansi
(b) The Shantung-Hopei
(c) Manchurian fields
(d) The South-western coal fields

70. Which is the largest coal province in Australia?
(a) Queensland (b) Victoria
(c) Tasmania (d) New South Wales

71. In India most of the coal is derived from:
(a) Gondwana rocks (b) Tertiary rocks
(b) Pre tertiary rocks (d) All the above

72. Which of the following is the most important coal field of Russia?
(a) Pechara basin (b) Kuznestsk basin
(c) Nothern Siberia (d) Eastern Siberia

73. Match the following and select the correct answer using the code given below:

Coal Field	**Country**
A. Ruhr region	1. USA
B. Kuzestask Basin	2. China
C. Appalachain	3. Germany
D. Shansi-Shensi	4. Russia

Code:

	A	B	C	D
(a)	3	2	1	4
(b)	4	3	2	1
(c)	3	4	1	2
(d)	2	4	3	1

74. Which one of the following is not a coal field of United kingdom?
(a) Lancahsire (b) Yorkshire
(c) Scottish lowland (d) London Basin

75. In which country are Karaganda coal-fields located?
(a) Ukraine (b) Kazakhstan
(c) Russia (d) South Korea

76. Which of the following countries has the largest coal reserves?
(a) USA (b) Russia
(c) China (d) India

77. Arrange the following countries in descending order according to coal reserve and select the correct answer using the code given below:
1. China 2. Russia
3. USA 4. India
Code:
(a) 3, 1, 4 and 2 (b) 3, 2, 1 ands 4
(c) 1, 3, 2 and 4 (d) 2, 4, 1 and 3

78. Which one of the following is the leading producer of crude oil?
(a) Iran (b) Russia
(c) Saudi Arabia (d) USA

79. Which one of the following is the leading consumer of petroleum?
(a) China (b) Germany
(c) Japan (d) USA

80. Arrange the producers of petroleum in descending order and select correct answer using the code given below:
1. Iran 2. Saudi Arabia
3. USA 4. Russia
Code:
(a) 3, 2, 1 and 4 (b) 4, 3, 2 and 1
(c) 2, 4, 3 and 1 (d) 1, 2, 3 and 4

81. Which one of the following is not a member of the OPEC (Oil Producing and Exporting Countries)?
(a) Mexico (b) Iraq
(c) Venezuela (d) Nigeria

82. Which one of the following is a leading producer of natural gas?
(a) USA (b) Canada
(c) Russia (d) Qutar

83. Which one of the following countries is not rich in petroleum resources?
(a) Iran (b) Qutar
(c) UAE (d) Morocco

84. Which one of the following is not known for the petroleum deposits?
(a) Mississippi delta
(b) Tigris delta
(c) Murray-Darling delta
(d) Azerbaijan

85. Which one of the following does not have oil deposits?
(a) France (b) Greace
(c) Germany (d) Azerbaijan

86. Which one of the following is not a member of the OPEC?
(a) Russia (b) Indonesia
(c) Angola (d) Algeria

87. Which one of the following does not have natural gas?
(a) Nigeria (b) Sudan
(c) Egypt (d) Algeria

88. **Assertion (A)** The age of fossil fuels is limited.

Reason (R) Fossil were made only in Carboniferous period.
(a) Both **A** and **R** are true and **R** is the correct explanation of **A.**
(b) Both **A** and **R** are true and **R** is not the correct explanation of **A.**
(c) **A** is true **R** is false
(d) **A** is false **R** is true

89. Which is not a major oil field of Saudi Arabia?
(b) Ghawar (b) Qatif
(c) Masjid-i-sulaiman (d) Abu Hadriya

90. The main oilfields of Iran is/are:
(a) Masjid-i-sulaiman (b) Aghajari
(b) Bahregan (d) All the above

91. In which country are Kirkuk, Mosul and Daura oil fields located?
(a) Iraq (b) UAE
(c) Bahrain (d) Kuwait

92. Which one of the following is the leading oil producing country of South America?
(a) Brazil (b) Venezuela
(c) Peru (d) Chile

93. China's position in world oil producing countries is:
(a) 3rd (b) 4th
(c) 5th (d) 6th

94. Arrange the following countries in descending order according to their oil reserves and select the correct answer using the code given below:
1. Saudi Arabia 2. Qatar
3. Russia 4. Iran
Code:
(a) 1, 3, 2 and 4 (b) 3, 4, 1 and 2
(c) 4, 3, 1 and 2 (d) 3, 4, 2 and 1

95. Among Iran, Qatar, Saudi Arabia and UAE, which is having highest natural gas reserves?
(a) Saudi Arabia (b) UAE
(c) Iran (d) Qatar

96. Where and when was the first hydro-power station started?
(a) United kingdom, 1850
(b) France, 1883
(c) Germany, 1892
(d) Spain, 1895

97. Which of the following physical features is not required for the generation of hydro-electricity?
(a) Abndant supply of water.
(b) Geological stability for construction of dams
(c) Fertile soil
(d) Suitable climate

98. Which of the continent is having the highest potentiality of hydro electricity generation?
(a) Asia (b) Africa
(c) South America (d) Europe

99. The highest producer of hydro-electricity in the world is:
(a) China (b) USA
(c) Canada (d) Brazil

100. On which river of Canada have major hydro electricity generation plants been developed?
(a) Winnipeg river
(b) Nipigon river and its tributasies
(c) Bridge river
(d) St. Lawrence river and its tributastes

101. Which country has the highest per capita electricity consumption?
(a) Canada (b) India
(c) Norway (d) USA

102. Identify the sites of high hydel potential in world:
(a) Rivers zambexi, Hwang he and Panama.
(b) Rivers Rhine and Nile
(c) Rivers St. Lawerence, Zaire & Tennessee.
(d) Rivers Zaire, Panama and Yangtse. A.

103. Consider the following statements:
   1. Natural Uranium contains 0.7 per cent of the fissile isotope U-235.
   2. Light water reactor at Kundankulam will use Low Enriched Uranium (LEU) which has 3–5 per cent of U-235.

   Which of these statements is/are true?
   (a) Only 1 (b) Only 2
   (c) Both 1 and 2 (d) Neither 1 or 2

104. Study the following statements and select the correct answer from the codes given below:
   1. The major component of natural gas is methane.
   2. The best quality coal is the lignite because of its low content.
   3. Being a clean fuel, Liquified Petroleum Gas (LPG) is now increasingly being used as an aerosol propellant and refrigerant.

   Which of the above statements is true?
   (a) Only 1 (b) Only 2
   (c) 1 and 3 (d) 1, 2 and 3

105. Identify the correct decreasing order of proven petroleum reserves in the world:
   (a) Saudi Arabia, UAE, Kuwait, Iraq
   (b) Iraq, UAE, Saudi Arabia, Kuwait
   (c) Iraq, Saudi Arabia, UAE, Kuwait
   (d) Saudi Arabia, Iraq, UAE and Kuwait

106. Which is considered as the largest hydropower project in the world?
   (a) Nigeria (b) Colorado
   (c) Aswan (d) Dniupur

107. Which one of the following is not an African hydro power project?
   (a) Victoria Falls
   (b) Kafue Dam
   (c) Dniepur
   (d) Cabora Bassa Dam

108. Where was the first nuclear power station built:
   (a) Chernoble (Russia)
   (b) Calder Hall (Britain)
   (c) Tarapur (India)
   (d) Tasmania (Australia)

Answer Key – SET 5

Que. No.	Answer	Que. No.	Answer	Que. No.	Answer	Que. No.	Answer	Que. No.	Answer
1	b	23	b	45	b	67	b	89	c
2	c	24	d	46	c	68	c	90	d
3	a	25	a	47	a	69	a	91	a
4	c	26	c	48	b	70	d	92	b
5	a	27	b	49	d	71	a	93	c
6	d	28	c	50	b	72	b	94	d
7	b	29	a	51	c	73	c	95	c
8	c	30	d	52	a	74	d	96	b
9	a	31	b	53	a	75	b	97	c
10	b	32	c	54	d	76	a	98	b
11	d	33	b	55	b	77	b	99	a
12	c	34	c	56	a	78	c	100	d
13	b	35	a	57	d	79	d	101	d
14	b	36	b	58	a	80	c	102	c
15	a	37	d	59	c	81	a	103	c
16	c	38	c	60	b	82	c	104	c
17	d	39	a	61	b	83	d	105	d
18	b	40	b	62	a	84	c	106	a
19	b	41	c	63	c	85	b	107	c
20	a	42	b	64	d	86	a	108	b
21	a	43	a	65	b	87	b		
22	c	44	b	66	b	88	c		

Set – 6
WORLD: Economic Geography

1. Which is not a characteristic of an industrial region?
   (a) Large Industrial complex in hierarchical order
   (b) Mining facilities
   (c) Large population engaged in industrial pursuits
   (d) Integration of some main industries with a group of subsidiary industries

2. Which is not an industrial region of the world?
   (a) North American region
   (b) European region
   (c) Industrial regions of China and Japan
   (d) Central African region

3. Which one of the following industrial regions is not correctly matched?
   (a) Rulr Basin – Germany
   (b) Po Basin – Italy
   (c) Silesia Basin – Romania
   (d) Saar Basin – France

4. Which one of the following is not an industrial region of USA?
   (a) Montreal region
   (b) Southern New England
   (c) Detroit region
   (d) Lake Michigan region

5. Which one of the following is not an industrial region of Canada?
   (a) Ontario & St. Lawerence Valley
   (b) Pittsburg region
   (c) Prairie region
   (d) Pacific coastal region

6. Match the following industrial regions and select the correct answer using the code given below:

   A. Lipzig — 1. France
   B. Cumberland — 2. USA
   C. Lorraine — 3. United Kingdom
   D. Detroit — 4. Germany

   Code:

	A	**B**	**C**	**D**
(a)	4	3	1	2
(b)	3	4	2	1
(c)	4	2	3	1
(d)	2	1	3	4

7. Which one of the following is a prominent industrial region of Europe?
   (a) Berlin industrial region
   (b) Greater London industrial region
   (c) Rhine industrial region
   (d) Paris Basin region

8. Which of the following statement is not true about industries in Japan?
   (a) Availability of hydroelectric power
   (b) Coastal location
   (c) Worldwide market for its products
   (d) Availability of raw material in the country

9. Which one of the following statements is correct?
   (a) The Great Lake region is famous for heavy industries, steel and automobiles.
   (b) Michigen state is well known for fruit production.
   (c) Indiana state has large deposits of iron ore.
   (d) Chicago is the third largest city of USA

10. Match List–I with List–II and select the correct answer using the code given below:

List–I (City)	**List–II (Industrial Region)**
A. Chelyabinsk	1. Central Industrial region
B. Irkutsk	2. Baikal region
C. Moscow	3. Volga region
D. Samara	4. Ural region

   Code:

	A	**B**	**C**	**D**
(a)	3	4	2	1
(b)	4	2	3	1
(c)	4	2	1	3
(d)	4	1	3	2

11. Which one of the following is not correctly matched?
    (a) Yokahana – Engineering goods
    (b) Tokyo – Printing and electrical goods
    (c) Kobe – Textile
    (d) Osaka – Iron and steel

12. Which one of the following sets represents the iron and steel centres?
    (a) Anshan, Wuhan, Nanchang, Hankow
    (b) Aushan, Tientsin, Wuhan, Nanchang
    (c) Nonafu, Tientsin, Shantug, Shanghai
    (d) Hankow, Tienstsin, Shantung, Shanghai

13. Match the List–I with List–II and select the answer using the code given below:

List–I (Cities)	List–II (Industry)
A. Vienna	1. Textiles
B. Havana	2. Steel
C. Chichago	3. Glass
D. Dundee	4. Cigars

Code:

	A	B	C	D
(a)	3	2	4	1
(b)	3	4	2	1
(c)	4	3	2	1
(d)	3	1	4	2

14. Leeds is well known for:
    (a) Woollen Textile (b) Cotton Textile
    (c) Iron smelting (d) Films

15. Philadelphia is well known for:
    (a) Dairy Industry
    (b) Silk Textile
    (c) Ship-building
    (d) Locomotives

16. Which one of the following is not correctly matched?
    (a) Houston – Petroleum industries
    (b) San Diego – Shipbuilding industry
    (c) Seattle – Aircraft industry
    (d) St. Louis – Textile industry

17 Which one of the following is not correctly matched?
    (a) Chicago – Agricultural machinery
    (b) Houston – Petrochemical
    (c) Piitsbug – Cotton Textile
    (d) San Diego – Ship building

18 Consider the following statements and select the correct answer using the code given below:
    1. The manufacturing of electric goods was started in 1880.
    2. Hydro-turbines and thermal generations are heavy industries.
    3. Heavy industries are located near the iron ore deposits.

    Code:
    (a) 1 and 2 are correct
    (b) 1 and 3 are correct
    (c) 2 and 3 are correct
    (d) 1, 2 and 3 are correct

19. Match List–I with List–II and select the correct answer using the code given below:

List–I	List–II
A. Ivanovo	1. Cotton Textile industry
B. Magnitogirsk	2. Engineering industry
C. Moscow	3. Iron and steel industry
D. St. Petersburg	4. Ship building industry

Code:

	A	B	C	D
(a)	4	3	2	1
(b)	1	3	2	4
(c)	1	2	3	4
(d)	3	2	1	4

20. Which one of the following is not correctly matched?
    (a) Central Rhine industrial region – France
    (b) Yorkshire region – UK
    (c) Nagoyo region – Japan
    (d) New England industrial region – USA

21. The Detroit region is US is famous for:
    (a) Iron and steel (b) Automobile
    (c) Petro chemicals (d) Ship building

22. Which is an example of a foot loose industry?
    (a) Electronic industry
    (b) Chemical industry
    (c) Fertilizer industry
    (d) Hydro power generation

23. Which region of USA is most important for automobile industry?
    (a) West-north central
    (b) Mid-Atlantic
    (c) East-north-central
    (d) South Atlantic

24. The food processing industries are considered as:
(a) Capital intensive
(b) Market oriented
(c) Raw material oriented
(d) Labour oriented

25. Textile industries are generally:
(a) Raw material oriented
(b) Labour oriented
(c) Transport oriented
(d) Market oriented

26. Which one of the following proposed the industrial location theory?
(a) Weber (b) A. Miller
(c) R. Johnston (d) Von Thunen

27. Which of the following are considered as bridge industries?
(a) Industries located near bridges
(b) Industries located between market and raw materials
(c) Industries located at transport nodes.
(d) Industries located between agricultural and mining areas.

28. If the processing involves weight loss, the industry will be:
(a) Raw material oriented
(b) market oriented
(c) Transport oriented
(d) Labour oriented

29. In which country was the first Jute mill estabilished?
(a) Bangladesh (b) India
(c) Scotland (d) Egypt

30. Weber's theory of industrial location is also called:
(a) Optimum profit theory
(b) Least cost theory
(c) Transport cost theory
(d) Labour cost theory

31. The location of Aluminium industry is influenced mostly by:
(a) Local market (b) Cheap labour
(c) Raw material (d) Availabily of power

32. D.M. Smith's Industrial theory is also known as:
(a) Maximum productivity theory
(b) Optimum profit theory
(c) Minimum cost theory
(d) Export linked theory

33. Moscow-Ivanovo belt is a major manufacturing region for:
(a) Petrochemical
(b) Chemical fertilizers
(c) Textile machinery
(d) Electrical equipments

34. Which one of the following country has a highly developed dairy industry?
(a) Denmark (b) Netherlands
(c) Germany (d) France

35. The most important industry of Scandinavian countries is:
(a) Textiles
(b) Heavy engineering
(c) Iron and steel
(d) Timber based industry

36. Most of the Japanese industries are based upon:
(a) Local coal resources
(b) Local market
(c) Local raw material
(d) Skilled labour

37. Which is called the manchester of Japan?
(a) Tokyo (b) Yokohama
(c) Osaka (d) Nagoyo

38. Match List–I with List–II and select the correct answer using the code given below:

List–I (Industrial significance)	**List–II (Locations)**
A. Large iron-ore deposits	1. Mesabi Range
B. Automobile capital	2. Pearl Harbour
C. Coal and steel city	3. Detroit
D. Naval station on Pacific coast	4. Pittsburgh

Code:

	A	B	C	D
(a)	3	4	1	2
(b)	1	4	3	2
(c)	1	3	4	2
(d)	1	3	2	4

39. Which is the leading centre of petro chemical production in Japan?
(a) Yokohama (b) Yokkaichi
(c) Hiroshima (d) Tokyo

40. The Ichinomiya in Japan is famous for:
(a) Textile (b) Automobile
(c) Silk industry (d) Woollen industry

41. The Hiroshima-Kobe industrial region is important for:
(a) Electronics
(b) Large iron and steel milles
(c) Ship-building
(d) Textile

42. The important industry of Australia is:
(a) Iron and steel (b) Chemical
(c) Automobiles (d) Textiles

43. Which Industry is most important in South-west Asia?
(b) Engineering goods (b) Iron and steel
(c) Petro-chemical (d) Automobiles

44. Which one of the following is a major centre of heavy industry in Japan?
(a) Kamaishi (b) Yawata
(c) Muroran (d) Kosaka

45. Consider the following statements:
1. Iron and steel industries developed around lake Michigen in USA due to the availability of good quality coal.
2. Pittsburgh-Youngstown area is the leading producer of steel in the USA.
3. The Lorraine area in France is significant for steel production.
4. The South Manchurian region accounts for nearly 60 per cent of China's pig iron production.

Which of the above statements are correct?
(a) 1, 3 and 4 (b) 1, 2 and 4
(c) 2, 3 and 4 (d) 2 and 3

46. Match List–I with List–II and select the correct answer using the codes given below:

List–I (Industries)	**List–II (Industrial centres)**
A. Automobile	1. Baltimore
B. Cotton Textile	2. Dallas
C. Iron and steel	3. New York
D. Ship-building	4. Proviedence
	5. Wheeling

Code:

	A	**B**	**C**	**D**
(a)	4	2	5	1
(b)	2	4	5	1
(c)	4	2	1	3
(d)	2	4	1	3

47. Match List–I with List–II and select the correct answer using the codes given below:

List–I (Russian Region)	**List–II (Industries)**
A. Central Industrial	1. Agro-based
B. Groznyy	2. Cotton textile
C. Volga	3. Machinery
D. Urals	4. Metallurgy
	5. Petroleum refining

Code:

	A	**B**	**C**	**D**
(a)	2	4	1	3
(b)	1	3	5	4
(c)	2	3	1	4
(d)	1	4	5	3

48. The leading producer of iron and steel in the world is:
(a) Japan (b) Russia
(c) USA (d) China

49. **Assertion (A)** The factors of iron and steel industry localization in Lake superior region are coal, iron and market.

Reason (R) Detroit is the largest steel consuming centre in the USA.

Code:
(a) Both **A** and **R** are true and **R** is the correct explanation of **A**.
(b) Both **A** and **R** are true and **R** is not the correct explanation of **A**.
(c) **A** is true **R** is false.
(d) **A** is false **R** is true

50. Which one of the following is the most important centre iron and steel industry in Germany?
(a) Rhenish-westphalia (b) Saxomy
(c) Siegerland Hessen (d) Krupp

51. Which is not an important iron and steel producing centre of UK?
(a) Lincolnshire (b) London
(c) Sheffield (d) Birmingham

52. Which of the following countries are first and second largest producers of cotton cloth in the world?
(a) USA – Russia
(b) China – Japan
(c) China – India
(d) Russia – Germany

53. Which one of the following is the leading country in the production of chemical products in the world?
(a) Germany (b) USA
(c) Russia (d) India

54. The largest producer of nitrogen fertilizer in the world is:
(a) China (b) India
(c) USA (d) France

55. The largest producer of phosphate fertilizers in the world is/are:
(a) India and France (b) Russia and Brazil
(c) USA and China (d) France and Belgium

56. The largest producer of potash fertilizers in the world is:
(a) Canada (b) China
(c) Russia (d) Belarus

57. The largest producer of wood pulp and newsprint paper in the world:
(a) USA (b) Canada
(c) Japan (d) Sweden

58. Which one of the following is not correctly matched?
(a) Seattle – Aircraft
(b) Orlando – Textile
(c) Buffalo – Food processing
(d) Phoenix – Precision instruments

59. Match List–I with List–II and select the correct answer using the code given below:

List–I (Industry)	**List–II (Location)**
A. Aircraft	1. Baltimore
B. Ship building	2. Seattle
C. Automobiles	3. Chicage
D. Agricultural machiners	4. Detroit

Code:

	A	**B**	**C**	**D**
(a)	2	1	4	3
(b)	2	1	3	4
(c)	3	4	1	2
(d)	3	4	2	1

60. The length of Trans-Canada Highway is:
(a) 8500 km (b) 7821 km
(c) 7235 km (d) 6980 km

61. The most important Highway of South America is:
(a) Pan-American Highway
(b) Brazilian Highway
(c) Lapaz to Bunos Aires Highway
(d) Racife to Montevideo Highway

62. First Railway line in the world was laid in:
(a) Western USA
(b) Federal Republic of Germany
(c) Russia
(d) North-East England

63. The Canadian Pacific Railway runs between:
(a) Montreal and Vancoures
(b) Edmonton and Halifex
(c) Halifex and Vancouver
(d) Ottawa and Prince Rupert

64. The Trans-Siberian Railway runes between:
(a) Moscow to Bukhaba
(b) Moscow to Vladivostok
(c) Bukhara to Temsk
(d) Moscow to Irkutsk

65. The Trans-Andean Railway connects:
(a) Buenos Aires with Valparaiso
(b) Lima with Santiago
(c) Buenos Airas with Sao Paulo
(d) Bogato with Lima

66. Which one of the following is not a tran-continental railway of USA?
(a) Union Pacific Railway
(b) Chicage NewYork Railway
(c) Great Northern Railway
(d) Southern Pacific Railway

67. **Assertion (A)** Cape to Cairo Railway was proposed to connmect North African city to Cape of Good Hope located in South Africa.
Reason (R) Cape to Cairo Raiwlay is still not fully completed.
Code:
(a) Both **A** and **R** are true and **R** is the correct explanation of **A**.
(b) Both **A** and **R** are true and **R** is not the correct explanation of **A**.
(c) **A** is true, **R** is false.
(d) **A** is false **R** is true

68. The famous port of Vladivostok is situated on the:
(a) White sea (b) Barents Sea
(c) Sea of Japan (d) Mediterranean Sea

69. The Rhine forms the busiest waterway of Europe because:
1. The river passes through one of the most industrialized regions of the world.
2. Other means of transport have not developed much here.
3. It passes through one of the most fertile regions of the continent.
4. The ancient monuments of this region attract large number of tourists.
Select the correct answer using code given below:
(a) Only 1 is correct
(b) 1 and 2 are correct
(c) 2 and 3 are correct
(d) Only 4 is correct

70. Which is the largest Railway system in the world?
(b) Trans – Canadian Railway
(b) Pertth-Brishane Railway
(c) Trans-US Railway
(d) Trans-Siberian railway

71. What is the function of 'locks' in Panama canal?
(a) Traffic control
(b) Raising the level of water
(b) Minimize shipping
(d) Avoid siling

72. River Amazon is navigable from its mouth inword by about:
(a) 1000 km (b) 500 km
(c) 1600 km (d) 1200 km

73. Consider the following statements and select the correct answer using the code given below:
1. Roads are characterisized by a very wide range of traffic.
2. Americans were the pioneer builders of metalled roads.
3. USA has the longest National Highways in the world.
Code:
(a) 1 and 2 are correct
(b) 1 and 3 are correct
(c) 2 and 3 are correct
(d) 1, 2 and 3 are correct

74. Consider the following statements and select the correct answer using the code given below:
(a) USA has the longest network of roads in the world.
(b) Waterways are cheaper than land transportation
(c) Roadways carry more traffic than any other means of transport.
Code:
(a) 1 and 2 are correct
(b) 2 and 3 are correct
(c) 1 and 3 are correct
(d) 1, 2 and 3 are correct

75. Which one of the following is not a waterway of Europe?
(a) Rhine (b) Kiel Canal
(c) St. Lawrence (d) Danube

76. The largest navigable river of United states is:
(a) Mississippi (b) Missouri
(c) St. Lawrence (d) Colorado

77. The most important waterway for navigation in China is:
(b) Sikiang (b) Yang-tse-kiang
(c) Pei-ho (d) Hwang-ho

78. **Assertion (A)** Most of the African rivers are navigable for a very limited length.

Reason (R) Most of the African rivers pass through rapis, falls and succession of cataracts.

Code:
(a) Both **A** and **R** are true and **R** is the correct explanation of **A**.
(b) Both **A** and **R** are true and **R** is not the correct explanation of **A**.
(c) **A** is true and **R** is false
(d) **A** is false **R** is true

79. **Assertion (A)** The North Atlantic route is the busiest of the ocean trade routes of the world.

Reason (R) This route connects western Europe and Eastern North America, the two highly developed regions of the world.
Code:
(a) Both **A** and **R** are true and **R** is the correct explanation of **A**.
(b) Both **A** and **R** are true and **R** is not the correct explanation of **A**.
(c) **A** is true and **R** is false
(d) **A** is false **R** is true

80. Why is the Panama canal famous?
1. It is a big canal.
2. It joins two major oceans
3. It lies in central Africa
4. It reduces the distance between the Eastern and Western coasts of USA.

Select the correct answer using the codes given below:
(a) 1, 2 and 4
(b) Only 1
(c) 1 and 2
(d) All the above

81. Match List–I with List–II and select the correct answer using the code given below:

List–I (Canals)	**List–II (Connects)**
A. Kiel canal	1. Erie to Ontario
B. Soo canal	2. North Sea Black Sea
C. Panama canal	3. Atlantic to Pacific Ocean
D. Welland canal	4. Siperior to Huron

Code:

	A	**B**	**C**	**D**
(a)	1	2	3	4
(b)	4	3	2	1
(c)	2	4	3	1
(d)	3	4	2	1

82. **Assertion (A)** Baltic sea remains open for international trade throughout the year.

Reason (R) Baltic sea lies in the warm temperature zone.
Code:
(a) Both **A** and **R** are true and **R** is the correct explanation of **A**.
(b) Both **A** and **R** are true and **R** is not the correct explanation of **A**.
(c) **A** is true and **R** is false
(d) **A** is false **R** is true

83. One will not pass through the Suez canal while going from Mumbai to:
(a) Suez (b) Alexandria
(c) Port Said (d) Benghazi

84. Consider the following statements and select the correct answer using the code given below:
1. Suez canal was opened in 1869.
2. The canal is owned and operated by Egypt.
3. The transit time for a ship is 24 hours.
4. The canal connects the Great Bitter Lake with the Red Sea.

Code:
(a) 1, 2 and 3 are correct
(b) 1, 3 and 4 are correct
(c) 2, 3 and 4 are correct
(d) 1, 2, 3 and 4 are correct

85. Examine the following statements and select the correct answer using the code given below:
1. The Canadian Pacific Railway was opened in 1885.
2. The Canadian Pacific Railway joins Halifax with Vancouver.
3. The Trans-Canadian Highway was completed in 1962.

Code:
(a) 1 and 2 are correct
(b) 2 and 3 are correct
(c) 1, 2 and 3 are correct
(d) Only 2 correct

86. Which one of the following is the most busy navigable river?
(a) Volga (b) Rhine
(c) Thames (d) Danube

87. Which one of the following statements is not correct?
(a) OB and Yenisey are not navigable
(b) Volga remains open throughout the year
(c) Amazon flows through sparsely populated areas
(d) The Orange river is not navigable

88. During a flight from Delhi to Tokyo the following are the landing airports:
1. Hongkong 2. Hanoi
3. Taipi 4. Bangkok

The correct sequence of these airports during an onward journey is:
(a) 1, 2, 3 and 4
(b) 4, 2, 1 and 3
(c) 3, 4, 1 and 2
(d) 4, 1, 2 and 3

89. Examine the following statements and select the correct answer using the code given below:
1. The Trans – Siberian railway is a double track.
2. The Trans – Siberian railway line connects Leningrad with Vladivostok.
3. Astrakhan and Odessa are the important junctions of Trans-Siberian railway.

Code:
(a) 1, and 2 are correct
(b) 1 and 3 are correct
(c) 2 and 3 are correct
(d) 1, 2 and 3 are correct

90. Which of the following countries will be served by proposed 'TAPI' natural gas pipeline?
(a) Kazakistan, Pakistan, India and Bangladesh
(b) Turkmenistan, Ajghanistan, Pakistan and India
(c) Afghanistan, Pakistan, India and Bangladesh
(d) Turkmenistan, Iran, Pakistan and India

Answer Key – SET 6

Que. No.	Answer	Que. No.	Answer	Que. No.	Answer	Que. No.	Answer	Que. No.	Answer
1	b	21	b	41	c	61	a	81	c
2	d	22	a	42	a	62	d	82	c
3	c	23	c	43	c	63	c	83	a
4	a	24	b	44	b	64	b	84	d
5	b	25	d	45	c	65	a	85	c
6	a	26	a	46	d	66	b	86	b
7	c	27	b	47	b	67	b	87	b
8	d	28	a	48	d	68	c	88	b
9	c	29	c	49	b	69	a	89	a
10	c	30	b	50	a	70	d	90	b
11	d	31	d	51	b	71	b		
12	b	32	b	52	c	72	c		
13	b	33	c	53	b	73	d		
14	a	34	a	54	a	74	a		
15	d	35	d	55	c	75	c		
16	d	36	d	56	d	76	a		
17	c	37	c	57	b	77	b		
18	a	38	c	58	c	78	a		
19	b	39	b	59	a	79	a		
20	a	40	d	60	b	80	a		